W9-CSF-217

ACS SYMPOSIUM SERIES **568**

Structure and Reactivity in Aqueous Solution

Characterization of Chemical and Biological Systems

Christopher J. Cramer, EDITOR
University of Minnesota

Donald G. Truhlar, EDITOR
University of Minnesota

Developed from a symposium sponsored
by the Division of Physical Chemistry
at the 207th National Meeting
of the American Chemical Society,
San Diego, California,
March 13–18, 1994

American Chemical Society, Washington, DC 1994

Library of Congress Cataloging-in-Publication Data

Structure and reactivity in aqueous solution: characterization of chemical and biological systems / Christopher J. Cramer, editor; Donald G. Truhlar, editor.

    p.   cm.—(ACS symposium series, ISSN 0097–6156; 568)

"Developed from a symposium sponsored by the Division of Physical Chemistry at the 207th National Meeting of the American Chemical Society, San Diego, California, March 13–18, 1994."

Includes bibliographical references and indexes.

ISBN 0–8412–2980–5

1. Solution (Chemistry)—Congresses.

I. Cramer, Christopher J., 1961–   . II. Truhlar, Donald G., 1944–
III. American Chemical Society. Division of Physical Chemistry.
IV. Series.

QD540.S77   1994
541.3'422—dc20   94–29430
                                               CIP

Foreword

THE ACS SYMPOSIUM SERIES was first published in 1974 to provide a mechanism for publishing symposia quickly in book form. The purpose of this series is to publish comprehensive books developed from symposia, which are usually "snapshots in time" of the current research being done on a topic, plus some review material on the topic. For this reason, it is necessary that the papers be published as quickly as possible.

Before a symposium-based book is put under contract, the proposed table of contents is reviewed for appropriateness to the topic and for comprehensiveness of the collection. Some papers are excluded at this point, and others are added to round out the scope of the volume. In addition, a draft of each paper is peer-reviewed prior to final acceptance or rejection. This anonymous review process is supervised by the organizer(s) of the symposium, who become the editor(s) of the book. The authors then revise their papers according to the recommendations of both the reviewers and the editors, prepare camera-ready copy, and submit the final papers to the editors, who check that all necessary revisions have been made.

As a rule, only original research papers and original review papers are included in the volumes. Verbatim reproductions of previously published papers are not accepted.

M. Joan Comstock
Series Editor

Contents

THE HYDROPHOBIC EFFECT

Preface

WATER IS A CRITICAL COMPONENT of the earth's surface and of the biological organisms upon it. All biochemical reactions take place in water, and water's enthalpic and entropic characteristics are unique. Consequently, aqueous solutions are a focal point for studies of structure and reactivity.

Participants in the symposium from which this book was developed came from around the globe to engage in stimulating exchanges highlighting recent research in this area. Speakers hailed from Australia, Belgium, Canada, France, Italy, the Netherlands, Spain, and the United States. The breadth of the experimental and theoretical chemistry discussed was reflected in the diverse affiliations of the speakers, with representatives from academia, industry, and government laboratories all presenting their most exciting results. Several sessions attracted audiences that overflowed the 200-seat capacity meeting room, a testimony to the high caliber of the science discussed.

This book represents a remarkable recording of those four days. Of the 28 invited and five contributed lectures, book chapters were developed from 24 of the former and two of the latter. In addition, one chapter derives from the work of a researcher in the field who was unable to travel to the symposium.

In an effort to keep this material fresh and exciting, we imposed draconian deadlines on all of our authors and peer reviewers and in general harassed them to the limits of human endurance. Thanks to their good humor and universal accommodation, this volume was completed a mere seven weeks after the end of the meeting! We take this opportunity to thank all of the participants in our symposium and we look forward to future gatherings devoted to this subject area.

CHRISTOPHER J. CRAMER
  AND DONALD G. TRUHLAR
Department of Chemistry and Supercomputer Institute
207 Pleasant Street, S.E.
University of Minnesota
Minneapolis, MN 55455–0431

May 12, 1994

Dedicated to

Sara and Stephanie

and

William and Matthew

Chapter 1

Structure and Reactivity in Aqueous Solution
An Overview

Christopher J. Cramer and Donald G. Truhlar

Department of Chemistry and Supercomputer Institute, University of Minnesota, 207 Pleasant Street, S.E., Minneapolis, MN 55455-0431

This introductory chapter provides a brief overview of the current state of the art in understanding and modeling structure and reactivity in aqueous solution. The contents of the chapters found in this book are discussed and areas where theory and experiment are working in tandem are highlighted.

Water is a remarkable substance. It covers two thirds of the Earth's surface, it makes up a large fraction of the total mass of biological organisms, it serves as the solvent in which essentially all biochemical reactions take place, and it's enthalpic and entropic characteristics repeatedly set it apart from other liquids (1). It is thus hardly surprising that chemists with an interest in the role that solvent plays in determining structure and reactivity devote themselves to aqueous solutions more than any other kind.

For theorists in particular this has been the case. Relative to modeling in the gas phase, it is only recently that theoretical techniques capable of simulating condensed phase chemistry have been developed. There are two complementary approaches which may be taken to modeling a solute embedded in a solvent. One is the quantum mechanical continuum approach (2-6) where the solvent is replaced by a continuous medium having the same dielectric constant as bulk liquid. This is a mean-field approach in which the solute polarizes the bulk medium, which back-polarizes the solute, etc.—a physical picture dating back to Onsager (7-9). In Chapter 2, the historical development of one such model is detailed by Tomasi, with special attention paid to algorithmic details. In addition, the present directions in which the model is expanding are detailed, including techniques to separate the continuum into fast and slow components for dynamical studies and methods for incorporating local anisotropy into the continuum model.

One significant drawback of any pure continuum approach is that the solvent bulk dielectric constant does not accurately describe the electric polarization field right up to the molecular "surface". Instead, the interaction of a solute with at least the first solvation shell is typically characterized by specific local energy components. One method which has been explored to address this is the assignment

0097–6156/94/0568–0001$08.00/0

of local surface tensions to specific regions of the molecular surface *(10-13)*. Chapter 3 by Storer *et al.* describes continuum solvation models incorporating such local effects. In addition, it provides specific details of the algorithms required to accurately account for dielectric screening, an important effect whereby the interaction of one portion of the solute with the surrounding continuum is mediated by the intervention of remaining portions of the solute. Finally, this chapter presents an example of using Specific Range Parameters for the modeling of particular chemical reactions, in particular for the study of the effect of solvation on the Claisen rearrangement, a reaction that several other chapters in this symposium also discuss (vide infra).

The interaction energy of a set of charges with a surrounding dielectric medium may be found by solution of the Poisson equation *(14-17)*. When the continuous charge density of a molecular solute is replaced by a collection of point charges, the solution of the Poisson equation becomes more facile; however, there are ambiguities involved in modeling a continuous charge density with point charges. There are also inherent ambiguities in defining the dielectric boundary between the solute and the solvent. In Chapter 4 Lim *et al.* discuss the use of calculated quantum mechanical densities for the evaluation of these quantities, and they present results for the solvation free energies of a number of ions and organic molecules. Tawa and Pratt present a similar Poisson equation derived formalism in Chapter 5, where they calculate the aqueous solvent-induced potential of mean force for the dissociation of sodium chloride and for nucleophilic substitution and addition reactions. Tawa and Pratt compare their calculations to other models where solvent has not been replaced by the mean-field continuum.

This alternative approach, i.e., representing the solvent in its discrete, molecular form, allows investigation of the microscopic details of solvation. It has been the case so far, though, that the number of solvent molecules that must be included to adequately model bulk solvation for a given solute is so large that a quantum mechanical treatment of the system has been prohibitively difficult. Instead, the solute-solvent and solvent-solvent interactions have been modeled classically using empirical force fields *(18-24)* or—very recently—a combination of such a force field with terms representing polarization of the water molecules. By using such a force field and following the trajectory of the system over time *(25-28)*, it is possible to calculate dynamical properties or time-averaged equilibrium properties, or Monte Carlo methods *(22,29)* may be used to obtain equilibrium properties. Clementi and Corongiu discuss in Chapter 7 the step-by-step development of a force field that has been designed to reproduce numerous static and dynamical properties of flexible water over a sizable temperature range. Nyberg and Haymet analyze in Chapter 8 a more complete model for liquid water in which the force field permits dissociation (accompanied by ionization) of individual water molecules. They compare their results to other models with respect to predicting the pH of liquid water. Finally, in Chapter 22 Van Belle *et al.* present calculations with polarizable water molecules (vide infra).

Nonequilibrium solvation can be critical for many processes, e.g., electron transfer *(30)* and spectroscopy *(31)*. If the effects of solvation can be modeled as an extra, general coordinate on a solute potential energy surface, it becomes straightforward to extend dynamical studies from the gas-phase into solution *(32-35)*. Using an approach which permits the separation of equilibrium and nonequilibrium components of aqueous solvation, Garrett and Schenter examine in Chapter 9 the radical addition of hydrogen atoms to benzene. In particular, they use variational transition state theory *(36)* to calculate kinetic isotope effects (KIEs) and explore solvent-induced changes in the KIEs for deuterium and muonium. Ando and Hynes also adopt this generalized coordinate approach in Chapter 10, where they consider the aqueous ionization of hydrochloric acid. In particular, they assess the importance of nonequilibrium microsolvation in the proton transfer from HCl to a water molecule and the subsequent separation of the two ions. This same system is also studied by Rivail *et al.* in Chapter 11 using a quantum mechanical continuum model with either one or two explicit water molecules. In addition to providing this stimulating study which the reader may compare to the work of Ando and Hynes, Rivail *et al.* also explore the hydrolysis of formamide in aqueous solution, again focusing on the details of those specific water molecules that are not part of the bulk solvent but are instead involved in the reaction itself. A third subject addressed in this chapter is a comparison of the properties of an isolated water molecule, a water molecule in the water dimer, and water in the bulk.

These various methods for modeling aqueous solvation provide considerable flexibility to researchers probing specific chemical problems. One area of particular interest is understanding how aqueous solvation affects organic reactions *(37,38)*. In Chapter 12, Bertrán *et al.* use various solvation models to study the effect of solvation on several organic reactions, paying particular attention to the location of the transition state along the solvated reaction coordinate and the degree to which nonequilibrium solvation effects must be included in solvation modeling.

Ultimately, some of the most interesting organic reactions occurring in aqueous media are those involving biological macromolecules. In this latter area, Warshel has explored many models for aqueous solvation *(27,39)* as it affects enzyme-mediated reactions. In Chapter 6, Warshel and Chu summarize their experience with numerous macroscopic (e.g., continuum), microscopic (e.g., discrete), and microscopic/macroscopic hybrid water models. In particular, they consider the balance between quantum and classical mechanics strategies and examine some of the algorithmic details of model implementation.

One possibility discussed by Warshel involves the replacement of explicit solvent molecules with pseudopotentials. Gordon and co-workers have also been active in this area *(40)*—their approach divides microsolvated clusters into a "solute" (potentially supermolecular) and a set of water molecules with which it interacts. The solute is treated quantum mechanically; the "spectator" region is modeled using pseudopotentials developed to accurately reproduce electrostatic, polarization, and exchange repulsion interactions in prototypical systems.

The marriage of quantum mechanical and classical mechanical treatments, where the former is applied to the solute and the latter to the solvent, is also a subject of much interest *(27,41-44)*. In Chapter 15, Gao applies such a strategy to the Claisen rearrangement, already discussed in Chapter 3, and to the Menshutkin nucleophilic substitution reaction. In addition, Gao extends a classical water model into the supercritical regime in order to examine the change in the potential of mean force for ion-pairing in sodium chloride. Tucker and Gibbons also consider supercritical water in Chapter 14, where they consider the hydrolysis of anisole in supercritical solvent. They illustrate that clustering of the solvent is an important phenomenon such that local dielectric constants may deviate significantly from the bulk value above the critical point.

The selectivity of biologically important reactions, like those mediated by enzymes, depends in part on the ability of macromolecules to recognize the structure of their often complicated substrates (45). As such, it is of considerable interest to examine how aqueous solvation affects conformational equilibria. Venanzi et al. consider this issue for a diuretic acylguanidine, amiloride, in Chapter 18. They additionally extend their studies into the regime of molecular recognition and present results for the hydrolysis of phenyl acetate in aqueous solution both as a free substrate and as a guest in a β-cyclodextrin host.

The contribution by Wilcox et al., chapter 19, also considers molecular recognition, and it includes a description of their synthetic design of cyclic polyaromatic receptors capable of transporting hydrophobic substrates into aqueous solution. In particular, they provide microscopic and thermodynamic analyses of host-guest interactions and illustrate the synergy between synthesis, spectroscopy, and molecular modeling in their experimental design.

Gajewski and Brichford, in Chapter 16, also combine experimental data and statistical modeling in an examination of the effects of solvation on the Claisen rearrangement. Their combination of kinetic isotope effect measurements and factor analysis using a variety of solvent descriptors provides a unique perspective which may be compared to the theoretical modeling of Chapters 3 and 15. Severance and Jorgensen provide a fourth perspective on the effect of solvation on the Claisen rearrangement. In Chapter 17, they employ Monte Carlo statistical mechanics simulations (29) with a classical water model. In particular, they consider in detail the effects of multiple conformational minima for the reactant allyl vinyl ether and examine the microscopic solvation of the gas-phase reaction coordinate. Comparison of these four studies of the Claisen rearrangement gives rise to a detailed understanding of the reaction as it occurs in water and moreover serves to illustrate the individual strengths and weaknesses of the various methods employed. This comparison is summarized in Chapter 3.

In Chapter 20, Breslow considers the effect of aqueous solvation on a different pericyclic process, the Diels-Alder reaction. This chapter discusses the experimental rates for several Diels-Alder reactions in the presence of various salts designed to tighten or disrupt internal water structure. In addition, other organic reactions and molecular recognition events are examined under the same conditions. Blokzijl and Engberts, in Chapter 21, offer additional experimental insights into aqueous acceleration of the Diels-Alder reaction by comparing inter- and intramolecular variants of the cycloaddition. They examine in detail the enthalpic and entropic components of the acceleration and identify both the reduction in hydrophobic surface area and the possibility of enhanced solute-solvent hydrogen bonding in the transition state as being critical to the observed rate accelerations.

As part of their explanation for aqueous acceleration of the Diels-Alder reaction, Breslow and Blokzijl and Engberts invoke the tendency of non-polar solutes to associate in order to minimize their exposed surface area in aqueous solution *(46-48)*. In Chapter 22, Van Belle *et al.* examine this hydrophobic interaction in detail using a simulation model that includes the polarization of discrete water molecules. They examine in particular the potential of mean force for the association of two methane molecules in water and find no evidence for a solvent-separated minimum, in disagreement with several prior studies that did not account for polarizability of the water molecules. Another particularly interesting result of their study is the reduced dipole moment of water molecules within the first two solvation shells of the methane solutes. This result has important implications for the dielectric continuum models discussed elsewhere in this volume. In Chapter 23, Hermann examines the interaction of hydrocarbons in aqueous solution using a decomposition of the free energy of solvation into a cavity-surface-tension contribution and a contribution calculated from configurationally averaged solute-solvent interaction potentials. In addition, comparisons between explicit-solvent molecular dynamics calculations and the results from transferable fragment solvent-distribution functions are offered.

The hydrophobic effect manifests itself for situations other than hydrocarbon-water and hydrocarbon-hydrocarbon interactions. Importantly, it appears to play an important role in dictating the folding and higher-order structure of proteins *(49,50)*. Understanding and predicting the structure and dynamics of proteins is of great interest. In Chapter 13, Gai *et al.* provide experimental details of the aqueous photophysics of 7-azaindole, a potentially useful surrogate chromophore for tryptophan which would permit the observation of short-time scale dynamics when incorporated into enzymes. Scheraga discusses in Chapter 24 theoretical approaches to predicting protein structure. In particular, continuum solvation techniques employing terms depending on either solvent shell volume or solvent-accessible surface area are presented; calculated and measured thermodynamic values are compared for the interaction of individual amino-acid residues and generic organic functionalities. Using molecular simulations, Ben-Naim also considers the interaction of specific protein residues in Chapter 25. This chapter concludes that solvent-induced attraction between hydrophilic groups can be stronger than that between hydrophobic groups, suggesting that surface tension models which are based on the positive values observed for hydrocarbons may be insufficiently flexible to model the solvation of more complex functional groups.

Chapter 26 describes the modeling of a different kind of biopolymer in water, namely DNA; Beveridge *et al.* describe a full nanosecond simulation of a dodecamer double helix in aqueous solution and analyze the effect of solvation and of complexation with a repressor-operator protein on the structural dynamics. They find a number of structures which exhibit metastability over periods of hundreds of picoseconds, suggesting that the dangers of short simulation times for biomolecular systems are significant. In Chapter 27 Pohorille and Wilson also explore a large biologically relevant system. They use molecular simulation to study the interaction of a model dipeptide with a glycerol 1-monooleate bilayer. The degree to which the dipeptide penetrates the interfacial region between the solvent and the bilayer is explored, as are the differential dynamics of the dipeptide surrounded by water compared to inside the bilayer.

Finally, Chapter 28 provides an additional perspective on an interfacial system; Benjamin uses molecular simulation to analyze the dynamics of electron transfer from liquid water to an adjoining 1,2-dichloroethane phase. The overall relaxation dynamics in response to a photochemically induced electron transfer event are found to depend on each of the two solvents and their different respective time scales.

In summary, aqueous solvation plays an important role in diverse systems of great chemical importance. This volume attempts to provide as wide a perspective as possible (in a four-day symposium) in this regard. We have reached a stage where theory and experiment may be used in conjunction, complementing each other's strengths and weaknesses, so as to illuminate chemical phenomena at a greater level of detail than would be possible from either one alone. We sincerely hope that some of the examples of this synergy that are provided in this book will stimulate further collaborations in this regard.

Literature Cited

(1)   Reichardt, C. *Solvents and Solvent Effects in Organic Chemistry*; VCH: New York, 1990.

(2)   Rivail, J.-L.; Rinaldi, D. *Chem. Phys.* **1976**, *18*, 233.

(3)   Tapia, O. In *Quantum Theory of Chemical Reactions*; R. Daudel, A. Pullman, L. Salem and A. Viellard, Eds.; Reidel: Dordrecht, 1980; Vol. 2; p. 25.

(4)   Miertus, S.; Scrocco, E.; Tomasi, J. *Chem. Phys.* **1981**, *55*, 117.

(5)   Tomasi, J.; Bonaccorsi, R.; Cammi, R.; Olivares del Valle, F. J. *J. Mol. Struct. (Theochem)* **1991**, *234*, 401.

(6)   Cramer, C. J.; Truhlar, D. G. In *Reviews in Computational Chemistry*; K. B. Lipkowitz and D. B. Boyd, Eds.; VCH: New York, 1994; Vol. 6; in press.

(7)   Onsager, L. *Chem. Rev.* **1933**, *13*, 73.

(8)   Kirkwood, J. G. *J. Chem. Phys.* **1934**, *2*, 351.

(9)   Onsager, L. *J. Am. Chem. Soc.* **1936**, *58*, 1486.

(10)   Reynolds, J. A.; Gilbert, D. B.; Tanford, C. *Proc. Natl. Acad. Sci., USA* **1974**, *71*, 2925.

(11)   Hermann, R. B. *Proc. Natl. Acad. Sci., USA* **1977**, *74*, 4144.

(12)   Ooi, T.; Oobatake, M.; Nemethy, G.; Scheraga, H. A. *Proc. Natl. Acad. Sci., USA* **1987**, *84*, 3086.

(13)   Cramer, C. J.; Truhlar, D. G. *J. Comput.-Aid. Mol. Des.* **1992**, *6*, 629.

(14)   Warwicker, J.; Watson, H. C. *J. Mol. Biol.* **1982**, *174*, 527.

(15)   Bashford, D.; Karplus, M.; Canters, G. W. *J. Mol. Biol.* **1988**, *203*, 507.

(16)   Davis, M. E.; McCammon, J. A. *Chem. Rev.* **1990**, *90*, 509.

(17)   Honig, B.; Sharp, K.; Yang, A.-S. *J. Phys. Chem.* **1993**, *97*, 1101.

(18)   Momany, F. A.; McGuire, R. F.; Burgess, A. W.; Scheraga, H. A. *J. Phys. Chem.* **1975**, *79*, 2361.

(19)   Lifson, S.; Hagler, A. T.; Dauber, P. *J. Am. Chem. Soc.* **1979**, *101*, 5111.

(20)   Berendsen, H. J. C.; Potsma, J. P. M.; van Gunsteren, W. F.; Hermans, J. In *Intermolecular Forces*; B. Pullman, Ed.; Reidel: Dordrecht, 1981.

(21)   Brooks, C. R.; Bruccoleri, R. E.; Olafson, B. D.; States, D. J.; Swaminathan, S.; Karplus, M. *J. Comp. Chem.* **1983**, *4*, 187.

(22)   Clementi, E. In *Structure and Dynamics: Nucleic Acids and Proteins*; E. Clementi and R. H. Sarma, Eds.; Adenine Press: New York, 1983; p. 321.

(23)   Weiner, S. J.; Kollman, P. A.; Nguyen, D. T.; Case, D. A. *J. Comp. Chem.* **1986**, *7*, 230.

(24)   Jorgensen, W. L.; Tirado-Rives, J. *J. Am. Chem. Soc.* **1988**, *110*, 1657.

(25)   Brooks, C. L.; Karplus, M.; Pettit, B. M. *Adv. Chem. Phys.* **1989**, *71*, 1.

(26)   Warshel, A. *Computer Modeling of Chemical Reactions in Enzymes and Solutions*; Wiley-Interscience: New York, 1991.

(27)   Åqvist, J.; Warshel, A. *Chem. Rev.* **1993**, *93*, 2418.

(28)   Whitnell, R. M.; Wilson, K. R. In *Reviews in Computational Chemistry*; K. B. Lipkowitz and D. B. Boyd, Eds.; VCH: New York, 1993; Vol. 4; p. 67.

(29)   Jorgensen, W. L. *Acc. Chem. Res.* **1989**, *22*, 184.

(30)   Marcus, R. A. *Annu. Rev. Phys. Chem.* **1964**, *15*, 155.

(31)   Karelson, M. M.; Zerner, M. C. *J. Phys. Chem.* **1992**, *96*, 6949.

(32)   Lee, S.; Hynes, J. T. *J. Chem. Phys.* **1988**, *88*, 6863.

(33)   Hanggi, P.; Talkner, P.; Borkovec, M. *Rev. Mod. Phys.* **1990**, *62*, 251.

(34)   Truhlar, D. G.; Schenter, G. K.; Garrett, B. C. *J. Chem. Phys.* **1993**, *98*, 5756.

(35)   Basilevsky, M. V.; Chudinov, G. E.; Newton, M. D. *Chem. Phys.* **1994**, *179*, 263.

(36)   Truhlar, D. G.; Garrett, B. G. *Annu. Rev. Phys. Chem.* **1984**, *35*, 159.

(37)   Grieco, P. A. *Aldrichim. Acta* **1991**, *24*, 59.

(38)   Li, C.-J. *Chem. Rev.* **1993**, *93*, 2023.

(39)   Warshel, A.; Levitt, M. *J. Mol. Biol.* **1976**, *103*, 227.

(40)   Jensen, J. H.; Day, P. N.; Gordon, M. S.; Basch, H.; Cohen, D.; Garmer, D. R.; Krauss, M.; Stevens, W. J. In *Modelling the Hydrogen Bond*; D. A. Smith, Ed.; American Chemical Society Symposium Series: Washington, DC, 1994; in press.

(41)   Field, M. J.; Bash, P. A.; Karplus, M. *J. Comp. Chem.* **1990**, *11*, 700.

(42)   Gao, J. *J. Phys. Chem.* **1992**, *96*, 537.

(43)   Stanton, R. V.; Hartsough, D. S.; Merz, K. M. *J. Phys. Chem.* **1993**, *97*, 11868.

(44)   Ten-no, S.; Hirata, F.; Kato, S. *Chem. Phys. Lett.* **1993**, *214*, 391.

(45)   Warshel, A.; Åqvist, J.; Creighton, S. *Proc. Natl. Acad. Sci., USA* **1989**, *86*, 5820.

(46)   Hermann, R. B. *J. Phys. Chem.* **1972**, *76*, 2754.

(47)   Huque, E. M. *J. Chem. Ed.* **1989**, *66*, 581.

(48)   Blokzijl, W.; Engberts, J. B. F. N. *Angew. Chem., Int. Ed. Engl.* **1993**, *32*, 1545.

(49)   Nemethy, G.; Scheraga, H. A. *J. Chem. Phys.* **1962**, *36*, 3401.

(50)   Kellis, J. T.; Nyberg, K.; Sali, D.; Fersht, A. R. *Nature* **1988**, *788*,

RECEIVED June 6, 1994

MODEL DEVELOPMENT

Chapter 2

Application of Continuum Solvation Models Based on a Quantum Mechanical Hamiltonian

J. Tomasi

Department of Chemistry and Industrial Chemistry, University of Pisa, Via Risorgimento 35, 56126 Pisa, Italy

This chapter discusses areas of study of chemistry in solution for which continuum models may be used profitably. A distinction among types of projects is introduced. The projects may be computational applications, in which existing computer codes are used to get numerical values of a desired property, or they may be methodological studies, addressed to the implementation of further elaboration of the basic procedure. Solvation energy, reaction mechanisms, energy derivatives, and local and large scale anisotropies are the main topics considered here. The exposition is mainly based on the past experience of our group, with the inclusion of some recent developments.

The main characteristic of continuum models of solvation is their simplicity in the description of the solvent structure. This quality is quite appealing for several reasons, in particular the possibility of extending such models to treat a large number of processes and systems, the ease of interpretation of experimental and computational results, and efficiency in routine calculations. The simplicity is tempered in quantum mechanical versions with a more accurate representation of the solute by means of ab initio or semi-empirical methods. We may draw from these general statements some indications about the most promising ways of using continuum models.

For this purpose we shall consider some specific topics where modelling and elaboration of efficient computer codes have different importance: the evaluation of the solvation energy, the description and interpretation of chemical reactions, and the elaboration of more detailed continuum descriptions of the solvent. These examples, to which more could be added, also show when and how quantum continuum methods are useful. This choice has been suggested by our personal experience, and we

0097–6156/94/0568–0010$08.00/0
© 1994 American Chemical Society

will mainly rely on the use of the polarized continuum model (PCM), a computational procedure we proposed years ago (*1*) and are continuously refining, as some previews in the following pages will show.

Models and routine computational codes

Modelling of solvent effects via continuous models dates back to the beginning of the molecular description of solutions and continues into current years. Einstein (*2*) in 1906 used a continuous model to describe transport properties of solutes (continuous models are not limited to electrostatic effects); the fundamentals of the theory of ionic solutions are given by the 1920 continuum model of Born (*3*); the basic aspects of electron transfer reactions are described by the 1956 theory of Marcus (*4*) invoking solvent fluctuations in a continuum model; finer aspects of similar reactions may be interpreted using a quantum description of the fast component of the continuum electrostatic polarization, as Gehlen et al. suggested in 1992 (*5*). These few examples, to which many others could be added, show that the persistence in the use of continuum models is accompanied by a remarkable versatility of the basic concept that surely will continue being exploited in the future.

The efficiency of some recent semi-empirical methods in routine computation of solvation energies of medium size molecules (*6*) is well documented and accompanied by a satisfactory quality of the results. The same methods may be used to describe reactions in solution: the thermodynamic balance of a reaction may be well reproduced (*6*), but description and interpretation of the reaction mechanism is limited by the approximations of the semi-empirical approach.

Similar considerations hold for semi-classical computational codes. These codes are not based on quantum calculations in solution, but they may rely on quantum calculations in vacuo to get the necessary parameters. Extended Born methods yeld fairly accurate solvation energies (*7*). Further extensions and refinements of the semi-classical codes will profit from analyses of continuum quantum results.

The Role of ab initio Continuum Solvation Methods.   We shall consider ab initio continuum methods as the primary object of this paper. There are two reasons for this choice.

The first has a methodological character. A sound strategy to test qualities, defects, and potentialities of a model consists in examining the output of the model at its best, in the most complete and detailed form, and then reducing it to a more manageable level when the essential features to be preserved are well ascertained. Shortcuts are not advisable; they may lead to wrong conclusions. The theoretical chemistry literature is rich in wrong statements based on a hurried examination of a model; some of these statements refer to continuum solvation models. When the usual homogeneous continuum model is considered, the description of the solute charge distribution and of the mutual solute-solvent interaction effects should be done at the highest possible level of accuracy to detect limits and potentialities of the approach. Analogous considerations hold when the homogeneous medium is replaced by other continuum distributions.

The second reason is related to the intrinsic superiority of ab initio quantum methods in describing fine details of molecular charge distributions. For some problems a detailed ab initio description is absolutely necessary.

The Solvation Free Energy

One of the basic problems in the study of solutions is the development of models able to describe their thermodynamic properties, for example the free energy of solvation, ΔG_{sol}, of neutral solutes at infinite diluition. From the application of ab initio continuum methods we may state the following points:

1) Calculations of ΔG_{sol} (and of related quantities) with continuum models are able to yield results within the range of experimental errors.

2) These calculations require the use of a well shaped cavity. The use of cavities with simple shapes can be accepted only under limited conditions.

3) Electrostatic (Coulomb and polarization), dispersion, cavitation, and repulsion terms are all necessary: $\Delta G_{sol} = \Delta G_{el} + G_{dis} + G_{cav} + G_{rep}$. The relative importance of these terms may be related to the bulk properties of the solvent and to the molecular properties of the solute. In making comparisons among solutes of the same class some terms may be neglected, but this choice must be based on the information derived from the decomposition of complete ab initio results.

4) The extra energy effects due to local variations induced in the solvent distribution by the solute (cybotactic changes) are of limited entity.

5) The contributions due to terms related to the vibrational, rotational and translational solute partition functions are not decisive for almost rigid solutes. Some corrections due to large-amplitude motions (hindered internal rotations, out-of-plane deformations) and to zero point contributions (stretching of M-H groups making hydrogen bonds with solvent molecules) may be easily introduced, when necessary.

The semi-empirical procedures we have mentioned satisfy points 2, 3 and 4. The analysis of ab initio results justifies their formulation and suggests some minor improvements.

Numerical Results. We report here the results of a linear regression analysis of computed against experimental ΔG_{hyd} values. The computed values refer to the PCM ab initio and to the AMSOL-SM2 semi-empirical procedures. The data reported in the Table are the coefficients of the regression line: $\Delta G_{hyd}(exp) = a \, \Delta G_{hyd}(comp) + b$ (Kcal/mol), the regression coefficient R, the standard error σ and the number n of cases within each set. Set 1 refers to a sample of chemically similar compounds (esters) all described at the 6-31G* SCF level with geometry optimisation in water (8). Set 2, presented here for the first time, it is not chemically homogeneous (neutral organic solutes with heteroatoms) and is computed using the same 6-31G* basis set with geometries optimised in vacuo. Some of the "experimental" values are drawn from the empirical formulas elaborated by Cabani et al.(9). Set 3 includes all the neutral polar solutes used by Cramer and Truhlar in their calibration procedure (6), but not the charged ones. The last set has been also computed by Cramer and Truhlar, but it was not used in the calibration. These results show that ab initio calculations may reach chemical accuracy and that semiempirical values are of a good level.

Table I. Model Predictions against Experimental Data

Set	Method	a	b	R	σ	n
1	PCM	1.03	0.33	0.999	0.015	16
2	PCM	0.98	0.35	0.910	0.452	102
3	AMSOL	0.90	-0.02	0.876	1.207	117
4	AMSOL	0.98	0.62	0.965	0.796	7

Good results are also obtained with other semiempirical procedures. The results are somewhat scattered and we refer to a still unpublished review by Cramer and Truhlar (Cramer C.J.; Truhlar, D.G. *Reviews in Computational Chemistry* 1994, in press) for more information. Note, however, that this field is in rapid evolution and that the number of methods and the quality of the results are increasing rapidly.

The reduction of the model may be carried further. Even semiclassical models give results of appreciable quality (see again the review by Cramer and Truhlar quoted above). A sequence of approximations starting from the quantum description and ending with atomic charges may be a guide to check this reduction of the model without shortcuts (10).

The continuum methods compare well with other approaches. The first systematic comparison between ab initio continuum and MD based free energy perturbation calculations of ΔG_{hyd} performed by Orozco, Jorgensen and Luque (11) gives an average error of 0.8 kcal/mol for the continuum PCM procedure and 1.5 kcal/mol for the MD-FEP technique with respect to the experimental values.

Of course many refinements may be introduced in this picture, but details are not essential here: we conclude that computationally very convenient continuum methods may be used to obtain gas-liquid and liquid-liquid transfer thermodynamical properties.

Chemical Reactions with the Continuum Model

The evaluation of solvation energy is but a tiny part of the topics facing theoretical chemistry in solution. We shall consider now a more challenging subject, the description of chemical reactions. Here the ab initio formulation of the continuum model has an important role.

Let us summarise the formal set-up of the approach. Continuum methods are able to give an evaluation of the free energy G of the solute in solution (not to be confused with ΔG_{sol}) as a function of the nuclear coordinates {R} of the solute (as "solute" we consider the bare reacting molecules, supplemented when necessary with a restricted number of solvent molecules playing an active role in the reaction). The $G(R)$ surface corresponds to the $E(R)$ surface defined in vacuo. We apply to the $G(R)$ surface (or, better, to the set of relevant $G(R)$ surfaces) the same concepts originally derived for reactions in vacuo. First, we define geometry, energy and electronic structure of reactants, products, intermediates, and saddle points. Then, we define the reaction coordinate and the portions of the energy surface near the reaction path necessary for dynamical studies of the reaction. The chemical interpretation of the mechanism will be based on a scrutiny of the solute wave function, to be performed with suitable techniques.

This simplified version of a rather formidable problem (a quantum study of a system composed of a large number of molecules) is corroborated by preliminary tests on the model. In particular the partition of the whole system into a "solute" and a medium is supported by the success in describing the energy profile of several significant reactions.

This scheme must be supplemented when certain dynamical effects of the solvent are considered. It is known that in many important classes of reactions the dynamics cannot be properly described by relying on the $G(R)$ surface alone: the typical case is the outer-sphere electron transfer model of Marcus (4), in which the dynamics is carried by a solvent coordinate alone, without intervention of the geometric coordinates of the solute. We are thus compelled to extend our definition of energy hypersurface $G(R)$ by including some extra dynamical coordinates {S}, and to use, when and where necessary, a more general function $G(R+S)$. This enlargement of the space is not equivalent to the addition of more solvent molecules in the "solute".

We have thus recognised at least three important problems: the correct evaluation of $G(R)$ and of its critical points, the description of the electronic structure at some significant points, and the definition of the additional {S} subspace, supplemented by the protocols for the use of $G(R+S)$

The Analysis of the G(R) Surface. Long experience in the evaluation of $E(R)$ surfaces tells us that semiempirical methods give, in the most favourable cases, nothing more than a first order guess. Ab initio calculations, of good quality, are necessary.

The analysis of the $G(R)$ surface is made easier by the use of the gradient of $G(R)$, grad $G(R)$, supplemented by the diagonalization of the Hessian matrix $H(R)$. The calculation of grad G and of H must be performed according to the conditions set in points 2 and 3 above, i.e. using a suitably shaped cavity and including in $G(R)$ all the necessary contributions: $G_{el}(R) + G_{dis}(R) + G_{cav}(R) + G_{rep}(R)$. Current ab initio continuum programs are not equipped for the analytical evaluation of gradG and H at this level of accuracy. We present here the computational scheme we have recently elaborated in the framework of the PCM (Cammi, R.; Tomasi, J., *J. Chem. Phys.*, May 1994).

Iterative and Direct PC Method. The original iterative PCM version is not suited to get analytical derivatives of G_{el} (the other terms of G are simpler to manage and may be treated separately). It is more convenient to resort to a matrix formulation of the electrostatic problem, exploiting the fact that in the PCM the solute is effectively replaced by a charge distribution σ on the cavity surface, and that this surface is divided into a finite number of tesserae. An accurate elaboration of the matrix-PCM has been done by the Sakurai group (*12*). We have elaborated a similar procedure, more computationally effective and suitable to compute the first and second derivatives of G_{el} with respect to parameters α and α, β, later indicated by G^{α} and $G^{\alpha\beta}$ respectively. This preliminary step highlights some points that deserve mention.

In the iterative procedure the Schrödinger equation

$$|(H^0 + V_\sigma)\Psi> = E|\Psi> \tag{1}$$

may be solved with the traditional variational techniques because V_σ at each step is fixed. Here H^0 is the Hamiltonian of the solute in vacuo, V_σ is the electrostatic component of the solute-solvent interaction operator, and $|\Psi>$ is the solute wave function. The free energy is expressed in terms of the final values of E, $|\Psi>$, and V_σ to which nuclear contributions must be added (*1*):

$$G = <\Psi|H^0 + V_\sigma|\Psi> -1/2 <\Psi|V_\sigma|\Psi> + V_{NN} + 1/2\, U_{N\sigma}. \tag{2}$$

Here V_{NN} is the intrasolute nuclear repulsion term also present in vacuo calculations and $U_{N\sigma}$ is the interaction between the solute nuclei and the apparent surface charge σ.

In the direct methods simultaneously optimizing the solute and solvent charge densities, the functional to be minimized is not the mean value of $H^0 + V_\sigma$, but rather the free energy fuctional G (*13*). When the problem is recast at the Hartree-Fock level, with expansion over a finite basis, the minimization of G is reduced to the solution of a pseudo-HF equation (*12*):

$$\mathbf{F'C} = \mathbf{ESC} \tag{3}$$

with

$$\mathbf{F'} = \mathbf{h'} + \mathbf{G'(P)} = (\mathbf{h} + 1/2(\mathbf{J} + \mathbf{Y})) + (\mathbf{G(P)} + \mathbf{X(P)}). \tag{4}$$

Here $\mathbf{h}$ and $\mathbf{G(P)}$ are the usual HF one-electron and two-electron integrals matrices, $\mathbf{P}$ is the density matrix, $\mathbf{J}$, $\mathbf{Y}$ and $\mathbf{X(P)}$ collect one-and two-electron integrals involving interactions with the surface apparent charges. (To be more precise, $\mathbf{J}$ describes the interactions between the elementary solute electron charge distributions and the surface charges having as origin the solute nuclear charges, $\mathbf{Y}$ describes the interactions between the solute nuclei and the surface charges having as source the elementary solute electronic distributions, X describes the interactions between elementary electron distributions and the surface charges they generate).

The expression of F' is different from that of the Fock matrix related to equation (1) and applied in the iterative procedure. The two approaches are formally equivalent (Cammi, R.; Tomasi, J. *J. Comp. Chem.*, to be published) and must give the same values for G and for the coefficients C of the wave function.

The Renormalization of the Surface Charges. This formal equivalence is numerically verified only when a renormalization of the surface charge σ (or of the point charges q which describe σ) is introduced in the computational scheme. The integrated value of σ must in fact satisfy a simple relationship with the total charge Q_M of the solute:

$$\int \sigma(s)ds = (\varepsilon-1)/\varepsilon \ Q_M. \tag{5}$$

Similar relationships hold for the electronic and nuclear components of the surface charge ($\sigma = \sigma^e + \sigma^N$) having as sources the solute electron and nuclear charge distributions ($Q_M = Q^e_M + Q^N_M$). These conditions are not satisfied by continuum quantum calculations (one reason is that a portion of the electronic charge distribution is spread out of the cavity, by definition). The effect of this lack of normalization is different in the iterative and direct methods and results without renormalization may differ. On this basis it has been stated that there are two alternative pictures for the description of the solvation energy. Actually, there is only one picture, and the choice is between two alternative but equivalent computational methods.

The introduction of a suitable renormalization permits one to recover in actual computations the equivalence of the two methods. In addition, the renormalization permits to exploit the formal equivalence between J and Y (the first, as said, describes the interactions of the elementary electronic solute charge distributions with σ^N, the second describes the interactions of nuclei with σ^e) that is lost in un-normalized calculations, with a further reduction of the computational times. With this reformulation the ab initio direct method is almost as fast as the iterative method for solutes of small size.

Analytical Derivatives with the PC Method. Surface charge renormalization plays an even more important role in the analytical determination of derivatives. Without renormalization no meaningful analytical derivatives may be computed in the ab initio apparent surface charge methods.

The expression of G, obtained using equation (3), is:

$$G = trPh' + 1/2 \, trPG'(P) + V'_{NN} \tag{6}$$

where $V'_{NN} = V_{NN} + 1/2 \, U_{NN}$ collects nuclear repulsions and nuclei-σ^N contributions to G. From this definition of G we may derive formal expressions for its first and second derivatives with respect to the cartesian components α and β of the solute nuclear coordinates:

$$G^\alpha = trPh'^\alpha + 1/2\ trPG'^\alpha(P) - trS^\alpha W + V'^\alpha_{NN} \qquad (7)$$

$$G^{\alpha\beta} = tr\ P^\beta h'^\alpha + tr\ P(h'^{\alpha\beta} + 1/2\ G'^{\alpha\beta}(P)) - tr\ S^{\alpha\beta}W - trS^\alpha W^\beta + V'^{\alpha\beta}_{NN} \qquad (8)$$

where S is the overlap matrix and $W = PF'P$. Some formal relationships established by Frisch et al. (*14*) have been exploited here.

The expression of the matrix elements of Y and X (J is no longer necessary) contains the q^e charges describing σ^e, that in the matrix formulation are given by:

$$q^e = -AD^{-1}E^e_n \qquad (9)$$

where q^e is a column vector whose length is equal to the number of tesserae, A is a square diagonal matrix collecting the areas of the tesserae, D is a square, non-symmetric matrix collecting the interaction operators between surface point charges, and E^e_n is a column vector collecting the normal components of the electric field, computed at the center of each tessera. A similar expression is employed for the q^N charges, necessary to compute U_{NN}. Note that D only depends on geometric factors (shape and size of the cavity, location of the centers of the tesserae).

The formal expressions (7) and (8), that are valid for each type of parameter, are greatly simplified when α and β are specified. We shall limit ourselves to parameters describing nuclear Cartesian coordinates.

In calculating derivatives we consider two significant options: derivatives at fixed cavity and full derivatives including cavity surface contributions. Here again the main goal of this analysis is to make available the data needed to elaborate simpler computational procedures, according to the philosophy stated at the beginning of this paper. It is in fact possible, for example, that gradient-driven procedures for geometry optimization and for reaction path definition can tolerate different approximations in the definition of grad G.

Derivatives at fixed cavity are simpler to compute. The elaboration of the formal expressions leads one to recognise that all the integrals one needs are of a simple nature and are present in ab initio computational packages. The final expressions have a computer demand comparable to that required for the corresponding calculations of E^α and $E^{\alpha\beta}$ in vacuo.

Derivatives with Surface Cavity Contributions. For the calculation of complete derivatives, including cavity surface contributions, we cannot exploit the recent developments in the calculation of molecular surface derivatives, because we need expressions for single surface elements. We recall that the cavity is given by the union of spheres centred on the atoms A, B... of the solute, with radii R_A, R_B... etc. A derivative with respect to a geometric coordinate α does not involve a change of the radii. When α is a Cartesian coordinate of atom A, the derivative involves the surface of the A sphere, and its intersection with the other spheres. The cavity surface contributions to the derivative are thus of two different types: the first regards the derivative of the position of all the tessera centres belonging to sphere A (and thus of the pertinent elements of the **D** matrix); the second regards the shape and size of the fragmental tesserae lying at the separation of sphere A with other spheres. Our algorithms reduce all the cases in the definition of derivatives of fragmental tesserae to two basic cases, for which compact analytical formulas have been given. To do it we have been compelled to abandon the finite formalism we introduced several years ago (*15*) to define shape area and position of the fragmental tesserae and that can be found in the GEPOL program (*16*).

Analysis of the Electron Distribution along the Reaction Coordinate. Analysing the shape of the G(R) surface is not sufficient to describe the static aspects of a reaction mechanism. Chemists desire to understand why the surface (or surfaces) for a given reaction has the shape shown by the calculations. This is another point that needs quantum continuum models.

   Methods elaborated for the analysis of reactive interactions in vacuo may be used here, applying the changes suggested by the higher complexity of the problem. Only methods based on a partition of the whole system into subunits will be considered here. The subunits may be of very different types: VB structures, molecular orbitals of the whole system or of some molecular fragments, atoms in molecules, localized orbitals, geminals, etc. A few general point may be stressed:

   1) The cavity must be well modelled.
   2) Surface charge renormalization may play an important role.
   3) The quality of the wave function is probably even more important than in vacuo

To satisfy these requirements, a reconsideration of the procedures now in use for the evaluation of the solvation energy of simple molecules is necessary. Let us consider a simple aspect of the general problem. When two molecules A and B interact to give the initial A·B complex from which the true reaction starts, there is a partial desolvation problem that must be modelled with care. It must be noted that in solution every new interaction always replaces another interaction: this does not occur in vacuo.

   An Example of Analysis. The number of detailed studies on reactions in solution is for the moment quite limited. We quote as an example a recent study on the catalytic influence of solvent on the aldol condensation reaction (Coitiño, E.L.; Tomasi, J.; Ventura, O.N. *J. Chem. Soc. Faraday Trans.*, in press). We have utilized here a localized orbital (LO) picture to examine in detail the almost contemporary mechanisms of bond formation, bond disruption and electron shifts occurring in distinct parts of the "solute". The analysis highlights what processes are favoured, or disfavoured, by the solvent and how a single water molecule acts as a catalyst. For this purpose we have employed several numerical indexes related to the LO description of chemical groups we elaborated several years ago to study systems in vacuo. These indexes are related to the polarity of the LOs, to their degree of localization, to their decomposition into generalized atomic hybrid orbitals, to the first and second moment of the LO description of chemical groups, etc. The results are satisfactory, but this investigation is spurring us to elaborate new tools for the analysis, such as the directional derivatives of the LO indexes starting from the Transition State, and the introduction of a partitioning of the reaction field acting on solute groups into contributions due to the catalyst and to the active and not active groups of the solute.

The Use of Solvent Dynamical Coordinates. In the sketch of a general strategy to study chemical reactions in solution, we indicated the possibility (or necessity) of expanding the {R} space to a {R+S} space including dynamical solvent coordinates.

As above noted, considering extra coordinates {S} is not equivalent to adding more solvent molecules in the "solute"; it is preferable to leave the latter option to other methods, better equipped to deal with the dynamics of a system composed by a large number of molecules, for example Molecular Dynamics.

The use of continuum models based on the apparent surface charge approach opens the way to other possibilities of defining the {S} coordinates. We are referring here to the proposals made by the Basilevsky group (*17*) supplemented by other tentative models advanced by our group (*18,19*). The most promising definitions are related to parameters already existing in the continuum models, i. e. the radii of the spheres defining the cavity and the charges on the tesserae. We have already partitioned the set of surface charges $\mathbf{q}$ into two components: $\mathbf{q} = \mathbf{q}^e + \mathbf{q}^N$. We may introduce other partitions, for example into slow and fast components, $\mathbf{q} = \mathbf{q}_{slow} + \mathbf{q}_{fast}$ (*20*), related to relaxation processes occurring in solution with different specific times. The slow component is related to the rotational relaxation modes of the solvent that are easily identified as pertinent dynamical coordinates. In a similar way the position of the cavity surface with respect to the nuclei may be used to define analogous dynamical coordinates.

A couple of simple examples may illustrate this statement. Let us consider a sudden change of the solute charge distribution (for example electronic excitation or photoionization): $\mathbf{q}_{fast}$ follows the solute change without appreciable time delay, $\mathbf{q}_{slow}$ dictates the dynamics of the solvent reorganization process via re-orientation of solvent molecules. As a second example, let us suppose that in a certain reactive process the motion of some atoms is faster than the slow solvent relaxation: here again it may be important to consider the time evolution of $\mathbf{q}_{slow}$ and of the cavity shape, for istance in the form of a delay with respect to the solute transformations (*19*).

Other partitionings of $\mathbf{q}$ may be adopted. Suppose a certain dynamical process can be described through the combination of two different electronic configurations: $\Psi(t) = c_1(t) \Psi_1 + c_2(t)\Psi_2$ (Ψ_1 and Ψ_2 may be, for example, a couple of covalent and charge transfer structures: AB and A^+B^-). We may define two sets of surface charges $\mathbf{q}_1$ and $\mathbf{q}_2$ generated by the solute charge distributions $|\Psi_1|^2$ and $|\Psi_2|^2$, and we may use them in dynamical calculations (*17*).

Note that similar definitions of {S} coordinates may be obtained using continuum quantum methods not based on the apparent charges. For example, the expression of the slow component of the polarization vector,

P_{slow}, in terms of diabatic state contributions has been employed many times by the Hynes group (21). A general definition of a complete set of {S} coordinates has been recently proposed by Truhlar et al. (22). In this method the solvent dynamical coordinates for a given solute geometry R are defined in terms of the polarization vector P obtained at a different appropriate solute geometry R'.

Solvent Fluctuations. The same approach may also be used to describe dynamical processes due to solvent fluctuations. As an example we quote a still unpublished attempt to describe the famous outer sphere electron transfer process (Cossi M.; Persico M.; Tomasi J. *J. Mol Liquids*, in press). The process is driven, as Marcus has shown (4), by an appropriate solvent fluctuation. This fluctuation may be modelled using the parameters we have introduced. In our tentative model the fluctuation has been represented by a change in the shape of the two cavities containing the two ions.

Solvent fluctuations always take place. They may be introduced in the model by a suitable change of the surface charge (in this case the computed solvation energy is not a free energy and an average over fluctuations must then be introduced) (18).

It may be added that dynamical coordinates are often concerned with only a small portion of the solvent surrounding the solute. The dynamical coordinates related to the apparent charge and to the cavity may describe local and directional effects: it is then sufficient to consider a portion of the q_{slow} charges or the radii of a limited number of atomic spheres.

These first attempts may be modified and improved in later work. The large variety of dynamical phenomena taking place in thermal and photochemical reactions requires the formulation of many other specialized models, some of which are already under development.

In this section we have discussed some extensions of the continuum model, with the aim of treating dynamical problems, in the perspective of introducing more "chemistry" in the modelisation. The "chemistry", i.e. a level of description able to account for the microscopic features that characterise the specific behaviour of different chemical systems, is given, at this stage of our development, by the ab initio continuum description. The progress of the research will show if simpler strategies are possible and what features, not considered until now, are necessary to improve our understanding of the dynamics of chemical events.

Anisotropies in Solution.

A discussion of problems where the use of continuum quantum models is advisable and of problems where further development of this approach may improve our theoretical tools cannot be confined to chemical reactions occurring in dilute solutions and to the evaluation of standard ΔG_{hyd} values.

Among the various problems deserving a mention we have selected the question of anisotropies, because it provides both examples of useful application of the existing methods and cases where further progress in the development of the model is within reach.

The anisotropies may be classified into two broad classes, local and large-scale anisotropies. Local anisotropies may be due both to random fluctuations (a subject we have already considered when dealing with dynamical aspects of reactions) and to permanent solute effects. We stated before that these effects may be discarded in the evaluation of ΔG_{sol}, but actually the model is stressed when the solute bears a large net charge. In this case it is better to introduce a field-dependent dielectric constant $\varepsilon(E)$.

New Computational Models. The model may be built up by introducing a set of nonoverlapping dielectrics, each with a different constant value of ε. Classical models consisting of concentric spheres at increasing values of ε give fair results for small ions. The description of the medium in terms of regions of different but constant ε has been generalized by Bonaccorsi et al. (*23*) and by Hoshi et al. (*12*) to the quantum model with arbitrary definitions of the boundaries among regions. These formulations extend the applicability of the model from ΔG_{sol} to G and permit one to treat an important class of large scale anisotropies.

Another possible option is to introduce an $\varepsilon(E)$ or an $\varepsilon(r)$ function in the quantum model. We have recently elaborated a model that presents some methodologically significant points (Cossi, M.; Mennucci, B.; Tomasi, J., unpublished). The introduction of a function ε depending on the electric field or on the distance involves giving up both the concept of apparent surface charges σ_b and the related Boundary Element Method (BEM) in favour of a diffuse apparent charge distribution ρ_b and a Finite Element Method (FEM). FEM approaches applied to diffuse apparent charges are not computationally convenient when the usual step definition of the dielectric function is used ($\varepsilon=1$ inside the cavity, ε=constant outside).

Both iterative and direct solutions of the variational problem are still possible, as well as a matrix formulation of the computational model. The variable ε formulations, also viable for neutral solutes, open new perspectives in the treatment of solvated ions; in particular the definition of the cavity boundary and the dependence of the apparent charges on the solute charge Q_M may be more properly dealt with.

Last, we quote a new computational model, addressed to describing solvent effects in a medium with a tensorial permittivity (this is an example of large-scale anisotropy, occurring in liquid crystals). In this case we have used a combination of apparent surface charges <u>and</u> volume charges, $\sigma_b + \rho_b$ (Mennucci, B.; Cossi, M.; Tomasi, J., unpublished). The computational model is a combination of BEM and FEM procedures used for the first time, at the best of our knowledge, in molecular problems. Both iterative and direct procedures may be adopted. The examination of results obtained with the iterative procedure, accompanied by a decoupling of σ_b and ρ_b contributions to the interaction operator, gives interesting hints for the treatment of the dynamics in such media: a not yet faced problem.

Confined Liquids. Another important kind of large-scale anisotropies concerns confined liquids. In our opinion, physical phenomena and chemical reactions occurring at a boundary surface (liquid/liquid, liquid/solid, liquid/gas,), in the crossing of a membrane, and in other limited systems can be treated with continuum models, in competition with more computer demanding simulation methods. The experience gained thus far (see, e.g. refs. (24,25)) indicates that for many systems and processes an ab initio quantum continuum description may be replaced by simpler methods, including semiclassical ones.

The variety and complexity of the phenomena that might be the object of study is too large to fully cover here. We quote, just as an example, the study of catalytic effects in host-guest reactions recently performed by the Sakurai group (26). In this field, even more than in other fields considered above, the possibility of devising models based on the continuum solvation approach and exploiting chemical intuition is large and appealing.

Conclusions

The concise exposition of the themes we have selected cannot be considered a review of these subjects, which are now studied with continuum approaches in many laboratories. We have drawn material from our past experience, and supplemented it with previews from our Laboratory to highlight some points that can be summarized as follows.

The impact of continuum methods in computational chemistry will increase in the near future. There are different ways of using these methods:

1) Computational packages for the evaluation of specific properties (the calculation of the solvation free energy is the example chosen here) are available and the reliability of the results is increasing. Their use may be competitive with other methods, at a lower computational cost.

2) Modelling of phenomena occurring in solution may exploit several features of continuum models. We have suggested some developments (related to aspects of chemical reactions and to the occurrence of anisotropies in the solution) that are within reach, but several others could be added. Here the use of the approach is based on the physical insight and ingenuity of the researcher. No general rules can be given.

Ab initio quantum methods play a pivotal role in both fields. They represent a full realization of the model and may be used to validate less expensive procedures (solvation energy, some aspects of the reactions, phenomena related to anisotropies). Ab initio quantum methods allow the analysis of complex events with the help of a wide set of tools, some of which have never been tested to study reactions in solution. The number of tools may be increased exploiting the information given by continuum calculations. The definition of the most appropriate tools, and their extension at the semi-empirical and even at the semi-classical level is a worthwhile line of investigation.

Literature Cited

(1) Miertus, S.; Scrocco, E.; Tomasi, J. *Chem. Phys.* **1981**, *55*, 117.
(2) Einstein, A. *Ann. Phys.* **1906**, *19*, 289.
(3) Born, M. *Z. Physik*, **1920**, *1*, 45.
(4) Marcus, R. A. *J. Chem. Phys.* **1956**, 24, 966.
(5) Gehlen, J. N.; Chandler, D.; Kim, H.J.; Hynes, J.T. *J. Phys. Chem.* **1992**, *96*, 1748.
(6) Cramer, C. J.; Truhlar, D. G. *J. Comp.-Aided Molec. Design* **1992**, *6*, 629.
(7) Jean-Charles, A.; Nicholls, A.; Sharp, K.; Honig, B.; Tempczyk, A.; Hendrickson, T. F.: Still, W. C. *J. Am. Chem. Soc.* **1991**, *113*, 1454.
(8) Tomasi, J. *Int J. Quant. Chem. Q.B.S.* **1991**, *18*, 73.
(9) Cabani, S.; Gianni, P.; Mollica,V.; Lepori, L. *J. Solut. Chem.* **1981**, *10*, 563.
(10) Tomasi, J.; Alagona, G.; Bonaccorsi, R.; Ghio, C.; Cammi, R. In *Theoretical Models of Chemical Bonding*; Editor, Maksic, Z., Springer: Berlin, Germany 1991; Vol. 3; pp 545-614.
(11) Orozco, M.; Jorgensen, W.L.; Luque, F.J. *J. Comp. Chem.* **1993**, *14*, 1498.
(12) Hoshi, H.; Sakurai, M.; Inoue,Y.; Chûjô, R. *J. Chem. Phys.* **1987**, *87*, 1107.
(13) Yomosa,S. *J.Phys. Soc. Japan* **1974**, *36*, 1655.
(14) Frisch, M.; Head-Gordon, M.; Pople, J. *Chem. Phys..* **1990**, *141*, 189.
(15) Pascual-Ahuir, J. L.; Silla, E.; Tomasi, J.; Bonaccorsi, R. *J. Comp. Chem.* **1987**, *8*, 778.
(16) Pascual-Ahuir, J.L.; Silla, E.; Tomasi, J.; Bonaccorsi, R. GEPOL 87 **1987**, QCPE program no. 554.
(17) Basilevski, M.V.; Chudinov, G.E.; Napolov, D.V. *J. Phys. Chem.* **1993**, *97*, 3270.
(18) Bianco,R.; Miertus, S.; Persico, M.; Tomasi, J. *Chem. Phys.* **1992**, *168*, 281.
(19) Aguilar, M.A.; Bianco, R.; Miertus,; Persico, M.; Tomasi, J. *Chem. Phys.* **1993**, *174*, 397.
(20) Aguilar, M.A.; Olivares del Valle, F.J.; Tomasi,J. *J. Chem. Phys.* **1993**, *98*, 7375.
(21) Kim, H.J.; Hynes, J.T. *J. Phys. Chem.* **1990**, *94*, 2736.
(22) Truhlar, D.G.; Schenter, G.K.; Garrett, B.C. *J. Chem. Phys.* **1993**, *98*, 5756.
(23) Bonaccorsi, R.; Scrocco, E.; Tomasi, J. *Int. J. Quant. Chem.* **1986**, *29*, 717.
(24) Bonaccorsi, R.; Ojalvo, E.; Palla, P.; Tomasi, J. *Chem. Phys.* **1990**, *143*, 245.
(25) Bonaccorsi, R.; Floris, F.; Palla, P.; Tomasi, J. *Termochim.. Acta* **1990**, *162*, 213.
(26) Furuki, T.; Hosokawa, F.; Sakurai, M.; Inoue, Y.; Chûjô, R. *J. Am. Chem. Soc.* **1993**, *115*, 2903.

RECEIVED June 14, 1994

Chapter 3

Solvation Modeling in Aqueous and Nonaqueous Solvents

New Techniques and a Reexamination of the Claisen Rearrangement

Joey W. Storer[1], David J. Giesen[1], Gregory D. Hawkins[1],
Gillian C. Lynch[1], Cristopher J. Cramer[1], Donald G. Truhlar[1],
and Daniel A. Liotard[2]

[1]Department of Chemistry and Supercomputer Institute, University
of Minnesota, 207 Pleasant Street, S.E., Minneapolis, MN 55455−0431
[2]Laboratoire de Physico-Chimie Theorique, Université de Bordeaux 1,
351 Cours de la Liberation, 33405 Talence Cedex, France

This chapter presents an overview of recent improvements and extensions of the quantum mechanical generalized-Born-plus-surface-tensions (GB/ST) approach to calculating free energies of solvation, followed by a new treatment of solvation effects on the Claisen rearrangement. The general improvements include more efficient algorithms in the AMSOL computer code and the use of class IV charge models. These improvements are used with specific reaction parameters to calculate the solvation effect on the Claisen rearrangement both in alkane solvent and in water, and the results are compared to other recent work on this reaction.

1. Introduction

The kinds of reaction pathways that one typically encounters in aqueous solution often differ qualitatively from those in the gas phase or in nonpolar solvents, even when water does not play a structural or catalytic role in the chemical reaction. Much of the controlling influence of the solvent in aqueous chemistry can be understood in terms of the thermodynamic solvation parameters of reactants, products, and transition states. Thus, as is so often the case in chemistry, a reasonable starting point for a quantitative understanding of chemical processes occurring in aqueous solution is the thermochemistry (1,2). Chemical equilibria are controlled by free energies, and rates—according to transition state theory (3)—are controlled by free energies of activation, so the free energies of hydration are the central quantities in the thermochemistry. This chapter is primarily concerned with the free energy of aqueous solvation, also called the free energy of hydration. We also consider the free energy of solvation in a nonpolar condensed medium, in particular hexadecane.

0097−6156/94/0568−0024$09.08/0
© 1994 American Chemical Society

Our discussion is focused on a class of quantum mechanical–continuum dielectric models, called the SMx models *(4-10)*. The original versions were SM1 *(4)*, SM1A *(4)*, SM2 *(5)*, and SM3 *(6)*, a newer version is SM2.1 *(8)*, and a version under development is SM4A *(9)*. These versions are all for water solvent. Parameterized versions for hexadecane solvent, called SM4C *(10)* and SM4A *(9)* (the latter having the same coulomb radii as the water SM4A model, and hence sharing the same name), are also under development. These models are concerned with transfer of a solute from the gas-phase into dilute solution. They are based on a quantum mechanical treatment of the internal electronic structure of a solute combined with a treatment of the solvent modeled as a nonhomogeneous continuum. The nonhomogeneity is very simple, consisting of an environment-specific first hydration shell superimposed on a homogeneous bulk dielectric medium. The interaction of the solute with the dielectric continuum is treated by classical electrostatics using partial charges on solute atoms, calculated either by Mulliken analyses *(11-13)* of neglect-of-diatomic-overlap *(14-17)* (NDDO) electronic wave functions or by new class IV charge models *(18)* based on semiempirical mappings of quantum mechanically derived partial charges.

The general framework of the SMx models is described extensively elsewhere, especially in overviewing *(7,19,20)* the SM2 and SM3 models, so this basis is reviewed only briefly here. This is done in Section 2, which also briefly reviews the difference of the aqueous models from the models (SM4C *(10)*, SM4A *(9)*) developed for hexadecane. Section 3 reviews recent algorithmic improvements *(8)* in the implementation of these models in the AMSOL code *(21)*. Section 4 reviews a new charge model *(18)*, which is used in the SMx models for $x = 4$.

Section 5 presents our first example of a new approach to solvation modeling, namely the use of parameters for a specific reaction or a specific range of solutes. This is called specific-reaction or specific-range parameterization (SRP) to distinguish it from using general parameterizations such as SM2.1 or SM4A, which have compromise parameters designed to model a broad range of solutes with a single parameter set. By targeting a specific range of solutes or reactions, the new approach allows faster parameterization and more accurate semiempirical predictions for the targeted systems. Section 6 presents an application of the SRP parameters of Section 5 to the Claisen reaction and compares the results to other recent work.

2. The SM2 and SM3 Models

In the SM2 and SM3 solvation models, we calculate the standard-state solvation free energy ΔG_S^o as a sum of two terms

$$\Delta G_S^o(aq) = \Delta G_{ENP}(aq) + G_{CDS}^o(aq) \tag{1}$$

where

$$\Delta G_{ENP}(aq) = \Delta E_{EN}(aq) + G_P(aq). \tag{2}$$

$\Delta E_{EN}(aq)$ is the change in the internal electronic kinetic and electronic and nuclear coulombic energy of the solute upon relaxation in solution, which is driven by the favorable electric polarization interaction with the solvent. $G_P(aq)$ is the electric polarization free energy, including both this favorable solute-solvent interaction and the unfavorable change in solvent molecule-solvent molecule interactions. Finally, G_{CDS}^o is the cavitation-dispersion-solvent-structural free energy.

In $\Delta G_{ENP}(aq)$, the solvent is treated as a continuum dielectric with bulk properties, and ΔG_{ENP} itself is the net favorable resultant free energy change due to bulk volume electrostatic effects. The solute internal, solvent internal, and solute-solvent effects are treated self-consistently by including the thermodynamic electrostatic effect as an operator in a self-consistent-field (SCF) semiempirical molecular orbital (MO) calculation on the solute, and this MO calculation is carried out by either of two popular parameterized models, namely Austin Model 1 (AM1) *(16)* or Parameterized Model 3 (PM3) *(17)*.

The second term in eq. 1, G_{CDS}^o, accounts for deviations from the first term due to the fact that the molecules in the first hydration shell of the solvent do not behave in the same way as the bulk dielectric. Thus, it includes the free energy of cavity formation, the short-range solute-solvent dispersion forces, and solvent-structure changing effects such as hydrogen bonding, the tightening of the first hydration shell around hydrophobic solutes, and the different degree of solvent polarization in the first hydration shell as compared to the bulk.

We approximate $G_P(aq)$ by a version of the generalized *(22-25)* Born *(26)* equation. This is a generalization of Born's treatment of a monatomic ion immersed in a dielectric medium. The generalization requires a form for two-center interactions in the dielectric medium and for treating the screening of some parts of the solute from the dielectric by other parts of the solute. Our treatment of these aspects of the generalization is based on the work of Still et al. *(27)*. The generalized Born equation is given by

$$G_P(aq) = -\frac{1}{2}\left(1 - \frac{1}{\varepsilon}\right)\sum_{i,i'} q_i q_{i'} \gamma_{ii'}, \tag{3}$$

where ε is the solvent dielectric constant, q_i is the net atomic partial charge on atom i of the solute, and $\gamma_{ii'}$ is a one-center ($i = i'$) or two-center ($i \neq i'$) coulomb integral. In the SMx models, these integrals are given by

$$\gamma_{ii'} = \{r_{ii'}^2 + \alpha_i \alpha_{i'} C_{ii'}(r_{ii'})\}^{-1/2}, \tag{4}$$

where α_i is the radius of the Born sphere associated with atom i, $r_{ii'}$ is the interatomic distance between atoms i and i', $C_{ii'}(r_{ii'})$ is given by

$$C_{ii'}(r_{ii'}) = C_{ii'}^{(0)}(r_{ii'}) + C_{ii'}^{(1)}(r_{ii'}), \tag{5}$$

where

$$C_{ii'}^{(0)} = \exp(-r_{ii'}^2/d^{(0)}\alpha_i\alpha_{i'}), \tag{6}$$

$d^{(0)}$ is an empirically optimized *(7)* constant equal to 4, and $C_{ii'}^{(1)}$ is given by

$$C_{ii'}^{(1)} = \begin{cases} d_{ii'}^{(1)} \exp\left(-d_{ii'}^{(2)} \middle/ \left\{1 - \left[r_{ii'} - r_{ii'}^{(1)} \middle/ r_{ii'}^{(2)}\right]^2\right\}\right), & \left|r_{ii'} - r_{ii'}^{(2)}\right| < r_{ii'}^{(2)} \\ \\ 0, & \text{otherwise.} \end{cases} \tag{7}$$

The constant $d_{ii'}^{(1)}$ is a new semiempirical element first introduced in SM1 *(4)* and is nonzero only for O–O and N–H interactions.

For the monatomic case, there is only a one-center term $(i = i' = 1)$, and α_i is set equal to a semiempirically determined intrinsic Born radius, ρ_i, where

$$\rho_i = \rho_i^{(0)} + \rho_i^{(1)}\left[-\frac{1}{\pi}\arctan\frac{q_i + q_i^{(0)}}{q_i^{(1)}} + \frac{1}{2}\right] \tag{8}$$

and where $q_i^{(1)}$ has been fixed at 0.1. Notice that the intrinsic Born radius depends on the atomic charge q_i, which is an integer for monatomic solutes. However, we also use eq. 8 as a starting point for polyatomic solutes, in which case q_i is not an integer. In the multicenter case, α_i is determined numerically, following the dielectric screening model introduced by Still et al. *(27)*. In this procedure α_i is chosen so that the Gp derived as in a monatomic case is equal to the Gp obtained by integrating about the atom the difference in the electronic free energy density fields of the charge distribution isolated in a vacuum and immersed in the dielectric solvent. Thus we calculate

$$\alpha_i^{-1} = \int_{\rho_i}^{\infty} \frac{dr}{r^2} \frac{A(r,\{\rho_i\})}{4\pi r^2} \tag{9}$$

where $A(r,\{\rho_{i'}\})$ is the numerically determined exposed surface area of a sphere of radius r centered at atom i, i.e., that area *not* included in any spheres centered around other atoms when those spheres have radii given by the set $\{\rho_{i'}\}$. (A sphere with such a radius is called a Born sphere in Section 3.) This area times dr is the part of the volume of the shell from r to $r + dr$ around atom i that is *in* the solvent.

The E, N, and P terms are obtained from the density matrix **P** of the aqueous-phase SCF calculation as

$$G_{ENP}(aq) = \frac{1}{2}\sum_{\mu\nu} P_{\mu\nu}\left(H_{\mu\nu} + F_{\mu\nu}\right) + \frac{1}{2}\sum_{i,i'\neq i} \frac{Z_i Z_{i'}}{r_{ii'}} \tag{10}$$

where **H** and **F** are respectively the one-electron and Fock matrices, μ and ν run over valence atomic orbitals, and Z_i is the valence nuclear charge of atom i (equal to the nuclear change minus the number of core electrons). The Fock matrix is given by

$$F_{\mu\nu} = F_{\mu\nu}^{(0)} + \delta_{\mu\nu}\left(1 - \frac{1}{\epsilon}\right)\sum_{i',\mu'\in i'} \left(Z_{i'} - P_{\mu'\mu'}\right)\gamma_{ii'}, \quad \mu \in i \tag{11}$$

where $\mathbf{F}^{(0)}$ is the gas-phase Fock matrix, and $\delta_{\mu\nu}$ is a Kronecker delta. $\mathbf{F}^{(0)}$ and $\mathbf{H}$ are the same as for gas-phase calculations *(14-17)*. The density matrix is determined self-consistently in the presence of solvent. Then

$$\Delta G_{ENP}(aq) = G_{ENP}(aq) - E_{EN}(g) \qquad (12)$$

The remaining contribution to the free energy of solvation beyond $\Delta G_{ENP}(aq)$ for *all* SM*x* models is calculated from

$$G^o_{CDS}(aq) = \sum_{\lambda=1}^{\Lambda} \sum_{i'} \left\{ \sigma_{i'}^{(0)}(\lambda) + \sigma_{i'}^{(1)}(\lambda) \left[F_{i'}\left(b^{(i')}\right) \right] \right\} A_{i'}\left(\beta_{i'}(\lambda), R_S(\lambda), \{\beta_i\}\right) (13)$$

where $\sigma_{i'}^{(0)}$ and $\sigma_{i'}^{(1)}$ are atomic surface tension parameters, λ is defined below, $\Lambda = 1$ for water and $\Lambda = 2$ for hexadecane, $F_{i'}$ is a function of the bond orders $b^{(i')}$ to atom i', and $A_{i'}(\beta_{i'}(\lambda), R_S(\lambda), \{\beta_i\})$ is the solvent-accessible surface area for atom i'. The latter is defined as the exposed surface area of atom i', which equals the exposed surface area of the atom-centered sphere with radius

$$\beta_{i'}(\lambda) = R_{i'} + R_S(\lambda) \qquad (14)$$

where $R_{i'}$ is the van der Waals radius of atom i', and $R_S(\lambda)$ is a solvent radius, taken for water as 1.4 Å. Exposed area is defined in this step as area that is *not* contained in any of the other atomic spheres when they also have radii given by eq. 14; this is why $A_{i'}$ depends on the full set of $\{\beta_i\}$. Notice that $A_{i'}$ represents the area of a surface through the first solvation shell around atom i', and thus, for a continuum solvent, it is proportional to the average number of solvent molecules in the first hydration shell *(28-30)*. In this model then, dispersion, solute-solvent hydrogen bonding, and disruption of solvent-solvent hydrogen bonding, as in the hydrophobic effect, are all assumed proportional to the number of waters that fit in the first hydration shell. This was the original motivation for making the nonhomogeneous part of the contribution to the first-hydration-shell free energy proportional to solvent-accessible surface area.

The bond orders are defined as elements of the covalent bond index matrix *(31)* as follows:

$$\left(b^{(i')}\right)_i = B_{ii'}, \qquad (15)$$

Note that the precise form of the function $F_{i'}$ in eq. 13 depends on which solvation model is under consideration. In the SM4-SRP model presented in section 5, all $\sigma_i^{(1)}$ are defined to be zero, so that no specification of $F_{i'}$ is required. This is a simplification relative to SM2, SM2.1, and SM3.

The essential physics behind the assumption that eq. 13 can represent the hydrophobic effect is consistent with a recent simulation study *(32)* of the hydration entropy of inert gases, which are often taken as atomic examples of hydrophobic solutes. This study showed that the solute-water orientational correlations are essentially confined to the first hydration shell and, within about 15%, are proportional to the number of water molecules in the first hydration shell. This is very encouraging although, of course, a more complete understanding of the first hydration shell must also take into account hydration enthalpies and the changes that occur for polar solutes. In a recent paper, Alary *et al.* (33) studied the hydration structure around a protein complex by molecular dynamics including 9154 explicit H_2O molecules. They found that the arrangement of first-shell water molecules strongly depends on the local nature of the protein surface and that the perturbation of the water structure by the protein essentially does not extend beyond one water layer. They concluded that the hydration structure in the immediate vicinity of any residue might be derived from data on this residue as a solute. All these conclusions are consistent with the present semiempirical approach in which the local surface tension, accounting for deviations from bulk solvent behavior in the first hydration shell, depends on the local character of the solute surface and is parameterized on the basis of small solutes.

For water, we use only one solvent radius, i.e., $\Lambda = 1$. In modeling solvation energies in hexadecane, on the other hand, we have used $\Lambda = 2$, with $R_S(1)$ representing the short range of the dispersion forces and $R_S(2)$ the long-range structural distortion of the solvent by the solute. This is a critical distinction between water and hexadecane. The water molecule, being small, is close enough to the solute for all its atoms to interact directly, whereas in a larger solute, dispersion affects only the closest subunits to the solute. Where $\Lambda = 2$, we may denote the solvent-accessible surface area calculated with $\lambda = 1$ as the area of the "dispersion shell" and solvent-accessible surface area calculated with $\lambda = 2$ as the area of the "structural solvation shell." For water both shells are taken as the traditional first solvation shell.

The parameters in the SM*x* models are adjusted to reproduce experimental data (34-37) for free energies of solvation. In parameterizing SM1, SM1a, SM2, and SM3, a specific approximately converged quadrature scheme was used for eq. 9, and the parameters made up for systematic deviations of these quadratures from converged ones. As discussed further below, the SM2.1 and SM4-type schemes are parameterized using well-converged quadratures.

3. Algorithms

A critical computational step that occurs repeatedly in the SM*x* solvation models, as just reviewed, is the calculation of the exposed surface area of an atom-centered sphere in the presence of several other spheres, centered at the other atoms of the solute. This step occurs in two contexts. First it occurs in the calculation of the electrostatic energy, where the singled-out sphere is either the Born sphere of one of the atoms or one of the enlarged spheres that occur as quadrature nodes in eq. 9, and the other spheres are the Born spheres of the other atoms. Second it occurs in the calculation of solvent-accessible surface areas of eq. 13 where the spheres are all van der Waals spheres augmented by the half-width, $R_S(\lambda)$, of the dispersion shell or structural solvation shell. In order to perform the calculations efficiently, we need an efficient algorithm to calculate these exposed surface areas and an efficient quadrature scheme for eq. 9. An efficient scheme for the latter will minimize the number of exposed surface areas to be calculated. Another way to minimize this number is by employing efficient SCF convergers since the dielectric screening calculation is embedded in the SCF process.

We next summarize some recent improvements we have made that speed up the exposed-surface-area calculation and that minimize the number of such calculations required.

Exposed Surface Area Calculations. The AMSOL code allows a choice of 3 algorithms: DOTS, GEPOL, and ASA.

DOTS. The DOTS algorithm is discussed elsewhere *(7)*. It has been greatly speeded up since the first released version of AMSOL.

GEPOL. The GEPOL algorithm has also been discussed in detail elsewhere *(38,39)*.

ASA. The ASA algorithm *(8)* is based on an analytic calculation of each exposed surface area A_i. It has seven steps for a given sphere i:

(1) Using triangular inequalities on interatomic distances and sphere radii, we set up a list $L^{(1)}$ of spheres that intersect sphere i. If sphere i is completely embedded in another sphere or not intersected by any other sphere, A_i is either 0 or 4π times the radius squared, and steps 2-7 are not needed.

(2) We note that each sphere j that intersects i defines a spherical cap $SC_j^{(i)}$, which is the portion of the surface of i that is contained in j. For each sphere j that intersects sphere i, we calculate the cosine of the angle between the unit vector from the center I of i to the center J of j and a vector from I to any point on the circle of intersection of i and j.

(3) We define a connectivity matrix $C^{(i)}$ whose element $C_{jk}^{(i)}$ is true if $SC_j^{(i)}$ intersects $SC_k^{(i)}$ and is otherwise false. Any spherical cap embedded entirely in another (larger) one or buried by two spherical caps is deleted from consideration at this stage.

(4) We calculate the buried surface area from isolated spherical caps and then delete them from $L_i^{(1)}$, making a new shorter list $L_i^{(2)}$.

(5) For each pair j, k of spheres in $L_i^{(2)}$ that intersect on the surface of i, we calculate unit vectors from I to the intersection points. We make a list of all such points that are not interior to any spherical cap $SC_l^{(i)}$ where $l \neq$ j, k. These are called free junction points. Spherical caps that are buried by more than two other spherical caps are detected at this stage and deleted from $L_i^{(2)}$, making a shorter list $L_i^{(3)}$. Also at this stage we check for junction points that are nearly shared by the circular intersections with i of more than two spheres. If these are detected, the radius of sphere i is increased by a factor of 1×10^{-11} to avoid roundoff instabilities in a later step. In this step, special attention was paid to avoiding unnecessary calculations of trigonometric functions and to re-using information from previous steps where possible to save work.

(6) For each spherical cap in $L_i^{(3)}$, we select a free junction point from its circular boundary as a starting point. The angles from the starting point to the center of the circular intersection to each of the other free junction points are then calculated and sorted by magnitude. There is an odd number of such angles. The smallest such angle is either the vertex of a spherical polygon or the vertex of a spherical cap slice. If the smallest angle is due to a spherical cap slice, then the sum of the odd angles

minus the sum of the even angles can be used to determine the contribution to A_i from spherical cap slices. Otherwise, 2π minus this quantity is used.

(7) Using $C^{(i)}$, we then count the number of spherical polygons, calculate the interior angles of each junction point vertex and add these to the sum of the interior angles of the intersection-centered vertices for the spherical polygons, the sum of the unused quantities in step 6. With this information, the contribution to A_i from spherical polygons can be calculated. Finally, by subtracting the surface area embedded in spherical caps, spherical cap sections, and spherical polygons, we obtain the exposed surface area of sphere i in a set of spheres.

Radial Quadratures for Dielectric Screening. The new algorithm for calculating the integral on the right side of eq. 9 begins by identifying a radius R for a sphere centered at I (the nucleus of atom i) such that a sphere of this radius would engulf all the Born spheres of the other atoms. Then we write

$$\frac{1}{\alpha_i} = \int_{\rho_i}^{R} dr \frac{a(r)}{r^2} + \frac{1}{R} \tag{16}$$

where

$$a(r) = \frac{A(r,\{\rho_k\})}{4\pi r^2}. \tag{17}$$

Next we define a sequence of quadrature segments $[b_{m-1}, b_m]$ by

$$b_m = \begin{cases} \rho_i, & m = 0 \\ b_{m-1} + T_1(1+F)^{m-1}, & m = 1,2,3,\ldots,M \end{cases} \tag{18}$$

where T_1 is determined by a two-step process. Starting with an input value of T_1 we calculate the minimum value of M such that $b_M \geq R$. Then for this M, we set

$$b_M = R \tag{19}$$

and recalculate T_1 to make this relation precisely true. Then we approximate $a(r)$ by

$$a^{(m)}(r) = c_0^{(m)} + c_1^{(m)}r \qquad b_{m-1} \leq r \leq b_m \tag{20}$$

by equating $a(r)$ to the numerically calculated values at the segment boundaries. This yields

$$\frac{1}{\alpha_i} = \frac{a(b_o)}{\rho_i} + \sum_{m=1}^{M-1} \frac{a(b_m) - a(b_{m-1})}{b_m - b_{m-1}} \left[\ln b_m - \ln b_{m-1}\right] \tag{21}$$

This algorithm yields good accuracy with less steps than were required by the previous one. One has two choices: (i) speed up the code greatly with systematic errors in the integration similar to those in SM2 and SM3 or (ii) speed up the code slightly with greatly improved convergence of the integrations. We chose the latter option as the basis of our parameterization for SM2.1 and the various SM4-type models. The default values of T_1 and F for SM2.1 and SM4-type models are $T_1 = 0.15$ Å and $F = 0.20$, but in cases where speed is of the essence the user may opt for lower levels of convergence. The user may also run in fixed-M mode in cases where strict continuity of the results as a function of geometry or initial guess is important.

SCF Convergers   As is well known, self-consistent fields governing the motions of electrons in molecular orbital calculations are achieved numerically by an iterative process. In gas-phase calculations the field in question is supplied by the nuclei and the other electrons. For molecules in solution, the free energy of solvation also contributes to the field, i.e., to the Fock operator (25). In devising an iterative strategy (which is *always* more of an art than a science), we should keep in mind that an update of the solvation terms is more costly than several SCF cycles with frozen solvation terms, with the precise value of "several" increasing with the average value of M in Section 3. Therefore an efficient code will need to be able to freeze or unfreeze the updates to the solvation terms in the Fock operator by some adaptive procedure. One must balance costs, though, against the counterconvergent nature of making an update after several cycles with frozen solvent if that update is too abrupt. Versions 4.0 and later of AMSOL involve damped updates at adaptively selected cycles. If the SCF converges without problems, damping is rapidly decreased. If problems are encountered, then, at first the damping factor is increased. If this does not cure the bad convergence, a damped level shift option is employed as the method of last resort.

4. New Charge Model

The traditional approach to semiempirical molecular orbital theory has placed all the parameters in the Hamiltonian, or equivalently in the Fock operator. Then, having obtained a wave function, it is used and analyzed by the same methods as if it were an *ab initio* wave function (e.g., Mulliken analysis (11-13), electrostatic potential fitting (40-42), etc.). We have recently proposed a new class of charge models that use a different philosophy (18). In particular an additional semiempirical layer is inserted between the semiempirical wave function and the physical observables inferred from it. This is accomplished by mapping partial charges obtained from the semiempirical wave function onto a new set of charges, q_k^{CM}, with the parameters of the mapping itself obtained from a general or specific parameterization. Like partial charge models based on fitting discrete charge representations to electrostatic potentials calculated from the full continuous wave electronic probability densities, the CM charge models make up for the errors inherent in using a discrete set of partial charges, but unlike electrostatic fitting charges, they also make up for the deficiencies (42c) of the electronic structure level and basis set employed because the parameters are fit to experimental data.

In the new charge models, the mapping occurs in two stages. The semiempirical Mulliken charge, $q_k^{(0)}$, for a given atom, k, is first adjusted by a parameterized change Δq_k,

$$q_k = q_k^{(0)} + B_k \Delta q_k, \tag{22}$$

In the second step of the mapping, the partial charges are readjusted to force the total charge on the molecule or ion to be the proper integral value (zero for neutrals). This is done by shifting charge locally between each atom whose charge has been adjusted and the atoms to which it has nonzero bond order. The final partial charge is then

$$q_k^{CM} = q_k^{(0)} + B_k \Delta q_k - \sum_{k' \neq k} B_{kk'} \Delta q_{k'} \tag{23}$$

where $B_{kk'}$ is the covalent bond index *(31)* between atoms k and k´. The sum of the bond orders from atom k to all other atoms is B_k. The following sum serves to ensure the charge is conserved in the mapping.

$$B_k = \sum_{k' \neq k} B_{kk'}. \tag{24}$$

where B_k is the sum of the covalent bond indices of atom k to all other atoms.

In work published elsewhere *(18)* we present new general parameterizations of such charge models, based on a large database of ions and neutrals. In the present study we use an early version parameterized only against neutral molecules. In the version we use here, $q_k^{(0)}$ is obtained by Mulliken analysis of the AM1 wave function, and Δq_k is obtained by

$$\Delta q_k = \begin{cases} -0.0283 & k = O \\ 0.1447 \sum_{k'} B_{kk'} & k = H, \, k' = O \\ 0 & k = C \end{cases}, \tag{25}$$

The parameters in eq. (25) were obtained from a database including 12 alcohols, 8 esters and lactones, 16 aldehydes and ketones, 9 acids, and 10 ethers. The root mean square (RMS) error between dipole moments derived from the new point charges and experimental dipole moments is 0.23 Debye for these 55 compounds. Point-charge dipole moments calculated from the unmodified AM1 Mulliken charges for the same set have an RMS error of 0.48 Debye. A comparison between dipole moments derived from the new point charges and those from the ab initio HF/6-31G* density or electrostatic potential for 7 representative members of the set (methanol, methyl formate, formaldehyde, cyclopropanone, acetic acid, dimethyl ether, and tetrahydrofuran) showed errors (compared to experiment) of 0.29 Debye for the dipoles from CM charges, 0.31 Debye for the dipoles from the Hartree-Fock density, and 0.31 Debye again for the dipoles corresponding to charges obtained from Hartree-Fock electrostatic potential. The electrostatic fitting was carried out by the ChelpG method *(42a,b)*.

Of course, as for any semiempirical method, there is the possibility that unphysical results will be obtained for compounds dissimilar to the test set used in development, but since the algorithm is linear and the test set contains diverse elements, such problems are not expected to be serious. Figure 1 compares the partial atomic charges for one example, benzaldehyde, which was not in the data set used for parameterization, as calculated at a variety of levels. The excellent perform-

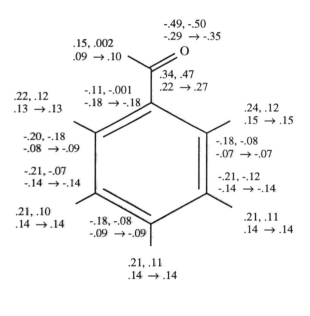

Experiment	$\mu = 2.8$	
Mulliken, ChelpG	4.0, 3.5	
AM1 → AM1-CM	2.6 → 2.9	

Figure 1. Partial charges and dipole moments (in Debye) for benzaldehyde. Next to each atom are shown four partial charges. Row 1 contains partial charges calculated by Mulliken analysis (given first) and ChelpG analysis of the HF/6-31G* wave function at the HF/6-31G* optimized geometry. Row 2 contains partial charges calculated by Mulliken analysis or the new charge model form the AM1 wave function at the AM1 optimized geometry. Below the molecule is given the experimental dipole moment, and below that are given dipole moments calculated from each of the four sets of partial charges described above in the same order.

ance of the new charge model for this molecule, as judged by its comparison to both the experimental dipole moment *(43)* and that derived from the MP2/6-31G* continuous density, illustrates the robust nature of the mapping. Of particular note is the effect of the mapping on the carbonyl bond dipole, which is considerably increased relative to the AM1 Mulliken charge result. The polarization is still larger for the HF/6-31G* Mulliken or electrostatic-potential fitted charges, but these methods both lead to charges that overestimate the dipole moment derived therefrom.

The utility of the more accurate charges yielded by the new charge model is discussed below in the context of solvation modeling.

We note that in SM1, SM1a, SM2, SM3, and SM2.1, the electrostatic term, ΔG_{ENP}, suffers deficiencies from the inaccuracy of the AM1 and PM3 charge distributions. These are absorbed into the parameterized surface tensions. Solvation models based on charges obtained from the new kind of charge model, on the other hand, are called SM4-type solvation models, e.g., SM4A, SM4C, SM4-SRP, and are expected to be less prone to such systematic deficiencies.

5. Specific Reaction Parameters for the Claisen Rearrangement

The critical step in the advancement of semiempirical molecular orbital theory from a promising approach studied by theoretical chemists to an everyday tool in the laboratory was the development of general parameterizations, such as CNDO/2 *(44)*, INDO/S *(45)*, MINDO/3 *(46)*, etc. The current working-horse general parameterizations are AM1 *(16)* and PM3 *(17)*. General parameterizations are not, however, the only way in which semiempirical molecular orbital theory can be useful. For example, one may want to study a particular class of compounds for which the general parameterization is known to have systematic errors. Alternatively, one may not have sufficient experimental data or enough time available to calibrate a general parameterization for a particular solvent at a given level of theory. In such cases, a parameterization for a specific range of compounds or an individual reaction might be useful. We call this an SRP model, as mentioned in the introduction. SRP-type models have been used for gas-phase studies of reaction dynamics *(47-49)*; we apply them here to solvation. The original goal of developing SRP models is to provide a convenient and flexible way of studying specific reactions or ranges of compounds, but another possible use would be to illuminate the *nature* of the deficiencies of general models.

In this section we present SRP solvation model in both water and hexadecane as an example of the new approach. The organic reaction in this case is the Claisen rearrangement, a [3,3] sigmatropic shift which converts an allyl vinyl ether into a γ,δ-unsaturated aldehyde (Figure 2). The reaction itself will be discussed more fully in the next section. Here, we first select a data set, and then we select a functional form for the SRP parameterization. The SRP solvation model that is presented in this section will be denoted SM4-SRP where the 4 denotes that it is based on CM-type charges, as explained at the end of Section 4. More specifically, we use the AM1 Hamiltonian and CM charges so an unnecessarily complete but more descriptive name is AM1-CM-SM4-SRP, just as SM2 may be called AM1-SM2 to emphasize the Hamiltonian.

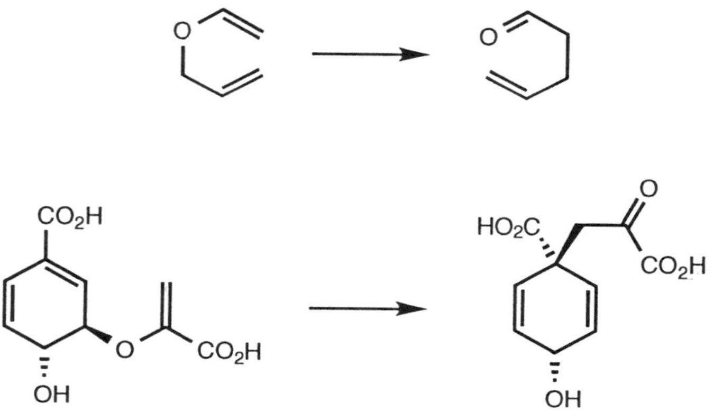

Figure 2. The Claisen rearrangement converts allyl vinyl ether to 4-pentenal. Also shown is the Claisen rearrangement which converts chorismic acid into prephenic acid, and important step in the biosynthetically important shikimic acid pathway.

An appropriate data set for determining SRP parameters for the Claisen rearrangement should include alkanes, alkenes, six-membered rings, ethers, aldehydes, and conjugated double bonds, since these are the functional groups involved in the reactants, products, and transition state. Within this range, we choose a representative set of molecules based on the availability *(34-37)* of experimental free energies of solvation. The number of parameters to be determined was reduced to a minimum by setting all $d_{ii}^{(1)} = 0$. Parameters $\sigma_i^{(1)}$ for the SM4-SRP model are defined to be zero. The optimal values of the SRP parameters are provided in Table 1, and the performance of the SRP models for the parameterization test set is detailed in Table 2. The hexadecane SRP model has an encouragingly small overall root-mean-square (RMS) error of 0.5 kcal/mol. The water SRP model is good at predicting the hydrophobic effect for hydrocarbons; it exhibits somewhat larger errors for the {C,H,O} compounds, with a tendency to underestimate the solvation free energies of ethers while overestimating those for aldehydes. Similar behavior was noted for SM2 *(5,7)*.

6. Solvation Effects on the Claisen Rearrangement

The Claisen rearrangement is an electrocyclic reaction converting an allyl vinyl ether into an unsaturated aldehyde or ketone via a [3,3] sigmatropic shift (Figure 2).

Table 1. Optimized parameters for the Claisen SM4-SRP models (distances are in Ångstroms, charges in electronic charge units, and surface tensions in cal mol^{-1} Å^{-2})

Parameter	Water			n-Hexadecane		
	H	C	O	H	C	O
$\rho_k^{(0)}$	0.590	1.798	1.350	0.590	1.798	1.350
$\rho_k^{(1)}$	1.283	0.000	0.000	1.283	0.000	0.000
$q_k^{(0)}$	−0.300	0.000	0.000	−0.300	0.000	0.000
R_k	1.200	1.700	2.000	1.200	1.700	2.000
$\sigma_k^{(0)}(1)$	8.886	−6.910	−6.424	−48.000	−56.200	−42.900
$\sigma_k^{(0)}(2)$				16.700	16.700	16.700
$R_S(1)$	1.400	1.400	1.400	2.000	2.000	2.000
$R_S(2)$				4.900	4.900	4.900

Table 2. Calculated SM4-SRP and experimental free energies of solvation (kcal/mol) for hydrocarbons, aldehydes, and ethers

	Water				n-Hexadecane			
	Calculated				Calculated			
	ENP	CDS	Total	Expt.	ENP	CDS	Total	Expt.
Hydrocarbons								
ethane	−0.05	1.38	1.33	1.83	−0.03	−0.60	−0.63	−0.66
propane	−0.03	1.66	1.62	1.96	−0.02	−1.34	−1.36	−1.42
butane	0.01	1.93	1.94	2.08	0.01	−2.08	−2.07	−2.19
pentane	0.06	2.20	2.26	2.38	0.03	−2.82	−2.79	−2.94
2–methylpropane	0.00	1.82	1.82	2.32	0.00	−1.99	−1.99	−1.91
hexane	0.12	2.47	2.59	2.49	0.06	−3.55	−3.49	−3.63
2–methylpentane	0.10	2.28	2.38	2.88	0.05	−3.29	−3.24	−3.45
cyclohexane	0.04	2.25	2.30	1.23	0.02	−2.64	−2.62	−3.96
heptane	0.17	2.75	2.92	2.62	0.09	−4.29	−4.20	−4.32
2,4–dimethylpentane	0.16	2.41	2.57	2.88	0.08	−3.87	−3.79	−3.87
octane	0.23	3.02	3.25	2.89	0.12	−5.03	−4.91	−5.01
neopentane	0.02	1.94	1.97	2.50	0.01	−2.53	−2.52	−2.47
benzene	−2.12	0.88	−1.24	−0.87	−1.03	−2.71	−3.74	−3.81

Continued on next page

Table 2. Continued

| | Water | | | | n-Hexadecane | | | |
| | Calculated | | | | Calculated | | | |
	ENP	CDS	Total	Expt.	ENP	CDS	Total	Expt.
toluene	−2.11	1.23	−0.88	−0.89	−1.02	−3.35	−4.37	−4.55
o–xylene	−2.15	1.50	−0.65	−0.90	−1.01	−4.00	−5.01	−5.36
m–xylene	−2.10	1.58	−0.52	−0.84	−1.05	−3.87	−4.92	−5.26
p–xylene	−2.08	1.59	−0.49	−0.81	−1.00	−4.00	−5.00	−5.25
RMS error			**0.42**				**0.37**	
{H,C,O} cmpds								
dimethyl ether	−2.19	0.83	−1.36	−1.90	−1.62	−0.39	−2.01	−1.67
diethyl ether	−1.74	1.74	0.00	−1.63	−1.44	−1.15	−2.59	−2.47
dipropyl ether	−1.39	2.31	0.92	−1.15	−1.35	−1.88	−3.23	−3.09
diisopropyl ether	−1.36	2.10	0.74	−0.53	−1.41	−2.62	−4.03	−3.77
dibutyl ether	−1.25	2.85	1.60	−0.83	−2.11	−3.13	−5.24	−5.43
anisole	−3.55	0.84	−2.71	−2.44	−1.29	−3.36	−4.65	−4.59
tetrahydropyran	−1.96	1.56	−0.40	−3.12	−1.26	−4.10	−5.36	−5.26
dioxane	−3.59	0.85	−2.74	−5.05	−1.23	−4.83	−6.06	−5.96
phenetole	−3.32	1.31	−2.01	−2.21	−1.20	−5.57	−6.77	−6.68
acetaldehyde	−5.22	0.28	−4.94	−3.50	−0.74	−0.78	−1.52	−1.28
propanal	−4.57	0.70	−3.87	−3.44	−0.63	−2.49	−3.12	−2.80
butanal	−4.30	0.97	−3.33	−3.18	−0.49	−3.97	−4.46	−4.07
pentanal	−4.26	1.24	−3.02	−3.03	−0.50	−3.67	−4.17	−3.48
benzaldehyde	−5.79	0.19	−5.60	−4.02	−0.42	−5.44	−5.86	−5.45
hexanal	−4.17	1.52	−2.65	−2.81	−1.52	−3.45	−4.97	−5.35
heptanal	−4.13	1.79	−2.34	−2.67	−0.71	−2.18	−2.89	−4.07
octanal	−4.06	2.06	−2.00	−2.29	−1.35	−1.72	−3.07	−3.84
nonanal	−4.01	2.33	−1.68	−2.08	−1.42	−4.30	−5.72	−5.64
E–2–hexenal	−5.68	0.53	−5.15	−4.63				
E–2–butenal	−5.62	0.41	−5.21	−4.22				
E,E–2,4–hexadienal	−5.26	1.03	−4.23	−3.68				
E–2–octenal	−5.16	1.57	−3.59	−3.44				
RMS error			**1.25**				**0.59**	

The reaction has demonstrated synthetic utility *(50-52)*. The Claisen rearrangement has the additional distinction of being one of the few electrocyclic reactions which has been demonstrated to occur in a biosynthetic pathway; in particular, the rearrangement of chorismic acid to prephenic acid, catalyzed by the enzymatic influence of chorismate mutase, plays a role in the shikimic acid pathway of primary plant metabolism *(53-55)* This has prompted the development of antibodies capable of catalyzing the Claisen rearrangement *(56,57)*.

The Mechanism of the Claisen Rearrangement.   Studies of the Claisen rearrangement have focused on several different mechanistic aspects of the reaction. One question of interest has been the degree of C–O bond breaking/C–C bond making which characterizes the transition state as inferred from analysis of kinetic isotope effects in the rearrangement *(58-63)*. Experimental results suggest that C–O bond breaking is significantly advanced over C–C bond making, leading to a transition state which involves two comparatively weakly coupled three-heavy-atom fragments *(58,59,62)*. Although early computational studies at the MNDO level were not in agreement with such an analysis *(60)*, more reliable levels of theory, including ab initio RHF calculations *(61)*, AM1 semiempirical calculations *(64)*, and MCSCF calculations *(63)*, all support the experimental interpretation.

An additional question of interest with respect to the Claisen transition state has been the extent to which it is characterized by interfragment charge separation and/or biradicaloid character. Figure 3 illustrates some of the limiting possibilities. Transition state **i** corresponds to the situation predicted by MNDO calculations *(60)*: C–C bond making is advanced over C–O bond breaking, and the overall character of the transition state is biradicaloid, that is, there are strong through-bond interactions

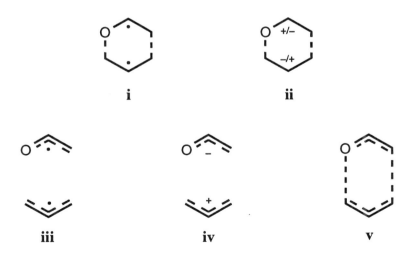

Figure 3. Limiting resonance structures which may be used to represent the transition state of the Claisen rearrangement.

between the oxallylic and allylic radical fragments. Alternatively, transition state **ii** corresponds to a similar situation with respect to relative bond making/breaking, but the fragments are more zwitterionic-like—although this structure is presented for completeness, it has never been suggested to be germane to the Claisen rearrangement.

Transition state **iii** corresponds to the case in which the interfragment coupling is considerably weaker (i.e., bond breaking is now advanced over bond making) with the character of the TS being close to a diradical (i.e., the net charges on the two fragments are near zero). The distinction between a transition state which is a "diradical" and one which has "biradicaloid" character is that there is no significant interaction between the radical centers in the former, but there is in the latter. Transition state **iv** resembles **iii**, except that the character of the TS is zwitterionic, being well described as a combination of an enolate anion and an allylic cation. Finally, transition state **v** refers to an "aromatic" transition state. This term was coined by Dewar to refer to a pericyclic transition state which (i) has roughly equal bond making and bond breaking and (ii) enjoys significant resonance stabilization from the mixing of the various molecular orbitals playing a significant role in the rearrangement even though the interfragment separation is large compared to the biradicaloid case **i** *(65)*. Since the aromatic MO's are delocalized, such a transition state may or may not be zwitterionic (and therefore synonymous with **iv**), but it is not a diradical.

Experimental studies have attempted to investigate the most likely character of the transition state by examination of substituent and solvent effects on the rate of the rearrangement. Based on rate accelerations observed in polar solvents, and sensitivity to substituent positioning, several groups have advocated the dipolar transition state **iv/v** *(66-70)*. However, Gajewski and co-workers have been reluctant to interpret the data similarly, and have suggested that a transition state **v** with quite limited dipolarity is more reasonable *(69,71)*.

Gas-phase modeling has been unable to provide a definitive answer since, at least for the parent allyl vinyl ether, the transition state structure proves extremely sensitive to the level of theory employed. At the RHF/6-31G* level of theory, the transition state has an interfragment distance (defined as the average of the breaking and forming C–O and C–C bonds, respectively) of about 2.1 Å. The enolate and allyl fragments have charges of –0.34 and 0.34 electrons, respectively, as analyzed from Mulliken population analysis *(61,63)* . Optimization at the MP2/6-31G* level gives a transition state structure which is slightly more biradicaloid in character, and kinetic isotope effects based on calculated harmonic frequencies are in poor agreement with experiment *(63)*. At the MCSCF level, on the other hand, correlating six electrons in six orbitals delivers a transition state structure which is much closer to transition state **iii**, with the distance between fragments increased to roughly 2.3 Å and the charge separation reduced to ± 0.16 electrons on each fragment *(63)*. The "true" transition state structure in the gas phase may be a little tighter than the MCSCF one due to dynamical electron correlation.

Calculations at the semiempirical AM1 level, by contrast, led Dewar to suggest that both biradicaloid (**i**, interfragment separation about 1.7 Å) and aromatic (weakly dipolar **v**, interfragment separation about 1.8 Å) transition states exist, corresponding to different reaction paths *(64,72)*. This analysis relies, however, on a somewhat dubious interpretation *(72)* of the molecular hypersurface. Although a similar bifurcation of reaction path was observed from MCSCF computational results obtained for the Cope rearrangement *(73,74)*, which is the all carbon analog of the Claisen (i.e., the oxygen atom is replaced by a methylene group), inclusion of dynamical correlation suggests there to be only a single reaction path *(75)*.

A conservative interpretation of all of these results taken together is that the nature of the Claisen rearrangement transition state is quite sensitive to external perturbations, whether those perturbations be in the form of structural modifications of the parent allyl vinyl ether fragments or in the form of differences in environmental conditions such as introducing or changing a solvent. With this caveat in mind, however, it remains interesting to consider what kinds of interactions with the surrounding solvent give rise to the phenomenologically observed rate accelerations of Claisen rearrangements in solvents of increasing polarity *(66,68-71)*. A number of computational studies employing quite different techniques have been undertaken with this goal in mind *(76-80)*, and we devote the remainder of this section to their comparison and critical analysis.

Modeling the Claisen Rearrangement in Solution. Two of the present authors performed the first computational study of the effect of solvent on the Claisen rearrangement, employing a GB/ST-type model, in particular SM2 *(76)*. In addition to the parent system, allyl vinyl ether, this study also considered all possible substitutions of a methoxy group for a proton in the molecule. Aqueous acceleration of the Claisen rearrangement was accounted for by electric polarization and first-hydration-shell hydrophilic effects. Polarization was found to dominate, although the relative magnitudes and even the signs of the two effects were quite sensitive to the substrate substitution pattern. Hydrophobic acceleration, from the joining together of the two terminal methylene groups in the transition state, was found to be accelerating as well, but only worth about 0.3 kcal/mol in every case. The SM2 calculations indicated that solvent polarization in the parent and methoxy-substituted systems tended to be most responsive to the oxygen atoms because of their smaller volume and moderately large charges. Geometrical changes on going from the reactant conformation to the transition state which sequestered charges of opposite sign into well separated regions of space contributed to rate acceleration, while the converse was true when charges of opposite sign were forced into proximity. Although the *intra*fragment charges were consistent with the enolate/allyl cation view of the transition state, i.e., negative charge buildup at either end of the former fragment and positive charge buildup at either end of the latter, the *inter*fragment charge separation was fairly small at only 0.14 electrons from Mulliken population analysis (Table 1). While this charge reorganization was related to the increase or decrease in overall dipole moment on progressing to the transition state, consideration of only the dipole moment proved insufficient to reproduce the distributed-monopole results of the SM2 model. Because competing axial and equatorial substitution patterns and competing chair and boat transition states involved different spatial arrangements of the critical charges, it was shown that energetic discrimination between competing reaction routes delivering stereochemically different products could be substantially changed in water in comparison to the gas phase or to nonpolar solvents.

The SM2 calculations led to a rate acceleration of 3.5 on going from the gas phase to aqueous solution at 298 K. Such quantitative predictions form the SM2 model are less reliable than the qualitative insights for two reasons: first, biradicals (and often zwitterions) require configuration interaction (CI) for an accurate representation of the electronic wave function, and the AM1 electronic structure level employed in the SM2 calculations does not allow for this; second, the AM1 Mulliken charges used in SM2 are not quantitatively accurate.

Severance and Jorgensen (77) analyzed the available experimental results and suggested that a more accurate rate acceleration would be on the order of 1000 at 348 K. These authors calculated the gas-phase reaction coordinate at the HF/6-31G* level for the Claisen rearrangement, and then solvated structures chosen periodically along that gas-phase path using Monte-Carlo simulations (81) employing the 4-Point Transferable Intermolecular Potentials (TIP4P) water model (82), Optimized Potentials for Liquid Simulations (OPLS) van der Waals parameters (83), and electrostatic-potential-derived HF/6-31G* atomic partial charges in order to generate an aqueous reaction coordinate. The simulations suggested a rate acceleration of 644 at 298 K, in good agreement with the extrapolated factor of about 1000. Analysis of the simulations revealed enhanced hydrogen bonding to oxygen in the transition state (1.9 average total number of hydrogen bonds) by comparison to the starting ether (0.9 average). This enhanced hydrogen bonding arises in part from increased accessibility of the oxygen atom in the transition state and also from its increased charge; in the gas-phase calculations from which the partial charges were taken, the oxygen becomes more negative by 0.07 electrons (80).

The AM1-SM2 quantum mechanical continuum model and the HF/6-31G*-OPLS-TIP4P explicit water simulation agreed nicely on one point of particular interest. As noted in a short commentary that summarized them (84), "Both studies show that reorganization of the charge distribution is more subtle and not in accord with an ion pair-like transition state. Rather polarization occurs within the enolate moiety of the transition state, rendering the oxygen more negative." Moreover, both studies indicated the bulk of the acceleration to be associated with improved solvation of the oxygen atom portion of the molecule.

Gao recently reported a study of the Claisen rearrangement which complements the earlier two in an extremely interesting way (78). Taking the same gas-phase reaction coordinate as Severance and Jorgensen, Gao performed combined QM/MM (85) simulations in which the solute was modeled using the AM1 Hamiltonian and OPLS van der Waals parameters, with the quantum mechanical Hamiltonian perturbed by surrounding explicit water molecules treated classically using the TIP3P (82) water model. This approach thus includes electronic polarization of the solute, together with explicit simulation of the solvent. The rate acceleration (gas phase → aqueous solution) calculated by Gao was 368 at 298 K. Gao emphasized several points of agreement between his study and those of the earlier authors. As before, charge buildup at oxygen was identified as the primary contributor to rate acceleration. In addition, little interfragment charge separation was observed in either the gas-phase or the solution electronic structures. Moreover, solute polarization in the transition state was calculated by Gao to account for 35% of the rate acceleration, in good agreement with the SM2 results. Although the study of Severance and Jorgensen does not explicitly account for solute polarization, charges derived from HF/6-31G* calculations appear to be systematically too large, and thus to some extent they mimic intrinsic incorporation of polarization effects.

Specific Reaction Parameter Modeling of the Claisen Rearrangement in Solution. The sensitivity of the rate acceleration to atomic partial charges and as well the sensitivity of the transition state structure to level of theory prompted us to investigate more closely the dependence of calculated rate acceleration on these factors within the SMx series of models. We have therefore calculated the rate accelerations for the aqueous Claisen rearrangement using both the SM2 and SM4-SRP models, in each case employing three choices of transition state geometry: (i) self-consistently optimized (in solution) chair transition states, (ii) the HF/6-31G* gas-phase transition state, and (iii) the MCSCF/6-31G* gas-phase transition state. The partial charges calculated for these different possibilities are presented in Table 3, and the accelerations are listed in Table 4. Several points warrant further discussion.

(i) First, it is obvious that looser transition states give rise to greater acceleration. This acceleration derives partly from larger atomic partial charges, with effects at oxygen predominating, and partly from greater accessibility of hydrophilic sites to solvent. It is especially interesting to note the synergistic effect of using both better transition state structures and a better charge model: the accelerations calculated by SM2 and SM4-SRP models are essentially identical at the semiempirically optimized transition state structure, which has the reacting fragments too tightly coupled. As the transition state structure becomes looser, the SM4-SRP model predicts steadily enhanced acceleration over the SM2 model, although the latter model itself predicts roughly 80-fold acceleration on going from the semiempirically optimized TS structure to the MCSCF gas-phase TS structure. Since the true transition state structure is probably looser than the HF/6-31G* one, and the looser transition state gives rise to larger aqueous acceleration, calculations based on the HF/6-31G* wavefunction and gas-phase reaction path should not give the quantitatively correct acceleration.

(ii) For the tight semiempirical case, it is interesting to note that bringing the methylene termini closer together in the transition state structure gives rise to a hydrophobic acceleration in water; however, this same change in structure corresponds to a solvo*philic decel*eration in hexadecane. In the latter case, the effect is sufficient to slow the reaction down relative to the gas phase! As expected, this effect becomes smaller in magnitude for both solvents in the looser transition states, in particular by 0.1 kcal/mol for water and 0.2 kcal/mol for hexadecane.

(iii) If the polarization free energies were to be calculated using the gas-phase partial charges, it is clear that water as solvent would give about twice the polarization as hexadecane—this follows from the difference in the dielectric constants of the two solvents and how that difference modifies the pre-factor on the right-hand side of eq. 3. This analysis suggests, using simple transition state theory and temporarily ignoring the small differences in $\Delta G_{CDS}^{o,\ddagger}$, that the rate acceleration in water

Table 3. Partial charges (including attached hydrogens) in Claisen chair transition state

TS geometry	C-1	C-2	O-3	C-4	C-5	C-6	Charge separation[a]
Mulliken gas-phase charges[b]							
AM1	0.01	0.10	−0.25	0.20	−0.17	0.10	0.14
HF/6-31G*[c]	−0.14	0.41	−0.61	0.34	0.00	0.00	0.34
MCSCF/6-31G*[d]	−0.04	0.35	−0.47	0.10	0.06	0.01	0.16
CM gas-phase charges							
AM1	0.01	0.14	−0.30	0.22	−0.17	0.10	0.15
HF/6-31G*	−0.06	0.18	−0.38	0.23	−0.11	0.13	0.26
MCSCF/6-31G*	−0.09	0.18	−0.41	0.25	−0.09	0.14	0.32
SM2 water charges[e]							
SM2	0.03	0.10	−0.28	0.21	−0.18	0.13	0.15
HF/6-31G*	−0.06	0.14	−0.37	0.25	−0.12	0.17	0.29
MCSCF/6-31G*	−0.12	0.14	−0.43	0.30	−0.10	0.22	0.41
SM4-SRP water charges[e]							
SM4-SRP	0.04	0.13	−0.36	0.22	−0.18	0.13	0.19
HF/6-31G*	−0.06	0.18	−0.45	0.26	−0.12	0.19	0.33
MCSCF/6-31G*	−0.12	0.18	−0.52	0.32	−0.11	0.24	0.46
SM4-SRP hexadecane charges[e]							
SM4-SRP	0.02	0.10	−0.27	0.21	−0.17	0.11	0.15
HF/6-31G*	−0.06	0.14	−0.35	0.24	−0.11	0.15	0.27
MCSCF/6-31G*	−0.10	0.14	−0.40	0.27	−0.10	0.18	0.36

[a] Absolute value of total charge on the enolate/allyl fragments. [b] All geometries fully optimized, all charges at the same level of theory. [c] Reference *(61)*. Note that this report *(61)* typographically interchanged the charges for the chair TS and the starting ether, as made clear in reference *(63)*. Those errors were included in Table III of reference *(76)*. [d] Reference *(63)*. [e] SM*x* geometry optimized in solution, all others are frozen gas-phase structures.

Table 4. Solvent-induced rate acceleration of the Claisen rearrangement at 298 K

Solvation Model	Solvent	TS geometry	Rate Acceleration[a]
SM2	water	SM2	3.5[b]
		HF/6-31G* [c]	21
		MCSCF/6-31G* [d]	270
SM4-SRP	water	SM4-SRP	3.4
		HF/6-31G*	46
		MCSCF/6-31G*	1400
SM4-SRP	n-hexadecane	SM4-SRP	0.7
		HF/6-31G*	2.3
		MCSCF/6-31G*	11

[a] Gas phase to solution. [b] Reference *(76)*. [c] Gas-phase geometry from reference *(61)*. [d] Gas-phase geometry from reference *(63)*.

should be roughly the square of that in hexadecane. However, self-consistent *additional* polarization of the electronic structure by the solvent reaction field is *also* more favorable in water, and this effect accounts for roughly 10-fold more aqueous acceleration than would otherwise be expected. Such a prediction cannot be made using classical models with fixed atomic charges. The net result from the SRP models is a ratio of 124:1 for the rate of the Claisen rearrangement of allyl vinyl ether in water as compared to hexadecane. This is in excellent agreement with an experimentally observed ratio of 214:1 for the rates of related sets of Claisen rearrangements measured in water and cyclohexane *(71)*.

(iv) The self-consistent-field SM4-SRP charge separation between fragments is found to be 0.46 electrons at the MCSCF transition state structure. This value is somewhat larger than the 0.34 electron separation derived from electrostatic potential fitting of HF/6-31G* charges as used by Severance and Jorgensen in their Monte Carlo simulations. All of the computational studies, then, point to a moderate degree of interfragment charge separation; however, it is the *intra*fragment charge organization which is critical to the calculated rate accelerations.

(v) Finally, it is worth emphasizing that the above results are extremely useful in terms of understanding the mechanism of the Claisen rearrangement in the condensed phase, but are probably no better than semiquantitative in terms of predicting absolute rate accelerations. It is clear that accounting for dynamical correlation *and* for the effect of solvent on the reaction coordinate itself may significantly alter the structure of the transition state. Of course, these two effects should be in opposite directions, dynamical correlation favoring a tighter structure and solvation a looser one, so there might be a fortuitous cancellation of errors in this instance.

In closing this section, it is appropriate to compare these results to the quite different approach used by Gajewski *(79)*, who has employed factor analysis to correlate the rate acceleration of the Claisen rearrangement in a wide variety of solvents against a series of empirical solvent descriptors. Gajewski suggests that solvent hydrogen bond donating ability is the factor most closely correlated with rate acceleration. This analysis clearly agrees with the Monte Carlo simulations of Severance and Jorgensen and of Gao, where enhanced hydrogen bonding is explicitly observed for transition state structures relative to reactants. Although the continuum model results presented here obviously do not include an explicit representation of the solvent, they are quite consistent with all of the other studies. In particular, the enhanced hydrogen bonding observed in the simulations is due primarily to the increased charge and increased accessibility of the oxygen atom in the Claisen transition state structure, and the continuum model finds rate acceleration to be similarly dependent on these two factors. Of course, it is not at all trivial to separate hydrogen bonding into electrostatic and non-electrostatic components, nor is it easy to separate electrostatic effects into hydrogen bonding and non-hydrogen-bonding components, and the extent to which those portions of experimental free energies of solvation assignable to hydrogen bonding effects are incorporated into the ENP and CDS terms of the SM*x* models is not clear. Thus, although the SM*x* results cannot be interpreted quantitatively in terms of hydrogen-bonding effects, they are not inconsistent with such effects dominating.

7. Summary and Concluding Remarks

We have presented an overview of recent improvements and extensions of the quantum mechanical generalized-Born-plus-surface-tensions SM*x* solvation models, and we illustrated a new way to treat reaction-specific solvation effects using the Claisen rearrangement as an example. General improvements discussed included more efficient algorithms in the AMSOL computer code for calculating analytical solvent accessible surface areas, for evaluating quadratures related to dielectric screening, and for solving the solvent-modified SCF equations. In addition, we have illustrated the use of class IV charge models. All of these improvements were used to develop a set of specific reaction parameters for calculating the effect of solvation on the Claisen rearrangement both in hydrocarbon solvent and in water. Calculated rate accelerations were in excellent agreement with the best available experimental estimates and also with other modeling approaches.

Acknowledgments.   The authors are grateful to Ken Houk for providing a preprint of reference *(63)*. This work was supported in part by the National Science Foundation.

Literature Cited
(1)   Daudel, R. *Quantum Theory of Chemical Reactivity*; Reidel: Dordrecht, 1973.
(2)   Reichardt, C. *Solvents and Solvent Effects in Organic Chemistry*; VCH: New York, 1990.
(3)   Kreevoy, M. M.; Truhlar, D. G. In *Investigation of Rates and Mechanisms of Reactions, Part I*; 4th ed.; C. F. Bernasconi, Ed.; Wiley: New York, 1986; p. 13.
(4)   Cramer, C. J.; Truhlar, D. G. *J. Am. Chem. Soc.* **1991**, *113*, 8305.
(5)   Cramer, C. J.; Truhlar, D. G. *Science* **1992**, *256*, 213.
(6)   Cramer, C. J.; Truhlar, D. G. *J. Comp. Chem.* **1992**, *13*, 1089.
(7)   Cramer, C. J.; Truhlar, D. G. *J. Comput.-Aid. Mol. Des.* **1992**, *6*, 629.
(8)   Liotard, D. A.; Hawkins, G. D.; Lynch, G. C.; Cramer, C. J.; Truhlar, D. G. to be published.
(9)   Storer, J. W.; Giesen, D. J.; Cramer, C. J.; Truhlar, D. G. to be published.
(10) Giesen, D. J.; Storer, J. W.; Cramer, C. J.; Truhlar, D. G. to be published.
(11) Mulliken, R. S. *J. Chem. Phys.* **1935**, *3*, 564.
(12) Mulliken, R. S. *J. Chem. Phys.* **1955**, *23*, 1833.
(13) Mulliken, R. S. *J. Chem. Phys.* **1962**, *36*, 3428.
(14) Pople, J. A.; Segal, G. A. *J. Chem. Phys.* **1965**, *43*, S129.
(15) Dewar, M. J. S.; Thiel, W. *J. Am. Chem. Soc.* **1977**, *99*, 4899.
(16) Dewar, M. J. S.; Zoebisch, E. G.; Healy, E. F.; Stewart, J. J. P. *J. Am. Chem. Soc.* **1985**, *107*, 3902.
(17) Stewart, J. J. P. *J. Comp. Chem.* **1989**, *10*, 209.
(18) Storer, J. W.; Giesen, D. J.; Cramer, C. J.; Truhlar, D. G. *J. Comp. Chem.* submitted for publication.
(19) Cramer, C. J.; Truhlar, D. G. In *Reviews in Computational Chemistry*; K. B. Lipkowitz and D. B. Boyd, Eds.; VCH: New York; Vol. 6; in press.
(20) Cramer, C. J.; Truhlar, D. G. In *Theoretical and Computational Chemistry: Solute/Solvent Interactions*; P. Politzer and J. S. Murray, Eds.; Elsevier: Amsterdam; Vol. 2; in press.
(21) Cramer, C. J.; Lynch, G. C.; Hawkins, G. D.; Truhlar, D. G. *QCPE Bull.* **1993**, *13*, 78.
(22) Hoijtink, G. J.; de Boer, E.; Van der Meij, P. H.; Weijland, W. P. *Recl. Trav. Chim. Pays-Bas* **1956**, *75*, 487.
(23) Peradejordi, F. *Cahiers Phys.* **1963**, *17*, 343.
(24) Jano, I. *Compt. Rend. Acad. Sci. Paris* **1965**, *261*, 103.
(25) Tapia, O. In *Quantum Theory of Chemical Reactions*; R. Daudel, A. Pullman, L. Salem and A. Viellard, Eds.; Reidel: Dordrecht, 1980; Vol. 2; p. 25.
(26) Born, M. *Z. Physik* **1920**, *1*, 45.
(27) Still, W. C.; Tempczyk, A.; Hawley, R. C.; Hendrickson, T. *J. Am. Chem. Soc.* **1990**, *112*, 6127.
(28) Hermann, R. B. *J. Phys. Chem.* **1972**, *76*, 2754.
(29) Hermann, R. B. *J. Phys. Chem.* **1975**, *79*, 163.
(30) Hermann, R. B. *Proc. Natl. Acad. Sci., USA* **1977**, *74*, 4144.
(31) Armstrong, D. R.; Perkins, P. G.; Stewart, J. J. P. *J. Chem. Soc., Dalton Trans.* **1973**, 838.
(32) Lazaridis, T.; Paulaitis, M. E. *J. Phys. Chem.* **1994**, *98*, 635.
(33) Alary, F.; Durup, J.; Sanenjouand, Y.-H. *J. Phys. Chem.* **1993**, *97*, 13864.
(34) Hine, J.; Mookerjee, P. K. *J. Org. Chem.* **1975**, *40*, 287.

(35) Cabani, S.; Gianni, P.; Mollica, V.; Lepori, L. *J. Solution Chem.* **1981**, *10*, 563.

(36) Abraham, M. H.; Whiting, G. S.; Fuchs, R.; Chambers, E. J. *J. Chem. Soc., Perkin Trans. 2* **1990**, 291.

(37) Zhang, Y.; Dallas, A. J.; Carr, P. W. *J. Chrom.* **1993**, *638*, 43.

(38) Pascual-Ahuir, J. L.; Silla, E. *J. Comp. Chem.* **1990**, *11*, 1047.

(39) Silla, E.; Tubon, I.; Pascual-Ahuir, J. L. *J. Comp. Chem.* **1991**, *12*, 1077.

(40) Rauhut, G.; Clark, T. *J. Comp. Chem.* **1993**, *14*, 503.

(41) Alhambra, C.; Luque, F. J.; Orozco, M. *J. Comp. Chem.* **1994**, *15*, 12.

(42) (a) Chirlian, L. E.; Francl, M. M. *J. Comp. Chem.* **1987**, *8*, 894. (b) Breneman, C. M.; Wiberg, K. B. *J. Comp. Chem.* **1990**, *11*, 361. (c) Merz, K. M. *J. Comp. Chem.* **1992**, *13*, 749.

(43) Liptag, W. In *Excited States*; E. C. Lim, Ed.; Academic Press: 1974; Vol. 1; p. 196.

(44) Pople, J. A.; Segal, G. A. *J. Chem. Phys.* **1966**, *44*, 3289.

(45) Ridley, J. E.; Zerner, M. C. *Theor. Chim. Acta* **1973**, *32*, 111.

(46) Bingham, R. C.; Dewar, M. J. S.; Lo, D. H. *J. Am. Chem. Soc.* **1975**, *97*, 1294.

(47) Gonzales-Lafont, A.; Truong, T. N.; Truhlar, D. G. *J. Phys. Chem.* **1991**, *95*, 4618.

(48) Viggiano, A. A.; Paschkewitz, J.; Morris, R. A.; Paulson, J. F.; Gonzales-Lafont, A.; Truhlar, D. G. *J. Am. Chem. Soc.* **1991**, *113*, 9404.

(49) Liu, Y.-P.; Lu, D.-H.; Gonzalez-Lafont, A.; Truhlar, D. G.; Garrett, B. C. *J. Am. Chem. Soc.* **1993**, *115*, 7806.

(50) Claisen, L. *Chem. Ber.* **1912**, *45*, 3157.

(51) Rhoads, S. J.; Raulins, N. R. *Org. React.* **1975**, *22*, 1.

(52) Ziegler, F. E. *Chem. Rev.* **1988**, *88*, 1423.

(53) Haslam, E. *The Shikimate Pathway*; Wiley: New York, 1974.

(54) Ganem, B. *Tetrahedron* **1978**, *34*, 3353.

(55) Bartlett, P. A.; Johnson, C. R. *J. Am. Chem. Soc.* **1985**, *107*, 7792.

(56) Hilvert, D.; Carpenter, S. H.; Nared, K. D.; Auditor, M.-T. M. *Proc. Natl. Acad. Sci., USA* **1988**, *85*, 4953.

(57) Jackson, D. Y.; Jacobs, J. W.; Sugasawara, R.; Reich, S. H.; Bartlett, P. A.; Schultz, P. G. *J. Am. Chem. Soc.* **1988**, *110*, 4841.

(58) Gajewski, J. J.; Conrad, N. D. *J. Am. Chem. Soc.* **1979**, *101*, 6693.

(59) McMichael, K. D.; Korver, J. L. *J. Am. Chem. Soc.* **1979**, *101*, 2746.

(60) Dewar, M. J. S.; Healy, E. F. *J. Am. Chem. Soc.* **1984**, *106*, 7127.

(61) Vance, R. L.; Rondan, N. G.; Houk, K. N.; Jensen, F.; Borden, W. T.; Komornicki, A.; Wimmer, E. *J. Am. Chem. Soc.* **1988**, *110*, 2314.

(62) Kupczyk-Subotkowska, L.; Saunders, W. H.; Shine, H. J.; Subotkowski, W. *J. Am. Chem. Soc.* **1993**, *115*, 5957.

(63) Yoo, H. Y.; Houk, K. N. *J. Am. Chem. Soc.* submitted for publication.

(64) Dewar, M. J. S.; Jie, C. *J. Am. Chem. Soc.* **1989**, *111*, 511.

(65) Dewar, M. J. S.; Ford, G. P.; McKee, M. L.; Rzepa, H. S.; Wade, L. E. *J. Am. Chem. Soc.* **1977**, *99*, 5069.

(66) Burrows, C. J.; Carpenter, B. K. *J. Am. Chem. Soc.* **1981**, *103*, 6983.

(67) Burrows, C. J.; Carpenter, B. K. *J. Am. Chem. Soc.* **1981**, *103*, 6984.

(68) Coates, R. M.; Rogers, B. D.; Hobbs, S. J.; Peck, D. R.; Curran, D. P. *J. Am. Chem. Soc.* **1987**, *109*, 1160.

(69) Gajewski, J. J.; Jurayj, J.; Kimbrough, D. R.; Gande, M. E.; Ganem, B.; Carpenter, B. K. *J. Am. Chem. Soc.* **1987**, *109*, 1170.

(70) Grieco, P. A. *Aldrichim. Acta* **1991**, *24*, 59.

(71) Brandes, E.; Grieco, P. A.; Gajewski, J. J. *J. Org. Chem.* **1989**, *54*, 515.

(72) Dewar, M. J. S.; Jie, C. *Acc. Chem. Res.* **1992**, *25*, 537.

(73) Dupuis, M.; Murray, C.; Davidson, E. R. *J. Am. Chem. Soc.* **1991**, *113*, 9756.

(74) Houk, K. N.; Gustafson, S. M.; Black, K. A. *J. Am. Chem. Soc.* **1992**, *114*, 8565.

(75) (a) Borden, W. T. In *Abstracts of the 207th National Meeting of the American Chemical Society*; American Chemical Society: Washington, DC, 1994; ORGN 129. (b) Hrovat, D. A.; Morokuma, K.; Borden, W. T. *J. Am. Chem. Soc.* **1994**, *116*, 1072.

(76) Cramer, C. J.; Truhlar, D. G. *J. Am. Chem. Soc.* **1992**, *114*, 8794.

(77) Severance, D. L.; Jorgensen, W. L. *J. Am. Chem. Soc.* **1992**, *114*, 10966.

(78) Gao, J. *J. Am. Chem. Soc.* **1994**, *116*, 1563.

(79) Gajewski, J. J.; Brichford, N. L. found elsewhere in this volume.

(80) Severance, D. L.; Jorgensen, W. L. found elsewhere in this volume.

(81) Jorgensen, W. L. *Acc. Chem. Res.* **1989**, *22*, 184.

(82) Jorgensen, W. L.; Chandrasekhar, J.; Madura, J. D.; Impey, R. W.; Klein, M. L. *J. Chem. Phys.* **1983**, *79*, 926.

(83) Jorgensen, W. L.; Tirado-Rives, J. *J. Am. Chem. Soc.* **1988**, *110*, 1657.

(84) Houk, K. N.; Zipse, H. *Chemtracts — Org. Chem.* **1993**, *6*, 51.

(85) Gao, J. *J. Phys. Chem.* **1992**, *96*, 537.

RECEIVED May 24, 1994

Chapter 4

Solvation Free Energies from a Combined Quantum Mechanical and Continuum Dielectric Approach

Carmay Lim, Shek Ling Chan, and Philip Tole

Protein Engineering Network of Centres of Excellence, Departments of Molecular and Medical Genetics, Biochemistry, and Chemistry, MSB 4388, University of Toronto, 1 King's College Circle, Toronto, Ontario M5S 1A8, Canada

We propose a new method to compute solvation free energies by combining quantum mechanical and continuum dielectric methods. Instead of using atom-based partial charges and a dielectric boundary derived from atomic radii to solve the Poisson or Poisson-Boltzmann equation, the solute charge distribution and the dielectric boundary are obtained from an electronic wavefunction determined using quantum mechanical methods. This results in several advantages over conventional applications of the continuum dielectric method as well as molecular dynamics and Monte Carlo simulation techniques. First, the input charge distribution is treated more accurately than that in the conventional application which uses atomic partial charges determined by non-unique procedures. Second, the electronic solute wavefunction provides a better description of the solute shape and thus, reduces the error in determining the dielectric boundary compared to a geometric dielectric boundary defined by arbitrary effective Born radii of the solute atoms. Third, the new method is generally a cost-effective means of obtaining relative free energies compared to free energy perturbation methods, which require parameterization and intensive cpu-time.

During the past decade, a large variety of methods have been developed to treat solvent effects on the static and dynamic properties of molecules. These methods fall generally into two classes depending on their treatment of the solvent molecules; viz., "microscopic" methods that treat solvent molecules and their interactions with the solute explicitly, and "macroscopic" methods that treat solvent as a continuous medium characterized by a dielectric constant. The "macro-

0097–6156/94/0568–0050$08.00/0

scopic" methods have a distinct advantage over the "microscopic" methods, especially for large systems, in speeding up computation considerably by not taking solvent molecules into account explicitly.

Under the class of "microscopic" methods, one of the earliest approaches was to employ quantum-mechanical methods to study the solute and a limited number of solvent molecules (*1*). This approach does not yield the energetics associated with the long-range bulk solvent forces and simply increasing the number of solvent molecules becomes computationally impractical.

A breakthrough for modeling solution chemistry came with the development of condensed-phase molecular dynamics and Monte Carlo simulation procedures using empirical energy potentials. The procedure (*2*) for modeling solution reactions involves (i) determination of the minimum energy reaction path in vacuum via ab initio calculations (*3*), (ii) determination of the solute-solvent non-bonded parameters as a function of the reaction coordinate by fitting to ab initio results of the reacting complex with a water molecule in various orientations, (iii) determination of the corresponding potential of mean force in solution from a series of fluid simulations using free energy statistical perturbation or thermodynamic integration techniques (*4*).

Recently, condensed-phase molecular dynamics and Monte Carlo methodology based on combined quantum and molecular mechanical potentials have been developed to simulate reactions in solution *directly*. Various combinations have been made: ab initio/Car-Parrinello methods with molecular dynamics (*5*) and semiempirical techniques with molecular dynamics (*6*) or Monte Carlo (*7*). These various approaches have in common that the system is partitioned into quantum mechanical and molecular mechanical regions and the combined quantum/molecular mechanical potential and forces are calculated at each step for use in classical dynamics simulations.

An approach bridging "microscopic" and "macroscopic" methods has been proposed whereby a quantum mechanical solute is surrounded with a grid of point dipoles representing the average solvent polarization, and solvent outside this region is treated as a dielectric continuum (*8*). This approach has been further simplified by incorporating continuum solvent models (e.g., a generalized Born term (*9*) or a sum of surface element terms (*10*)) in the solute Hamiltonian and carrying out self-consistent field calculations. The continuum solvent model can also be employed on its own to compute solvation free energies by solving a finite-difference Poisson equation (*11,12*); it has been used to estimate solvent effects on gas-phase activation free energy profiles (*13, 14*).

Unfortunately, the "microscopic" and "macroscopic" methods described above suffer from the disadvantage that extensive parameterization is required; e.g., parameters for van der Waals and electrostatic solute-solvent interactions in simulations using empirical energy potentials (*2*), parameters for van der Waals solute-solvent interactions in simulations based on combined quantum and molecular mechanics potentials (*5*), parameters for the effective Born radius and accessible surface tension in quantum mechanical calculations employing a semiempirical Hamiltonian augmented by a generalized Born term (9). However, experimental data are not always available to calibrate the parameters; this is often the case for short-lived transition states or reaction intermediates. Furthermore, parameterization against ab initio calculations on the reacting complex with a single water molecule may not yield sufficiently accurate parameters. "Microscopic" methods have the further disadvantage that free energy perturbation calculations are computationally demanding as simulations must run long enough to permit adequate sampling of the relevant configuration space. If there are multiple minima with significant intervening barriers, the problem may be difficult to detect and overcome. Moreover, a simulation with combined quantum mechanical/molecular mechanical potentials takes two to three times longer than a corresponding simulation using empirical energy functions.

Here, a new method for computing solvation free energies is proposed. The new procedure combines quantum mechanical and continuum dielectric methods by employing the quantum mechanical charge density *directly* in solving the finite-difference Poisson or Poisson-Boltzmann equation for the electrostatic contribution to the solvation free energy. To test the reliability of the new method, it was employed to calculate solvation free energies of monatomic ions and polyatomic molecules for which experimental data are available for comparison. These test calculations serve to calibrate (i) the basis set and level of quantum mechanical calculation required to yield the most accurate charge density, and (ii) the cut-off value of the charge density for determining the dielectric boundary. The method is outlined in the next section. This is followed by a discussion of the results of monatomic ions and polyatomic molecules. The performance of the method and its potential application to model reactions in solution are discussed in the last section.

Method

The continuum dielectric method sets up a grid in space and solves for the electric potential at each grid point given a molecule residing in a dielectric medium. In the standard continuum dielectric model (11, 12) the charge distribution on the grid is determined from partial atomic charges, and the dielectric boundary is derived from atomic radii of the solute atoms. The essence of the new procedure is to employ directly a charge density grid generated from quantum mechanical calculations and to derive the dielectric boundary from this electron distribution. The grid of electron charge densities was generated from the solute wavefunction using the ab initio program GAUSSIAN 92 (15). For monatomic ions, the basis sets employed were 6-31G*/6-31+G* for first and second row elements and STO-3G* for third row elements. Unless stated otherwise, the geometries for polyatomic ions were fully optimized at the Hartree Fock (HF) level with the 6-31G* basis, and the charge density grids were computed at the second-order Møller-Plesset perturbation (MP2) level.

The grid for the continuum dielectric calculation was checked to ensure that the dielectric boundary was at least 1.5 Å from the grid edge. Unless stated otherwise, the grid-spacing (or grid-eye) was $0.2\,\text{Å} \times 0.2\,\text{Å} \times 0.2\,\text{Å}$; it was the same as that used in the ab initio calculations so that the electron density output of GAUSSIAN 92 could be fed directly into the input of the continuum dielectric program. As GAUSSIAN 92 gave the electron densities at the grid points, when the density at each grid point was multiplied by the grid-eye volume and subsequently summed, the result might deviate from the total electron charge by as much as 10%. To correct this, the nuclear charges were scaled down so as to preserve the net charge of the molecule/ion. For example, if $-Q$ is the total electron charge of the molecule/ion, $-q$ is the sum of the electron charge on the grid, and N is the total nuclear charge, then all nuclear charges are scaled by a factor of $(N-Q+q)/N$; this preserves the ratio of the charges on the nuclei. The scaled nuclear charges were then distributed between the eight nearest grid points by the procedure of Wachspress (*16*).

The physical boundary of the molecule was defined using the electron density distribution: all grid points with an electron density above a certain cutoff value were considered to be inside the molecule. The low dielectric region was then defined as the space inaccessible to contact by a 1.4 Å solvent sphere rolling over the physical boundary. Thus, the definition of the dielectric boundary is similar to that in the conventional application of the continuum dielectric method. The difference lies in the definition of the physical boundary of the molecule, which is determined by a set of atomic radii in the conventional application. Here, the need to calibrate the atomic radii for different atom types is bypassed and replaced by a single cutoff value for the determination of the physical molecular boundary. The low dielectric region was assigned a dielectric constant of 1, whereas the high dielectric region was characterized by a dielectric constant of 1 for vacuum and 80 for water.

The numerical procedure for solving the electric potential on the grid is the same as that in conventional applications. The electric potential ϕ at every grid point was initialized using Coulomb's Law with a relative permittivity of the solvent medium. Then Poisson equation was solved for the values of ϕ on the grid by an over-relaxation algorithm (*17*). The work W of assembling the electric charges is given by

$$W = \sum_i \frac{1}{2} q_i \phi_i$$

where q_i is the charge associated with the ith grid point and ϕ_i is the electric potential at that point; the summation goes over all the grid points. The difference in the work calculated in vacuum and solution yields the electrostatic part of the solvation free energy.

Since each solvation free energy calculation required the difference between the work term in vacuum and solution, it took two iterative cycles for one free energy calculation. For each iterative cycle, 900 steps were performed; the maximum

Table I. Effect of Diffuse Functions on the Electrostatic Solvation Free Energy

| Species | Grid Size | Cut-off | Solvn free energy (kcal/mol) | |
| | | | Charge Density Basis | |
	Å^3	$e/\text{Å}^3$	$6\text{-}31\text{+}G^{*a}$	$6\text{-}31G^{*a}$
Na^+	51x51x51	0.0002	101	106
Na^+	51x51x51	0.0002	100^b	106^b
$CH_3NH_3^+$	71x71x71	0.0005	62.6	68.7
ImH^+	47x97x97	0.0001	75.8	78.0
$(OH)^-$	51x51x51	0.0025	74.6	88.8
$(OH)^-$	51x51x51	0.0045	81.2	94.0
CH_3COO^-	61x61x61	0.0045	81.7	86.2
CH_3NH_2	61x61x61	0.0015	11.0	8.2
CH_3NH_2	71x71x71	0.0005	7.9	6.1
CH_3COOH	61x61x61	0.0015	2.5	3.2

[a]The charge densities were obtained from a single point MP2 calculation corresponding to HF/6-31G* geometries.
[b]The charge densities were computed at the HF level.

change in the magnitude of the potential at a grid point in the last step dropped to the order of several micro-volts. The cpu time required for each iterative cycle is strongly dependent on the linear dimension of the grid; if the linear dimension of the grid is doubled, the number of points in the grid is eight times the original. On an IBM 340, each iterative cycle took about 2 to 3 hours for a 50 Å×50 Å×50 Å grid and 5 to 6 hours for a 60 Å×60 Å×60 Å grid.

Results and Discussion

Dependence on Charge Densities. The effect of diffuse functions in evaluating the charge densities on the electrostatic solvation free energy is summarized in Table I. Table I shows that for charged species, the electrostatic solvation free energies computed with MP2/6-31+G* charge densities are generally smaller than those calculated with MP2/6-31G* charge densities. This is because the 6-31+G* electronic charge densities of the ions are more spread out than the 6-31G* values, hence the same charge density cutoff defines a physical boundary that is further away from the nucleus; i.e., a larger effective Born radius, and thus a smaller solvation free energy.

Dependence on Charge Density Cut-off. The dependence of the electrostatic solvation free energy on the cut-off value employed for the charge densities is summarized in Table II for monatomic ions and Table III for polyatomic molecules. The computed and experimental solvation free energies correspond

Table II. Effect of Charge Density Cut-off on the
Electrostatic Solvation Free Energy: Monatomics

Species	Basis[a]	Solvation free energy (kcal/mol)				
		Charge density cut-off $(e/Å^3)$				
		0.5^{-4}	2^{-4}	5^{-4}	10^{-4}	expt[b]
Li^+	6-31+G*	−121	−136	−152	−[c]	−(115−124)
Be^{2+}	6-31+G*	−607	−721	−780	−[c]	−563
Na^+	6-31+G*	−92	−101	−108	−[c]	−(90-98)
Mg^{2+}	6-31+G*	−421	−476	−511	−[c]	−443
Na^+	STO-3G*	−[c]	−82	−92	−101	−(90-98)
Mg^{2+}	STO-3G*	−[c]	−353	−400	−445	−443
K^+	STO-3G*	−[c]	−65	−72	−79	−(74-75)
Ca^{2+}	STO-3G*	−[c]	−284	−327	−362	−363

[a]The charge densities were obtained from a single point MP2
calculation corresponding to HF/6-31G* geometries;
a 50 Å×50 Å×50 Å grid was used for the dielectric calculations.
[b]Experimental values are taken from Latimer W. M.;
Pitzer, K. S.; Slansky, C. M.; *J. Chem. Phys.* **1939**, *7*, 108;
R. Gomer, G. Tryson *J. Chem. Phys.* **1977**, *66*, 4413;
H. L. Friedman and C. V. Krishnan *Water: A Comprehensive
Treatise*, Eds: F. Franks, Plenum Press, NY **1973**, *3*, 1.
[c]Value has not been computed for this cut-off.

to a common standard state of 1 M for both gas and solution. As the charge
density cut-off increases, the computed solvation free energy increases. This can
be rationalized using the Born formula for the heat of hydration W (*18*),

$$W = -\frac{q^2}{2r}\left(1 - \frac{1}{\epsilon_o}\right)$$

which shows that the solvation free energy of a monatomic ion with charge q
depends inversely upon the Born radius r. Thus, as the cutoff value of the
electron density is increased, the volume defined as inside the ion decreases; i.e.,
the effective Born radius decreases, and so the solvation free energy increases.
In general, the same variation in cutoff has a smaller effect on the solvation free
energy of a polyatomic ion than a monatomic ion. This is because the volume
of a polyatomic ion is generally bigger than that of a monatomic ion and hence,
for the same absolute difference in the positioning of the physical boundary, the
effect on the overall volume mass is smaller for polyatomic ions.

Comparison with experiment. Although experimental solvation free energies
of neutral species may be directly measured, "experimental" values of charged

Table III. Effect of Charge Density Cut-off on the Electrostatic Solvation Free Energy: Polyatomics

Species	Basis[a]	Grid Size	Solvation free energy (kcal/mol)		
			Cut-off R (e/Å^3)		
		Å^3	R$_1^b$	R$_e^c$	expt[d]
H_3O^+	6-31G*	61^3	−109	−104	−104
PH_4^+	3-21G*	61^3	−77	−72	−73
$CH_3NH_3^+$	6-31G*	61^3	−72	−69	−70
$(OH)^-$	6-31G*	51^3	−101	−94	−(96–106)
CH_3COO^-	6-31G*	61^3	−98	−86	−(77–82)
$(SH)^-$	6-31G*	61^3	−87	−78	−76
CH_3COOH	6-31G*	61^3	−4.1	−2.3	−6.7
H_2O	6-31G*	51^3	−10.3	−6.4	−6.3
CH_3NH_2	6-31G*	61^3	−10.6	−6.1	−4.6
$(SH)_2$	6-31G*	51^3	−6.5	−3.5	−0.7

[a]The charge densities were obtained from a single point MP2 calculation corresponding to HF/6-31G* geometries.
[b]R$_1$ equals 0.0025, 0.0100, and 0.0030 for cations, anions and neutrals, respectively.
[c]R$_e$ equals 0.0015, 0.0045, and 0.0005 for cations, anions and neutrals, respectively.
[d]Experimental values are taken from
R. G. Pearson *J. Am. Chem. Soc.* **1986**, *108*, 6109.
A. Ben-Naim, Y. J. Marcus *J. Chem. Phys.* **1984**, *81*, 2016.

species are commonly determined <u>indirectly</u> from the experimental values of the solvation free energy ΔG_s of the corresponding neutral species and the proton, the gas-phase proton affinity or basicity (ΔG_{gas}) and solution pK_a or free energy (ΔG_{sln}) using Scheme I:

$$
\begin{array}{ccccc}
 & \Delta G_{gas} & & & \\
BH(g) & \rightarrow & B(g) & + & H^+(g) \\
-\Delta G_s(BH) \downarrow & & \downarrow -\Delta G_s(B) & \downarrow -\Delta G_s(H^+) & \\
BH(s) & \rightarrow & B(s) & + & H^+(s) \qquad \text{(Scheme I)} \\
 & \Delta G_{sln} & & &
\end{array}
$$

Since experimental gas basicities are accurate to only 2–5 kcal/mol, and there are at least five different experimental values for the proton solvation free energy –(254 to 261) kcal/mole with a spread of 7 kcal/mol, the uncertainty in the "experimental" solvation free energies of charged species is at least ± 5 kcal/mol. On the other hand, the computed solvation free energies are subject to errors due to uncertainties in the charge densities and cut-off employed and the neglect of the non-polar contribution to the solvation free energy.

For the monatomic ions in Table II, a 0.0010 $e/\text{Å}^3$ cutoff of the MP2/STO-3G* charge densities gives solvation free energies in excellent agreement with experiment, whereas a 0.5×10^{-4} $e/\text{Å}^3$ cutoff of the MP2/6-31+G* charge densities yields solvation free energies that are within 7% of the experimental values. The STO-3G* basis requires a larger cut-off than the 6-31+G* basis as the latter results in a larger solvation free energy than the former for the same cutoff (Table II). For the polyatomic molecules in Table III, a charge density cut-off of 0.0015 $e/\text{Å}^3$ for cations, 0.0045 $e/\text{Å}^3$ for anions, and 0.0005 $e/\text{Å}^3$ for neutrals, yields solvation free energies in accord with experiment. As expected, the agreement with experiment for neutral species is not as good as the charged ions since for neutral molecules, the non-polar contribution (around 1–2 kcal/mol) is no longer negligible relative to the electrostatic component of the solvation free energy.

Concluding Remarks

Applications. The present method can be applied to study reactions in aqueous solutions by computing the solvation free energies of reaction complexes. Thus, if X and Y are stationary points on the gas-phase ab initio potential energy surface and $\Delta A_{gas}^{\ddagger}$ is the free energy difference from ab initio calculations, the present method can be used to obtain the solvation free energies of X and Y and thus, $\Delta A_{sln}^{\ddagger}$ from Scheme II.

$$
\begin{array}{ccc}
 & \Delta A_{gas}^{\ddagger} & \\
X(gas) & \longrightarrow & Y(gas) \\
 & & \\
-\Delta A_s(X) \downarrow & & \downarrow -\Delta A_s(Y) \quad \text{(Scheme II)} \\
 & & \\
X(sln) & \longrightarrow & Y(sln) \\
 & \Delta A_{sln}^{\ddagger} &
\end{array}
$$

By piecing together consecutive $\Delta A_{sln}^{\ddagger}$, a qualitative free energy profile is obtained in solution. Note that for methods which require parameterization, there are generally no experimental data available on short-lived reaction intermediates/transition states to calibrate parameters.

The above procedure automatically yields the separate gas-phase and solvation contributions to the observed solution activation free energy. If the X and Y species involved have gas-phase activation and solvation free energy errors that are similar and cancel, the proposed method may yield reasonably accurate solvation and solution activation free energy *differences*. In such cases, the outcome of two competing reaction pathways in solution and substituent effects may be predicted. The present method can readily be applied to study reactions in organic solvents as well as the ionic-strength and temperature dependence of reaction rates.

Advantages. The present methodology is not only versatile (as described above), it also possesses several advantages over existing methods, in principle. First, since the charge density is directly employed in the calculations, the need to fit solute partial charges empirically and errors due to ambiguous ways of extracting partial charges (e.g., Mulliken population analysis) are eliminated. Second, as the dielectric boundary is determined by a single charge density cut-off value, the need to fit the Born radii empirically (9, 11) for various atom types (e.g., C, H, O, N, P, S) and errors due to arbitrary ways of defining the solute Born radii (e.g., in terms of van der Waal's, ionic or covalent radii) are eliminated. Third, if the number of grid-points considered is not excessive, the new method is a cost-effective means of obtaining relative free energies compared to expensive free energy perturbation methods. Note that the limitations in the quantum mechanical calculations; e.g., the basis set size and CPU requirements, are inherent in all existing methods.

Potential Improvements Accuracy may be improved if the charge density outside the physical boundary of the molecule is set to zero and the nuclear charge rescaled accordingly, as described in the Method section. Another interesting avenue is to explore means of using the ab initio electrostatic potential grid directly instead of the quantum mechanical charge density distribution in solving for the electrostatic potentials in vacuum and in solution; however, the charge density grid will still be needed to compute the electrostatic self-energy (see equation in Method section). This may speed up convergence as well as potentially decrease the dependence of the electrostatic self-energy on the accuracy of the charge density.

Acknowledgments

This work was supported by the Protein Engineering Network of Centres of Excellence of Canada.

Literature Cited

(1) Goldblum, A.; Perahia, C.; Pullman, A. *FEBS Lett.* **1979**, *91*, 213.

(2) Madura, J.; Jorgensen, W. L. *J. Am. Chem. Soc.* **1986**, *108*, 2517–2527.

(3) Hehre, W. J.; Radom, L.; Schleyer, P. v.R.; Pople, J. A. *Ab Initio Molecular Orbital Theory.* John Wiley and Sons, (1986).

(4) Beveridge, D. L.; DiCapua, F. M. *Annu. Rev. Biophys. Biophys. Chem.* **1989**, *18*, 431–492.

(5) Field, M. J. *J. Phys. Chem.* **1991**, *95*, 5104–5108.

(6) Field, M. J.; Bash, P. A.; Karplus, M. *J. Comp. Chem.* **1990**, *11*, 700–733.

(7) Gao, J. *J. Phys. Chem.* **1992**, *96*, 537–540.

(8) Luzhkov, V.; Warshel, A. *J. Comp. Chem.* **1992**, *13*, 199–213.

(9) Cramer, C. J.; Truhlar, D. G. *J. Computer-Aided Molecular Design* **1992**, *6*, 629–666.

(10) Ford, G. P.; Wang, B. *J. Am. Chem. Soc.* **1992**, *114*, 10563–10569.

(11) Lim, C.; Bashford, D.; Karplus, M. *J. Phys. Chem.* **1991**, *95*, 5610–5620.

(12) Gilson, M. K.; Honig, B. H. *Proteins: Structure, Function and Genetics* **1988**, *4*, 7–18.

(13) Lim, C.; Tole, P. *J. Phys. Chem.* **1992**, *114*, 7245.

(14) Tole, P.; Lim, C. *J. Phys. Chem.* **1993**, *97*, 6212.

(15) Frish, M. J.; Head–Gordon, M.; Trucks, G. W.; Foresman, J. B.; Schlegel, H. B.; Raghavachari, K.; Robb, M.; Binkley, J. S.; Gonzalez, C.; Defrees, D. J.; Fox, D. J.; Whiteside, R. A.; Seeger, R.; Melius, C. F.; Baker, J.; Martin, R. L.; Kahn, L. R.; Stewart, J. J. P.; Topiol, S.; Pople, J. A. Gaussian Inc.; Pittsburgh, PA 15213, USA.

(16) Wachspress, E. I. *Iterative Solution of Elliptic Systems*, Prentice Hall, Englewood Cliffs, (1966).

(17) Press, W. H.; Flannery, B. P.; Teukolsky, S. A.; Vetterling, W. T. *Numerical Recipes. The Art of Scientific Computing*; Cambridge University Press, Cambridge, (1986).

(18) Born, M. Z. *Z. Phys.* **1920**, *1*, 45.

(19) Chan, S. L.; Lim, C. *J. Phys. Chem.* **1994**, *98*, 692.

RECEIVED April 25, 1994

Chapter 5

Tests of Dielectric Model Descriptions of Chemical Charge Displacements in Water

Gregory J. Tawa and Lawrence R. Pratt

Theoretical Division, Los Alamos National Laboratory, Los Alamos, NM 87454

A dielectric model of electrostatic solvation is applied to describe potentials of mean force in water along reaction paths for: a) formation of a sodium chloride ion pair; b) the symmetric S_N2 exchange of chloride in methylchloride; and c) nucleophilic attack of formaldehyde by hydroxide anion. For these cases simulation and *X*RISM results are available for comparison. The accuracy of model predictions varies from spectacular to mediocre. It is argued that: a) dielectric models are physical models, even though simplistic and empirical; b) their successes suggest that second-order perturbation theory is a physically sound description of free energies of electrostatic solvation; and c) the most serious deficiency of the dielectric models lies in the definition of cavity volumes. Second-order perturbation theory should therefore be used to refine the dielectric models. These dielectric models make no attempt to assess the role of packing effects but for solvation of classical electrostatic interactions the dielectric models sometimes perform as well as the more detailed *X*RISM theory.

An important quality of water as a solvent is its ability to stabilize ions and polar molecules. Since displacement of electric charge is often central to chemical reactivity, water is a special solvent for chemical reactions in solution. An accurate molecular theory of water participation in chemical reactions in aqueous solution has not been established but a few possibilities are available. The range of theoretical approaches includes simulation calculations, integral equation theories, and dielectric models.

Comparison of the predictions of dielectric models with those of other methods in cases where thermal precision in solvation free energies is possible is the goal of this paper. For several molecular complexes we obtain thermally accurate solutions of the

0097–6156/94/0568–0060$08.00/0

governing macroscopic Poisson equation. The particular examples are chosen because of the availability of results from alternative methods. The comparison of dielectric model results with data from simulation calculations or with XRISM (*1*) results should teach us about the utility of the dielectric model and about fruitful directions for the discovery of better theories. Since the goal is unambiguous comparisons that test the performance of the dielectric model, the example problems are not discussed for their own sake.

Dielectric models of electrostatic solvation free energy

The dielectric model we apply is physically viewed as follows (*2, 3*). Attention is focused on a solute of interest. A solute volume is defined on the basis of its geometry. Partial charges describing the solute electric charge distribution are positioned with respect to this volume. For liquid water under the most common conditions it is known that the van der Waals volume of the molecule is a satisfactory choice for the molecular volume (*4, 5*). But that is coincidental and elaborations of the theory will have to consider more general possibilities. For the present applications, it is essential that the defined solute volume permit disconnection when the solute fragments are widely separated. We define the solute volume as the volume enclosed by spheres centered on solute atoms. The solvent is idealized as a continuous dielectric material with dielectric constant ε. The value $\varepsilon = 77.4$, appropriate to water at its triple point, is used everywhere below. The solvent is considered to be excluded from the solute volume and that region is assigned a dielectric constant of one, $\varepsilon = 1$.

Methods. The equation to be solved for the model is

$$\nabla \bullet \varepsilon(\mathbf{r})\nabla\Phi(\mathbf{r}) = -4\pi\rho_f(\mathbf{r}) \tag{1}$$

where $\rho_f(\mathbf{r})$ is the density of electric charge associated with the solute molecule, the function $\varepsilon(\mathbf{r})$ gives the local value of the dielectric constant, and the solution $\Phi(\mathbf{r})$ is the electric potential. To solve this equation, we first cast it as an integral equation, *e.g.*

$$\Phi(\mathbf{r}) = \Phi_0(\mathbf{r}) + \int_V G_0(\mathbf{r},\mathbf{r}')\left(\frac{\nabla'\varepsilon(\mathbf{r}')}{4\pi\varepsilon(\mathbf{r}')}\right)\bullet \nabla'\Phi(\mathbf{r}')d^3r' \ . \tag{2}$$

Here $G_0(\mathbf{r},\mathbf{r}')$ is the Green function for the Poisson equation with $\varepsilon(\mathbf{r}) = 1$ and $\Phi_0(\mathbf{r})$ is the electrostatic potential for that case. It is assumed that all the charges of $\rho_f(\mathbf{r})$ are positioned in regions where $\varepsilon(\mathbf{r}) = 1$. This equation is correct both for a localized distribution $\rho_f(\mathbf{r})$ and zero boundary data on a surface everywhere distant and for periodic boundary conditions on a cell of volume V. $G_0(\mathbf{r},\mathbf{r}')$ is different in those two cases. This equation is not the only such form that can be solved and more general considerations can be helpful. But we do not pursue those issues here.

The integrand of equation 2 is concentrated on the interface between the solute volume and the solvent. We can then use boundary element ideas to solve it *(6-11)*. The principal novelty in our numerical methods is that we use a sampling method based upon quasi-random number series *(12)* to evaluate the surface integral rather than more specialized methods for that evaluation. Advantages of our method are that it facilitates systematic studies of numerical convergence and exploitation of systematic coarse-graining. More specific discussion of numerical methods can be expected at a later date.

With the solution of equation 2 in hand we obtain the desired potential of mean force as

$$W = U + \left(\frac{1}{2}\right)\int_V \rho_f(\mathbf{r})(\Phi(\mathbf{r}) - \Phi_0(\mathbf{r}))d^3r \qquad (3)$$

where U is the static energy in the absence of the solvent. Since for the present examples $\rho_f(\mathbf{r})$ is a sum of partial charges, the integral in equation 3 is a sum over those partial charges.

Results

a) Pairing of sodium and chloride ions in water. Dielectric model results for the $Na^+ \cdots Cl^-$ potential of average force in water are shown in Figure 1. The radii used were those recommended by Rashin and Honig *(5)*. See also Pratt, *et al.* (Pratt, L. R., Hummer, G., and Garcia, A. E., *Biophys. Chem.,* in press). These results agree with those of Rashin who studied a similar dielectric model *(13)*. Rashin assumed a somewhat different solute volume and his predicted potential of mean force displayed a more prominent barrier to escape from the contact minimum. Those results were similar to XRISM *(14)*. The dielectric model results are surprising in showing minimum free energies both at ion contact and at a larger distance that indicates a solvent-separated pairing. Although surprising, these results are not in quantitative agreement with simulation calculations of solvation free energies of ion pairing (Hummer, G., Soumpasis, D. M., and Neumann, M., *Mol. Phys.,* in press). Most importantly, the contact minimum is much deeper than the simulation results. We note also that the simulation results were obtained for finite concentrations of NaCl. Thus, the large distance behavior of that potential of mean force is influenced by ionic screening. The dielectric models and this XRISM result conform to the asymptotic variation expected at infinite dilution.

A particular concern over recent years has been sensitivity of the molecular results for potentials of mean force to modelled intermolecular interactions *(15)*. Figure 1

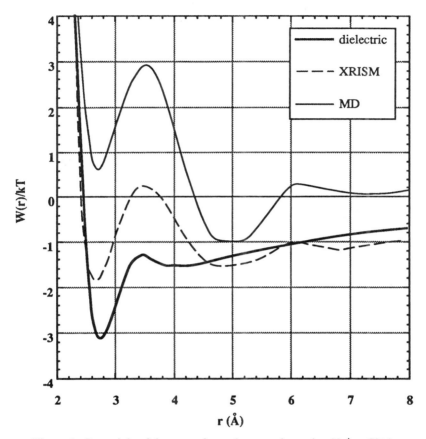

Figure 1. Potentials of the mean forces between ion pairs $Na^+ \cdots Cl^-$ in water. The XRISM results are redrawn from Reference *14,* and the MD results are redrawn from Pratt, *et al.* (Pratt, L. R., Hummer, G., and Garcia, A. E., *Biophys. Chem.,* in press). Those original MD results are to be found in Hummer, *et al.* (Hummer, G., Soumpasis, D. M., and Neumann, M., *Mol. Phys.*, in press).

does not attempt to give the wide range of results that have been obtained. However, to the extent that such sensitivity is considered important, it highlights a fundamental deficiency of the dielectric models. Those molecular details are not present in the dielectric models as currently applied. The dielectric models do depend on the dielectric constant of the solvent but for many applications that dependence is not decisive because of the high values of the dielectric constant that are typically relevant.

b) Symmetric S_N2 chloride exchange in methyl chloride. Following the efforts of Chandrasekhar, *et al. (16)*, this example has been taken as a theoretical model of reactions in solutions over recent years *(17-18)*. We used radii of R_H = 1.00Å, R_C = 1.85Å, and R_{Cl} = 1.937Å, and partial charges of Chandrasekhar, *et al. (16)* (Figure 2) and Huston, *et al. (18)* (Figure 2 inset). As Figure 2 shows, the agreement between dielectric model and simulation results is as satisfactory as the agreement between XRISM and simulation. As discussed in the paper of Huston, *et al. (18)*

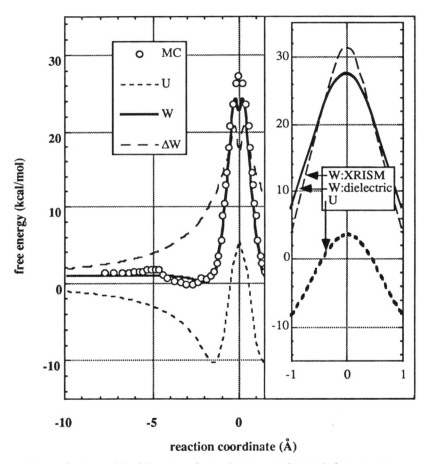

Figure 2. Potential of the mean force along a reaction path for symmetric S_N2 replacement of the Cl^- ion in CH_3Cl. $\Delta W \equiv W - U$. See References *16* and *18*.

the original assignment of partial charges of the reacting complex in the neighborhood of the barrier leads to the prediction of the notch in the potential of the mean force by the XRISM theory. This is also true of the dielectric model, as is seen in Figure 2. Huston, *et al.* (18) reanalyzed this modeling of solute-solvent interactions and proposed alterations of the original parameterization. The results for this alternative description of the barrier region are shown in the inset.

c) **Nucleophilic attack of** HO^- **on** H_2CO. This example has been previously studied as a prototype of an initial step in the formation and destruction of the peptide unit (19, 20). We used radii of $R_{O(hydroxide)} = 1.65$Å, $R_C = 1.85$Å, $R_H = 1.0$Å, and $R_{O(carbonyl)} = 1.60$Å, and the partial charges and geometries of Madura and Jorgensen (19). The results are shown in Figure 3. The predictions of the dielectric model are qualitatively similar to those of the Monte Carlo simulation and of the XRISM theory, particularly in predicting the existence of a solvation barrier prior to contact of the reacting species.

The results of the dielectric model are not quantitatively accurate in this case, however. It seems clear that adjustment of the cavity radii at each geometry could bring the model results into agreement with the simulation data. Recently, a new proposal was given for achieving a physically valid parameterization of the model on the basis of additional molecular information (Pratt, L. R., Hummer, G., and Garcia, A. E., *Biophys. Chem.,* in press); that proposal is discussed below. As an observation preliminary to attempts to parameterize the dielectric model on more basic information, we demonstrate here that reasonable radii can achieve excellent agreement with the data. To do this we adjusted the radii of the hydroxide oxygen to reproduce at a coarse level of calculation the simulation results at a few points along the potential of mean force shown, used a simple interpolation for radii at other points, and observed the variation of the hydroxide oxygen radius. The adjusted radii are shown in Figure 4 and the results for the potential of mean force are the circles of Figure 3. This type of comparison can be misleading. Agreement just as good would be achieved even if the model result were physically wrong, *e.g.*, if we had mistakenly done the calculation for acetaldehyde rather than formaldehyde. However, the typical rough expectation for cavity radii for these models is between the van der Waals radius and the distance of closest approach for solvent molecular centers. To the extent that the radii are treated as truly adjustable, the model has more than enough flexibility to match the data with reasonable changes in those radii.

Discussion

The importance of the dielectric models is that they provide free energies of solvation for cases that are not accessible by alternative methods. The free energies obtained from the dielectric model are quadratic functionals of the charge distribution of the solute. It has been pointed out previously (Pratt, L. R., Hummer, G., and Garcia, A. E., *Biophys. Chem.,* in press) that this indicates that the dielectric models correspond to a

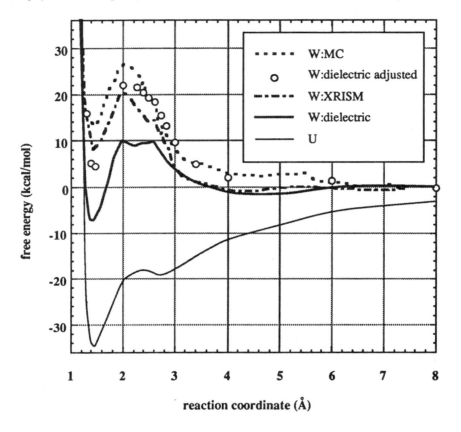

Figure 3. Potential of the average force along a reaction path for nucleophilic attack of H_2CO by HO^- according to the dielectric model, simulation *(19),* and XRISM theory *(20).* The circles are the results of the dielectric model with empirical adjustment of the radius of the hydroxide oxygen. See the text and Figure 4.

modelistic implementation of second-order perturbation theory for the excess chemical potential of the solute. Thus, the successes of dielectric models suggest that second-order thermodynamic perturbation theory is a physically sound theory for the desired solvation free energy due to electrostatic interactions. But in addition to the 'second-order' limitation, dielectric models also drastically eliminate molecular detail of the solvation structures. This detail can be restored by implementing second-order thermodynamic perturbation theory on a molecular basis. The fundamental formula for that approach is

$$\Delta\mu \approx \Delta\mu_0 + \left\langle \sum_{C=constituents} \varphi(C) \right\rangle_0 - \left(\frac{\beta}{2}\right)\left\langle \left(\sum_{C=constituents} \varphi(C) - \left\langle \sum_{C'=constituents} \varphi(C') \right\rangle_0 \right)^2 \right\rangle_0 . \qquad (4)$$

This approximation has been discussed previously by Levy, *et al. (21).* Here the subscript '0' indicates quantities obtained for the reference system in which no electrostatic interactions are expressed between the solution constituents and a designated solute molecule. The $\varphi(C)$ are the electrostatic potential energies of interaction between the constituent C of the solution and that solute. Several aspects of this molecular approach should be noted. First, it requires knowledge of $\Delta\mu_0$, the solvation free energy when electrostatic interactions are neglected. This is not supplied by the dielectric models. Second, the molecular approach includes a term linear in the charges; this involves the *potential at zero charge* induced by short ranged forces. This term is generally

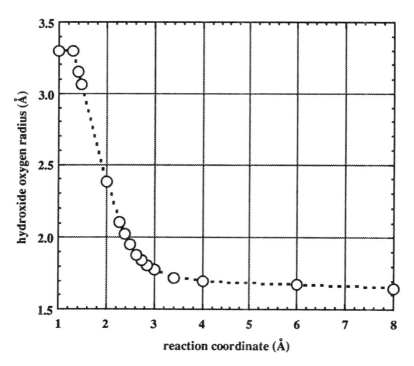

Figure 4. Variation of the hydroxide-oxygen radius adjusted to fit the data for the potential of mean force in example **c**. See Figure 3.

present, non-zero, dependent upon molecular geometry and thermodynamic state, but the dielectric models assume that it vanishes. Third, the second-order term corresponds to the quantities usually obtained from dielectric models but this formula avoids the classic empirical adjustments of cavity radii. From the perspective of a continuum approach, this second order term incorporates nonlocality of the polarization response of the solution.

In view of the corresponding molecular theory, a natural way to improve dielectric model results is first to obtain molecular results for the second-order term to better establish the solute volume on a proper molecular basis. Cavity radii thus determined will be dependent upon the thermodynamic state just as empirical cavity radii must generally be considered functions of thermodynamic state. For example, evaluation of enthalpies by temperature differentiation should include derivatives of the cavity radii with respect to temperature *(22)*. After definition of that solute volume is better controlled, an alternative source of information on the leading two terms must be developed. At this level, the cavity radii are *independent* of charges and charge distributions of the solute. Finally, the importance of succeeding terms in the perturbation theory must be assessed (Pratt, L. R., Hummer, G., and Garcia, A. E., *Biophys. Chem.*, in press, Rick, S. W., and Berne, B. J., *J. Am. Chem. Soc.*, in press). Those succeeding terms are likely to be especially troublesome for circumstances where composition fluctuations are physically important, *e.g.*, for mixed solvents.

We note again that the reactions and models studied here were chosen solely because of the availability of molecularly detailed simulation data and of results of integral equation theory. Since testing of the dielectric model by comparing its predictions to simulation and integral equation results is the objective, the same model systems must be treated by the various methods. Solute geometries and partial charges must be accepted for the test. Because of the simplicity of the dielectric model, however, we do not expect that conclusions regarding the utility of the model for description of charge displacement in aqueous solutions would be significantly changed if more elaborate models of these reactions were available. For example, the generalization of equation 4 to apply when the solute-solvent electrostatic interactions are described with the aid of higher-order multipole moments of the solute charge distribution is straightforward.

Conclusions

To the extent that the dielectric model is physically sound, second-order thermodynamic perturbation theory should provide an accurate description of free energies due to electrostatic interactions between the solute and the solution. Second-order thermodynamic perturbation theory restores molecular detail of the solvation structures that is discarded when the dielectric model is used. More fundamentally, second-order thermodynamic perturbation theory identifies the potential at zero charge that is neglected in the dielectric model.

Despite their simplistic character, dielectric models provide a physically sound description of chemical charge displacements in water. Because of these qualities they can be helpful where only rough but physical results are required; for example, they might be expected to provide serviceable umbrella functions *(23)* for more accurate molecular calculations of free energies along reaction paths of the sort considered here. It should be recognized, however, that stratification of the reaction coordinate is typically more important *(24, 25)*.

Considered directly the dielectric models are not reliably accurate for thermal level energy changes. A large part of the unreliability of the dielectric model predictions is surely due to the assignment of cavity radii; the predictions of the model are sensitive to those parameters, they clearly ought to vary along a reaction path, and the determination of proper values for the radii comes from outside the model. Although the dielectric model does not attempt to assess the importance of packing effects on solvation properties, it is sometimes of comparable accuracy to the more detailed XRISM theory for treatment of electrostatic contributions. The dielectric model also has the advantage of being simple and physical when used for those purposes.

Acknowledgments. We are grateful for helpful discussions with Drs. J. Blair, S.-H. Chou, P. Leung, D. Misemer, J. Stevens, and K. Zaklika of 3M Corporation on the topics of solvation and reaction chemistry in solution. LRP thanks Gerhard Hummer and Angel E. Garcia for helpful discussions and acknowledges partial support for this work from the Tank Waste Remediation System (TWRS) Technology Application program, under the sponsorship of the U. S. Department of Energy EM-36, Hanford Program Office, and the Air Force Civil Engineering Support Agency. This work was also supported in part by the US-DOE under LANL Laboratory Directed Research and Development funds.

Literature Cited

1.   Rossky, P. J., *Ann. Rev. Phys. Chem.* **1985**, 36, 321.
2.   Rashin, A. A., *J. Phys. Chem.* **1990**, 1725.
3.   Honig, B., Sharp, K., and Yang, A.-S., *J. Phys. Chem.* **1993**, 97, 1101.
4.   Latimer, W. M., Pitzer, K. S., and Slansky, C. M., *J. Chem. Phys.* **1939**, 7, 108.
5.   Rashin, A. A., and Honig, B., *J. Phys. Chem.* **1985**, 89, 5588.
6.   Pascual-Ahuir, J. L., Silla, E., Tomasi, and J., Bonaccorsi, *J. Comp. Chem.* **1987**, 8, 778.
7.   Zauhar, R. J., and Morgan, R. S., *J. Mol. Biol.* **1985**, 186, 815; *J. Comp. Chem.* **1988**, 9, 171.

8. Rashin, A. A., and Namboodiri, K., *J. Phys. Chem.* **1987**, 91, 6003.
9. Yoon, B. J., and Lenhoff, A. M., *J. Comp. Chem.* **1990**, 11, 1080; *J. Phys. Chem.* **1992**, 96, 3130.
10. Juffer, A. H., Botta, E. F. F., van Keulen, A. M., van der Ploeg, A., and Berendsen, H. J. C., *J. Comp. Phys.* **1991**, 97, 144.
11. Wang, B., and Ford, G. P., *J. Chem. Phys.* **1992**, 97, 4162.
12. Hammersley, J. M., and Handscomb, D. C., *Monte Carlo Methods;* Chapman and Hall: London, 1964; pp 31-36.
13. Rashin, A. A., *J. Phys. Chem.* **1989**, 93, 4664.
14. Pettitt, B. M., and Rossky, P. J., *J. Chem. Phys.* **1986**, 84, 5836.
15. Dang, L. X., Pettitt, B. M., and Rossky, P. J., *J. Chem. Phys.* **1992**, 96, 4046.
16. Chandrasekhar, J., Smith, S. F., and Jorgensen, W. L., *J. Am. Chem. Soc.* **1985**, 107, 154.
17. Chiles, R. A., and Rossky, P. J., *J. Am. Chem. Soc.* **1984**, 106, 6867.
18. Huston, S. E., Rossky, P. J., and Zichi, D. A., *J. Am. Chem. Soc.* **1989**, 111, 5680.
19. Madura, J. D., and Jorgensen, W. L., *J. Am. Chem. Soc.* **1986**, 108, 2517.
20. Yu, H.-A., and Karplus, M., *J. Am. Chem. Soc.* **1990**, 112, 5706.
21. Levy, R. M., Belhadj, M., and Kitchen, D. B., *J. Chem. Phys.* **1991**, 95, 3627.
22. Roux, B., Yu, H.-A., and Karplus, M., *J. Phys. Chem.* **1990**, 94, 4683.
23. Valleau, J. P, and Torrie, G. M., in *Statistical Mechanics, Part A: Equilibrium Techniques;* Berne, B. J., Ed.; Modern Theoretical Chemistry; Plenum: New York, 1977, Vol. 5; pp 178-182.
24. Hammersley, J. M., and Handscomb, D. C., *Monte Carlo Methods;* Chapman and Hall: London, 1964; pp 55-57.
25. Kalos, M. H., and Whitlock, P. A., *Monte Carlo Methods Volume I: Basics;* Wiley-Interscience: New York, 1986; pp 112-115.

RECEIVED April 5, 1994

Chapter 6

Calculations of Solvation Free Energies in Chemistry and Biology

A. Warshel and Z. T. Chu

Department of Chemistry, University of Southern California, Los Angeles, CA 90089–1062

Different approaches for the calculation of solvation free energies are considered. The performance of classical solvation models is discussed comparing macroscopic models, simplified microscopic models, and all-atom microscopic models. The incorporation of these different classical models in quantum mechanical solute Hamiltonians is considered, discussing both in the molecular orbital and Valence-bond approaches. New pseudopotential and density functional approaches that represent the solvent on a more quantum mechanical level are also discussed. An effective way of performing free energy perturbation calculations with quantum mechanical solvation models is outlined. Finally, the implementation of different solvation models in macromolecular simulations is reviewed.

Many biological processes can be described as solvation processes where the interaction between the protein water system and the given active group is considered formally as solvation free energy [1, 2, 3]. Thus it is crucial to be able to estimate solvation energies in any attempt to gain quantitative understanding of biological processes. Of course, the need to evaluate solvation effects is also crucial in quantitative studies of chemical reactions in solutions.

The advance in computer power in the last two decades has led to enormous progress in calculations of solvation free energies [1-36]. However, some early attempts involved strategies that were not so successful with hindsight. For example, the use of the supermolecule approach (e.g. [7, 6]) has reflected the assumption that the same quantum mechanical approach used to study a single small molecule can also be used to study several molecules and eventually solutions. Unfortunately, solvation effects involve cooperativity between many solvent molecules and studies that include only one or two solvent molecules do not provide a direct information about solvation free energies. Similarly, the progress in using continuum models in quantitative studies has been rather slow due to the persistent use of cavity prescriptions. The use of such models

0097–6156/94/0568–0071$08.72/0

remained confined for many years to qualitative applications of uncalibrated reaction-field models. This reflected the slow realization that the use of explicit solvent models (which was impossible before the emergence of computers) could overcome the main stumbling blocks in quantitative evaluation of electrostatic energies.

In retrospect, it seems clear that the proper strategy for early computations of solvation energies should have involved modeling of solvation effects using dipolar solvent models. In fact, such a strategy was quite effective in providing semiquantitative results in the computer evaluation of solvation free energies [15, 14] Nevertheless, the recent advance in computer hardware has made it possible to evaluate solvation free energies by different approaches and probably one day it will be possible to do this by *ab initio* supermolecule approaches. This might happen, however, after many of the important problems already have been solved.

In this chapter, we will consider some aspects of the field of computer studies of solvent effects. The first section will outline various classical solvent models and consider their performance. The second section will review the incorporation of solvent models in quantum mechanical calculations and the third section will consider briefly the calculation of absorption spectra in solutions. Finally the application of solvation models to studies of macromolecules will be reviewed in the fourth section.

Classical Solvent Models

Any formal treatment of solvation problems should consider the entire solute-solvent system quantum mechanically. However, such calculations are impractical at the present time and it is essential to consider approximate treatments. This will be done below, starting with classical solute-solvent models and then describing treatments that represent the system on a more quantum mechanical level.

One way to consider solvation problems is to represent the Born-Oppenheimer potential surface of the solute-solvent system by "classical" models. In fact, viewing the solvent as a dielectric continuum is the granddaddy of all classical models. Such models date back to the pioneering work of Born [37], Kirkwood [38] and Onsager [39]. However, attempts to obtain quantitative solvation energies for molecules of arbitrary shape are much more recent.

In developing various classical models for solvation studies one can consider the three main options illustrated in Fig. 1. Such models cover the entire range from explicit all atom model to simplified microscopic models (that replace the explicit solvent by dipoles) and finally to macroscopic models that consider a collection of dipoles as a polarized continuum. In considering the three classes of models, it is important to realize a point that is frequently overlooked; the continuum models represent the roughest approximation despite the fact

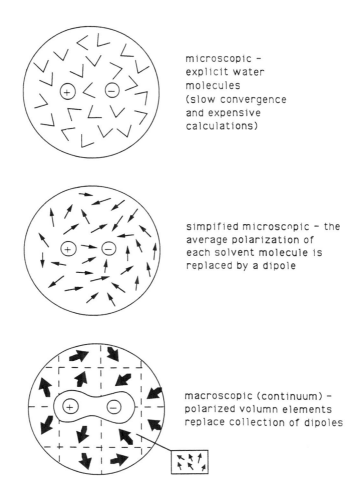

microscopic –
explicit water
molecules
(slow convergence
and expensive
calculations)

simpified microscopic – the
average polarization of
each solvent molecule is
replaced by a dipole

macroscopic (continuum) –
polarized volumn elements
replace collection of dipoles

Figure 1. Illustrating the three main options for representing the solvent in computer simulation approaches. The microscopic model uses detailed all-atom representation and evaluates the average interaction between the solvent residual charges and the solute charges. Such calculations are expensive. The simplified microscopic model replaces the time average dipole of each solvent molecule by a point dipole, while the macroscopic model is based on considering a collection of solvent dipoles in a large volume element as a polarization vector (Reproduced with permission from reference 49. Copyright 1993 John Wiley.)

that they are most closely related to the available electrostatic literature. The actual performance and limitations of the different models will be considered below.

Macroscopic models. The use of the Born's and Onsager's macroscopic models has provided an extremely useful tool for qualitative considerations of many solvation problems (e.g. [40, 41]) However, serious problems are associated with the quantitative use of the Onsager's model. That is, this model expresses the solvation energy of a dipolar solute by [39]

$$\Delta G_{sol} = -166 \frac{\mu^2}{b^3} \frac{(2\epsilon - 2)}{(2\epsilon + 1)} \qquad (1)$$

Where ΔG is given in kcal/mol, μ is the dipole moment of the solute (in $\overset{\circ}{A}$ and atomic charge unit), b is the cavity radius (in $\overset{\circ}{A}$) and ϵ is the medium dielectric constant. Unfortunately, it is very hard to use a simple rule of thumb to assess the effective value of b that gives the correct solvation energy. In fact, the correct effective value of b for an ion pair (composed of ions of the same radius) in high dielectric solvents is (see Appendix A of [42])

$$b^3 \simeq R^2 \bar{a}[2(1 - \bar{a}/R)] \qquad (2)$$

where $\bar{a}$ is the effective cavity radius of the ions and R is the internuclear separation distance. This value has not been used in early studies and is only valid when ϵ is large. Nevertheless, reasonable estimates of b are used in some recent studies (e.g. [26])

The problems associated with the shape of the solute and the corresponding cavity size have been solved in part with the emergence of discretized continuum models [23, 24, 43, 44, 45, 46] but this progress has occured in a relatively late step of the game, after the development of more explicit models [15].

An important advance in continuum studies has emerged with the systematic use of the generalized Born model [47, 48]. It is instructive to note, however, that this treatment is not based on an analytical expression for the energetics of a system of nonoverlapping cavities (which is still unavailable) but is simply the result of the implicit use of the Coulomb's law for charges in a dielectric medium. That is, the energy of a collection of charges is given (in kcal/mol) by [17]

$$\Delta G = \sum_{ij} 332 \frac{Q_i Q_j}{r_{ij}\epsilon} - 166 \sum_i \frac{Q_i^2}{a_i}(1 - \frac{1}{\epsilon}) = \sum_{ij} 332 \frac{Q_i Q_j}{r_{ij}} + \Delta G_{sol} \qquad (3)$$

where a_i is the effective Born radius of the ith atom. In a medium of high dielectric constant ($\epsilon \gg 1$), one obtains [17]

$$\Delta G_{sol} \simeq -\sum_{ij} 332 Q_i Q_j / r_{ij} - 166 \sum_i Q_i^2 / a_i \qquad (4)$$

This equation, is basically the so called "generalized Born" equation (in the limit of a large ϵ) and it simply reflects the almost complete compensation of the vacuum Coloumb's energy by the change in solvation energy which is the main feature of high dielectric media. The simple physics of this compensation effect provides an accurate estimate of the relevant solvation energy in aqueous solution. However, the corresponding generalized Born equation is not expected to work well in heterogeneous environments (like proteins), where the above compensation effects are not so complete.

Simplified Microscopic Models. Although the continuum models are based on an authoratative literature of more than a century, they involve enormous conceptual problems when applied to charges in nonhomogeneous environments (e.g. macromolecules) or to basic microscopic problems. These difficulties and the entire dielectric problem disappear (at least in principle) when the solvent is described explicitly [14, 17]. However, true all-atom models with complete phase space averaging were far too expensive to be used in the early stage of the introduction of computer methods for solvation studies. The realization of this point led to the introduction of simplified solvent models [1]. The most widely used of these models is the Langevin dipoles (LD) model [14, 17]. This model simulates the solvent effect by using a grid of point dipoles that represent the average polarization of the solvent molecules at the corresponding sites by Langevin-type point-dipoles. The point-dipoles interact self consistently with the solute charges and with the field from each other. The LD model has been parameterized to reproduce observed solvation energies of different ions as well as the distance dependence of the average field-dipole interaction, obtained from MD simulations of the corresponding all-atom model. The current version of the model, which is implemented in the program POLARIS [49] includes a field-dependent hydrophobic term and a van der Waals term [49]. The corresponding parameter set involves group-charges which were calibrated to reproduce the observed solvation energies of a large "training set" of molecules and ions as well as the corresponding observed dipole moments. This parameter set, which is incorporated in POLARIS 3.22, gives quite accurate solvation energies as illustrated in Table I and Figs. 2 and 3.

Other early simplified models include the Surface Constrained Soft Sphere Dipole (SCSSD) model [15] which uses explicit Stockmayer-type solvent model and introduced spherical boundary conditions to force the finite simulation system to reproduce the polarization expected from an infinite system [15, 50].

It is perhaps instructive to point out that the conceptual simplicity of the LD and SCSSD model and their rapid convergence, have provided quite reliable results for classical and quantum mechanical problems long before the emergence of similar studies using continuum or all-atom models.

Table I. POLARIS solvation free energies (in water) for a bench mark of neutral molecules[a]

#	name	dGexp	dGcal	dGexp-cal
1	alanine-dipeptide	-16.96	-15.80	-1.15
2	p-nitrophenol	-11.92	-10.11	-1.81
3	m-nitrophenol	-10.14	-9.45	-0.69
4	N-methylacetamide	-10.20	-10.65	0.44
5	1-methyl-3-nitrobenzene	-3.45	-4.90	1.45
6	1-methyl-2-nitrobenzene	-3.59	-3.90	0.32
7	nitrobenze	-4.11	-3.99	-0.12
8	2-nitropropane	-3.14	-2.21	-0.93
9	1-nitropropane	-3.34	-3.41	0.07
10	nitroethane	-3.71	-4.29	0.58
11	chlorobenzene	-1.12	-1.41	0.29
12	chloromethane	-0.56	-0.36	-0.19
13	benzoic acid methyl ester	-4.28	-4.41	0.13
14	hexanoic acid methyl ester	-2.49	-3.15	0.66
15	propionic acid 2-methylethyl ester	-2.22	-2.05	-0.17
16	butanoic acid methyl ester	-2.83	-2.40	-0.43
17	propionic acid ethyl ester	-2.80	-2.88	0.09
18	acetic acid 2-methylethyl ester	-2.64	-1.75	-0.90
19	acetic acid propyl ester	-2.85	-1.86	-0.99
20	propionic acid methyl ester	-2.93	-3.15	0.22
21	acetic acid ethyl ester	-3.09	-3.23	0.13
22	acetic acid methyl ester	-3.31	-3.15	-0.16
23	formic acid ethyl ester	-2.64	-2.62	-0.02
24	butanoic acid	-6.35	-5.93	-0.42
25	propionic acid	-6.47	-6.20	-0.27
26	acetic acid	-6.70	-6.93	0.24
27	octanal	-2.29	-3.60	1.31
28	heptanal	-2.67	-2.36	-0.31
29	hexanal	-2.81	-3.42	0.61
30	pentanal	-3.03	-3.11	0.08
31	butanal	-3.17	-3.04	-0.14
32	proponal	-3.44	-3.75	0.31
33	acetaldehyde	-3.50	-3.98	0.48
34	acetophenone	-4.58	-3.79	-0.80
35	2,4-dimethyl-3-pentanone	-2.74	-2.39	-0.34
36	4-heptanone	-2.92	-2.53	-0.39
37	2-heptanone	-3.04	-3.48	0.44
38	4-methyl-2-pentanone	-3.06	-2.66	-0.40
39	2-hexanone	-3.29	-3.62	0.33
40	3-methyl-2-butanone	-3.24	-3.01	-0.23
41	3-pentanone	-3.41	-3.36	-0.05
42	2-pentanone	-3.53	-3.05	-0.48
43	2-butanone	-3.63	-3.06	-0.58
44	2-propanone	-3.97	-3.43	-0.54

Table I. Continued

45	pyridine	-4.70	4.41	-0.28
46	trimethylamine	-3.24	-1.78	-1.46
47	dibutylamine	-3.33	-4.48	1.15
48	dipropylamine	-3.66	-4.85	1.19
49	diethylamine	-4.07	-4.31	0.25
50	dimethylamine	-4.28	-4.92	0.64
51	hexylamine	-4.03	-4.96	0.94
52	pentylamine	-4.09	-4.91	0.81
53	butylamine	-4.29	-4.01	-0.28
54	propylamine	-4.39	-4.88	0.49
55	ethylamine	-4.50	-4.89	0.39
56	methylamine	-4.56	-4.68	0.12
57	dibutylether	-0.83	-1.25	0.42
58	diisopropylether	-0.53	-0.53	0.00
59	dipropylether	-1.15	-1.79	0.63
60	2-methoxy-2-methylpropane	-2.21	-0.93	-1.28
61	1-ethoxypropane	-1.81	-1.29	-0.52
62	2-methoxypropane	-2.00	-2.32	0.32
63	1-methoxypropane	-1.66	-2.38	0.72
64	diethyl ether	-1.63	-1.63	-0.01
65	dimethyl ether	-1.89	-2.21	0.32
66	1,1-dinethylethyl phenol	-5.92	-5.94	0.02
67	4-methylphenol	-6.13	-6.04	-0.09
68	2-methylphenol	-5.87	-5.30	-0.57
69	phenol	-6.61	-6.09	-0.52
70	cycloheptanol	-5.48	-4.94	-0.54
71	cyclohexanol	-5.47	-6.16	0.69
72	cyclopentanol	-5.49	-5.58	0.09
73	4-heptanol	-4.00	-4.51	0.51
74	2,3-dimethyl-2-butanol	-3.91	-3.11	-0.80
75	3-methyl-1-butanol	-4.42	-5.29	0.87
76	1-pentanol	-4.47	-5.66	1.18
77	1-butanol	-4.71	-6.03	1.31
78	1-propoanol	-4.82	-5.88	1.05
79	ethanol	-5.01	-5.11	0.10
80	methanol	-5.11	-5.94	0.83
81	biphenyl	-2.64	-1.39	-1.25
82	benzene	-0.87	-0.67	-0.20
83	hexane	2.48	2.19	0.30
84	2-methylbutane	2.38	2.69	-0.31
85	pentane	2.33	1.03	1.30
86	2-methylpropane	2.32	2.41	-0.09
87	butane	2.08	1.59	0.49
88	propane	1.95	2.16	-0.20
89	ethane	1.83	1.99	-0.16
90	methane	2.00	2.40	-0.40

Continued on next page

Table I. Continued

91	aspatic acid	-8.60	-9.73	1.13
92	serine	-7.00	-7.89	0.89
93	histidine	-12.20	-13.20	1.00
94	arginie	-12.90	-8.42	-4.48
95	asparagine	-11.70	-11.10	-0.60
96	lysine	-6.30	-6.06	-0.24
97	aniline	-4.90	-5.80	0.90
98	toluene	-0.90	-1.20	0.30

(a) Energy in kcal/mol

All-atom solvent models. With the emergence of supercomputers it became possible to study solvation problems with all-atom solvent models. In such models one uses classical force field for the solute-solvent interactions (sometimes with induced dipole terms [14, 16, 50, 51] and preferably with proper boundary conditions (see below)). With this description and MD or MC simulation methods, it is possible to use Free Energy Perturbation (FEP) [52] methods to evaluate solvation free energies. Such calculations are based on a "charging" process using a potential of the form [16]

$$\epsilon_m = \epsilon_1(1 - \lambda_m) + \epsilon_2\lambda_m \qquad (5)$$

where ϵ_1 and ϵ_2 are the potential surfaces of the uncharged and charged states, respectively. Changing λ_m from zero to one in n steps gives the charging free energy by [52]

$$\Delta G(Q = 0 \rightarrow Q = 1) = \sum_{m=1}^{m=n-1} -RT \ln\{< \exp[-(\epsilon_{m+1} - \epsilon_m)/RT] >_m\} \qquad (6)$$

A typical example of such a calculation that involves an adiabatic charging of Na^+ in water [50] is given in Fig. 4. The figure also demonstrates that the results obtained by a Langevin dipole model are very similar to those obtained by the all-atom model (see ref. [50] for more details).

The earliest FEP studies of solvation free energies were reported in 1982 [16, 22]. The interest in this approach has increased significantly following the work of Jorgensen [27] that involved a simple test case with good convergence. The FEP approach provides now the most popular tool for microscopic calculations of solvation free energies e.g. [30, 31, 32].

solvation free energy(gas->solution)

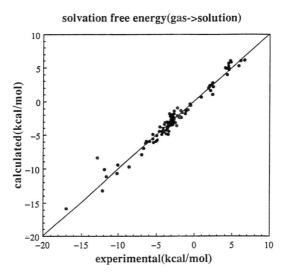

Figure 2. The correlation between the observed solvation energies and the corresponding calculated results obtained using the POLARIS group charges. All the molecules considered are not charged (see Table I).

POLARIS vs. Experiments

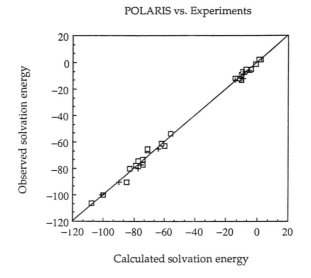

Figure 3. The correlation between observed solvation energies and the corresponding polaris results for a set of charged molecules.

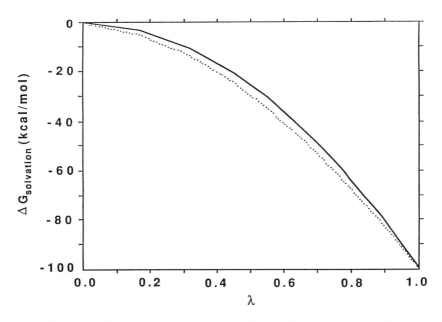

Figure 4.   Showing the free energy as a function of the parameter λ for an adiabatic charging of a Na^+ ion in water. The dotted curve was obtained with a Langevin Dipole model (see [50] for more details). (Reproduced with permission from reference 50. Copyright 1989 American Institute of Physics.)

Despite the great popularity of FEP calculations, they still involve major problems when applied to absolute solvation energies. Many workers have tried to avoid these problems by focusing on the less challenging task of evaluating changes in solvation energies ($\Delta\Delta G$) (e.g. $Na^+ \to K^+$) rather than absolute solvation energies. However, in order to obtain consistent parameters for inter-molecular interaction, it is essential to be able to calculate abosolute solvation energies (ΔG_{sol}). Such calculations are very sensitive to boundary conditions and to the cutoff radius used, and standard periodic boundary conditions do not give reliable results (see below).

Crucial problems with boundary conditions and long-range interactions. Computer simulations of solvation processes should take into account a system with an infinite number of solvent molecules. In the case of pure solvents (without a solute) one can use periodic boundary conditions to simulate a finite system as a part of an infinite system. However, this approach does not provide such a good approximation for studies of solvated ions. That is, as is illustrated in Fig. 5, the symmetry around a charge is clearly not periodic (a very informative discussion of the problem associated with the use of periodic boundary conditions is given in a recent work of Aqvist [53]). Thus it is imperative to use nonperiodic boundaries. Such a treatment has been introduced in the early SCSSD model [15] and led to the development of the surface constrained all-atoms solvent (SCAAS) model [17, 50]. This model uses a spherical simulation region surrounded by a surface region and a bulk region. The surface region involves radial and angular constraints that force the polarization of a finite simulation region to follow the polarization expected from an infinite system. Related models that emphasize the thermal fluctuations of the surface region but did not introduce polarization constraints, have also been developed [54]. The SCAAS model allows one to get a consistent estimate of the bulk contribution (which is difficult to evaluate consistently when one uses periodic boundary conditions) and to get relatively fast convergence with a relatively small number of solvent molecules.

Another problem that has not received sufficient attention is the cutoff of the long-range electrostatic interactions. Customary cutoffs of $10\mathring{A}$ lead to overpolarization of the solvent molecules and the overestimation of the corresponding solvation energies. Our effort to overcome this problem involves the development of the generalized Ewald method [55] and the simpler but more effective Local Reaction field(LRF) method [56]. The latter method provides an excellent approximation for the energetics obtained without any cutoff, while using as small a cutoff as $7\mathring{A}$. The LRF method also gives much more stable FEP results than those obtained with standard cutofff methods [56]. Other methods that can help in solving the long range problem have been reported [57] but the performance of these methods has not been examined in systematic calculations of solvation free energies.

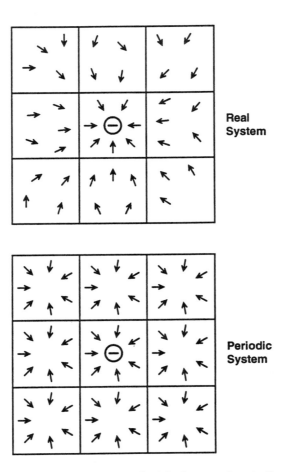

Figure 5. Showing the problems associated with the use of periodic boundary conditions for studies of solvation of ions.

Quantum mechanical solvation models

The classical approaches considered above cannot be used when the solvent field has a major effect on the solute wave function. This includes, of course, chemical reactions and in particular charge separation processes. Here it is essential to represent quantum mechanically at least a part of the system. Different attempts to develop quantum mechanical solvation models will be reviewed below.

Supermolecular approaches. Seemingly, the most straightforward approach to quantum mechanical solvation problems is to treat the solute and the solvent quantum mechanically as one large supermolecule. Unfortunately, it is impractical to treat sufficient number of molecules vquantum mechanically and it is clearly impossible to explore sufficiently large phase space in any attempt to evaluate solvation free energies. Supermolecular calculations that involve only few solvent molecules were reported by several workers [e.g. [6, 58]] but the results of such studies could not be converted to solvation free energies in solutions.

Reaction field models. The incorporation of continuum solvent models in quantum chemical calculations has been explored in the early work of Klopman [4] who considered the problem in a heuristic way and provided a useful expression for the solvation contribution to the energetics of a given solute charge distribution. Subsequent studies (e.g. [9, 10, 11, 12, 20] considered the solute in a spherical cavity located inside a continuum dielectric medium. The corresponding potential from the solvent was expressed in terms of the solute charge distribution and incorporated into the solute Hamiltonian. While such approaches seem quite rigorous (especially when applied self-consistently [9, 11]), they depend on the *third power* of the cavity radius, which is basically an adjustable parameter whose actual value is not given in a reliable way by the available prescriptions (see first section). Thus the calculations may involve substantial errors and can only be considered as rough estimates. It is possible that the elegant form of the macroscopic treatments led some to overlook the bottom line, which is the ability to predict reasonable solvation free energies. A model that cannot give reliable solvation energies should not be used for quantitative quantum mechanical studies of processes in solutions regardless of its formal justification.

Recent attempts to estimate the effective cavity size [10, 26] led to some progress and more quantitative results were obtained by the use of descretized continuum methods [21, 36]. The use of a well parameterized generalized Born model also led to significant progress [48], illustrating again the crucial need to parameterize and validate any solvation model [15].

In general, we feel that cavity models have led to very significant conceptual progress (in particular with regards to the general trend in solvation energies) but probably held back the quantitative progress of the field. With digital computers, it is much simpler to overcome all the conceptual problems associated with continuum treatment by using explicit dipolar models. This is particularly true with regard to the incorporation of induced dipoles in excited state calculations (see below) and to the treatment of solvent fluctuations.

MO/Explicit-Solvent Models. The emergence of computers allows one to overcome the long standing problems associated with continuum formulations and to represent the solvent by explicit dipolar models. The realization of this point led quite early [14, 15] to quantitative quantum mechanical solvation calculations. The corresponding models were based on a hybrid treatment that treated the solute quantum mechanically while considering the solvent classically. The simplest way to understand this hybrid concept is to think of the solute-solvent system as a supermolecule by writing the SCF matrix as

$$\begin{vmatrix} \mathbf{F}^{SS} & \mathbf{F}^{Ss} \\ \mathbf{F}^{Ss} & \mathbf{F}^{ss} \end{vmatrix} \tag{7}$$

where S and s designate solute and solvent respectively and the $\mathbf{F}$ matrix elements are given in [1]. The assumption that the solvent and the solute AO's are orthogonal to each other, implies, within the CNDO approximation, that $\mathbf{F}^{Ss} = 0$. Treating the complete $\mathbf{F}$ matrix within the CNDO-type all valence electron approximation and freezing the MO coefficients of the solvent orbitals one obtains

$$\mathbf{F}^S_{\mu\mu} = (\mathbf{F}^S_{\mu\mu})_0 - \sum_{B \in s} q_B \gamma_{AB} \quad \mu \in A \in S \tag{8}$$

$$\mathbf{F}^S_{\mu\mu} = (\mathbf{F}^S_{\mu\nu})_0$$

where $(\mathbf{F}^S)_0$ stands for the SCF matrix of the isolated solute molecule, A and B are, respectively, solute and solvent atoms, q_B are the solvent atomic charges and γ_{AB} is the two-electron repulsion integral, which accounts here for the solvent influence on the solute electronic structure.

Taking eq. (8) and approximating γ_{AB} by e^2/r_{AB}, we obtain for any solvent configuration

$$\mathbf{F}^S_{\mu\mu} = (\mathbf{F}^S_{\mu\mu})_0 - \sum_B e^2 q_B/r_{AB} = (\mathbf{F}^S_{\mu\mu})_0 - U_A \tag{9}$$

where $\mu \in A$ and U_A designates the total electrostatic potential from the solvent atoms at the site of atom A. This equation can be used in the more general case where the solvent electrons are polarized by the field of the solute.

Basically, the effect of the solvent is to modify the ionization energy of the solute atoms and thus to change its electronic structure. The new charges of the solute atoms interact with the solvent which in turn changes its polarization. The resulting polarization of the solvent modifies the solute charges and this leads to a new solvent polarization and so on until self-consistency is acheived. When this procedure is implemented with MD or energy minimization and with non-polarizable all-atom solvent models, the self-consistency problem is simplified. In such case it turns out that the solute electronic wave function is determined uniquely for any instantaneous solvent structure.

Although hybrid quantum/classical MO studies of solvation processes were reported quite early (i.e. the LD study of ref. [14] and a MINDO/SCSSD study of the $H_2O + H_2O \rightarrow H_3O^+ + OH^-$ autoionization reaction in water [15]), they have only become popular in recent years [29, 59, 34, 35, 33, 42]. The popularity of these approaches reflects the fact that they provide a practical and quite rigourous way for evaluating solvation effects by semiempirical quantum mechanical software packages.

In considering hybrid models, it is important to mention studies that evaluated the charges of the reacting system by gas-phase *ab-initio* calculations and then solvated this system in classical all-atom solvent models (e.g. ref. [28]). Although such studies were clearly important in demonstrating that one can reproduce the observed energetics of S_N2 reactions in solutions, they were somewhat inconsistent since they neglected the coupling between the solute wave function and the solvent polarization. Such approaches cannot deal properly with reactions that involve large charge separation (e.g. the above mentioned autoionization of water or the heterolytic bond cleavage reactions discussed in ref. [60]).

The EVB model. Although the above mentioned MO models are very appealing, they still involve some problems. This includes the difficulties in implementing such approaches in rigorous FEP treatments, the need to perform configuration-interaction treatments in studies of bond breaking reactions and the fact that the semiemprical integrals used in current semiempirical treatments are not calibrated for studies of solution reactions. Perhaps the best way to capture the physics of bond-breaking reactions in solutions is offered by the Valence Bond (VB) approach. The semiemperical integrals in this approach can be calibrated easily by forcing the reaction surface to reproduce the observed energies of the reactant and products at infinite separation and the effect of the solvent can be included in a rather unique way. The incorporation of classical solvent models in the VB Hamiltonian has been introduced in the Empirical Valence Bond (EVB) method [1, 60, 61]. This treatment has been used extensively in studies of the energetics and dynamics of chemical reactions in solution and proteins [1]. Closely related approaches are now emerging

[62, 63, 64] indicating the powerful physics of this method. More details on the EVB methods and its extensive applications are given in ref. [19].

Representing the solvent quantum mechanically. The above mentioned hybrid quantum/classical methods capture a major part of the effect of the solvent on the solute wave function. However, some apsects of solvation processes are not accounted for. These include charge transfer from the solute to the solvent and other related effects. Progress in this direction has started only recently with the emergence of an *ab-initio* solvation study that represented the solvent by a pseudopotential [65] (studies of solvated electrons, e.g. ref. [66], are also relevant to this issue). The pseudopotential approach involves *ad-hoc* assumptions about the nature of relevant parameters. A more promising strategy may be provided by the recently developed Frozen Density Functional Theory (FDFT) [67, 68]. This method considers the entire solute-solvent system as a supermolecule in the DFT formulation. However the electronic density of the solvent molecules is kept frozen, thus allowing for efficient calculations of the energy of solvated molecules.

Ab-initio and semiempirical FEP calculations. Proper microscopic calculations of solvation free energies require an extensive average over the solvent configurations. Although this can be accomplished with efficient free energy perturbation approaches, it is still quite challenging to combine such calculations with quantum mechanical solvation models. The first incorporation of quantum mechanical and FEP calculations has been introduced in EVB studies [18]. In fact, the EVB provides a natural way for FEP calculations using mapping potentials of the form [60, 18]

$$\epsilon_m = \sum_i \lambda_i^m \epsilon_i \qquad (10)$$

where the ϵ_i are the different resonance structures that describe the given reaction.

The implementation of FEP calculations with MO models is much less straightforward. Attempts to use gas phase *ab-initio* surfaces as the basis for FEP studies in solutions [28] are somewhat inconsistent (see the section on MO/Explicit-Solvent Models). More recent attempts [59] used the coordinates of the solute in their reactants and product states for the FEP mapping, thus introducing artificial constraints. Basically, it is difficult to find simple mapping parameters within the MO formulation; for example the Bond order cannot be used as a mapping parameter since its value changes with the fluctuations of the solvent [60].

A general way for FEP calculations with semiempirical and *ab-initio* MO treatments has been introduced recently [42, 65, 68]. This approach is based

on the use of the EVB surface as a reference state for the FEP/MO calculations. The evaluation of the MO free energy only involves calculations of
the free energy associated with changing the potential from the EVB to the
MO representation in a few points along the reaction path. Furthermore, this
approach does not require the evaluation of the *ab-initio* forces since the trajectories involve the EVB forces. Such an approach has been used in AM1
calculations of solvation energies and most significantly in *ab-initio* pseudopotential calculations [65] and in FDFT calculations of solvation free energies
[67].

Solvent effects on electronic transitions

Trying to account for the the effect of the solvent on the spectrum of a given
molecule, one has to evaluate the change in the solute-solvent interaction upon
electronic excitation. This change may involve many factors such as dispersion, polarization, Coulombic and charge-transfer interactions. The polarization and Coulombic interactions are expected to have the dominant effect
on electronic transitions that involve large changes in the solute charge distribution. This is, in fact, the assumption made by McRay [69] and others
[70, 71, 72, 73, 41] in their pioneering studies of solvent effects on absorption
spectra. However, these early studies and more recent works (e.g. [74, 75, 76]),
were formulated on a phenomenological level and involved macroscopic continuum models with an arbitrarily defined cavity radius as an adjustable parameter. Such treatments, which might seem quite reasonable at first sight, lead
to major problems when one tries to evaluate the absolute value of the solvent effect (rather than the trend associated with changes of solvent polarity).
Furthermore, significant difficulties are involved in macroscopic treatments
of electronic excitation in polar solvents. In such cases the induced dipoles
of the solvent should adjust themselves to the charge distribution of the excited solute, while still responding to the polarization of the solvent permanent
dipoles (which are kept at their ground state orientations). While macroscopic
treatments of this problem are available for cases where the solute charge distribution can be represented by the dipole approximation, the treatment of
more general charge distribution presents a significant challenge. This challenge becomes more serious when the solvent polarization in the ground state
involves saturation effects.

Early attempts to evaluate absorption spectra of solvated molecules on a
quantitative level has emerged with the use of microscopic polar models [15,
17, 77]. More recent studies [78] along that line include a systematic study that
used both the LD model and the SCAAS all-atom model. The latter model was
used with a systematic MD simulation and with the dispersed polaron model

[79] that provided a rigorous way to obtain line shapes of solvated molecules [78]. Important attempts to use microscopic models in evaluating the spectra of solvated models were reported in refs.[80, 81] and many more studies are expected to emerge in the near future. Microscopic studies of the spectra of proteins were also reported [77].

Modeling Solvated Macromolecules

The importance of electrostatic energies in macromolecules have been pointed out frequently (e.g. [2, 3, 17, 25, 82, 83, 84]). However, it is not so widely accepted that the evaluation of the "solvation" of active groups in proteins might provide the best way of correlating structure and function [3]. Nevertheless, recent progress in calculations of solvation energies of macromolecules have opened up new possibilities for exploring the structure and function of macromolecules. These options have been reviewed recently [17, 25, 85, 86] and we will only consider here several important points.

Macroscopic models. Macroscopic studies of electrostatic interactions in proteins dates back to the pioneering studies of Lingstrum Lang [87] and to the extremely influential model of Kirkwood and Tanford [88] that were formulated before the availability of X-ray structure of macromolecules. In retrospect, it appeared that the use of macroscopic models to describe macromolecules is even more problematic than the corresponding use in studies of small molecules. Not only that the value of the local dielectric constant could not be determined by the use of macroscopic concepts [89], but the proper treatment of polar environments around charged groups presented a major challenge that was not considered by the early macroscopic treatments (see [90]). In fact, all macroscopic studies of solvated macromolecules have involved major inconsistencies until as late as the mid eighties (see discussion in [85, 90, 91]) significantly after the emergence of consistent treatments by microscopic polar models [14, 17].

Recent years led to enormous progress in the implementation of macroscopic models in studies of macromolecules. Discretized continuum (DC) models (e.g. [23, 25]) provided a practical way to represent the solvent around the macromolecules and eventually started to consider the local polarity and to take care of self energies. The issue of the proper local dielectric is still largely misunderstood although this question can be addressed on a consistent microscopic level [89]. It is also hard to account properly for internal water molecules and to represent consistently the protein reorganization during a charging process although these two effects can be considered as a part of the protein dielectric constant.

Despite the problems involved with initial implementation of continuum models they are becoming very popular and quite reliable when applied consistently. These methods are proving repeatedly that the solvent around the protein can be represented by simple models which are frequently more reliable than some microscopic all-atom treatments with improper boundary conditions.

The PDLD model. As in the case of small molecule, it appears that the first model that captured the physics of electrostatic effects in macromolecules was a simplified microscopic model; the protein dipoles Langevin Dipoles (PDLD) model [14, 17]. This model considered explicitly the protein permanent and induced dipoles while representing the solvent by the Langevin dipoles model. The development of the PDLD model reflected the point of view that the "safest" way to treat electrostatic effects in proteins is to count all the interactions microscopically, thus avoiding the dielectric concept all together.

The PDLD method has been used extensively in evaluations of electrostatic energies in proteins. This included calculations of intrinsic pK_a's, redox potentials, enzyme catalysis, absolute binding energies, and penetration of ions through ion channels [49, 85]. The current version of the PDLD approach is implemented in the program POLARIS [49]. This version involves MD generation of protein configurations, which are then used in PDLD calculations. The recent version also includes an hydrophobic term which is corrected for the effect of the local field at the surface of the protein. In addition the POLARIS program includes a semimacroscopic version [49] where the effect of the protein relaxation (in response to changes of charges) and the effect of the induced dipoles are considered as a dielectric effect and the PDLD results are scaled accordingly. This model gives stable results and converges faster than the microscopic PDLD methods. In fact, the semimacroscopic model provides a connection between the PDLD and the discretized continuum models. The semimacroscopic model gives, for example, encouraging results in calculation of binding free energies [49].

All-atom microscopic models. The availability of powerful computers made it possible to use all-atom models in studies of electrostatic energies in solvated proteins. Impressive progress has occured in the field since the early FEP studies of electrostatic energies in proteins [18, 30, 92, 93, 94] and FEP calculations of solvated macromolecules are now growing exponentially (see ref [30, 31, 32]). FEP simulations have been applied to a wide range of properties of proteins including pK_a's [49], redox processes [49, 95, 96], ligand binding [30, 32], enzymatic reactions [1, 19, 97], ion channels [98, 99], and protein stability [100]. Despite this progress, there are still major problems with regards to convergence and boundary conditions (see below).

The importance of proper boundary conditions. Despite the progress in all-atom simulations, we are still far from the stage where a simple use of a standard computer package will give accurate results for solvation free energies in proteins. As in the case of small molecules, periodic boundary conditions do not reflect the correct symmetry of the field around proteins. If one likes to assume that the periodic treatment is reasonable, then each unit cell should include the entire protein. The Ewald method is not useful in this case, since the main problem is now associated with the treatment of the long range interactions within the enormous unit cell. Here the combination of the SCAAS boundary conditions and the LRF long-range corrections appears to provide a very effective toll (see ref. [49]).

Perhaps the most common error in evaluating electrostatic energies in macromolecules involves the treatment of ionizable residues while neglecting to include a sufficient number of solvent molecules. In fact, the electrostatic field of surface groups is almost completely screened by the solvent and it is much better to neglect the charge of this group than to treat them in their ionized state, while neglecting the dielectric effect of the solvent [101]. The same is true with regards to the so called "Helix dipoles" whose effect is largely overestimated without proper boundary conditions [102].

Because of the problems associated with incomplete microscopic treatments of electrostatic energies, it is important to compare the results of such calculations to those obtained with the PDLD or DC approaches. This point is nicely illustrated in a recent study of the absolute pK_a of the acidic groups of BPTI [56] and in calculations of the energetics of the charge transfer states in bacterial reaction centers [101].

Concluding Remarks

In this chapter we considered different solvation models and reemphasized the point that the solvent effects are the missing link between gas phase calculations and the evaluation of the properties of molecules in solutions. Apparently, the physics of the solute-solvent interaction is not so complicated and can be captured by models with different degrees of complexity. While each of these models may have its strength and weaknesses, it is becoming increasingly clear that a proper and self-consistent representation of the entire system in which the "solute" finds itself is essential for approaching quantitative success. Hence, seemingly rigorous treatments that only consider a limited part of the system or do not treat the solute-solvent coupling in an adequate way may yield somewhat irrelevant results. The importance of representing the complete system can hardly be overemphasized when dealing with macromolecules. Here the use of all-atom models do not always present the best choice, particularly when such approaches do not involve the implementation of proper boundary conditions.

It is important to point out that the most accurate results are obtained at present with models which are empirically calibrated using observed solvation energies. Despite the belief by some that all-atom models reproduce correct results because of an inherent physics, both continuum and all atom model involve as adjustable parameters the effective atom size (the van der Waals radius) and the residual charges. We are still not at a stage where *ab-initio* surfaces give accurate solvation energies. In fact, it is possible that only the inclusion of the charge-transfer interaction between the solute and the solvent will allow *ab-initio* solvation energies to be of quantitative predictive value.

The importance of hybrid quantum/classical models is becoming increasingly clear [19]. Such methods allow one to describe chemical processes in solution without attempting to treat the entire system quantum mechanically. It is quite possible that such approaches will provide sufficiently reliable tools for studies of most classes of chemical processes in solution and in solvated macromolecules. Yet it is clearly important to find new approaches that can also represent the solvent quantum mechanically. Some initial progress is being made in this direction [67, 68] but the main challenge is clearly ahead.

Acknowledgement

This work was supported by the National Institute of Health (Grant GM 24492) and by the ONR (Grant 0014-91-J-1318).

References

1. Warshel, A. *Computer Simulation of Chemical Reactions in Enzymes and Solutions* ; John Wiley & Sons: New York, 1991.
2. Warshel, A. *Proc. Natl. Acad. Sci., USA* **1978**, *75* , 5250.
3. Warshel, A. *Acc. Chem. Res.* **1981**, *14*, 284.
4. Klopman, G. *Chem. Phys. Lett.* **1967**, *1*, 200.
5. Yomosa, S. J. *Phys. Soc. Japan* **1973**, *35*, 1738.
6. Pullman, A.; Pullman, B. *Quart. Rev. Biophys.* **1975**, *7*, 506.
7. Pullman, A. *Environmental Effects on Molecular Structure and Properties*; Reidel: Dodrecht, 1976, p 149.
8. Rinaldi, D.; Rivali, J.-L. *Theor. Chim. Acta* **1973**, *32*, 57.
9. Rivali, J.-L.; Rinaldi, D. *Chem. Phys.* **1976**, *18*, 223.
10. Rinaldi, D.; Rivail, J.-L.; Rguini, N. *J. Comput. Chem.* **1992**, *13*, 675.
11. Tapia, O.; Goscinski, O.; *Mol. Phys.* **1975**, *29*, 1653.
12. Tapia, O.; Silvi, B. *J. Phys. Chem.* **1980**, *84*, 1646.
13. Tapia, O. *J. Mol. Struct.* **1991**, *226*, 59.
14. Warshel, A.; Levitt, M. *J. Mol. Biol.* **1976**, *103*, 227.
15. Warshel, A. *J. Phys. Chem.* **1979**, *83*, 1640.

16. Warshel, A. *J. Phys. Chem.* **1982**, *86*, 2218.
17. Warshel, A.; Russell, S.T. *Quart. Rev. Biophys.* **1984**, *17*, 283.
18. Warshel, A. In *Specificity in Biological Interactions*; Pontificiae Academiae Scientarum Scripta Varia: 1984, p. 60.
19. Aqvist, J.; Warshel, A. *Chem. Rev.* **1993**, *93*, 2523.
20. McCreery, J.; Christoffersen, R.E.; Hall, G.G. *J. Am. Chem. Soc.* **1976**, *98*, 7191.
21. Miertus, S.; Scrocco, E.; Tomasi, J. *Chem. Phys.* **1981**, *55*, 117.
22. Postma, J.P.M.; Berendsen, H.J.C.; Haak, J.R. *Faraday Symp. Chem. Soc.* **1982**, *17*, 55.
23. Warwicker, J.; Watson, H. *J. Mol. Biol.* **1982**, *157*, 671.
24. Gilson, M.K.; Honig, B.H. *Proteins* **1988**, *3*, 32.
25. Sharp, K.A.; Honig, B. *Ann. Rev. Biophys. Biophys. Chem.* **1990**, *19*, 301.
26. Katritzky, A.R.; Zerner, M.C.; Karelson, M.M. *J. Am. Chem. Soc.* **1986**, *108*, 7213.
27. Jorgensen, W.L.; Ravimohan, C. *J. Chem. Phys.* **1985**, *83*, 3050.
28. Chandrasekhar, J.; Smith, S.F.; Jorgensen, W.L. *J. Am. Chem. Soc.* **1985**, *107*, 2974.
29. Singh, U.C.; Kollman, P.A. *J. Comput. Chem.* **1986**, *7*, 718.
30. Kollman, P.A.; Merz, K.M., Jr. *Acc. Chem. Res.* **1990**, *23*, 246.
31. Beveridge, D.L.; DiCapua, F.M. *Ann. Rev. Biophys. Biophys. Chem.* **1989**, *18*, 431.
32. Straatsma, T.P.; McCammon, J.A. *Ann. Rev. Phys. Chem.* **1992**, *43*, 407.
33. Field, M.J.; Bash, P.A.; Karplus, M. *J. Comput. Chem.* **1990**, *11*, 700.
34. Gao, J. *J. Phys. Chem.* **1992**, *96*, 537.
35. Wang, B.; Ford, G.P. *J. Chem. Phys.* **1992**, *97*, 4162.
36. Bonaccorsi, R.; Cammi, R.; Tomasi, J. *J. Comput. Chem.* **1991**, *12*, 301.
37. Born, M. *Z. Phys.* **1920**, *1*, 45.
38. Kirkwood, J.G. *J. Chem. Phys.* **1934**, *2*, 351.
39. Onsager, L. *J. Am. Chem. Soc.* **1936**, *58*, 1486.
40. Shaik, S.S.; Schlegel, H.B.; Wolfe, S. *Theoretical Aspects of Physical Organic Chemistry. The Sn2 Reaction*; Wiley and Sons: New York, 1992.
41. Salem, L. *Electrons in Chemical Reactions: First Principles*; Wiley and Sons: New York, 1982.
42. Luzhkov, V.; Warshel, A. *J. Comput. Chem.* **1992**, *13*, 199.
43. Zauhar, R.J.; Morgan, R.S. *J. Comput. Chem.* **1988**, *9*, 171.
44. Rashin, A.A. *J. Phys. Chem.* **1990**, *94*, 1725.
45. Honig, G.; Sharp, K.; Yang, A.-S. *J. Phys. Chem.* **1993**, *97*, 1101.
46. Orrtung, W.H. *Annals of the New York Acad. Sci.* **1977**, *303*, 22.
47. Still, C.S.; Tempczyk, A.; Hawley, R.; Hendrickson, T. *J. Am. Chem. Soc.* **1990**, *112*, 6127.
48. Cramer, C.J.; Truhlar, D.G. *J. Am. Chem. Soc.* **1991**, *113*, 8305.
49. Lee, F.S.; Chu, Z.-T.; Warshel, A. *J. Comput. Chem.* **1993**, *14*, 161.
50. King, G.; Warshel, A. *J. Chem. Phys.* **1989**, *91*, 3647.
51. Kuwajima, S.; Warshel, A. *J. Phys. Chem.* **1990**, *94*, 460.
52. Valleau, J.P.; Torrie, G.M. In *Modern Theoretical Chemistry* ; Plenum Press: 1977, p. 169.

53. Aqvist, J. *J. Chem. Phys.* in press.
54. Brooks, C.L., III.; Karplus, M. *J. Chem. Phys.* **1983**, *79*, 6312.
55. Kuwajima, S.; Warshel A. *J. Chem. Phys.* **1988**, *89*, 3751.
56. Lee, F.S.; Warshel, A. *J. Chem. Phys.* **1992**, *97*, 3100.
57. Greengard, L.; Rokhlin, V. *Chem. Scr.* **1989**, *29A*, 139.
58. Jorgensen, W.L. *J. Am. Chem. Soc.* **1978**, *78*, 1057.
59. Bash, P.A.; Field, M.J.; Karplus, M.J. *J. Am. Chem. Soc.* **1987**, *109*, 8092.
60. Hwang, J.-K.; King, G.; Creighton, S.; Warshel, A. *J. Am. Chem. Soc.* **1988**, *110*, 5297.
61. Warshel, A.; Weiss, R.M. *J. Am. Chem. Soc.* **1980**, *102*, 6218.
62. Chang, Y.-T.; Miller, W. *J. Phys. Chem.* **1990**, *94*, 5884.
63. Kim, H.J.; Hynes, J.T. *Int. J. Quant. Chem.* **1990**, *24*, 821.
64. Kim, H.J.; Hynes, J.T. *J. Am. Chem. Soc.* **1992**, *114*, 10508.
65. Vaidehi, N.; Wesolowski, T.A.; Warshel, A. *J. Chem. Phys.* **1992**, *97*, 4264.
66. Schnitker, J.; Rossky, P.J. *J. Chem. Phys.* **1987**, *86*, 3462.
67. Wesolowski, T.A.; Warshel, A. *J. Phys. Chem.* **1993**, *97*, 8050.
68. Wesolowski, T.; Warshel, A. unpublished.
69. McRae, E.G. *J. Phys. Chem.* **1957**, *61*, 562.
70. Bayliss, N. *J. Chem. Phys.* **1950**, *18*, 292.
71. Amos, A.; Burrows, B.L. **1973**, *7*, 289.
72. Kosower, E. *An Introduction to Physical Organic Chemistry*; John Wiley and Sons: New York, 1968.
73. Mataga, N.; Kubota, T. *Molecular Interactions and Electronic Spectra*; Marcel Dekker: New York, 1970.
74. Karelson, M.; Zerner, M.C. *J. Am. Chem. Soc.* **1990**, *112*, 9405.
75. Karelson, M.M.; Zerner, M.C. *J. Phys. Chem.* **1992**, *96*, 6949.
76. Pappalardo, R.R.; Reguero, M.; Robb, M.A.; Frisch, M. *Chem. Phys. Lett.* **1993**, *212*, 12.
77. Warshel, A.; Lappicirella, A. *J. Am. Chem. Soc.* **1981**, *103*, 4664.
78. Luzhkov, V.; Warshel, A. *J. Am. Chem. Soc.* **1991**, *113*, 4491.
79. Warshel, A.; Hwang, J.-K. *J. Chem. Phys.* **1986**, *84*, 4938.
80. Blair, J.T.; Karsten, K.-J.; Levy, R.M. *J. Am. Chem. Soc.* **1989**, *111*, 6938.
81. Zeng, J.; Hush, N.S.; Reimer, J.R. *J. Chem. Phys.* **1993**, *99*, 1508.
82. Perutz, M.F. *Science* **1978**, *201*, 1187.
83. Hol, W.G. *J. Prot. Biophys. Mol. Biol.* **1985**, *45*, 149.
84. Harvey, S.C. *Proteins* **1989**, *5*, 78.
85. Warshel, A.; Åqvist, J. *Ann. Rev. Biophys. Biophys. Chem.* **1991**, *20*, 267.
86. Warshel, A. *Curr. Opinion Struct. Biol.* **1992**, *2*, 230.
87. Linderstrom-Lang, K. *C. R. Trav. Lab. Carlsberg* **1924**, *15*, 1.
88. Tanford, C.; Kirkwood, J.G. *J. Am. Chem. Soc.* **1957**, *79*, 5333.
89. King, G.; Lee, F.S.; Warshel, A. *J. Chem. Phys.* **1991**, *95*, 4366.
90. Warshel, A.; Russell, S.T.; Churg, A.K. *Proc. Natl. Acad. Sci., USA* **1984**, *81*, 4785.

91. Warshel, A. *Nature* **1987**, *333*, 15.

92. Wong, C.F.; McCammon, J.A. *J. Am. Chem. Soc.* **1986**, *108*, 3830.

93. Warshel, A.; Sussman, F.; King, G. *Biochemistry* **1986**, *25*, 8368.

94. Jorgensen, W.L. *Acc. Chem. Res.* **1989**, *22*, 184.

95. Langen, R.; Brayer, G.D.; Berghuis, A.M.; McLendon, G.; Sherman, F.; Warshel, A. *J. Mol. Biol.* **1992**, *224*, 589.

96. Warshel, A.; Parson, W.W. *Annu. Rev. Phys. Chem.* **1991**, *42*, 279.

97. Rao, S.N.; Singh, U.C.; Bash, P.A.; Kollman, P.A. *Nature* **1987**, *328*, 551.

98. Åqvist, J.; Warshel, A. *Biophys. J.* **1989**, *56*, 171.

99. Roux, B.; Karplus, M. *Biophys. J.* **1991**, *59*, 961.

100. Dang, L.X.; Merz, K.M.; Kollman, P.A. *J. Am. Chem. Soc.* **1989**, *111*, 8505.

101. Warshel, A.; Chu, Z.T.; Parson, W.W. *Photochem. Photobiol.* in press.

102. Åqvist, J.; Luecke, H.; Quiocho, F.A.; Warshel, A. *Proc. Natl. Acad. Sci.*, *USA* **1991**, *88*, 2026.

RECEIVED June 1, 1994

Chapter 7

Simulated Water Structure

E. Clementi and G. Corongiu

CRS4, Center for Advanced Studies, Research and Development
in Sardinia, P.O. Box 488, 09100 Cagliari, Italy
Université Louis Pasteur, 3 rue de l'Université, 67084 Strasbourg,
France

The ab-initio quantum mechanically derived potential for flexible water molecules of Nieser-Corongiu-Clementi, NCC-vib, is used for molecular dynamics simulations of liquid water in the temperature range 242-361 K for H_2O and 238-368 K for D_2O. Pair correlation functions, x-ray and neutron scattering structures, translational, librational, bending and stretching mode frequencies are found in good agreement with laboratory data. The simulated trajectories are used to obtain a model for liquid water.

Proposals on what liquid water is like have been and continue to be a major focus in this century physical-chemistry. We recall the Born proposal, later extended by Onsager, where water is described as a continuum material characterized by high dielectric constant; as it is known, this very simple model accounts for macroscopic properties, particularly solvation energy. For a structural characterization at the atomic level, we recall the proposals by Bernal and Fowlers, later extended by Sceats and Rice, where liquid water is essentially formed by tetrahedrally coordinated water molecules hydrogen bonded one to another. However, recent laboratory experiments support the hypothesis that in liquid water there are most likely more than one species, in equilibrium one with the other. In our work, summarized in this paper, we have attempted to determine and characterize these different species and to predict their probability distribution at different temperatures, from supercooled to hot water. We have used molecular dynamics trajectories as main tool; the computed trajectories have been obtained with a force field, which we have developed.

We will start by reviewing our force field, then we present the static properties obtained from the MD simulations (pair-correlation functions, x-ray and neutron diffraction intensities), and the dynamical properties (particularly the density of states). The good agreement with a very broad amount of laboratory experiments allows us to consider our computed trajectories as a reliable data bank, from which we can extract a model for liquid water.

0097–6156/94/0568–0095$08.00/0

The NCC potential

In the last few years many new potentials have been proposed to model water-water interactions. In the literature, two classes of many-body potentials have been developed to describe water-water interactions: (i) the class of empirical potentials (1-13), where the parameters in the analytic form of the potential energy are parametrized against experimental data, and (ii) the class of ab initio potentials, where the parameters are determined on the basis of quantum mechanical calculations on small clusters of water molecules (14-20). The first type of ab initio potentials is of pair-wise nature without correlation corrections (14), the second type adds the electron correlation corrections (15-17), and finally the third type introduces many body effects, either as 3- (18) or as 4-body (19) corrections or as polarization (20). An alternative, presently limited to a few water molecules and short simulated times, is the direct use of quantum mechanics in computing the forces at each time step (21). Most simulations obtained with empirical potentials discuss pair-correlation functions, thermodynamic quantities, and, but less frequently, x-ray and neutron diffraction data. Dynamical properties are very seldom analyzed and, to our knowledge, no temperature dependence study has been discussed. For this reason, we will compare our results with experimental data rather than with previous simulations.

This paper reviews the existing numerical simulation based on the ab initio polarizable potential (20) which has been developed by Nieser, Corongiu and Clementi (NCC). Parts of this paper follow on from the original papers (20, 22-28), to which we refer the interested reader for further information. The good agreement between the experimental and simulated data found in the liquid and solid phases for static and dynamic properties suggests that a computer "substance" very similar to real water, based only on ab initio quantum mechanical calculations, has indeed been created.

The NCC potential has been parametrized against quantum mechanical calculations a the MP4 level for water dimers and at the HF level for water trimers (20). It is composed of two contributions: a two-body potential, $V_{two\text{-}body}$, and a polarization term, E_{pol}. The polarization effects in the NCC potential are represented by a point polarizability along each of the two OH bonds. An intramolecular term, V_{intra}, has been added to the NCC potential (22) to take into account the intramolecular degrees of freedom. The potential is therefore composed of three terms:

$$V_{NCC} = V_{two-body} + V_{pol} + V_{intra} \tag{1}$$

where the $V_{two\text{-}body}$ is an additive term, which takes care of the repulsion between atoms and their electrostatic interactions; V_{pol} is a polarization term expressed as

$$V_{pol} = -\frac{1}{2} \sum_{i=1}^{N_m} \sum_{\lambda=1}^{2} \vec{\mu}_{i_\lambda}^{ind} \vec{E}_{i_\lambda}^{q} \tag{2}$$

where λ is an index running over the polarizability points of each molecule, $\vec{E}_{i_\lambda}^{q}$ is the electric field generated by the charges q on molecule i at the λ position, N_m is the number of molecules in the system, and $\vec{\mu}_{i_\lambda}$ is given by

$$\vec{\mu}_{i_\lambda}^{ind} = \vec{\alpha}_{i_\lambda}\left(\vec{E}_{i_\lambda}^q + \sum_{k \neq i}^{N_m} \sum_{\nu = 1}^{2} \tilde{T}_{i_\lambda} k_\nu \vec{\mu}_{k_\nu}^{ind} \right) \tag{3}$$

with $\tilde{T}$ the dipole-dipole matrix and $\bar{\alpha}_{i_\lambda}$ the polarizability tensor. The value of charges q and of the point polarizability tensor α are obtained by the parametrization procedure (20). The inter-molecular potential yields room temperature thermodynamical properties in nice agreement with laboratory data (for details see Refs. 20 and 22). In summary, the computed vaporization energy is 10.9 Kcal/mole, which reduces to 10.2 ± 0.3 Kcal/mole when quantum corrections are included; the corresponding experimental value is 9.9 Kcal/mole. The average dipole moment for water in the liquid is 2.8 Debye, which is the correct value to obtain the dielectric constant experimentally observed. The simulated specific heat is 73.5 J/(mole K) to be compared with the experimental value of 74.9 J/(mole K). The computed diffusion coefficient is 0.25×10^{-5} cm^2/s to be compared with the experimental value of 0.24×10^{-5} cm^2/s. The NMR relaxation time from our simulation is computed as 1.8 ± 0.2 ps, the experimental value is 2.0 ps. The computed values of the low and high frequency sound modes are 1288 and 3200 m/s, respectively, to be compared with the experimental values of 1390 ± 100 and 3310 ± 350 m/s, respectively. The computed value of the pressure fluctuates strongly around the value of -1180 atm $\pm$ 470 atm, however, this simulated value is algorithm dependent, and we refer the interested reader to Ref. 22 for more details.

The intra-molecular potential, V_{intra}, is also derived from ab-initio computations (29). It is expressed, up to quartic terms, in functions of the three internal coordinates of the water molecule, the changes in the OH bond lengths and in the OHO bond angle. The equilibrium bond length and bond angle predicted by the potential, for one water molecule in the gas phase, are 0.9576 Å and 104.59° respectively, to be compared with the experimental values (30) of 0.9572 Å and 104.5°.

Molecular dynamics simulations in the liquid phase have been performed with samples of 1000 or 512 water molecules at several temperatures (24), and at a density of 0.997g/cm³. Simulations in the solid phase (hexagonal ice, I_h) have been performed with 432 molecules (26, 27). The simulations have been carried out at constant energy and constant density (NVE ensemble). Periodic boundary conditions have been applied to simulate the infinite system, and to handle the electrostatic interactions we have used the reaction field technique. The self-consistent set of induced dipole moments is calculated at each step by an iterative method, as explained in detail in Ref. 20. A sixth-order Gear predictor corrector method was used to integrate the equations of motion with a time step of 0.125 fs. Readers are referred to the original papers for further information.

The liquid phase: x-ray and neutron diffractions and pair correlation functions

To stress the reliability of our interaction potential we compare in Figs. 1a and 2a, respectively, the experimental X-ray (31a) and neutron (32) intensity data at $T = 298K$ with the same quantities calculated from the MD simulated data at $T = 305K$. Notice the overall satisfactory agreement between simulated and experimental data, which is the best thus far reported, including even those based on interaction potentials fitted to experimental data.

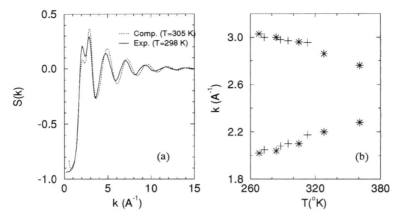

Figure 1. (a) X-ray scattering intensity. (b) Temperature dependence of the first two peak positions on $S(k)$. Reprinted with permission from Ref. 28. Copyright 1993 Elsevier Science Publishers B.V.

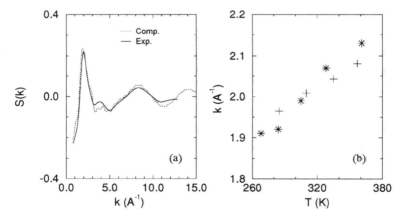

Figure 2. (a) Neutron structure factor. (b) Temperature dependence of the position of the first peak in the neutron $S(k)$. Reprinted with permission from Ref. 28. Copyright 1993 Elsevier Science Publishers B.V.

Characteristic quantities in the X-ray scattering are the position and intensity of the first two peaks, which are known to vary drastically with temperature (*31*). The position of the first and second structural peaks in S(k) as a function of temperature, for both experimental (crosses) (*31*) and simulated (stars) data are reported in Fig. 1(b). The MD data show that, for temperatures above the supercooled region, both the intensity (not reported) and the position of the first peak grow almost linearly with temperature, whereas there is a decrease for the second peak; eventually, at higher temperatures than those considered here, the two peaks merge into a single one. Thus, the NCC data agree well, qualitatively and quantitatively, with the experimental results, on the shape of the $S(k)$ function as well as in its temperature variation. As for the X-ray data, we find a quantitative agreement also for neutron diffraction data (Fig. 2a). In the structure factor, the position of the first peak is temperature dependent and shifts towards low k values on cooling the sample, approaching values characteristic of the amorphous low density ice. In Fig. 2b we compare the computed (stars) and the experimental (crosses) position of the first peak as a function of temperature.

The neutron diffraction MD simulated data are analyzed in Figs. 3 and 4. We first recall the definition of the total structure function, $S(k)$:

$$S(K) = S_d(k) + S_m(k),\qquad (4)$$

where the quantity $S_d(k)$ is related to the intermolecular scattering caused by the correlation of atoms in different molecules and $S_m(k)$ is the contribution due to the atomic correlation within the same molecule. The calculated $S(k)$ curves in Fig. 3 are obtained from the MD simulated $g_{\alpha\beta}(r)$ intramolecular parameters; the only empirical information used are the coherent scattering lengths (*32*). In Fig. 3 we report the total structure function, $S(k)$, for water with 99.75%, 67.89, 35.79 and 0.01% of deuterium content. Two sets (*32, 33*) of laboratory neutron scattering data are reported and compared with the MD simulated scattering intensity. The agreement is very satisfactory, especially with the data of Ref. 33.

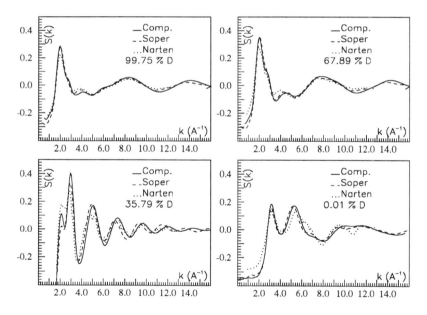

Figure 3. Total neutron scattering intensities for samples with different D_2O content.

In Fig. 4 we move to an even more detailed comparison with experimental data, this time considering the partial structure functions. Unlike the total structure functions, these three functions, $h_{OO}(k)$, $h_{OH}(k)$, and $h_{HH}(k)$, are related only to individual pair correlations and thus should provide crucial tests of various water models. The general agreement between the simulated and experimental $h_{OO}(k)$, $h_{OH}(k)$ and $h_{HH}(k)$ is good and our data is closer to the results of Ref. 33.

The simulated $H(k)$ for the samples with different deuterium content in function of temperature are reported in Ref. 24.

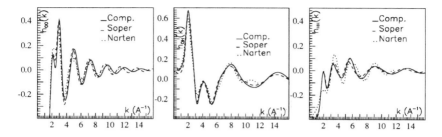

Figure 4. Computed and experimental partial neutron scattering intensities.

In Fig. 5 are compared the simulated oxygen-oxygen and oxygen-hydrogen pair correlation functions, $g_{OO}(r)$ and $g_{OH}(r)$, with those obtained from X-ray and neutron diffraction experiments by Narten (*31, 34*) and from neutron diffraction by Soper (*33*), both at a temperature of 298K. There are notable differences in the two experimental sets of data. The MD data agree much better with Soper's data.

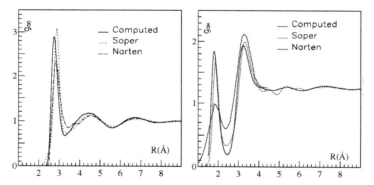

Figure 5. Oxygen-oxygen and oxygen -hydrogen radial distribution functions. Adapted from Ref. 28.

The $g_{OO}(r)$ values at six different temperatures are reported in Fig. 6 (left inset). We note the existence of radial positions where the value of the radial distribution function does not change. The same property is found for the $g_{OH}(r)$ and $g_{HH}(r)$ functions (see Ref. 24). Indeed, the difference between any pair of temperatures is zero at exactly the same points. This property was observed experimentally by Bosio *et al.* (*31b*). These authors made accurate measurements of the oxygen radial distribution function for two different temperatures T_1 and T_2, chosen such that the density ρ of water is the same for both temperatures; this is possible owing to the existence of a maximum in the function $\rho(T)$. By subtracting the respective radial distribution functions, they obtained "isochoric temperature differential functions" displaying the remarkable property that only the heights of the peaks and valleys (but not their positions) depend on temperature.

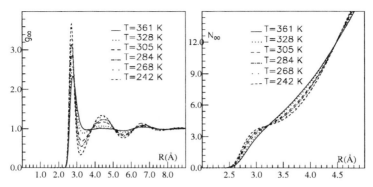

Figure 6. Left: Oxygen-oxygen radial distribution functions: Temperature dependence. Right: Coordination numbers, $N_{OO}(r, T)$ at different temperatures. Adapted from Ref. 24.

The existence of a point invariant for a change in temperature is also found in the integral of the radial distribution function, i.e. in the number of neighbors found inside a sphere centered on a selected atom (Fig. 6 right insert). The properties of invariance shown by the $g(r)$ functions, and by its integral, speak strongly in favor of a decomposition of the local environment in two large groups. Following Ref. 35 we assume, as a first approximation, that in the liquid there are two species A and B, with "temperature-dependent" concentrations $x_A(T)$ and $x_B(T)$ and with $(x_A + x_B = 1)$. Then the total radial distribution function g_T is related to the radial distribution functions g_A and g_B of species A and B by the relation

$$g_T(r, T) = x_A(T)g_A(r) + x_B(T)g_B(r) \qquad (5)$$

With this functional choice, the points where $g_A(r) = g_B(r)$ are temperature invariant. In Ref. 35 it is shown that Eq. (5) is satisfied also by the radial distribution functions calculated over the inherent structures. This means that the existence of points invariant for a change of temperature in $g(r)$ is due to the temperature dependent underlying structure of the liquid. Thus the existence of such points cannot be accounted for by invoking "trivial" thermal disorder broadening of the main peaks in $g(r)$. The fortuitous crossing of the different temperature is ruled out by the existence of more than one invariant point, not only in $g(r)$ but also in the integral of $g(r)$, up to about 7Å (see right insert of Fig. 6). From Eq. 5 we have

$$N_{OO}(r) = \int g_T(r,T)r^2 dr = x_A(T)\int g_A(r)r^2 dr + x_B(T)\int g_B(r)r^2 dr \qquad (6)$$

In the same way the isochoric temperature differential function is given by

$$g_T(r,T_2) - g_T(r, T_1) = [x_B(T_2) - x_B(T_1)][g_B(r) - g_A(r)] \qquad (7)$$

By using Eq. (7), we can interpret the changes in temperature of Fig. 6 as an effect of the decreasing concentration of the B species on cooling. It is important

to stress that the two species, A and B, are characterized by a variety of geometries whose probability distribution is expressed by $g_A(r)$ and $g_B(r)$. Also, being a
liquid mixture of A and B, a cross term in the expression for the radial distribution function would be expected (Eq. 5). The fact that this term is not needed for
distances shorter than 7Å, suggests that the local structure imposed by the central
molecule extends up to that distance.

Simulated density of states

We recall that the Fourier transform of the velocity autocorrelation functions

$$\phi(\omega) = \frac{m}{\pi k_b T} \int_0^\infty dt < \mathbf{v}(0) \cdot \mathbf{v}(t) > e^{-i\omega t} \qquad (8)$$

is the density of states, or spectral density.

Fig. 7 collects the results in the frequency range $0 - 4500$ cm^{-1} at $T = 305K$ for
liquid H_2O (left), which shows frequency shifts relative to both liquid D_2O (right)
and to the gas phase. The gas phase frequencies are indicated with dotted lines in
the two insets. In the figure, the oxygen atom contribution is separated from that
of the hydrogen (or deuterium) atom. Moving from low to high frequency, the
main bands have been assigned, on the base of laboratory experiments, to the
translational, librational, bending, and stretching modes, respectively.

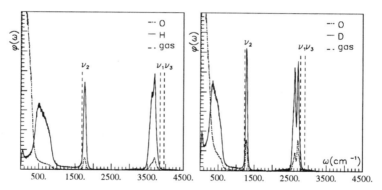

Figure 7. Left: Oxygen and hydrogen power spectrum from simulated H_2O
at T = 305 K. Right: Same but for D_2O.

Notice that at low frequencies the spectrum is dominated by the contribution from
the oxygen atom trajectories, while at high frequencies the hydrogen atoms dominate with the deuterium at intermediate frequencies (this is expected on the base
of the H/D relative masses). The low frequency region is dominated by the *intermolecular* water-water interactions, while the high frequency region is dominated
by the *intramolecular* motions for a single molecule of water.

In the following, we shall first provide an overall view of the density of state in
the liquid, up to 4500 cm^{-1}, and then we shall comment band by band in detail.

The gas phase values of Fig. 7 have been computed with *normal mode* analysis.
By comparing the computed gas phase values with the corresponding exper-

imental values (*36, 37*), it is found that the three vibrational frequencies (the bending, v_2, and the stretching modes symmetric, v_1, and antisymmetric, v_3) are overestimated by 90, 189 and 199 cm^{-1}, respectively for H_2O and by 55, 110, and 113 cm^{-1} for D_2O (*36*). These differences arise because the experimental values account for both the harmonic and anharmonic contributions to the vibrations, whereas the normal mode analysis is limited to the harmonic part only. It is note-worthy in this respect to recall that experimental values (*37*) have estimated the *harmonic* contribution for H_2O to be 1649, 3832, 3943, cm^{-1}, for v_2, v_1 and v_3 respectively not far from the computed values of 1685, 3846, 3955 cm^{-1}. In Fig. 7, left inset, the band centered at ~ 1750 cm^{-1} is the intramolecular *bending* mode, while those at 3626 and 3694 cm^{-1} are the intramolecular OH *stretching*, symmetric and asymmetric, respectively. Comparing these frequencies with those of the single molecule gas phase (vertical dotted lines), we obtain a ~ 65 cm^{-1} up shift for v_2 and down shifts of -220 and -261 cm^{-1} for v_1 and v_3 respectively. These shifts should be compared with the experimental values (*38-40*) of 55 cm^{-1} for v_2, and ~ -300 cm^{-1} for the stretching modes (experimentally v_1 and v_3 are not well resolved). Therefore, from the liquid simulation we obtain bending and stretching bands which substantially deviate from the free molecule and show characteristics typical of the liquid state. This is an indication that the anharmonicity of the intermolecular *H*-bond interactions are reliably represented by our model.

The *bending mode*, found in the liquid phase at 1650 cm^{-1} in Ref. 39, should be compared with our value of ~ 1750 cm^{-1} for $T = 305$ K. In our simulations the position of the bending mode is overestimated by about 100 cm^{-1}, and the stretching by about 250 cm^{-1}. These overestimations are about as large as the anharmonic effects discussed above for the gas phase.

The *librational, bending and stretching* D_2O spectra in the right inset of Fig. 7, are essentially very similar, in appearance, with those of H_2O except that for D_2O the bands are narrower and more pronounced (see, for example, the very good separation between v_1 and v_3, with a split of about 90 cm^{-1}).

Concerning the temperature dependency of the *translational, librational, bending and stretching* modes for both H_2O and D_2O it has been observed that only *the bending mode is essentially temperature independent*, for the other modes, shifts are obtained from our simulations. For details see Ref. 25.

A model for liquid water

Since the pioneering works of Röntgen (*41*) and Bernal and Fowler (*42*) there have been many attempts to explain the unusual properties of water in terms of simple models, which can be subdivided *grosso modo* in two broad categories: The "continuum" and the "mixture" models.

The "continuum model" assumes that the dominant structure of the liquid is a "locally tetrahedral continuous hydrogen-bonded network", (*43, 44*) whereas the "mixture models" propose (*45*) the existence of a thermodynamic equilibrium between two or more different aggregates of water molecules. Among the recent proposed "mixture models", we recall those presented in Refs. 46-50. We recall in addition that a detailed analysis of Raman spectra by Walrafen *et al.* (*39*) sup-ports the hypothesis of two components: the HB and the NHB, corresponding to four-hydrogen bonded and 3-hydrogen bonded water molecules, respectively.

Our MD simulations support the hypothesis of an equilibrium between water molecules hydrogen bonded to two, three, four and five water molecules, in agreement with previous proposal (23, 47); in particular, we present an analysis, where we predict specific spectral features for the different species present in liquid water.

Analysis carried out on the stored trajectories revealed (23) that liquid water can be viewed as composed of "clusters" of different sizes and with different probabilities of being present in the liquid, depending also from the temperature. We recall that a "cluster" can be characterized either by its oxygen atom positions or by the position of both the oxygen and the hydrogen atoms; we have placed quotation marks on the term cluster, to emphasize that a cluster in the liquid is different from a cluster in vacuo. Indeed, the former relates to *bulk water*, the latter to *surface structures*. A cluster in the liquid is defined as an association of (n+1) water molecules, i.e. a central water molecule coordinated, or hydrogen bonded (HB), to n; alternatively, if we refer specifically to the central molecule, we talk of a bi-, three-, n-coordinated water molecule (n=2, n=3, etc.). In Fig. 8 we present four clusters. The label A, B, C, and D identifies the central molecule in each cluster, i.e. the "solvated water molecule". The solvated molecules A and B are examples of tetra-coordination. Notice that A is coordinated to four water molecules (ice-like structure) via four hydrogen bridges (two between the lone pair electrons of A and hydrogen atoms of two solvating water molecules and two between the hydrogen atoms of A and the oxygen atoms of two additional solvating water molecules). In B the situation is different, since we have one hydrogen atom of the solvated molecule B bridging *two oxygen atoms* of two solvated water molecules (*bifurcated hydrogen bond*); two additional water molecules are bridging the lone pairs of B, as for A. Molecules C and D are examples of three-coordinated water molecules: C has a *free lone pair*, D has a *free hydrogen atom*.

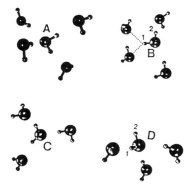

Figure 8. A and B are examples of tetra-coordinated water molecules.   C and D are examples of water molecules three-coordinated. Reprinted with permission from Ref. 51. Copyright 1993 Elsevier Science Publishers B.V.

An analysis (23) carried out by considering *only oxygen atoms*, revealed that at low temperature n=4 is more frequent than n=5, whereas n=5 is more probable at high temperature (see Fig. 9 left insert).

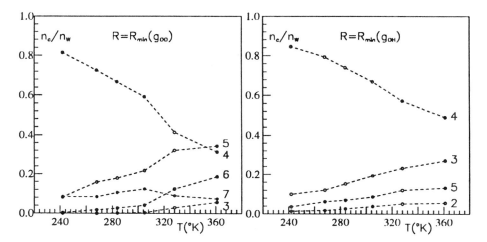

Figure 9. Distribution of water molecules with coordination number from 2 to 7. Left: O-O coordination number within a sphere with $R=R_{min}$ of g_{OO}. Right: O-H coordination number within a sphere with $R=R_{min}$ of g_{OH}.

Three-, hexa-and epta-coordination exist, but less abundantly. Instead, by considering both the oxygen and hydrogen atoms, then the analysis revealed that, at all the temperatures, the tetra-coordination is the most important, with the three- and penta-coordinations becoming more and more abundant as the temperature rises (see Fig. the right inset of Fig. 9). The lifetime of the tetra-coordinated water molecules is the longest; from the analysis we found in addition smaller and larger clusters, but the corresponding populations are small and the lifetime vanishingly short. The resulting overall picture of liquid water is that of a very dynamical "macromolecular" system, where clusters of different size and structure coexist in different subvolumes of the liquid and each has characteristic lifetimes and specific temperature dependencies.

For the liquid at T=305 K, we report the density of states of the hydrogen atoms belonging to the water molecules bi-, three, tetra- and penta-coordinated, defining the coordination number by considering both oxygen and hydrogen atoms. Thus, we do not analyze the spectrum in terms of the entire liquid, but we generate the spectra corresponding to different types of solvated water molecules, each type with its specific coordination.

In Fig. 10 we report the density of states in the region 0-4500 cm^{-1} obtained by considering the hydrogen atoms belonging to water molecules with n=2, 3, 4, and 5. The first band extends up to 1000 cm^{-1} and corresponds to the librational motions, the second one to the bending mode (v_2), and the last one to the stretching modes, symmetric (v_1) and asymmetric (v_3). The dotted vertical lines refer to frequencies for one single water molecule in vacuo, computed with the same potential we have used to study the liquid. We have labelled as "total" the previously reported (see the left inset of Fig. 7) spectrum obtained by considering the entire sample of the liquid without any subdivision into clusters. In this spectrum the modes v_2, v_1, and v_3 occur at ~ 1756, ~ 3626, ~ 3694 cm^{-1}, respectively. The spectra labelled as "2", "3", etc. refer to water molecules with n=2, n=3, etc.

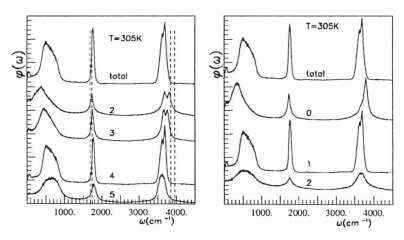

Figure 10. Left: Density of states of the hydrogen atoms. "Total": full liquid water sample; "2": only bi-coordinated water molecules; "3": only three-coordinated water molecules; "4": only tetra-coordinated water molecules; "5": only penta- coordinated water molecules. Right: "Total": full liquid water sample; "0": hydrogen atoms not hydrogen bonded to any oxygen atom; "1": hydrogen atoms hydrogen bonded to one oxygen atom; "2": hydrogen atoms hydrogen. Adapted from Ref. 51.

In general, by comparing this set of data, notable differentiating features can be noted at all frequencies. In particular, in the librational region, the water molecules bi-coordinated show a maximum at ~380 cm^{-1}, which blue shifts at ~450 cm^{-1} for the three-coordinated and to ~510 cm^{-1} for the tetra-coordinated. Following this first maximum a shoulder starts to appear at high frequencies for the three-coordinated water molecules, and it becomes more evident for the tetra-coordination. The shoulder develops into a broad maximum, centered at ~650 cm^{-1} (extending up to 1000 cm^{-1} for the penta-coordinated water molecules). The superposition of these partial bands yields the curve reported as "total".

Walrafen *et al.* (*39*) assuming for the tetra-coordinated water molecules a C_{2v} symmetry as in ice (i.e. four nearest neighbors with four hydrogen bonds equal in length, angle, etc., two for its two protons and two lone pairs) report the libration around the C_2 axis of A_2 symmetry at 550 cm^{-1}, the in-plane libration of B_2 symmetry at 425-450 cm^{-1}, and the out-of-plane libration of B_1 symmetry at 720-740 cm^{-1}. From Fig. 10 we notice for the tetra- coordinated water molecules a maximum at ~510 cm^{-1} and a shoulder at ~ 730 cm^{-1}; it should be noted that one could assume under the first broad maximum another gaussian distribution at lower frequencies.

We have already pointed out that in our analysis tetra-coordination does not necessarily implies one water molecule with four neighbors as in ice (as for A in Fig. 8). In our definition tetra-coordination can arise from many and different configurations: for example, either the one considered above (however with a *distribution* of HB lengths and angles), or one with the lone pairs coordinated to two water molecules and one hydrogen atom hydrogen bonded to two different oxygen atoms (bifurcated hydrogen bond) and the second hydrogen atom free, etc. It is reasonable to assume that the first type (the ice-like structure) is the most probable. These possible combinations *can coexist* for all the distributions of coordinations given in Fig. 9, preserving, however, the total number of HBs for each combination. In Fig. 8 water molecules C and D provide examples for two different combinations of three-coordination.

Comparing the spectra corresponding to different coordinations in the bending region, we observe small variations in frequency; for example the penta-coordinated water molecules show a maximum at ~1780 cm^{-1}, which progressively shifts towards lower frequencies (towards the gas phase value) as the coordination number decreases reaching a value of ~1735 cm^{-1} for the bi-coordination.

More evident are the shifts in the stretching region. Starting from the penta-coordination and decreasing the coordination number, we notice a clear shift towards higher frequencies (moving towards the gas frequencies) pointing out the presence of an increasing percentage of water molecules with one hydrogen atom unbounded. Notice also that for all the coordinations, except for n=5, we find very well distinct peaks for the symmetric and asymmetric stretching modes.

To provide additional details on the role played by the "free" hydrogen atoms of the solvated water molecules, we have attempted the following analysis. We have subdivided the water molecules in the liquid into different distributions, each one being characterized by the number (0, 1, 2) of hydrogen bonds for a given hydrogen atom. In particular, when for one water molecule one hydrogen atom is free (i.e. it has not HB to any molecule) it belongs to group "0", when it is HB to a water molecule (i.e. pointing towards the oxygen lone pair), it belongs to group "1", when HB to two water molecules it belongs to group "2", independently from the second hydrogen atom, for which the same kind of analysis is carried out. In Fig. 8, examples for 0, 1, and 2 are, respectively, the hydrogen atom 2 of molecule D, the hydrogen atom 2 of molecule B, and the hydrogen atom 1 of molecule B. By averaging over all the hydrogen atoms belonging to a given group and over all the simulation time steps, we obtain the data reported in the right inset of Fig. 10. Again, the data labelled as "total" refers to the entire liquid sample. Those hydrogen atoms with no HB show in the librational region a maximum at ~350 cm^{-1}, in the bending region a maximum at ~1735 cm^{-1} and in the stretching region a peak at ~3790 cm^{-1}. Those hydrogen atoms with one HB (the most probable type) show in the librational region a maximum at ~530 cm^{-1} and a shoulder, which covers a gaussian with a maximum at ~730 cm^{-1}. In the bending region the maximum occurs at ~1770 cm^{-1} and finally, in the stretching region the symmetric and asymmetric modes occur at ~3630 cm^{-1} and ~3690 cm^{-1}, respectively. The hydrogen atoms with two HBs (bifurcated hydrogen bond) show in the librational region a broad maximum centered at ~550 cm^{-1}; the bending mode occurs at ~1760 cm^{-1} and in the stretching region a non-split maximum is present at ~3670 cm^{-1}. Note that the population, at T=305 K, of the 0, 1, and 2 distributions are vastly different, i.e. large for 1, very small for 2 and small for 0.

Conclusions

The NCC-vib interaction potential has been obtained by fitting *ab initio data* rather than experimental data. This potential represents the latest stage of an evolution, which started with a Hartree-Fock potential (*14*), after several systematic refinements (*15-19*), has brought to the MCY model. As one can expect, this evolution takes advantage both of the increased performance of computational means and of advances of theoretical nature. The present potential has accurately reproduced much structural and dynamical data, predicted overall trends and very detailed features observed in infrared, Raman and neutron scattering experiments, both for the liquid and the solid phases. The MD trajectories have been used as "experimental data" to develop a descriptive model for liquid water, which is much richer of quantitative features than simply assuming essentially a network of tetrahedrally oriented water molecules. We have proposed the model of a dynamical liquid, which can be represented as a temperature dependent mixture of "clusters" of different size and lifetime, with well defined probability. Individual water molecules break off from one cluster to build up another cluster. The hydrogen bond is energetically affected by the temperature, as we have reported (see Ref. 23) and by its belonging to a given type of cluster. Its O-O distance and O-H-O angle varies with temperature. There is a complex distribution of lifetimes for the hydrogen bond in liquid water, even for a given temperature, since it depends also on the cluster to which it belongs. Typically we can say that a hydrogen bond in liquid water can last up to a few picoseconds. Work is in progress to elucidate further aspects of liquid water using a recently completed molecular dynamics code, where the forces are obtained at each time step using Density Functional Theory and Gaussian basis sets (51).

The computations presented in this work are very computer intensive tasks, both in the generation of the NCC potential and in performing the molecular dynamics simulations and corresponding analysis. We recall that quantum chemistry (needed for constructing the ab initio potentials) and molecular dynamics simulations are areas particularly well suited to parallel computers. In this context we recall that computational chemists use more and more from clusters of workstations to parallel systems like the IBM-SP1, the Crsy T3D, or the CM5 of the Thinking Machine Corporation. It is worth noting that the European Communities Commission has established a special program to parallelise scientific codes, key to the industry, under the ESPRIT-III Programme. In the area of computational chemistry there are two projects, coordinated by Smith System Engineering Limited, UK, whose task is the porting to parallel platforms of computer codes both in quantum chemistry and molecular dynamics modelling.

Acknowledgement It is a pleasure to acknowledge financial support from the "Regione Autonoma della Sardegna".

References

1. Rahman, A.; Stillinger, F.H. *J. Chem. Phys.* **1971**, *55*, 157.
2. Lemberg, H.L.; Stillinger, F.H. *J. Chem. Phys.* **1975**, *62*, 1667.
3. Stillinger, F.H.; Rahman, A. *J. Chem. Phys.* **1978**, *68*, 666.
4. Watts, R.O. *Chem. Phys.* **1977**, *26*, 367.
5. Reimers, J.R.; Watts, R.O. *Mol. Phys.* **1984**, *52*, 357.
6. Coker, D.F.; Watts, R.O. *J. Phys. Chem.* **1987**, *91*, 2513.
7. Barnes, P.; Finney, J.L.; Nicholas, J.D.; Quinn, J.D. *Nature* **1979**, *282*, 459.
8. Jorgensen, W.L. *J. Am. Chem. Soc.* **1981**, *103*, 335.
9. Jorgensen, W.L. *J. Chem. Phys.* **1982**, *77*, 4156.
10. Berendsen, H.J.C.; Postma, J.P.M.; van Gunsteren, W.F.; Hermans, J. *Intermolecular Forces*, B. Pullman, Ed., Reidel, Dordrecht, (1981).
11. Teleman, O.; Jönsson, B.; Engström, S. *Mol. Phys.* **1987**, *60*, 193.
12. Cieplak, P.; Kollman, P.; Lybrand, T. *J. Chem. Phys.* **1990**, *92*, 6755.

13. Sprik, M. *J. Chem. Phys.* **1991**, *95*, 6762.
14. Popkie, H.; Kistenmaker, H.; Clementi, E. *J. Chem. Phys.* **1973**, *59*, 1325.
15. Lie, G.C.; Clementi, E. *J. Chem. Phys.* **1975**, *62*, 2195.
16. Matsuoka, O.; Clementi, E.; Yoshimine, M. *J. Chem. Phys.* **1976**, *64*, 1351.
17. Carravetta, V.; Clementi, E. *J. Chem. Phys.* **1984**, *81*, 2646.
18. Clementi, E.; Corongiu, G. *Int. J. Quant. Chem. Symp.* **1983**, *10*, 31.
19. Detrich, J.H.; Corongiu, G.; Clementi, E. *Chem. Phys. Letters* **1984**, *112*, 426.
20. Nieser, U.; Corongiu, G.; Clementi, E.; Kneller, G. R.; Bhattacharya, D. *J. Phys. Chem.* **1990**, *94*, 7949.
21. Laasonen, K.; Sprok, M.; Parrinello, M. *J. Chem. Phys.* **1993**, *99*, 9080.
22. Corongiu, G. *Int. J. Quantum Chem.* **1992**, *44*, 1209.
23. Corongiu, G.; Clementi, E. *J. Chem. Phys.* **1993**, *98*, 2241.
24. Corongiu, G.; Clementi, E. *J. Chem. Phys.* **1992**, *97*, 2030.
25. Corongiu, G.; Clementi, E. *J. Chem. Phys.* **1993**, *98*, 4984.
26. Sciortino, F.; Corongiu, G. *J. Chem. Phys.* **1993**, *98*, 5694.
27. Sciortino, F.; Corongiu, G. *Mol. Phys.* **1993**, *79*, 547.
28. Clementi, E.; Corongiu, G.; Sciortino, F. *J. Mol. Struct.* **1993**, *296*, 205.
29. Bartlett, R.; Shavitt, I.; Purvuis, G.D. *J. Chem. Phys.* **1979**, *71*, 281.
30. Benedict, W.S.; Gailar, N.; Plyler, E.K. *J. Phys. Chem.* **1956**, *24*, 1139.
31. (a) Narten, A.H.; Levy, H.A. *J. Chem. Phys.* **1971**, *55*, 2263.
    (b) Bosio, L.; Chen, S.H.; Teixeira, J. *Phys. Rev. A.* **1983**, *27*, 1468.
32. Thiessen, W.E.; Narten, A.H. *J. Chem. Phys.* **1982**, *77*, 2656.
33. Soper, A.K.; Phillips, M.G. *Chem. Phys.* **1986**, *107*, 47. Soper, A.K.; Silver, R.N. *Phys. Rev. Lett.* **1982**, *49*, 471.
34. Narten, A.H. *J. Chem. Phys.* **1972**, *56*, 5681.
35. Sciortino, F.; Geiger, A.; Stanley, H.E. *Phys. Rev. Lett.* **1990**, *65*, 3452.
36. Herzberg, G. *Molecular Spectra and Molecular Structure II. Infrared and Raman Spectra of Polyatomic Molecules* Van Nostrand, Princeton, NJ, 1945.
37. Califano, S. *Vibrational States*, Wiley, London, 1976.
38. Ratcliffe, C.I.; Irish, D.E. *J. Chem. Phys.* **1982**, *86*, 4897.
39. Walrafen, G.E.; Hokmabadi, M.S.; Yang, W.-H. *J. Chem. Phys.* **1988**, *92*, 2433.
40. Walrafen, G.E.; Hokmabadi, M.S.; Yang, W.-H. *J. Chem. Phys.* **1986**, *85*, 6964.
41. Röntgen, W.K. *Ann. Phys.* **1892**, *45*, 91.
42. Bernal, S.D.; Fowler, R.H. *J. Chem. Phys.* **1933**, *1*, 515.
43. Sceats, M.G.; Rice, S.A. In *Water: A Comprehensive Treatise*, Franks, F.; Ed.; Plenum Press, New York, NY, 1982, Vol. 7.
44. Henn, A.R.; Kauzmann, W. *J. Phys. Chem.* **1989**, *93*, 3770.
45. Frank, H.S. In *Water: A Comprehensive Treatise*, Franks, F.; Ed.; Plenum Press, New York, NY, 1982, Vol. 1.
46. Blumberg, R.L.; Stanley, H.E.; Geiger, A.; Mausbach, P. *J. Chem. Phys.* **1984**, *80*, 5230.
47. Nemethy, G.; Scheraga, H.A. *J. Chem. Phys.* **1962**, *36*, 3382.
48. Gill, S.J.; Dec, S.F.; Olofsson, G.; Wadsö, I. *J. Phys. Chem.* **1985**, *89*, 3758.
49. Grunwald, E. *J. Phys. Chem.* **1986**, *108*, 5819.
50. Benson, S.W.; Siebert, E.D. *J. Am. Chem. Soc.* **1992**, *114*, 4270.
51. G. Corongiu and E. Clementi *Chem. Phys. Letters* **1993**, *214*, 367.
52. Estrin, D.; Corongiu, G.; Clementi, E.; In *Methods and Techniques in Computational Chemistry*, Clementi, E.; Ed., Stef, Cagliari, Italy (1993).

RECEIVED July 25, 1994

Chapter 8

The Dissociation of Water
Analysis of the CF1 Central Force Model of Water

Anna Nyberg and A. D. J. Haymet[1]

School of Chemistry, University of Sydney, New South Wales 2006, Australia

The relative free energies of the solvated species $H^+_{(aq)}$ and $H_3O^+_{(aq)}$ are calculated for the CF1 model of water. The calculations lead to an upperbound for the pH of CF1 water. Comparison is made with other calculations for the dissociation of water.

Recently we have calculated the pH of the CF1 central force model of water.[1] The CF1 model is a slight modification of the central force model of Stillinger and Rahman,[2] designed to improve the pressure at 25 °C and 1.00 g cm^{-3}. For the CF1 model an upper bound to the pH is found to be 8.5±0.7. The model has a dielectric constant[1] of 69 ± 11.

Our first calculation used classical mechanics, and predicted the equilibrium concentration of 'loosely solvated' (defined below) species $H^+_{(aq)}$ and $OH^-_{(aq)}$, resulting from the dissociation $H_2O_{(\ell)} \rightleftharpoons H^+_{(aq)} + OH^-_{(aq)}$. Standard methods[3] were used to calculate the relative fraction of dissociated species. Since the extent of hydration of H^+ and OH^- in the CF1 model of water is not known (nor to our knowledge, is it known for any other model), this calculation of the pH in the CF1 model established an *upper bound*. Further stabilisation of the 'loosely solvated' species would lead to an increase in the total equilibrium concentration of $H^+_{(aq)}$. Within the CF1 model, a hydrogen species is defined to be 'loosely solvated' if both (i) the distance to the nearest oxygen species is greater than 1.2 Å and (ii) the distance to the nearest hydrogen species is greater than 1.8 Å.

---

[1]Corresponding author

0097–6156/94/0568–0110$08.00/0

Here we begin the calculation of the relative stability of tightly solvated species such as $H^+_{(aq)}$, $H_3O^+_{(aq)}$, $\ldots$,$H_9O^+_4\,{}_{(aq)}$ and similar $OH^-_{(aq)}$ species. For the CF1 model, 'tightly solvated' ions have all oxygen and hydrogen species connected by 'bonds', where an oxygen-hydrogen bond has a separation less than 1.2 Å, and a hydrogen-hydrogen bond has a separation less than 1.8 Å. Our ultimate goal is to calculate the total concentration of dissociated species. We begin with the $H_3O^+_{(aq)}$ ion, defined in the CF1 model to have all three OH distance less than 1.2 Å, and all three intramolecular hydrogen–hydrogen distances less than 1.8 Å.

CF1 Model of Water

Despite the well-known role of *p*H on the structure of proteins and activity of enzymes, only modest interest has been shown in the dissociation of water, with the notable exceptions of work by Stillinger,[4] Warshel,[5] and Bratos and co-workers.[6–9] The rigid models of water[10,11] used most frequently in computer simulations have zero H^+ ion concentration, since by construction they cannot address dissociation. Some flexible models also do not permit dissociation.[12–15]

The central force model of Stillinger and co-workers[2,16,17] consists of three pair potentials acting between fractionally charged hydrogen and oxygen species. There is a single Hamiltonian which describes both intra– and inter–molecular degrees of freedom. The central force (CF) potential energies, as revised in 1978 by Stillinger and Rahman[2] are:

$$V_{OO}(r) = \frac{144.538}{r} + \frac{26758.2C_1}{r^{8.8591}} - 0.25e^{-4(r-3.4)^2} - 0.25e^{-1.5(r-4.5)^2}, \quad (1)$$

$$V_{HH}(r) = \frac{36.1345}{r} + \frac{18}{1 + e^{40(r-2.05C_2)}} - 17e^{-7.62177(r-1.45251)^2},$$

$$V_{OH}(r) = -\frac{72.269}{r} + \frac{6.23403}{r^{9.19912}} - \frac{10}{1 + e^{40(r-1.05)}} - \frac{4}{1 + e^{5.49305(r-2.2)}},$$

where the values of the constants are $C_1 = C_2 = 1$. The hydrogen species have a fraction charge of approximately one-third, which should be regarded as an effective value arising from integrating out the many contributions to the total potential energy omitted from a two-body prescription. At the temperature $T = 25\,°C$ and density 1.00 g cm^{-3}, this model has a pressure of 3,540 bar, thousands of times the correct value.[18] The CF1 model[1] attempts to both preserve the useful properties of this model and correct the pressure. The same potential energies are used, but with the slightly different constants $C_1 = 0.9$ and $C_2 = 1/1.025$. With this change, the pressure for the CF1 model is 120 bar.

Two high peaks in CF1 pair correlation functions $g_{OH}(r)$ and $g_{HH}(r)$ correspond to species within the same molecule, and at 25 °C, these peaks arise solely from intra–molecular correlations. The CF1 model has also been studied by integral equation methods.[19,20]

The Free Energy Calculation

The Helmholtz free energy of solvation for the species H^+, OH^- and H_3O^+, all of importance in the dissociation of water, are calculated using gradual changes of the interaction potential between the species and the surrounding solvent. This method is called thermodynamic integration, and it has been used in calculations of the chemical potential of water[21,22] and free energy of hydration of molecules and ions by Jorgensen, Kollman and others.[22–25]

A parameter λ describes the path chosen between the initial and final state, and the change in free energy is

$$\Delta A^{ex} = A^{ex}(\lambda = 1) - A^{ex}(\lambda = 0) = \int_0^1 d\lambda \left\langle \frac{\partial \mathcal{H}(\mathbf{p}^N, \mathbf{q}^N, \lambda)}{\partial \lambda} \right\rangle , \qquad (2)$$

where $\mathcal{H}$ is the Hamiltonian of a system of N particles, and q are coordinates and p the corresponding momenta. The angle brackets denote an average over phase space, which is approximated by a (relatively short) time average from a molecular dynamics simulation. The Helmholtz free energy corresponds to the canonical ensemble, a choice that is implemented easily in molecular dynamics simulations. The path is chosen so that no phase transitions are encountered. For example, the pressure remains positive throughout the simulation. The computational details are the same as those used earlier.[1]

The change in free energy is calculated for the following processes:

Process 1. $H^+_{(aq)} \to \mathcal{N}$

Process 2. $H_2O_{(l)} \to OH^-_{(aq)} + \mathcal{N}$

Process 3. $H_3O^+_{(aq)} \to H_2O_{(l)} + \mathcal{N}$

where $\mathcal{N}$ denotes a non-interacting or null particle. A common feature of the calculations is that a hydrogen species disappears. The measured changes in free energy are summarized in Table 1.

To evaluate the consistency of the results we use closed thermodynamic cycles. The free energy is a state function, and the free energy of the final state is the same, independent of the path chosen. In order to obtain full cycles, we mutate a hydrogen species into a Lennard-Jones particle (denoted Lj) with similar size as a H-species ($\epsilon = 0.757$ kcal/mol and $\sigma = 1.2$ Å for interaction with O-species and no interaction with H-species), following the scheme:

$$H_2O + H^+{}_{(aq)} \quad \rightarrow \quad H_3O^+{}_{(aq)}$$

$$\uparrow \qquad\qquad\qquad\qquad \downarrow$$

$$H_2O + Lj_{(aq)} \quad \leftarrow \quad H_2OLj_{(aq)}$$

Note that in all these calculations $\Delta(PV)$ is small and hence the Helmholtz and Gibbs free energies are approximately equal.

Table 1: Changes in the Helmholtz free energy for the processes related to the dissociation of water.

Process			ΔA (kcal mol^{-1})	number of simulations
$H^+{}_{(aq)}$	$\rightarrow$	$\mathcal{N}$	-	0
$H_2O_{(l)}$	$\rightarrow$	$\mathcal{N} + OH^-{}_{(aq)}$	12.1	4
$H_3O^+{}_{(aq)}$	$\rightarrow$	$H_2O_{(l)} + \mathcal{N}$	46.1	5
$Lj_{(aq)}$	$\rightarrow$	$\mathcal{N}$	-	0
$H_2OLj_{(aq)}$	$\rightarrow$	$H_2O_{(l)} + \mathcal{N}$	-	0
$Lj_{(aq)}$	$\rightarrow$	H^+	-	0
$H_2O_{(l)}$	$\rightleftharpoons$	$H^+{}_{(aq)} + OH^-{}_{(aq)}$	-	0
$2H_2O_{(l)}$	$\rightleftharpoons$	$H_3O^+{}_{(aq)} + OH^-{}_{(aq)}$	34.0	4

The above free energy simulations are still in progress, and final estimates are therefore lacking. In order to keep the system neutral, a H-species appears at 10 Å from the place where the primary reaction with a vanishing H-species takes place. The accuracy of the free energies is not high at this stage, and further averaging is required, but the preliminary evidence is that in the CF1 model, the loosely solvated H-species has a free energy similar to the free energy of the hydronium ion, comparing values from PMF-calculations with the above value. The ion $H_5O_2^+$ is not stable in the expected tetrahedral configurations, with O-species sharing more than one H-species, since in the CF1 model the OO potential is highly repulsive for distances $r < 2$ Å.

The values measured in our simulations are excess free energy changes when a particle in liquid water is mutated into another type of particle or a null particle, and should not be confused with the absolute hydration free energy of gaseous H^+

$$H^+_{(g)} + H_2O_{(g)} \rightarrow H^+(H_2O)_{(g)}. \tag{3}$$

Values for the gas phase reaction are summarised by Pearson[26] and Hepler and Woolley,[27] who also give experimental values for free energy changes in solution, which are in the same range as our measured values (19.1 kcal mol^{-1} in the standard state, $[H^+]=[OH^-]= 1$ M).

Summary of pH calculation

For comparison below we summarise the earlier pH calculation of Nyberg et al.[1] This method uses the potential of mean force (PMF) $w(\mathbf{r}, \mathbf{r}')$ between two dissociating species, which is defined by

$$g(\mathbf{r}, \mathbf{r}') = \exp[-\beta w(\mathbf{r}, \mathbf{r}')] , \tag{4}$$

where $\mathbf{r}$ and $\mathbf{r}'$ are the positions of the particles for which the average force is calculated, and $\beta^{-1} = kT$. Using this definition, it can be shown that $w(\mathbf{r}, \mathbf{r}')$ is the quantity which – when differentiated – yields the average or mean force,

$$- \nabla_\mathbf{r} w(\mathbf{r}, \mathbf{r}') = < \mathbf{F}(\mathbf{r}, \mathbf{r}') > = < \mathbf{F}(|\mathbf{r} - \mathbf{r}'|) > \tag{5}$$
$$= -D^{-1} \int d\mathbf{r}_3 \cdots d\mathbf{r}_N \nabla_\mathbf{r} U_N(\mathbf{r}, \mathbf{r}', \mathbf{r}_3, ..., \mathbf{r}_N) \exp[-\beta U_N(\mathbf{r}, \mathbf{r}', \mathbf{r}_3, ..., \mathbf{r}_N)] ,$$

where $< \mathbf{F}(|\mathbf{r} - \mathbf{r}'|) >$ is the average force, U_N is the total N-body potential, and the denominator $D = \int d\mathbf{r}_3 \cdots d\mathbf{r}_N \exp[-\beta U_N(\mathbf{r}, \mathbf{r}', \mathbf{r}_3, ..., \mathbf{r}_N)]$.

Nyberg et al followed the dissociation of a single water molecule by 'pulling apart' the two hydrogen species in a single CF1 water molecule. The distance between the hydrogen species is denoted $R = r_{HH} = |\mathbf{r}_{H_a} - \mathbf{r}_{H_b}|$. The two oxygen – hydrogen distances $r_{OH_a} = |\mathbf{r}_O - \mathbf{r}_{H_a}|$ and $r_{OH_b} = |\mathbf{r}_O - \mathbf{r}_{H_b}|$ may take any values, and will assume on average the values which minimize the total system free energy, subject only to the single constraint that the hydrogen – hydrogen distance of one molecule is constrained to be R. The PMF as a function of the hydrogen – hydrogen separation R was obtained from

$$w(R) = w(R_o) - \int_{R_o}^R d\mathbf{r} \cdot < \mathbf{F}(r) > , \tag{6}$$

where $w(R_o)$ is a constant of integration, R_o is an arbitrary distance, and the potential is independent of the path of the integration. This quantity clearly depends on the projection of the average force, arising from all the surrounding water molecules, onto the hydrogen-hydrogen axis.

The hydrogen species concentration was calculated directly from the PMF using standard methods,[3]

$$[H^+] = \frac{\exp(-\beta w(\infty))}{N_A \int_0^{R_f} dr' \exp(-\beta w(r'))} , \tag{7}$$

where N_A is the Avogadro constant, R_f is the HH distance beyond which a water molecule is unambiguously dissociated (3 Å), and $w(\infty)$ has been approximated by the value of the PMF at the largest separation considered. This calculation yielded the value $pH = -\log_{10}[H^+] = 8.5 \pm 0.7$ where the uncertainty results from the uncertainty in the value of $w(\infty)$.

An Alternate Approach

We seek the equilibrium constant $K(p,T)$ at the pressure p and the temperature T, for the dissociation

$$H_2O \rightleftharpoons H^+ + OH^- . \tag{8}$$

The chemical potential for each species i may be written

$$\mu_i = \mu_i^{ref} + kT \ln(a_i/a_i^{ref}), \tag{9}$$

where a_i is the activity, a_i^{ref} is the activity in the reference state and μ_i^{ref} is the chemical potential in the reference state. For the reference states, we make the conventional standard state choices of 1 M for H^+ and OH^- ions, and 55.55 M for H_2O.

The change in Gibbs free energy at equilibrium is by definition

$$0 = \Delta G = \mu_{H^+} + \mu_{OH^-} - \mu_{H_2O} \tag{10}$$
$$= \Delta G^{ref} + kT \ln(a_{H^+} a_{OH^-}),$$

where we have assumed that the water activity is approximated by its value at its reference state. The product of activities may be written

$$a_{H^+} a_{OH^-} = \exp(-\beta \Delta G^{ref}). \tag{11}$$

The total change in free energy is the sum of the excess and ideal parts

$$\Delta G = \Delta G^{ex} + \Delta G^{ideal} = 2\Delta w + kT \ln \left(\frac{\Lambda_{H+}^3 \rho_{H+} \Lambda_{OH-}^3 \rho_{OH-}}{\Lambda_{H_2O}^3 \rho_{H_2O}} \right), \quad (12)$$

assuming that the excess part for concentrations less than or equal to the reference state does not depend on the concentrations; that is, that the dissociated water molecules do not interact, and moreover that the interaction between dissociated and associated water molecules is the same on the average.

From Equations (12) and (11), we obtain

$$a_{H+} a_{OH-} = [H^+][OH^-] = \frac{[H_2O]}{\Lambda_{H+}^3 N_A} \exp(-2\beta\Delta w), \quad (13)$$

where $[H_2O]$ is the standard state concentration, Λ_i the thermal wavelength of species i, and we have assumed that the thermal wavelengths for H_2O and OH^- are almost identical and cancel. Inserting all the values into Equation (13), again one obtains the value $pH = 8.5$ for CF1 water.

Comparison with an earlier calculation

Bratos and coworkers,[6-9] denoted GGB, have calculated the pH of water using (i) different empirical potentials, and (ii) a combination of particle insertion in MD simulations and approximate calculations of the different contributions to the excess free energy. Their model is semiclassical: translations and rotations are treated classically, and vibrations are treated quantum mechanically.

The Hamiltonian used by GGB is summarised here. Separate Hamiltonians are used for intra– and inter–molecular degrees of freedom. The internal degrees of freedom are described by experimental IR frequencies. For the interaction between water molecules, Lennard-Jones plus Coulomb potentials are used, with parameters from the simple point charge (SPC) model of Berendsen.[10] For the ion–water interactions, the same functional form is used, but with the addition of both molecular polarization terms and so-called 'Zundel polarisation'[28] terms modeling the polarizability of strong H-bonds. The Zundel polarizability terms are similar in form to the molecular polarization terms after multiplication with a phenomenological switching factor to turn off the Zundel polarization for high-energy configurations. No polarisation is included for the interactions between water molecules. The new parameters in the ion–water potentials were chosen to reproduce experimental gas phase energies and calculated geometries from quantum chemistry.

It is worthwhile to compare the two calculations of the pH. The GGB calculation is related to the alternate method described above, where the change in the Gibbs free energy is

$$\Delta G^{\text{ref}}(p,T) = \mu^{\text{ref}}_{\text{H}_3\text{O}+} + \mu^{\text{ref}}_{\text{OH}-} - 2\mu^{\text{ref}}_{\text{H}_2\text{O}} = kT \ln K(p,T), \qquad (14)$$

where μ^{ref} is a chemical potential at the standard state and K is the equilibrium constant. The chemical potential is written

$$\mu_i = \mu^{\text{ref}}_i + kT \ln a_i, \qquad (15)$$

where a_i is the activity at equilibrium. Note that the activity at the standard state is not included, as it was in Equation (9) above. The authors write that the reference state for the ions is chosen to be at infinite dilution, which is an unusual choice in our view. The resulting equilibrium constant is

$$K(p,T) = K^{id}(p,T)\exp[-\beta(\mu^{ex}_{\text{H}_3\text{O}} + \mu^{ex}_{\text{OH}-} - 2\mu^{ex}_{\text{H}_2\text{O}})], \qquad (16)$$

where μ^{ex}_i is the excess chemical potential of component i, and K^{id} is the ideal contribution to the equilibrium constant, which is calculated quantum mechanically using experimentally determined average IR frequencies to describe the intra–molecular vibrations.

The calculation of the excess chemical potential is performed using classical mechanics and a particle insertion method, in which ions are inserted into distributions of water molecules generated from molecular dynamics simulations. The calculated value of the pH is 6.5. Moreover, the pH is divided into contributions from the different terms in the potential energy. The authors conclude that polarization of water molecules and Zundel polarization together play the dominant role. Given the approximate nature of the potentials, particularly the terms describing the Zundel polarization, and the assumption that there is ion–water polarisation but no water–water polarization, this conclusion seems brave. The omission of water–water polarisation seems certain to lead to an over-estimate of the importance of ion–water polarisation. In any event, the CF1 model, which has a single Hamiltonian for water and ions, intra– and inter–molecular degrees of freedom, and no *explicit* polarisation, yields an equally acceptable description of water dissociation.

There are many applications for a dissociative model of water such as the CF1 model. For example, Booth *et al*[20] have studied CF1 water next to a charged surface, using an integral equation method. These calculations investigate the role of the molecular nature of water, omitted in the traditional continuum descriptions of the electrode / electrolyte interface, such as the Gouy-Chapman approximation, which treat the water as a structureless dielectric continuum.

Acknowledgments: This research was supported by the Australian Research Council (ARC) (grant No. A29131271). ADJH thanks Dr. David Smith (PNL) for many helpful conversations, and acknowledges (USA) NIH grant GM34668 in which this calculation was proposed.

References

[1]Nyberg, A.; Smith, D.E.; Zhang, Ling; Haymet, A.D.J. The dissociation of water: Molecular dynamics computer simulation of the CF1 central force model *J. Chem. Phys.* **submitted September 1993**.

[2]Stillinger, F.H.; Rahman, A. *J. Chem. Phys.* **1978**, *68*, 666.

[3]Ciccotti, G.; Ferrario, M.; Hynes, J. T.; Kapral, R. *Chem. Phys.* **1989**, *129*, 241.

[4]Stillinger, F.H.; David, C.W. *J. Chem. Phys.* **1978**, *69*, 1473.

[5]Warshel, A. *J. Phys. Chem.* **1979**, *83*, 1640.

[6]Guissani, Y.; Guillot, B.; Bratos, S. *J. Chem. Phys.* **1988**, *88*, 5850.

[7]Bratos, S.; Guissani, Y.; Guillot, B. in *Chemical Reactivity of Liquids — Fundamental Aspects*, edited by M. Moreau and P. Turq (Plenum, New York, 1987).

[8]Bratos, S.; Guillot, B.; Guissani, Y. in *Springer Proceedings in Physics*, edited by Davidovic, M.; Soper, A.K., (Springer-Verlag, Berlin, Heidelberg, 1989), Vol. 40.

[9]Guillot, B.; Guissani, Y.; Bratos, S. in *Synergetics, Order and Chaos*, edited by Velarde, M.G., (World Scientific Press, Singapore, 1987).

[10]Berendsen, H. J. C.; Postma, J. M. P.; Gunsteren, W. F.van; Hermans, J. in *Intermolecular Forces*, edited by Pullman, B., (D. Reidel, Dortrecht, Holland, 1981).

[11]Jorgensen, W.L.; Chandrasekhar, J.; Madura, J.D.; Impey, R. W.; Klein, M.L. *J. Chem. Phys.* **1983**, *79*, 926.

[12]Reimers, J. R.; Watts, R. O.; Klein, M. L. *Chem. Phys.* **1982**, *64*, 95.

[13]Reimers, J. R.; Watts, R. O. *Mol. Phys.* **1984**, *52*, 357.

[14]Bopp, P.; Jancso, G.; Heinzinger, K. *Chem. Phys. Lett.* **1983**, *98*, 129.

[15]King, J. F.; Rathore, R.; Lam, J. Y. L.; Guo, Z. R.; Klassen, D. F. *J. Am. Chem. Soc.* **1992**, *114*, 3028.

[16]Lemberg, H.L.; Stillinger, F.H. *J. Chem. Phys.* **1975**, *62*, 1677.

[17]Stillinger, F.H. *Adv. Chem. Phys.* **1975**, *31*, 1.

[18]Andrea, T. A.; Swope, W. C.; Andersen, H. C. *J. Chem. Phys.* **1983**, *79*, 4576.

[19]Duh, D.-M.; Perera, D.N.; Haymet, A.D.J. *J. Chem. Phys.* **1994**, submitted.

[20]Booth, M.J.; Duh, D.-M.; Haymet, A.D.J. *J. Phys. Chem.* **1994**, submitted.

[21]Quintana, J.; Haymet, A.D.J. *Chem. Phys. Lett.* **1992**, *189*, 273.

[22]Jorgensen, W.L.; Blake, J.F.; Buckner, J.K. *Chem. Phys.* **1989**, *129*, 193.

[23]Cieplak, P.; Kollman, P.A. *J. Am. Chem. Soc.* **1988**, *110*, 3734.

[24]Jorgensen, W.L.; Ravimohan, C.J. *J. Chem. Phys.* **1985**, *83*, 3050.

[25]Jorgensen, W.L.; Briggs, J.M. *J. Am. Chem. Soc.* **1989**, *111*, 4190.

[26]Pearson, R.G. *J. Am. Chem. Soc.* **1986**, *108*, 6109.

[27]Hepler, L. G.; Woolley, E. M. in *Water, A Comprehensive Treatise*, edited by Franks, F., (Plenum Press, New York, 1977), Vol. 3, Chap. 3.

[28]Zundel, G. in *The Hydrogen Bond*, edited by P. Schuster, G. Zundel and C. Sandorfy, (North-Holland, Amsterdam, 1976), Vol. 2, p. 363.

RECEIVED April 25, 1994

NONEQUILIBRIUM SOLVATION

Chapter 9

Nonequilibrium Solvation for an Aqueous-Phase Reaction

Kinetic Isotope Effects for the Addition of Hydrogen to Benzene

Bruce C. Garrett and Gregory K. Schenter

Molecular Science Research Center, Pacific Northwest Laboratory, Richland, WA 99352

Variational transition state theory with semiclassical tunneling corrections is applied to a model of H atom addition to benzene in the gas phase and aqueous solution. The model allows the separation of equilibrium (static) solvation effects on the free energy of activation from nonequilibrium (dynamic) solvation effects that enter through frictional terms. Using a classical mechanical treatment with this model, the static effect of the solvent on the equilibrium free energy of solvation is independent of the mass of the solute. Therefore, within a classical equilibrium solvation model, kinetic isotope effects are the same for gas and aqueous phases. Observations of changes of kinetic isotope effects (KIEs) upon solvation therefore indicate that dynamic solvent effects are important. The model calculations show that the nonequilibrium solvation effects are small for the hydrogen and deuterium addition reactions, but are large for the addition reaction of muonium (approximately one ninth the mass of H). These studies correctly account for the anomalous quenching of the Mu KIE by aqueous solution that has been observed experimentally by Roduner and Bartels (*Ber. Bunsenges. Phys. Chem.* **1992**, *96*, 1037).

Effects of nonequilibrium solvation on rates of activated chemical reactions in solution have been postulated for many years, yet their importance has not been well established. Theoretical considerations based on the Kramers (*1*) and Grote-Hynes (*2*) theories predict monatonically decreasing rate constants as a function of increasing solvent friction for activated bimolecular reactions. For these classical mechanical theories the decreases are typically much less than an order of magnitude for physically reasonable models. However, model studies have shown that when quantum mechanical effects are important, nonequilibrium solvation effects can be greatly enhanced (McRae, R. P.; Schenter, G. K.; Garrett, B. C., "Dynamic Solvent Effects on Activated Chemical Reactions II. Quantum Mechanical Effects", in preparation). Unfortunately, these effects are hard to confirm experimentally since changes in solvent friction are often accompanied by changes in the static (equilibrium) free energy of solvation that can greatly alter the reaction rate. Comparison of rates of reactions in gas phase and in solvents exhibit the importance of solvent effects on the reactions, but cannot help distinguish equilibrium from nonequilibrium solvation effects.

0097–6156/94/0568–0122$08.18/0

The study of the effect of solvation on kinetic isotope effects (KIEs) holds the promise of enabling us to validate the importance of dynamic (nonequilibrium) solvent effects. Classically, the free energy of solvation of a rigid solute is independent of the mass of the solute. If the free energy of solvation does not vary rapidly with internuclear geometry of the solute, then the shape of the intermolecular potential energy surface of the solute will not change severely and the static effect of the solvent on the equilibrium free energy of solvation will be approximately independent of the mass of the solute. In this case, if nonequilibrium solvation effects are unimportant, the KIE will be approximately independent of environment (*i.e.*, gas phase *vs.* solvent). Conversely, substantial changes in KIEs on going from the gas phase into solution will be attributable to dynamic solvent effects if the changes in internal vibrations of the solute upon solvation are small.

In a recent paper, Roduner and Bartels (*3*) have compared aqueous-phase experimental rate constants for the addition of hydrogen (H) (*3*), deuterium (D) (*3*), and muonium (Mu) (*4*) to benzene with the gas-phase rate constants for H (*5*), D (*5*), and Mu (*4*). Muonium is a positive muon-electron pair that behaves classically like H but is one-ninth the mass. The rate constant for the reaction in water is faster than the rate constant in gas phase by factors between 20 and 30 for D and 30 and 40 for H. Roduner and Bartels have shown that equilibrium solvation effects can account for this effect – the transition state complex is more stable in solvent than the separately solvated reactants. The transition state has a less positive free energy of solvation than reactants, thereby effectively lowering the activation free energy in aqueous solution relative to the gas-phase value. However, they find that equilibrium solvation cannot account for the much smaller increase in the rate constant for Mu (about a factor of three) and propose "that there may be a more fundamental, dynamic reason for the occurrence of such a pronounced mass effect." In this paper, we provide a theoretical study of the effects of nonequilibrium (dynamical) aqueous solvation on KIEs for H addition to benzene.

We employ a simplified model of the reaction that incorporates the features necessary to reproduce the rate constant dependence on temperature, isotope mass, and environment (*i.e.*, gas and aqueous phases). The dynamical influence of the solvent is treated by a harmonic bath linearly coupled to the solute reaction coordinate. In this model the classical equation of motion for the reaction coordinate is equivalent to the Generalized Langevin equation (GLE) (*6,7*). For a quadratic system (the potential along the reaction coordinate is a parabolic barrier), using a Langevin equation of motion to describe the dynamics leads to the well-known Kramers result (*1*), whereas using a generalized Langevin equation (GLE) leads to the Grote-Hynes result (*2*). These classical theories include the collective influence of the solvent on solute dynamics described in terms of frictional and stochastic forces. The usefulness of the GLE for describing the dynamics of activated chemical reactions in liquids has been established by comparisons with classical molecular dynamics calculations (*8-17*).

For the reaction of current interest, quantum mechanical effects are expected to be important, especially for the Mu isotope. Over the last several years there has been increased interest in understanding the influence of dissipative media on quantum mechanical effects, especially tunneling. In a recent paper (*18*), two approximate methods – variational transition state theory (VTST) with semiclassical corrections for quantum effects on reaction coordinate motion (*19-21*) and centroid density path-integral based quantum transition state theory (*22*) – were tested against accurate benchmark quantum mechanical calculations. Both approximate methods were found to provide reliable estimates of the rate constants for a model reaction in solution. In the present study we use the VTST approach to study the qualitative trends in the addition reaction of H to benzene.

Methods

General considerations of the methods for including both equilibrium and nonequilibrium solvation effects in VTST are discussed in a recent review (23). The authors of that work indicated how VTST could be applied to reactions in solution once the Hamiltonian for the system is defined. In this section, we review VTST methods that can be used for a generic Hamiltonian. The models for the reaction of H with benzene in gas phase and solution are presented, and the application of the VTST to the different models is outlined in the next section. The methods used for calculation of the gas-phase rate constant using variational transition state theory with semiclassical tunneling corrections are described in detail elsewhere (19-21).

We assume the following form for the Hamiltonian

$$H = \frac{p^2}{2\mu} + V(q) \tag{1}$$

where q and p are mass-scaled coordinates and conjugate momenta, respectively; μ is the reduced mass for reactants; and $V(q)$ is the potential energy surface. The VTST calculation begins with the definition of the reaction path as the minimum energy path (MEP); that is, the path of steepest descent from the saddle point to reactants and products in the mass-scaled coordinates. The reaction coordinate s along the MEP is the (signed) distance from the saddle point (negative on the reactant side and positive on the product side). Generalized transition state dividing surfaces are constrained to be orthogonal to the reaction path and are defined by their location s along the reaction coordinate. For a dividing surface at s, the generalized TST expression for the gas-phase canonical rate constant is given by (19,24,25)

$$k^{GT}(T,s) = \frac{k_B T}{h} \frac{Q^{GT}(T,s)}{\Phi^R(T)} \exp\left[-\beta V^{MEP}(s)\right]. \tag{2}$$

In this expression k_B is the Boltzmann's constant, T is the absolute temperature, h is Planck's constant, $\Phi^R(T)$ is the partition function of the reactants, $Q^{GT}(T,s)$ is the generalized transition state partition function for the bound modes orthogonal to the reaction path at s, $\beta = 1/k_B T$, and $V^{MEP}(s)$ is the potential evaluated on the MEP at s. Note that the usual symmetry factor has been omitted in equation 2. The canonical variational theory (CVT) rate constant is obtained by minimizing equation 2 with respect to s

$$k^{CVT}(T) = \min_s k^{GT}(T,s) = k^{GT}\left(T, s^{CVT}(T)\right) \tag{3}$$

where $s^{CVT}(T)$ is the location of the dividing surface that minimizes equation 2 at temperature T.

The partition function for the bound modes is computed quantum mechanically. The energy levels for the bound modes are computed using an independent normal mode approximation. Normal modes at the saddle point are obtained from the diagonalization of the matrix of second derivatives in mass weighted coordinates (the Hessian matrix). For points off the saddle point but on the minimum energy path, these bound modes are obtained from diagonalizing the Hessian matrix with the reaction coordinate motion projected out (26). Within the independent normal mode approximation, the quantized partition function is given by the product of partition functions for each mode:

$$Q^{GT}(T,s) = \prod_m q_m^{GT}(T,s) \tag{4}$$

where the partition function for mode m is given by

$$q_m^{GT}(T,s) = \sum_{n_m} \exp\left[-\beta \varepsilon_m^{GT}(s,n_m)\right] \tag{5}$$

and $\varepsilon_m^{GT}(s,n_m)$ is the generalized transition state energy level for level n_m in mode m.

The rate constant expression in equation 2 treats bound modes quantum mechanically, but the reaction coordinate motion is treated classically. A consistent route to include quantum mechanical effects on reaction coordinate motion is provided by the vibrationally adiabatic theory of reactions (27-30). Recent work (31) showing that quantized transition states globally control reactivity in the threshold region for a bimolecular reaction dramatically confirm the adiabatic model for tunneling through the transition state region. In the adiabatic model, reaction probabilities for each adiabatic state are obtained by considering the dynamics on the one-dimensional vibrationally adiabatic potential

$$V_a(s,\mathbf{n}) = V(s) + \sum_m \varepsilon_m^{GT}(s,n_m) \tag{6}$$

where the sum is over the bound vibrational modes of the generalized transition state at s, and the generalized transition state energy levels are the same as those used in the partition functions. The reaction probabilities $P^A(\mathbf{n},E)$ can then be thermally averaged to yield the rate constant. When reaction coordinate motion is treated classically, the adiabatic theory of reactions yields an expression for the thermal rate constant that is equivalent to that obtained from microcanonical variational theory (μVT) even though the approximations in the two theories are very different (24,32). Since the one-dimensional scattering problem can be treated quantum mechanically, a multiplicative tunneling correction factor for the adiabatic theory of reactions can be obtained, and the equivalency of μVT and adiabatic theory makes it consistent to use the same correction factor to account for the quantization of reaction coordinate motion in μVT.

At low temperatures, where tunneling corrections are most important, quantized systems tend to be in the ground state and tunneling through the adiabatic ground-state potential is adequate. As the temperature increases the tunneling correction factor tends to unity and the ground-state tunneling correction factor is still adequate. The ground-state tunneling correction factor for the CVT rate constant is defined by (33)

$$\kappa^{CVT/AG}(T) = \frac{\int\limits_0^\infty dE \, e^{-\beta E} \, P^A(\mathbf{n}=0,E)}{\int\limits_0^\infty dE \, e^{-\beta E} \, \theta\left[E - V_a\left(s^{CVT}(T),\mathbf{n}=0\right)\right]} \tag{7}$$

and the resulting quantum mechanical CVT rate constant is given by

$$k^{CVT/AG}(T) = \kappa^{CVT/AG}(T) \, k^{CVT}(T) . \tag{8}$$

In equation 7, $\theta(x)$ is a Heaviside step function that is zero for x<0 and 1 for x>0.

The adiabatic approximation is made in a curvilinear coordinate system, and although the potential term is simple, the kinetic energy term is complicated by factors dependent upon the curvature of the reaction path (26,27,34). For systems in which the curvature of the reaction path is not too severe, successful methods specify a tunneling path that 'cuts the corner' and shortens the tunneling length. The small-curvature semiclassical adiabatic ground-state (SCSAG) method (35,36) is one such method that has been extensively tested and shown to be valid for systems in which only one bound mode is coupled to the reaction coordinate. The centrifugal-dominant SCSAG approximation (CD-SCSAG) provides a more suitable approximation for systems with non-zero components of the reaction path curvature along several generalized normal modes (21). The CD-SCAG method reduces to the previous SCSAG method in the limit of one non-zero component of the reaction-path curvature.

Models

The focus of the current study is on qualitative trends of rate constant ratios (e.g., KIEs and ratios of aqueous-phase to gas-phase rate constants) and activation energies rather than on the accurate prediction of absolute rate constants. The approach is to develop simplified models that reproduce the qualitative trends for the gas-phase and for aqueous solvation for the heaviest isotopes (D and H) and then use this model to study the solution effects on the reaction with Mu.

Gas-Phase Model. In this section, we describe a simple model that predicts the correct qualitative trends in the H/D and Mu/H KIEs and activation energies for the gas-phase reactions of H, D, and Mu with benzene. The relevant gas-phase experimental data is presented in Table I and summarized below. (The aqueous-phase experimental data in Table I is discussed in the next section.) The gas-phase reaction of H with benzene has been well studied by Nicovich and Ravishankara (5), especially in the temperature range 298-470 K. Nicovich and Ravishankara deduced from their studies that abstraction of ring hydrogens could be ignored at temperatures below 1000 K leaving the addition of H to benzene to form the cyclohexadienyl radical as the only reaction channel. Thus we are interested in the reaction

$$H + C_6H_6 \xrightarrow{\quad k_{H,g} \quad} C_6H_7 \qquad\qquad (R_{H,g})$$

and its isotopic variant

$$D + C_6H_6 \xrightarrow{\quad k_{D,g} \quad} C_6H_6D. \qquad\qquad (R_{D,g})$$

Nicovich and Ravishankara provided absolute rate constants for temperatures from 298 to 470 K and a value of 4.3 kcal/mol for the activation energy in this range for reaction $R_{H,g}$ and absolute rate constants at 379, 633, and 788 K for reaction $R_{D,g}$. We extrapolate the rate constants for reaction $R_{H,g}$ to temperatures down to 286 K and interpolate them for intermediate temperatures using an Arrhenius fit to the experimentally observed results. Similarly, we extrapolate the rate constants for reaction $R_{D,g}$ at lower temperatures from an Arrhenius fit to the experimentally observed results. Using the interpolated and extrapolated rate constants, the H/D KIEs at temperatures below 379 K show only small variations from the value of 1.1 at 379 K.

Roduner et al. (4) have extended these studies to the reaction of the light muonium isotope

$$Mu + C_6H_6 \xrightarrow{\quad k_{Mu,g} \quad} C_6H_6Mu \qquad\qquad (R_{Mu,g})$$

over the temperature range 296 to 500 K. Compared to reaction $R_{H,g}$, the activation energy is significantly decreased to 1.6 kcal/mol and the rate constant is enhanced by a factor of 19 at room temperature. We extrapolate the Mu rate constants to temperatures down to 286 K and interpolate them for intermediate temperatures using an Arrhenius fit to the experimentally observed results. Using the interpolated and extrapolated rate constants for the H and Mu reactions, the Mu/H KIEs show much greater variation with temperature than the H/D KIEs. The smaller activation energy and greater variation of the KIE with temperature are indications of increased tunneling in the Mu reaction.

Table I. **Experimental rate constants (in units of 10^7 M^{-1} s^{-1}), activation energies (in kcal/mol), kinetic isotope effects, and ratios of aqueous-phase to gas-phase rate constants for the reaction of hydrogen atom isotopes with benzene in gas and aqueous phases**

	T(K)	Gas Phase			Aqueous Phase		
		D[a]	H[a]	Mu[b]	D[c]	H[c]	Mu[b]
k	286	2.0[d]	2.0[d]	58[d]	46±2	69±1	
	293	2.4[d]	2.5[d]	62[d]		86[e]	190±40
	296	2.5[d]	2.6[d]	65±2		93[e]	
	298	2.7[d]	3.4±0.5	65[e]	78±9	110±1	
	338	6.2[d]	6.6[e]	89±4		245[e]	
	368	10.4[d]	11.1[e]	109[e]		400±8	
	379	12.5±2.1	13.7±1.3	116[e]		510[d]	
E_{act}[f]		4.2	4.3	1.6		4.6	
k_X/k_H	286	1.0		29	0.7		
	293	1.0		25			2.2
	298	0.8		19	0.7		
	338	0.9		13			
	379	0.9		8.5			
k_{aq}/k_g	286				23	35	
	293					34	3.1
	298				29	32	
	368					36	
	379					37	

[a] Reference (*5*).
[b] Reference (*4*).
[c] Reference (*3*).
[d] Extrapolated from Arrhenius fit to experimental data.
[e] Interpolated from Arrhenius fit to experimental data.
[f] Activation energies obtained from least squares fits of experimental rate constants to Arrhenius equations.

The heats of formation at room temperature of H and benzene are 52.1 and 19.8 kcal/mol (*37*), and Nicovich and Ravishankara estimate the heat of formation of the cyclohexadienyl radical to be 45.7 kcal/mol. Thus the reaction is exothermic by 26.2 kcal/mol. Furthermore, as mentioned above, it has a relatively small barrier of about 4 kcal/mol (*5*). Hammond's postulate (*38*) leads us to expect that this reaction has an early barrier and that the saddle point resembles reactants more than products; that is, the distance for the closest C atom on the benzene to the addition H atom will be much longer than a normal CH bond in benzene (1.1 Å). The barrier will be asymmetric, falling off more rapidly towards the product and a realistic description of the potential only needs to be provided for a region extended from a short distance on the product side of the barrier out into the reactant entrance channel. For this entire region we expect the model of H atom addition to a rigid benzene molecule to give at least a qualitative description of the reaction.

The Hamiltonian for the gas-phase reaction $R_{X,g}$ (X=H,D,Mu) is written

$$H_{X,g} = \frac{p^2}{2\mu_X} + V_g(q) \tag{9}$$

where the gas-phase potential for the H atom motion relative to the frozen benzene is a function of three coordinates, $q=(s,x_1,x_2)$, and μ_X is the reduced mass for the hydrogen isotope (H, D, or Mu) relative to benzene; μ_X is well approximated by the mass of the hydrogen isotope. The masses used in the calculations are 1.008, 2.014, and 0.113 amu for H, D, and Mu, respectively. The distance from the H isotope to one of the carbon atoms is taken to be the reaction coordinate s, and the total gas-phase potential is taken to be the sum of a potential along the reaction path (RP) and two bending potentials

$$V_g(q) = V_g^{RP}(s) + \sum_{k=1}^{2} V_{bk}(x_k,s). \tag{10}$$

Since it was experimentally determined (5) that the gas-phase reaction is an association reaction (see above), the potential along the reaction coordinate is modeled by a Morse potential with a Gaussian term added to create a barrier

$$V_g^{RP}(s) = D_M\{1 - \exp[-\alpha_M(s - s_0)]\}^2 - D_M + V_G \exp[-\alpha_G(s - s_G)^2]. \tag{11}$$

For the two bending degrees of freedom, a simple harmonic form could be chosen in which the frequency depends of the reaction coordinate s. Because the bend frequencies change with the mass of the hydrogen isotope, different parametrizations of the bend frequencies would be needed for each isotope. Instead we choose a functional form for the bending potential that is independent of isotopic mass and we derive the bending frequencies from it. This requires only one parametrization of the bending potentials for all three hydrogen isotopes. The bending potentials are modeled by the functional form

$$V_{bk}(x_k;s) = v_k \left[\exp\left(-\alpha_k\sqrt{s^2 - x_k^2}\right) - \exp(-\alpha_k s) \right], \tag{12}$$

for the two modes (k=1,2). The bending coordinates x_1 and x_2 are defined in terms of bending angles Φ_1 and Φ_2 that are deviations from the minimum energy path both parallel and perpendicular to the plane of the benzene. These are defined as the angles between the vector from the H isotope to the C atom on the benzene and the vector from the H isotope to the C atom when the H isotope is on the minimum energy path, that is

$$x_k = s \sin\Phi_k, \tag{13}$$

for k=1,2. With this form for the bending potential, the kth bending frequency for reaction $R_{X,g}$ is defined by

$$\omega_{Xk}^2(s) = \frac{1}{\mu_X s^2} \left(\frac{\partial^2 V_{bk}}{\partial \Phi_k^2} \right)_s \Bigg|_{\Phi_k = 0} = \frac{\alpha_k v_k}{\mu_X s} \exp(-\alpha_k s), \tag{14}$$

for $k=1,2$. The fitting of the parameters of these potentials is presented below.

In this three-dimensional model, the generalized transition state rate constant (*cf.,* equation 2) for reaction $R_{X,g}$ is approximated by

$$k_{X,g}^{GT}(T,s) \approx \frac{k_B T}{h} \left(\frac{2\pi\hbar^2}{\mu_X k_B T} \right)^{3/2} \prod_{k=1}^{2} q_{Xk}^{GT}(T,s) \exp\left[-\beta V_g^{RP}(s) \right] \tag{15}$$

where $\hbar = h/2\pi$; $q_{Xk}^{GT}(T,s)$, $k=1,2$, are the partition functions for the two bending motions at the generalized transition state; and the minimum energy path is just the reaction coordinate s so that the potential along the MEP is given by $V_g^{RP}(s)$. The CVT rate constant is obtained by finding the optimum location that minimizes equation 15 with respect to s, $s_{Xg}^{CVT}(T)$. We have assumed that the rotational partition functions of the reactant (benzene) and the generalized transition state are nearly identical. This has been verified numerically by calculating the moments of inertia of the approximate complex at the saddle point and CVT geometries and comparing them with the moments of inertia of benzene; they differ by less than 2%. Of the 32 vibrational modes of the complex, only 2 are treated explicitly. The effects of variations of the ground-state energy levels with reaction coordinate for the other 30 vibrations are assumed to have no dependence on the mass of the hydrogenic reactant, and are implicitly included in the reaction coordinate potential $V_{RP}(s)$. The contributions of excited state energy levels in the partition functions will be small for individual partition functions (typically less than two near room temperature and only for low frequencies) and have small temperature dependence over the temperature range of the experiment. However, the neglect of the product of these contributions can cause the absolute rate constant to be in error by over an order of magnitude.

The tunneling correction factor for reaction $R_{X,g}$, $\kappa_{X,g}^{CVT/AG}(T)$, is computed using equation 7, where the tunneling probabilities are computed for the ground-state adiabatic barrier for reaction $R_{X,g}$. The adiabatic potential is approximated by

$$V_{X,g}^{AG}(s) = V_g^{RP}(s) + \sum_{k=1}^{2} \varepsilon_{Xk}^{GT}(s, n=0) \tag{16}$$

where the generalized transition state energy levels $\varepsilon_{Xk}^{GT}(s, n_k)$, $k=1,2$, are for the two bending modes. As noted above, the reaction coordinate potential is assumed to included the effect of the change in the ground-state energy levels between the reactants and generalized transition state location at s for the other 30 vibrational modes. The generalized transition state energy levels are the same as are used in computing the partition functions for the bending modes. Note that the bending vibrations at generalized transition states near the barrier correlate with zero frequency rotational motion of the reactants. In this study, the energy levels for the two bending vibrations are treated harmonically. Thus the partition functions are given by

$$q^{GT}_{Xk}(T,s) = \frac{1}{2\sinh(\hbar\omega_{Xk}(s)\beta/2)}, k=1,2 \qquad (17)$$

and the adiabatic ground-state potential reduces to

$$V^G_{X,g}(s) \approx V^{RP}_g(s) + \sum_{k=1}^{2} \frac{\hbar\omega_{Xk}(s)}{2}. \qquad (18)$$

The parameters of the potential are fitted to reproduce the experimental enthalpy of reaction and the H atom activation energy at room temperature, and to qualitatively reproduce the H/D and Mu/H kinetic isotope effects. The bend frequencies for the product (cyclohexadienyl radical) were chosen to be similar to H bend frequencies in benzene. The activation energies are computed by a two-point approximation at 298 and 300 K. The parameters for the gas-phase model are given in Table II and the computed activation energies and KIEs are compared with experimental ones in Table III. The activation energies are fitted well by the simple model. The KIEs have the correct qualitative trend (e.g., the Mu/H KIE is about a factor of 20 greater than the H/D KIE), but the magnitudes of the computed KIEs are too large. A more sophisticated model that includes the variation of other important degrees of freedom along the reaction coordinate is needed to accurately reproduce both the KIEs and activation energies.

Table II. Parameters of the gas-phase potential model

D_M (kcal/mol)	30.0
α_M (Å^{-1})	3.7
V_G (kcal/mol)	4.8
ω_G (cm^{-1})	700
α_G (Å^{-2})	2.0
s_G (Å)	2.7
v_1 (kcal/mol)	475
v_2 (kcal/mol)	755
ω_1 (cm^{-1})	1150
ω_2 (cm^{-1})	1450
α_1, α_2 (Å^{-1})	1.7
s_0 (Å)	1.1

Table III. Comparison of experimental and computed activation energies and kinetic isotope effects for the gas-phase reactions of hydrogen isotopes with benzene at room temperature

	CVT/ CD-SCSAG	Experimental
k_H/k_D	2.7	1.0
k_{Mu}/k_H	63	19
E_{Dact} (kcal/mol)	4.5	4.2
E_{Hact} (kcal/mol)	4.3	4.3
E_{Muact} (kcal/mol)	2.1	1.6

Equilibrium Solvation Model. The experimental rate constants, activation energies, and kinetic isotope effects for the aqueous-phase analogs of reactions $R_{X,g}$ (denoted R_{Xaq}), and ratios of aqueous-phase to gas-phase rate constants are summarized in Table I. The Mu reactions were studied by Roduner *et al.* (*4*) and the H and D reactions were studied and compared with the Mu reaction by Roduner and Bartels (*3*). They find that the rate constants are enhanced upon aqueous solvation by factors of 23-29 for D and 32-37 for H. The enhancements for Mu is much lower (about 3.1 at room temperature). In this section we describe an equilibrium solvation approach to describing the effect of aqueous solvation on these reactions.

In the equilibrium solvation approximation the solvent molecules are assumed to adjust instantaneously to motions of the solute. The effect of the solvent is included by its average effect on the solute potential energy function and only the solute degrees of freedom are treated explicitly. The equilibrium solvation Hamiltonian for the aqueous-phase reaction R_{Xaq} is written

$$H_{X,es} = \frac{\mathbf{p}^2}{2\mu_X} + V_{X,es}(T,\mathbf{q}) \tag{19}$$

where $V_{X,es}(T,\mathbf{q})$ is the equilibrium solvation potential for reaction R_{Xaq} (including the free energy of solvation) and it depends upon the temperature of the solvent. The equilibrium solvation potential is also written as a function of the three coordinates of the H atom relative to rigid benzene [*e.g.*, $\mathbf{q}=(s,x_1,x_2)$]. In this work we take the equilibrium solvation potential to be the sum of the gas-phase potential and the change in free energy of solvation in going from reactants to a location s along the reaction coordinate

$$V_{X,es}(T,\mathbf{q}) = V_g(\mathbf{q}) + \Delta\Delta G_{Xsolv}^{GT}(T,s). \tag{20}$$

The change in free energy of solvation, $\Delta\Delta G_{Xsolv}^{GT}(T,s)$, is obtained by extending the equilibrium solvation model of Roduner and Bartels (*3*).

Roduner and Bartels approximated the ratio of aqueous-phase to gas-phase rate constants using an equilibrium transition state theory model

$$\frac{k_{X,es}^{\ddagger}(T)}{k_{X,g}^{\ddagger}(T)} = \exp\left[-\Delta\Delta G_{Xsolv}^{\ddagger}(T)/RT\right] \tag{21}$$

with the free energy change between the saddle point and reactants defined by

$$\Delta\Delta G_{Xsolv}^{\ddagger}(T) = \Delta G_{solv}^{\ddagger}(T) - \Delta G_{solv}^{X}(T) - \Delta G_{solv}^{\phi}(T), \tag{22}$$

and $\Delta G_{solv}^{\ddagger}(T)$, $\Delta G_{solv}^{X}(T)$, and $\Delta G_{solv}^{\phi}(T)$ defined as the free energy of solvating the transition state, hydrogen isotope X (X=H, D, Mu), and benzene, respectively. They defined the free energy of solvation of species Y using a temperature dependent model based on the enthalpy and entropy of solvating species Y. Using this scheme they were able to quantitatively account for the temperature dependence of the ratio of aqueous-phase to gas-phase rate constants for H and to qualitatively reproduce the experimental results for D that were available over a much more narrow temperature range. In the present study we focus on the large discrepancy between this model and the Mu results. Since the aqueous-phase rate constant for the Mu reaction is available only near room temperature, we only need the room temperature free energies of solvation of the species.

The free energies of solvating benzene and the transition state at room temperature are taken from the model of Roduner and Bartels (*3*). In their work Roduner and Bartels assumed that the free energies of solvation for all three hydrogen isotopes were the same and approximated them from literature values for H_2. (Roduner and Bartels provided compelling arguments, based upon the similarity of the dispersion forces for H and H_2, for why the solvation free energy of H atom is well approximated by the solvation free energy for H_2.) The free energy of solvation of a solute is independent of the mass when computed using classical statistical mechanics. However, because of the light masses of the H isotopes (especially Mu) we include the possibility that the free energies of solvation may have a solute mass dependence when computed quantum mechanically. In the present study, as justified by Roduner and Bartels, the free energy of solvation of D at 298 K is estimated to be that for H_2. The difference in the free energy of solvation of H and D, and H and Mu are estimated from quantum mechanical calculations by Gai and one of the authors (*39*). All the equilibrium solvation parameters are listed in Table IV.

It is interesting to note that using these parameters, the change in free energy of solvation in going from reactants to the saddle point is approximately 2.2 kcal/mol, which yields a ratio of equilibrium solvation to gas-phase rate constant of 40 at 300 K, in good agreement with experiment for the H isotope. The classical free energy of solvation is independent of the mass of the H isotope, and thus the same ratio is obtained for the D reaction, which is also in qualitative agreement with experiment. This simple model based on the free energy change does not agree well with the experimental value for Mu of 3.1.

Table IV. Parameters of the equilibrium solvation model[a]

$\Delta\Delta G^{\ddagger}_{\text{Hsolv}}$ (kcal/mol)	-2.12
$\Delta\Delta G^{\ddagger}_{\text{Dsolv}}$ (kcal/mol)	-2.19
$\Delta\Delta G^{\ddagger}_{\text{Musolv}}$ (kcal/mol)	-3.32
α_{solv} (Å^{-1})	1.7
s^* (Å)	3.75

[a] Free energies are for room temperature

For a given temperature T, the change in free energy of solvation in going from reactants to a location s along the reaction coordinate is given in terms of the free energy change between the saddle point, and reactants by

$$\Delta\Delta G^{GT}_{\text{Xsolv}}(T,s) = -\frac{\Delta\Delta G^{\ddagger}_{\text{Xsolv}}(T)}{2}\left\{\tanh[\alpha_{\text{solv}}(s-s^*)]-1\right\}. \tag{23}$$

The generalized free energy of solvation is assumed to have a similar dependence on the reaction coordinate as the bend frequencies and the parameters α_{solv} and s^* are fitted to reproduce this behavior. These parameters are also listed in Table IV.

The equilibrium solvation potential can also be written in terms of an effective equilibrium solvation potential along the reaction coordinate and the bending potentials

$$V_{X,es}(T,\mathbf{q}) = V_{X,es}^{RP}(T,s) + \sum_{k=1}^{2} V_{bk}(x_k,s) \tag{24}$$

where

$$V_{X,es}^{RP}(T,s) = V_g^{RP}(s) + \Delta\Delta G_{Xsolv}^{GT}(T,s). \tag{25}$$

The equilibrium solvation potentials along the reaction coordinate for D, H, and Mu are compared with the reaction coordinate gas-phase potential in Figure 1. In this model we have chosen to include the effect of equilibrium solvation on the reaction path motion only and neglect the effect on the bending modes. As discussed above, all the species (reactants and transition state) are hydrophobic with positive free energies of solvation. As the hydrogen isotope bonds to the benzene molecule, the unfavorable solvation of the hydrogen atom is mitigated, thereby leading to the large increase in the rate constant upon solvation. The energetics for the hydrophobic interactions are determined largely by the size of the cavity created by the solute. We expect this to depend very weakly on the bending coordinate, especially at the transition state where the solvation energy is determined primarily by the size of the benzene molecule. We do not expect the neglect of equilibrium solvation on the bending modes to significantly affect the qualitative trends in our predictions of ratios of rate constants.

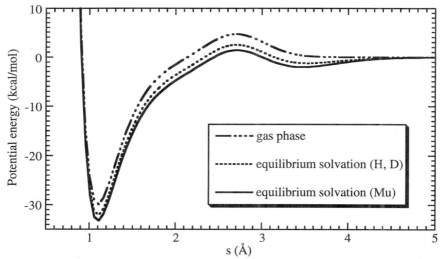

Figure 1. Comparison of the gas-phase and equilibrium solvation potentials along the reaction coordinate s.

For the equilibrium solvation Hamiltonian, the VTST rate constant with semiclassical tunneling correction can be obtained as described in the previous section. The generalized TST rate constant for the equilibrium solvation model is given by an expression identical to equation 15 except with $V_g^{RP}(s)$ replaced by $V_{X,es}^{RP}(T,s)$. The rotational motions of the transition state complex and benzene in the gas phase become low frequency vibrations or hindered rotations in solution. In the equilibrium solvation model they are treated as free rotations as in the gas phase, but the rotational partition functions for reactants and the transition state are again assumed to cancel. This is equivalent to assuming that the partition functions for the low frequency vibrations or hinder rotations for the reactants and transition state would also cancel for an explicit solvation model. The equilibrium solvation CVT rate constant for reaction R_{Xaq} is obtained from the generalized transition state one by locating the value of the reaction coordinate, $s_{X,es}^{CVT}(T)$, that minimizes the generalized transition state expression. Note that the contribution from the change in the equilibrium solvation free energy can move the optimum location from the optimum location found for the gas phase.

The tunneling correction factor for the equilibrium solvation Hamiltonian, $\kappa_{X,es}^{CVT/AG}(T)$, is computed using equation 7, where the tunneling probabilities are computed for the equilibrium-solvation ground-state adiabatic barrier. This adiabatic potential is given by an expression identical to equation 18 except with gas-phase potential $V_g^{RP}(s)$ replaced by the equilibrium solvation potential $V_{X,es}^{RP}(T,s)$. The use of this adiabatic potential may seem inconsistent, since the bound solute modes are treated quantum mechanically in the ground-state vibrationally adiabatic approximation while the solvent modes are treated classically and are not restricted to their ground state but are effectively thermally averaged. However, for the low frequency modes of the solvent, the ground-state vibrations are not representative of the those states that contribute most significantly to the thermally averaged rate constant. This is true because the density of state rises very rapidly in these low frequency modes. Using the thermally averaged potential of mean force represents a physically motivated method that includes the important energy levels in an average sense. This is discussed in more detail elsewhere (40).

The ratio of the aqueous-phase and gas-phase rate constants for the reaction of hydrogen isotope X with benzene is then approximated in the equilibrium solvation model by

$$\frac{k_{X,es}^{CVT/AG}(T)}{k_{X,g}^{CVT/AG}(T)} = \frac{\kappa_{X,es}^{CVT/AG}(T)}{\kappa_{X,g}^{CVT/AG}(T)} \prod_{k=1}^{2} \frac{\sinh\left[\hbar\omega_{Xk}\left(s_{X,g}^{CVT}(T)\right)\beta/2\right]}{\sinh\left[\hbar\omega_{Xk}\left(s_{X,es}^{CVT}(T)\right)\beta/2\right]}$$

$$\times \frac{\exp\left[-\beta V_{X,es}^{RP}\left(T, s_{X,es}^{CVT}(T)\right)\right]}{\exp\left[-\beta V_g^{RP}\left(s_{X,g}^{CVT}(T)\right)\right]}. \tag{26}$$

Note that this differs from the simple transition state theory expression given in equation 21 but reduces to that expression when tunneling is negligible or the tunneling factors cancel for the gas and aqueous phases, and the optimum locations of the transition states are the same for the gaseous and aqueous phases.

Nonequilibrium Solvation Model. The dissipative influence of solvent friction on the solute dynamics is modeled by linear coupling between harmonic bath oscillators and the solute coordinates. The Hamiltonian for this system is given by

$$H_{ns} = \frac{p^2}{2\mu} + V_{es}(T,q) + \sum_n \left[\frac{P_n^2}{2\mu} + \frac{1}{2}\mu\omega_n^2(Q_n - C_n q)^2\right] \tag{27}$$

where $\mathbf{Q}$ and $\mathbf{P}$ are the bath oscillator coordinate and momentum vectors, ω_n is the harmonic oscillator frequencies for bath coordinate n, and $\mathbf{C}_n$ is the vector of constants coupling the solute coordinates to bath coordinate n. The mass for the bath coordinates is arbitrary (changing it scales the coupling constants) and we choose it to be the same as the solute reduced mass for convenience. The classical equations of motion for the solute coordinates obtained from this Hamiltonian are equivalent to a generalized Langevin equation (6,7). The connection between the system Hamiltonian, equation 27, and a GLE is completed with the identification of the components of the friction tensor and random force vector ($\eta_{nm}(t)$ and $F_n(t)$, respectively) for the GLE in terms of the bath parameters ω_n and $\mathbf{C}_n$ in the Hamiltonian. The friction tensor is given by

$$\eta_{jk}(t) = \mu \sum_n C_{nj} C_{nk} \omega_n^2 \cos(\omega_n t). \tag{28}$$

where j and k are indices over the solute coordinates. For the GLE, the random force vector is a zero-centered Gaussian and it is related to the friction tensor through the second fluctuation-dissipation theorem (41).

For the purpose of calculating the rate constant, the most significant coupling of the solute to the bath is for the reaction coordinate. For the model considered here, the MEPs for gas phase and equilibrium solvation only contain the reaction coordinate motion, and coupling of the bath to the bending modes does not have a direct effect on the reaction coordinate motion. Therefore, in the present study we assume the heat bath is only coupled to the reaction coordinates, not the bending motion. In this case the nonequilibrium solvation Hamiltonian for reaction $R_{X,aq}$ is expressed as

$$H_{X,ns} = \frac{\mathbf{p}^2}{2\mu_X} + \frac{\mathbf{P}^2}{2\mu_X} + V_{X,ns}(T,\mathbf{q},\mathbf{Q}) \tag{29}$$

where the nonequilibrium solvation potential is given by

$$V_{X,ns}(T,\mathbf{q},\mathbf{Q}) = V_{X,es}^{RP}(T,s) + \sum_{k=1}^{2} V_{bk}(x_k,s) + \sum_n \frac{1}{2}\mu\omega_n^2(Q_n - C_n s)^2 . \tag{30}$$

The bath parameters C_n and ω_n are determined to obtain the best fit of an analytic expression for the friction kernel to the cosine expansion given in equation 28. The procedure used to fit the parameters is similar to that described elsewhere (42). The friction tensor reduces to a single component and we employ a Gaussian form

$$\eta(t) = \frac{\eta_0}{\sigma}\sqrt{\frac{2}{\pi}}\exp\left(-\frac{t^2}{2\sigma^2}\right) \tag{31}$$

where η_0 determines the strength of the friction and σ determines the time scale. The bath frequencies are distributed uniformly

$$\omega_n = \frac{\pi}{\tau}\left(n - \frac{1}{2}\right), \quad n=1,2,...,N \tag{32}$$

where τ is chosen to provide the best cosine expansion of the Gaussian friction kernel with a finite number of terms. The coupling coefficients are obtained from the cosine Fourier transform of the Gaussian friction kernel

$$C_n^2 = \left(\mu\omega_n^2\right)^{-1} \frac{2}{\tau} \int_0^\tau dt \, \cos(\omega_n t) \, \eta(t), \quad n=1,2,\ldots,N \tag{33}$$

which, on substitution of equation 31, with $\tau \gg \sigma$, yields

$$C_n^2 = \frac{2\eta_0}{\mu\tau\omega_n^2} \exp\left[-(\omega_n\sigma)^2 / 2\right]. \tag{34}$$

The bath parameters are determined by the physical parameters of the Gaussian friction, η_0 and σ. In principle the calculations should be converged with respect to the parameter τ and the number of oscillators. In previous calculations (18) we have shown that a value of $\tau=2\sigma$ and one oscillator provide correct qualitative trends, and these are the values used in the present study.

The solvent friction can be obtained from direct classical dynamical simulations of the force autocorrelation; however, consistent with our interest to study qualitative trends, we use estimates of the time scale σ and strength η_0 of the friction. The time scale for the force autocorrelation function is determined by motion of solvent molecules around a clamped solute and we approximate it by the time scale for water motion in the pure solvent given by the velocity autocorrelation function. In this study we take the half width at half maximum from molecular dynamics simulations (43). This approach gives a value of $\sigma=8.5 \times 10^{-15}$ seconds. The magnitude of the friction is harder to estimate and we will present results for a range of values for η_0. To estimate the range, we use the Stokes-Einstein relationship

$$\eta_0 = (\beta D)^{-1} \tag{35}$$

where D is the diffusion constant for H in water. The room temperature diffusion constant has recently been measured to be about 8×10^{-5} cm^{2}s^{-1} (Bartels, D. M., private communication). Note that this value agrees well with the diffusion constants for H_2 and He in water of 4.5×10^{-5} and 6.3×10^{-5} cm^{2}s^{-1}, respectively (37). For these values of the diffusion constant, values of η_0 less than 20 atomic units are consistent.

The procedure for the nonequilibrium solvation rate calculations parallels that for the calculations corresponding to the gas phase and equilibrium solvation; however, the calculation of the minimum energy path is performed for the nonequilibrium Hamiltonian (equation 29) in the extended space of the three solute coordinates (s,x_1,x_2) and bath oscillator coordinates. Another difference arises in the calculation of the reactant partition function that includes both the solute and solvent degree of freedom. For a unimolecular reaction, in which all modes are bound at reactants, if the solute modes are approximated harmonically, the classical mechanical reactant partition function for the nonequilibrium solvation Hamiltonian is independent of the coupling C_n (44). Therefore, the classical reactant partition function reduces to the product of the equilibrium solvation reactant partition function and the partition function for the coupled bath. This is the correct result since the reactant partition function is an equilibrium property and should not be affected by the nonequilibrium coupling between the solute and the bath. When the reactant partition function for the nonequilibrium solvation Hamiltonian is computed quantum mechanically, it is no longer independent of coupling, even for the unimolecular reaction. However, we do know that we want to compute the (equilibrium) reactant partition function that corresponds to zero coupling between the solute and nonequilibrium bath modes. Thus, the nonequilibrium solvation Hamiltonian with full coupling is used just for the interaction region and the uncoupled ($C_n=0$) nonequilibrium solvation Hamiltonian is used to describe the reactant region. Consistent ways to handle this problem in terms of a projector of the reactant region of phase space or using a coordinate-dependent friction that dies off in the reactant and product regions are discussed elsewhere (23).

The nonequilibrium solvation, generalized TST rate expression for reaction R_{Xaq} is approximated by

$$k_{X,ns}^{GT}(T,s) \approx \frac{k_B T}{h} \left(\frac{2\pi\hbar^2}{\mu_X k_B T}\right)^{3/2} \frac{Q_{X,ns}^{GT}(T,s)}{Q_{bath}(T)} \exp\left[-\beta V_{X,ns}^{MEP}(T,s)\right]. \tag{36}$$

The generalized transition state partition function $Q_{X,ns}^{GT}(T,s)$ is expressed in terms of the bound state energy levels for each mode m

$$Q_{X,ns}^{GT}(T,s) = \prod_m \sum_{n_m=0} \exp\left[-\beta \, \varepsilon_{X,ns,m}^{GT}(s,n_m)\right]. \tag{37}$$

The bound energy levels $\varepsilon_{X,ns,m}^{GT}(s,n_m)$ are different from their equilibrium solvation counterparts because of the different definition of the reaction path and the fact that there are also more modes corresponding to the bath oscillators. For the nonequilibrium solvation Hamiltonian, the generalized transition state partition function for the bound modes orthogonal to the reaction coordinate can be written as the product

$$Q_{X,ns}^{GT}(T,s) = \prod_{k=1}^{2} q_{Xk}^{GT}(T,s) \, Q_{Xbath}^{GT}(T,s) \tag{38}$$

where $q_{Xk}^{GT}(T,s)$ is given by equation 17 and $Q_{Xbath}^{GT}(T,s)$ is the generalized transition state partition function for the bath modes that are perturbed by their coupling to the reaction coordinate. In the uncoupled case, this partition function reduces to the following

$$Q_{bath}(T) = \prod_n \frac{1}{2\sinh(\hbar\omega_n(s)\beta/2)}. \tag{39}$$

The potential along the minimum energy path $V_{X,ns}^{MEP}(T,s)$ is for the nonequilibrium solvation path through the extended space of the solute and bath oscillator coordinates.

The tunneling correction factor $\kappa_{X,ns}^{CVT/AG}(T)$ is obtained using equation 7 and the prescription outlined above. The ground-state adiabatic potential is constructed from the potential along the MEP and the ground-state energy levels $\varepsilon_{X,ns,m}^{GT}(s,n_m=0)$. The ratio of the nonequilibrium to equilibrium rate constants provides a measure of the importance of nonequilibrium solvation on the rate constant and, for the reaction of hydrogen isotope X with benzene, this ratio is given by

$$\frac{k_{X,ns}^{CVT/AG}(T)}{k_{X,es}^{CVT/AG}(T)} = \frac{\kappa_{X,ns}^{CVT/AG}(T)}{\kappa_{X,es}^{CVT/AG}(T)} \prod_{k=1}^{2} \frac{\sinh\left[\hbar\omega_{Xk}\left(s_{X,es}^{CVT}(T)\right)\beta/2\right]}{\sinh\left[\hbar\omega_{Xk}\left(s_{X,ns}^{CVT}(T)\right)\beta/2\right]}$$

$$\times \frac{Q_{Xbath}^{GT}\left[T,s_{X,ns}^{CVT}(T)\right]}{Q_{bath}(T)} \frac{\exp\left\{-\beta V_{X,ns}^{MEP}\left[T,s_{X,ns}^{CVT}(T)\right]\right\}}{\exp\left\{-\beta V_{X,es}^{RP}\left[T,s_{X,es}^{CVT}(T)\right]\right\}}. \qquad (40)$$

Note that the location of the CVT transition state for equilibrium solvation and nonequilibrium solvation are not necessarily the same. The ratio of the nonequilibrium solvation and gas-phase rate constant is given by combining the equations 26 and 40, and this ratio gives the best estimate of the experimentally observed ratio of the aqueous-phase and gas-phase rate constants.

Results and Discussion

The ratios of equilibrium solvation to gas-phase rate constants computed using equation 26 are compared with experimental ratios of the aqueous-phase to gas-phase rate constants in Table V. Note that the equilibrium solvation results for H and D agree well with experimental results just like the simple model given by equation 21. This indicates that the optimum locations of the dividing surfaces are at the saddle points for these systems and that tunneling effects are nearly negligible. The 12% increase in the computed ratios for H relative to D is in good agreement with a 10% increase in the experimental ratio of aqueous-phase to gas-phase rate constants. The free energy of solvation of Mu is about 1 kcal/mol higher than for H and, as a result, the change in the free energy of solvation between the saddle point and reactants is more negative for Mu (see Table IV). This leads to the large increase (by about a factor of 3) in the Mu ratio compared to the H ratio. Although this effect is the opposite of that observed experimentally for the ratios of the aqueous-phase to gas-phase rate constants, it shows a consistent trend that as the hydrogen isotope become lighter it has a larger positive free energy of solvation and drives the reaction faster.

Table V. Comparison of computed and experimental ratios of aqueous-phase and gas-phase rate constants at room temperature

Isotope X	$\dfrac{k_{X,es}^{CVT/CD-SCSAG}}{k_{X,g}^{CVT/CD-SCSAG}}$	$\dfrac{k_{X,ns}^{CVT/CD-SCSAG}}{k_{X,g}^{CVT/CD-SCSAG}}$	Experimental $k_{X,ag}/k_{X,g}$
D	32	24	29
H	36	21	32
Mu	96	2.5	3.1

[a]   The nonequilibrium solvation calculations are for a value of $\eta_0=6$ au.

In Figure 2 we display the ratio of nonequilibrium solvation and equilibrium solvation rate constants, k_{ns}/k_{es} as a function of the strength of dynamical coupling, η_0. These ratios were calculated with the CVT/CD-SCSAG method using equations 26 and 40. From Figure 2, the strong dependence of the rate constant on nonequilibrium solvation effects for the case of the Mu isotope and weak dependence for the case of H and D is easily seen. As the coupling increases from 0 to 20, the Mu k_{ns}/k_{es} drops by more than three orders of magnitude while for the case of H and D isotopes, the ratio drops by less than an order of magnitude. If we look at the value of $\eta_0 = 6$ au, we find that we are able to reproduce the quantitative experimental trends (see Table V). Here, the dramatic suppression of the Mu rate when going from an equilibrium solvation description to a nonequilibrium description is required to reproduce the experimental trend. For $\eta_0 = 6$, the ratio k_{es}/k_{ns} is equal to 39 for Mu, 1.7 for H, and 1.4 for D. We see that the introduction of nonequilibrium solvation effects suppresses the rate.

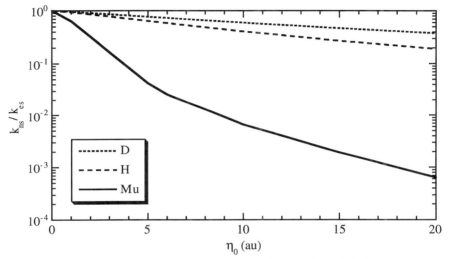

Figure 2. Variation of the ratio of nonequilibrium and equilibrium CVT/CD-SCSAG rate constants with the magnitude of the solvent friction.

Within our model we can separate the influence of the solvent on quantum mechanical tunneling by comparing estimates of tunneling factors, κ, and the rate with tunneling effects removed, k^{CVT}. Part of the large decrease of the Mu rate constant upon aqueous solvation arises from suppression of quantum mechanical tunneling. This is shown in Table VI. The CVT/CD-SCSAG tunneling correction factors are small for H and D and decrease only slightly from the gas-phase to the equilibrium and nonequilibrium solvation models. The changes in the Mu tunneling correction factors are much more pronounced. The decrease in the tunneling correction factor for Mu of about a factor of 2.5 in going from the gas-phase to the equilibrium solvation model arises largely because of the lowering and a slight broadening of the barrier for the equilibrium solvation potential (see Figure 1). The subsequent decrease of the Mu tunneling correction factors by a factor of 7.6 upon adding coupling to the bath in the nonequilibrium solvation model is much more pronounced. The quenching of tunneling by the solvent friction in this type of model has been discussed previously [for example, see the review by Hänggi *et al.* (*45*)], and has been discussed in the context of VTST calculations by McRae *et al.* (*18*).

Table VI. Contributions of tunneling to the computed room-temperature rate constants

Isotope X	$\kappa_{X,g}^{CVT/CD-SCSAG}$	$\kappa_{X,es}^{CVT/CD-SCSAG}$	$\kappa_{X,ns}^{CVT/CD-SCSAG}$
D	1.3	1.3	1.2
H	1.6	1.6	1.4
Mu	31.5	12	1.6

The nonequilibrium solvation calculations are for a value of $\eta_0=6$ au.

We also examine the effect of equilibrium and nonequilibrium solvation on the CVT rate constant that neglects tunneling. Ratios of the equilibrium solvation and gas-phase CVT rate constants, the nonequilibrium solvation and gas-phase CVT rate constants, and the nonequilibrium solvation and equilibrium CVT rate constants are shown in Table VII. The major contribution to the large values of the ratio of equilibrium solvation to gas-phase rate constants is the negative value of the change in solvation free energy in going from reactant to the transition state. The trends in the equilibrium solvation to gas-phase ratio in Table VII for the three isotopes reflect the increasing negative values for this free energy for the sequence D, H, and Mu as discussed above. In the nonequilibrium solvation model, the barrier heights at the saddle point for the reactions are the same as those for the equilibrium solvation model. Therefore, changes in the ratio of the nonequilibrium solvation and gas-phase rate constants relative to the ratio of the equilibrium solvation and gas-phase rate constants have at most a small dependence on changes in the energetic factor in the rate expression (*e.g.*, the energy in the Boltzmann factors of equation 40). Instead, the change in these ratios is caused by changes in the partition function $Q_{Xbath}^{GT}(T,s)$ for the bath modes that are perturbed by their coupling to the reaction coordinate. For Mu this effect decreases the ratio of rate constants by a factor of 5.2. This is comparable to the decrease of a factor of 7.6 in the Mu tunneling factor that was seen for the nonequilibrium solvation model relative to the equilibrium solvation model. These ratios are also summarized in the last two columns of Table VII as the ratios of the CVT rate constants and tunneling correction factors for the nonequilibrium and equilibrium solvation models.

Table VII. Contributions to the computed ratios of room-temperature rate constants

Isotope X	$\dfrac{k_{X,es}^{CVT}}{k_{X,g}^{CVT}}$	$\dfrac{\kappa_{X,es}^{CVT}}{\kappa_{X,g}^{CVT}}$	$\dfrac{k_{X,ns}^{CVT}}{k_{X,g}^{CVT}}$	$\dfrac{\kappa_{X,ns}^{CVT}}{\kappa_{X,g}^{CVT}}$	$\dfrac{k_{X,ns}^{CVT}}{k_{X,es}^{CVT}}$	$\dfrac{\kappa_{X,ns}^{CVT}}{\kappa_{X,es}^{CVT}}$
D	32	1.00	25	0.93	0.77	0.91
H	36	0.99	25	0.83	0.71	0.83
Mu	239	0.40	46	0.053	0.19	0.13

The nonequilibrium solvation calculations are for a value of $\eta_0=6$ au.

Concluding Remarks

In this work we have shown that nonequilibrium solvation effects of aqueous solvation can be substantial, causing a decrease in the aqueous-phase rate constant relative to the gas-phase rate constant by over an order of magnitude. The key to distinguishing nonequilibrium from equilibrium solvation effects was the examination of kinetic isotope effects or the change in the ratio aqueous-phase to gas-phase rate constants with change in the mass of the reacting solute molecules. When equilibrium solvation effects on internal vibrations of the solute are small, equilibrium solvation effects are independent of the mass of the solute, and differences in aqueous-phase and gas-phase KIEs are an indicator of important nonequilibrium effects.

Models for the gas-phase reaction and equilibrium and nonequilibrium aqueous solvation have been developed that reproduce qualitative trends in the activation energies and KIEs for the reaction of H and D with benzene. We find that the experimentally observed, anomalous quenching of the Mu/H KIE by aqueous solvation can be explained using a nonequilibrium solvation model that incorporates solvent friction effects. Both changes to quantum mechanical tunneling and partition functions for bound energies contribute about equally to the large nonequilibrium solvation effect on the Mu/H KIE.

The large nonequilibrium solvation effects seen in the present work are unusual in comparison to those predicted by classical mechanical theories such as Kramers (*1*) and Grote-Hynes (*2*) theories, or a recent nonequilibrium electric polarization model (*46*). The present study indicates that large nonequilibrium solvation effects are possible when quantum mechanical effects are important. This is true for the reaction studied here because of the light mass of one of the solutes, Mu, which is one-ninth that of hydrogen. This finding raises the question of whether these large effects will be seen for other, less exotic, systems such as proton, hydrogen atom, or hydride transfer reactions. For the addition reaction of the hydrogen isotopes to benzene the barrier height is quite low (less than 5 kcal/mol) and the barrier is broad enough that tunneling is only important for the lightest mass, Mu. Also, for this reaction, changes in the partition functions for the bound energies are more pronounced for the lighter mass. However, for other reactions with higher narrower barriers, we expect to see comparable nonequilibrium solvation effects for the heavier hydrogen isotopes (*e.g.,* H).

Acknowledgments. The authors wish to acknowledge Dr. David Bartels for many stimulating conversations and Dr. Michael Messina and Professor Donald G. Truhlar for critically reading the manuscript. This work was supported by the Division of Chemical Sciences, Office of Basic Energy Sciences, U.S. Department of Energy under Contract DE-AC06-76RLO 1830 with Battelle Memorial Institute, which operates the Pacific Northwest Laboratory.

Literature Cited

1. Kramers, H. A. *Physica (Utrecht)* **1940**, *7*, 284.
2. Grote, R. T.; Hynes, J. T. *J. Chem. Phys.* **1980**, *73*, 2715.
3. Roduner, E.; Bartels, D. M. *Ber. Bunsenges. Phys. Chem.* **1992**, *96*, 1037.
4. Roduner, E.; Louwrier, P. W. F.; Brinkman, G. A.; Garner, D. M.; Reid, I. D.; Arseneau, D. J.; Senba, M.; Gleming, D. G. *Ber. Bunsenges. Phys. Chem.* **1990**, *94*, 1224.
5. Nicovich, J. M.; Ravishankara, A. R. *J. Phys. Chem.* **1984**, *88*, 2534.
6. Ford, G. W.; Kac, M.; Mazur, P. *J. Math. Phys.* **1965**, *6*, 504.
7. Zwanzig, R. *J. Stat. Phys.* **1973**, *9*, 215.
8. Bergsma, J. P.; Reimers, J. R.; Wilson, K. R.; Hynes, J. T. *J. Chem. Phys.* **1986**, *85*, 5625.
9. Bergsma, J. P.; Gertner, B. J.; Wilson, K. R.; Hynes, J. T. *J. Chem. Phys.* **1987**, *86*, 1356.
10. Straub, J. E.; Borkovec, M.; Berne, B. J. *J. Chem. Phys.* **1988**, *89*, 4833.
11. Gertner, B. J.; Wilson, K. R.; Hynes, J. T. *J. Chem. Phys.* **1989**, *90*, 3537.
12. Zichi, D. A.; Ciccotti, G.; Hynes, J. T.; Ferrario, M. *J. Phys. Chem.* **1989**, *93*, 6261.
13. Ciccotti, G.; Ferrario, M.; Hynes, J. T.; Kapral, R. *J. Chem. Phys.* **1990**, *93*, 7137.

14. Benjamin, I.; Lee, L. L.; Li, Y. S.; Liu, A.; Wilson, K. R. *Chem. Phys.* **1991**, *152*, 1.
15. Benjamin, I.; Lee, L. L.; Li, Y. S.; Wilson, K. R. *J. Chem. Phys* **1991**, *95*, 2458.
16. Keirstead, W. P.; Wilson, K. R.; Hynes, J. T. *J. Chem. Phys.* **1991**, *95*, 5256.
17. Whitnell, R. M.; Wilson, K. R. In *Review in Computational Chemistry*; K. B. Lipkowitz and D. B. Boyd, Ed.; VCH: New York, 1993; Vol. 4; pp 67.
18. McRae, R. P.; Schenter, G. K.; Garrett, B. C.; Haynes, G. R.; Voth, G. A.; Schatz, G. C. *J. Chem. Phys.* **1992**, *97*, 7392.
19. Truhlar, D. G.; Isaacson, A. D.; Garrett, B. C. In *Theory of Chemical Reaction Dynamics*; M. Baer, Ed.; CRC Press: Boca Raton, FL, 1985; Vol. IV; pp 65.
20. Isaacson, A. D.; Truhlar, D. G.; Rai, S. N.; Steckler, R.; Hancock, G. C.; Garrett, B. C.; Redmon, M. J. *Comp. Phys. Comm.* **1987**, *47*, 91.
21. Lu, D. H.; Truong, T. N.; Melissas, V. S.; Lynch, G. C.; Liu, Y. P.; Garrett, B. C.; Steckler, R.; Isaacson, A. D.; Rai, S. N.; Hancock, G. C.; Lauderdale, J. G.; Joseph, T.; Truhlar, D. G. *Comp. Phys. Comm.* **1992**, *71*, 235.
22. Voth, G. A.; Chandler, D.; Miller, W. H. *J. Chem. Phys.* **1989**, *91*, 7749.
23. Garrett, B. C.; Schenter, G. K. *Int. Rev. Phys. Chem.* **1994**, (in press).
24. Garrett, B. C.; Truhlar, D. G. *J. Phys. Chem.* **1979**, *83*, 1079.
25. Garrett, B. C.; Truhlar, D. G. *J. Am. Chem. Soc.* **1979**, *101*, 4534.
26. Miller, W. H.; Handy, N. C.; Adams, J. E. *J. Chem. Phys.* **1980**, *72*, 99.
27. Marcus, R. A. *J. Chem. Phys.* **1966**, *45*, 4493.
28. Marcus, R. A. *J. Chem. Phys.* **1967**, *46*, 959.
29. Marcus, R. A. *J. Chem. Phys.* **1968**, *49*, 2610.
30. Truhlar, D. G. *J. Chem. Phys.* **1970**, *53*, 2041.
31. Chatfield, D. C.; Friedman, R. S.; Truhlar, D. G.; Garrett, B. C.; Schwenke, D. W. *J. Am. Chem. Soc.* **1991**, *113*, 486.
32. Garrett, B. C.; Truhlar, D. G. *J. Phys. Chem.* **1979**, *83*, 1052.
33. Garrett, B. C.; Truhlar, D. G.; Grev, R. S.; Magnuson, A. W. *J. Phys. Chem.* **1980**, *84*, 1730.
34. Marcus, R. A. *J. Chem. Phys.* **1964**, *41*, 610.
35. Skodje, R. T.; Truhlar, D. G.; Garrett, B. C. *J. Phys. Chem.* **1981**, *85*, 3019.
36. Skodje, R. T.; Truhlar, D. G.; Garrett, B. C. *J. Chem. Phys.* **1982**, *77*, 5955.
37. *Handbook of Chemistry and Physics*; 70 ed.; Weast, R. C., Ed.; CRC Press, Inc.: Boca Raton, FL, 1989.
38. Hammond, G. S. *J. Am. Chem. Soc.* **1955**, *77*, 334.
39. Gai, H.; Garrett, B. C. *J. Phys. Chem.* **1994**, (submitted).
40. Truhlar, D. G.; Liu, Y.-P.; Schenter, G. K.; Garrett, B. C., *J. Phys. Chem.* (in press).
41. Kubo, R.; Toda, M.; Hashitsume, N. *Statistical Physics II: Nonequilibrium Statistical Mechanics*; Springer-Verlag: New York, 1985.
42. Schenter, G. K.; McRae, R. P.; Garrett, B. C. *J. Chem. Phys.* **1992**, *97*, 9116.
43. Palmer, B. J.; Garrett, B. C. *J. Chem. Phys.* **1993**, *98*, 4047.
44. Pollak, E. *J. Chem. Phys.* **1986**, *85*, 865.
45. Hänggi, P.; Talkner, P.; Borkovec, M. *Rev. Mod. Phys.* **1990**, *62*, 251.
46. Truhlar, D. G.; Schenter, G. K.; Garrett, B. C. *J. Chem. Phys.* **1993**, *98*, 5756.

RECEIVED August 12, 1994

Chapter 10

Ionization of Acids in Water

Koji Ando and James T. Hynes

Department of Chemistry and Biochemistry, University of Colorado, Boulder, CO 80309-0215

The acid ionization of HCl in water is examined via a combination of electronic structure calculations and Monte Carlo computer simulation. The mechanism is found to involve: first, an activationless (or nearly so) motion in a solvent coordinate, which is adiabatically followed by the quantum proton, to produce a "contact" ion pair $Cl^- \text{-} H_3O^+$, which is stabilized by ~ 7 kcal/mol; second, motion in the solvent with a small activation barrier, as a second adiabatic proton transfer produces a "solvent-separated" ion pair from the "contact" ion pair in a nearly thermoneutral process.

Proton transfer reactions represent a central elementary process in chemistry and biochemistry.(1-7) In aqueous solution, the acid ionization

$$HA(aq) \rightleftharpoons A^-(aq) + H_3O^+(aq) \qquad (1)$$

is of particular importance in itself and also in connection with acid-base catalysis. (5-7) But despite its importance, the molecular mechanism of the solution phase reaction is unclear. Some reasons for this situation include (i) the quantum character of the proton motion, (ii) the complexity of the solution phase reaction, and (iii) the specificity of proton transfers in water. In this work, we address these problems on the basis of a realistic molecular modeling — *ab initio* quantum chemistry calculations of the potential energy surfaces and Monte Carlo simulations for the solution phase reaction — for HCl ionization in water.

The basic idea of our approach is as follows. First we note that the large polarity change in equation 1 suggests significant effects from the polar solvent. Indeed, acid ionization does not occur in the isolated 1:1 hydrogen-bonded complex with H_2O (8-10), indicating that the ion stabilization by the polar solvent water is essential for the reaction. The solvent fluctuation and reorganization are then

0097–6156/94/0568–0143$08.00/0

expected to be critical factors in determining both the rate and the mechanism of the reaction. This aspect (11-15) is similar to that for electron transfer reactions in polar solvents (16,17), for which the free energy curve/surface crossing induced by the fluctuation of the polar environment is key. For proton transfers, we postulate (and confirm) the following scheme, focusing for the moment and for simplicity solely on the first proton transfer between the HCl and a neighbouring water molecule: the proton potential and its asymmetry are modulated by the fluctuating polar environment, and the proton transfer occurs at the "crossing" point of the solvent configuration which gives a symmetric proton potential (Figure 1). The asymmetry modulation of the proton potential can be qualitatively comprehended in terms of a valence-bond picture of the electronic structure (18); the relative stability between the neutral diabatic state ($ClH\cdots OH_2$) and the ionic one ($Cl^- \cdots HOH_2^+$), which varies with the solvent polarization fluctuation, alters the asymmetry of the (electronically) adiabatic proton potential curve. Since attainment of the crossing point by solvent motion will generally require activation, the reaction coordinate is in the solvent and there can be a free energy barrier in this coordinate.

Second, we take account of the quantized nature of the proton motion, associated with the barrier height and shape of the proton potential. Depending on the strength of the hydrogen-bonding interaction, the height of the proton potential barrier and the position of the proton (vibrational) energy levels may vary in a way that controls the transfer mechanism. For "weak" hydrogen-bonding complexes where the barrier is high enough to hold several proton levels in either well, the proton transfer will be a quantum tunneling process.(11,12) In the case of "strong" hydrogen-bonding, where the proton double well has a low barrier such that the ground proton level is located above the barrier top, the proton transfer is no longer tunneling, although it is still quantum.

We call this last situation an "adiabatic proton transfer" (15): the proton wavepacket motion adiabatically follows the modulation of the proton potential from the reactant solvent configuration through the diabatic crossing point to the product state (Figure 1). The distance between the heavy particles, e.g. the oxygen atoms in an H_3O^+-H_2O arrangement, between which the proton transfers can be important in allowing this adiabatic pathway.

Finally, we consider a possible "Grotthuss mechanism" (7), which is quite specific for proton transfers in aqueous media: the solvent water molecules themselves can be involved in the process as proton acceptors. The first two proton transfer steps may be illustrated as in Scheme I:

$$ClH'\cdots H_2O\cdots H_2O \rightleftharpoons Cl^- \cdots H'OH_2^+ \cdots H_2O \rightleftharpoons Cl^- \cdots H'OH\cdots H_3O^+.$$
$$\quad\quad 1 \quad\quad\quad\quad\quad\quad 2 \quad\quad\quad\quad\quad\quad 3$$

Scheme I

One remarkable aspect is that a "proton relay" among water molecules may take place through the hydrogen-bonding network without need of large displacements of an individual proton. This mechanism is believed to be the origin of the anomalously high mobility of the proton in water and ice (*19,20*) and is also suggested (*5-7*) to play a role in various acid-base catalyses in aqueous solution, e.g. hydration-dehydration reactions, hydrolysis reactions and other prototropic changes. A natural and interesting question then would be whether the double (or multiple) proton transfer is "step-wise" or "concerted" in the case of HCl ionization.

The outline of this paper is as follows. After summarizing the computational methods in the next section, we discuss the results of the solvent free energies and the proton potentials, focusing on the above issues. A final section concludes. Full details of the calculations and many further aspects of the HCl ionization will be discussed in a more extensive article (*21*).

Computational Methods

Ab initio molecular orbital (MO) methods (*22*) are used (i) to optimize the nuclear geometries of small clusters, (ii) to determine the model potential functions used in Monte Carlo simulations and (iii) to compute the potential energy surfaces as a function of proton coordinates.

We first optimize the geometry of the reaction system **1** surrounded by eight water molecules, shown schematically in Figure 2. The eight external waters constitute the nearest neighbour solvation around the reacting molecules **1** — each of the three (HCl, H_2O_a and H_2O_b) is coordinated by four molecules. The analytic gradient method for the restricted Hartree-Fock (RHF) wavefunction (*23*) is employed with the 3-21G* basis set (*24,25*); the standard 3-21G set was augmented with d-polarization functions with the exponents of 0.8 and 0.75 on O and Cl atoms, respectively. Because of the practical difficulty in the full calculation of the whole system in Figure 2, the geometries of smaller four-hydrated clusters, designated by the thin curves in Figure 2, are optimized. Some hydrogen-bonds are constrained to be linear. For example, for structure **1**, the optimized heavy atom distances $Cl \cdots O_a$ (R_1) and $O_a \cdots O_b$ (R_2), and the Cl-H (r_1) and O_a-H (r_2) bond lengths are 2.96, 2.83, 1.33 and 0.98 Å respectively.

To determine the free energies in the solvent coordinate, the Monte Carlo simulations are carried out using standard procedures with Metropolis sampling, periodic boundary conditions and the canonical (constant NVT) ensemble.(*26*) Each cubic cell contains the reaction system **1**, **2** or **3** and 248 water molecules. The temperature is fixed at 298 K. The box length is 19.6 Å, so that the mass density of the system is 0.997 g/cm³. Each simulation starts with an equilibration run with 5×10^5 (or more) configuration generations. Additional runs are carried out to compute the free energy curves, which consist of several (4-8) sets of 1×10^6 configuration generations. The intermolecular interactions are spherically truncated at a cutoff distance of half the box length, referenced to the center-of-mass distance of each molecule.

Under certain solvent configurations identified below, we calculate the potentials for the proton transfers. These are computed with the 6-31G**(Cl+) basis set

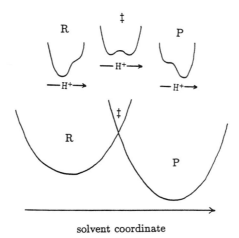

solvent coordinate

Figure 1. Schematic illustrations of (top) the evolution of the proton transfer potential and (bottom) the two (diabatic) free energy curves in the solvent coordinate.

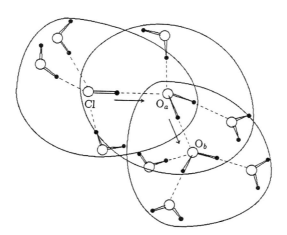

Figure 2. Schematic illustration of the primary reaction system $ClH\cdots OH_2\cdots OH_2$, with eight external waters displayed. See the text for explanation of the light line enclosures.

($24,25$) and the second order Møller-Plesset (MP2) approximation ($27,28$). The exponents of the d-polarization functions on O and Cl atoms are the same as those in the 3-21G* set, and the exponent of the p-polarization function on H atom is 1.0. A set of p-type diffuse functions with the exponent of 0.049 is also placed on the Cl atom to better describe the diffuse anion Cl⁻.(29) It is recognized ($27,28,30$) that the MP2 approximation with this basis set quality (double-zeta plus polarization and diffuse) generally provides reliable ground electronic state energies in systems where the wavefunction is well represented by a Hartree-Fock configuration. This is expected to apply for the present cases considering the small distance of the proton shifts (~ 0.6 Å) in the strongly overlapping hydrogen-bonded complexes. We have checked this by carrying out calculations with the single and double excitations configuration interaction (CI(SD)/3-21G**) method for the proton shift, HCl$\cdots$H$_2$O $\rightarrow$ Cl$^-\cdots$H$_3$O$^+$, with the equilibrium Cl$\cdots$O distance (2.96 Å) at 7 points of the proton displacement, and found that the Hartree-Fock configuration dominates (more than 95 %) at every point.

We determine model potential parameters to be used in the Monte Carlo simulations in the potential form

$$V_{int} = \sum_a \sum_b [\frac{q_a q_b}{r_{ab}} + 4\epsilon_{ab}\{(\frac{\sigma_{ab}}{r_{ab}})^{12} - (\frac{\sigma_{ab}}{r_{ab}})^6\}]. \qquad (2)$$

The electrostatic term here represents Coulomb interaction between point charges on each atomic site. The point charges on each atomic site of external waters are taken from the TIP3P model for water (31). The point charges on the reaction system **1-3** are determined so as to reproduce the electrostatic potentials at ~ 500 points around the trimer **1-3** computed from the electron density matrix of the RHF/6-31G**(Cl+) wavefunction. To include the solvent induced polarization of the solute, RHF/6-31G**(Cl+) wavefunctions of the system **1-3** are recalculated under the influence of eight externals reorganized to each charge distribution of the systems **1-3**. For the ion pair states **2** and **3**, the cluster structures optimized by using the model potential equation 2 with the charge distribution of the isolated **2** and **3** are used. The LJ parameters ϵ and σ in equation 2 are determined so as to approximate the interaction energies and the average heavy atom (Cl$\cdots$O and O$\cdots$O) distances for smaller clusters, HCl(H$_2$O)$_4$, Cl$^-$(H$_2$O)$_4$ and H$_3$O$^+$(H$_2$O)$_3$. The reference energies and geometries to be approximated are taken from MP2/6-31G**(Cl+) energy calculations and RHF/3-21G* geometry optimizations, respectively.

Results and Discussion

We first present the results for the first proton transfer **1→2** in Scheme I. Figure 3a displays the free energy curves in the solvent coordinate

$$\Delta E_{12} = V_1(\mathbf{S}; \mathbf{R}_1) - V_2(\mathbf{S}; \mathbf{R}_2), \qquad (3)$$

where $V_i(\mathbf{S}; \mathbf{R}_i)$ denotes the total potential energy including solute internal, solute-solvent interaction and solvent-solvent interaction energies, as functions of the solvent molecule configurations denoted as $\mathbf{S}$, at fixed solute coordinates $\mathbf{R}_i$. These curves are obtained by the Monte Carlo sampling described above, together with a free energy perturbation method ($32,33$) to access the thermally improbable regions of high free energy.

Solvent configurations corresponding to the intersection $\Delta E_{12} = 0$ in Figure 3a, are then used as configurations for which to generate the potential for the proton transfer in the fixed field of the solvent molecules. Figure 3b shows that this potential is symmetric, and that the calculated (from a one-dimensional Schrödinger equation) ground adiabatic state proton vibrational level is only very slightly below the proton barrier top. In fact it is argued elsewhere (15) that any simulation — such as the present one — which in effect treats the solvent electronic polarization classically and via effective charges on the solvent molecules, will give an <u>overestimate</u> of the proton barrier. Thus the true proton barrier will be lower than that displayed, and the proton vibrational level will be above it. This last feature signals an adiabatic proton transfer.

The calculations above were performed at an O-O distance of 2.83 Å between the two explicit waters in Scheme I. This is the equilibrium distance between two neutral water molecules in $(H_2O)_5$. When the above calculations are repeated at an O-O distance value of 2.56 Å — the H_3O^+-H_2O equilibrium distance in $H_3O^+(H_2O)_3$, the same basic picture emerges.

The picture for the first transfer $\mathbf{1}{\rightarrow}\mathbf{2}$ that emerges is then the following. The acid-base proton transfer is adiabatic, rather than tunneling, with the proton adiabatically following the slow solvent rearrangement to configurations with $\Delta E_{12} = 0$. This rearrangement occurs at a slight cost of free energy. The precise cost to reach the transition state in the solvent coordinate must be calculated as follows. We must determine the vibrational energy level for the proton in the reactant and at the transition state in ΔE_{12}, by solving the Schrödinger equation for the proton in its potential in those ΔE_{12} regions. The results are denoted by the small cross marks in Figure 2a. It is then seen that the process is nearly barrierless, $\Delta G_{12}^{\ddagger} \sim 0.1$ kcal/mol. In a similar way, a reaction free energy of $\Delta G_{12} \sim$-6.7 kcal/mol is estimated (cf. Figure 3a). This is about 70~80 % of the full experimental ($1,34,35$) free energy of ionization -8~-10 kcal/mol. Thus the adiabatic $\mathbf{1}{\rightarrow}\mathbf{2}$ transfer is an almost activationless process in the solvent coordinate and is markedly downhill in free energy. (By contrast, the $\mathbf{1}{\rightarrow}\mathbf{2}$ transfer is calculated to be uphill by $\sim$20 kcal/mol in the gas phase.) Since our estimated simulation uncertainty is only a few kcal/mol (21), we anticipate that the qualitative picture stated above is robust, i.e., the adiabatic first proton transfer is at most nearly activationless and the major part of the reaction exothermicity is gained in this first step.

Figure 4 displays the corresponding pictures for the second proton transfer $\mathbf{2}{\rightarrow}\mathbf{3}$. Here the O-O distance is taken to be 2.56 Å, which is the equilibrium distance for H_3O^+-H_2O and the appropriate solvent coordinate is now

$$\Delta E_{23} = V_2(\mathbf{S}; \mathbf{R}_2) - V_3(\mathbf{S}; \mathbf{R}_3) \tag{4}$$

in analogy to equation 3. It is seen that again the proton transfer is adiabatic. It is found (21) instead to be a nonadiabatic tunneling proton transfer if the O-O

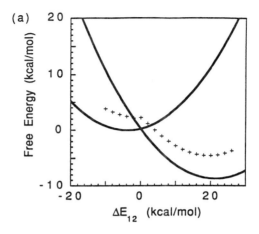

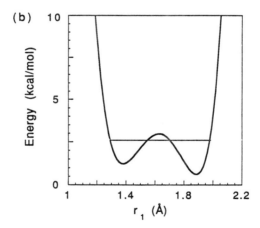

Figure 3. (a) The diabatic free energy curves in the solvent coordinate ΔE_{12}. The ground proton vibrational levels (+++) are shown. (b) Proton transfer potential evaluated at $\Delta E_{12} = 0$, with the ground vibrational level displayed. The variance of the calculated proton barrier height for 5 different independent solvent configuration samplings (all with $\Delta E_{12} = 0$) is 0.1 kcal/mol. (The variance of the average of the forward and reverse barrier heights is ~ 0.01 kcal/mol.)

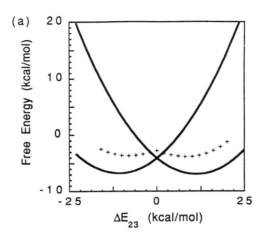

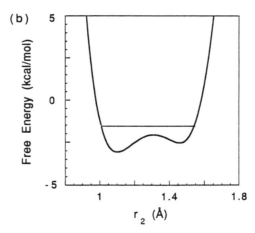

Figure 4. Same as Figure 3, but now for the second proton transfer.

separation is not allowed to "reorganize" from the H_2O-H_2O equilibrium value of 2.83 Å. Thus nuclear reorganization plays a key role in this second transfer.

Now, activation in the solvent is required, and an estimate of the barrier height which takes account of the quantized proton vibrational level is $\Delta G_{23}^{\ddagger} \sim 0.9$ kcal/mol. This second transfer is nearly thermodynamically neutral, with an estimated reaction free energy of ΔG_{23} of only $\sim$-0.2 kcal/mol. The overall experimental reaction free energy of HCl ionization in water is ($1,34,35$) around -8$\sim$-10 kcal/mol. We estimate (21) that the residual rise in free energy on going from **3** to the completely separated ions is only $\sim$0.6 kcal/mol in the high dielectric constant water solvent by a simple dielectric continuum argument.

Finally, the issue of concerted versus stepwise transfer of the protons in Scheme I can be examined by calculation of the free energy curves for the concerted double transfer, now in the solvent coordinate ΔE_{13}. It is found that the activation free energy for this reaction route is $\sim$6.2 kcal/mol (including the energy cost for O-O compression), which is noticeably higher than that for the stepwise transfers discussed previously. This strong bias against a concerted transfer can be qualitatively understood (21) in terms of the larger reorganization free energy for the solvent required for the large charge separation in **1**→**3**, versus the smaller reorganization costs for the less drastic charge separations in **1**→**2** and **2**→**3**.

Concluding Remarks

We have outlined here the methods and results of a study of the acid ionization of HCl in water. More extensive description, analysis and discussion will be presented elsewhere (21). The picture which emerges is that the HCl ionization is a consecutive set of adiabatic, non-tunneling, proton transfers. The first transfer involves an almost activationless solvent reorganization to "contact" Cl^--H_3O^+ ion pair, with an accompanying reaction free energy of $\sim$-6.7 kcal/mol. The second transfer is slightly activated in a solvent coordinate and is approximately thermoneutral. While no simulations have been performed for the remaining passage to the fully separated ions, a dielectric continuum estimate suggests that that process involves only a very small overall free energy change. We expect, however, that there can be small barriers in a solvent coordinate — much like that in the second step **2**→**3** above — along the pathway, as well as nuclear reorganization of O-O distances to assist the adiabatic proton transfer.

Since the solvent barriers for the two steps are either negligible or small, there could well be considerable barrier recrossing effects which would need to be taken into account in reaction rate constants (36-39). This feature would not, however, alter the energetic pathways above.

The considerations outlined here can also be employed to study the issue of HCl ionization at ice surfaces in connection with ozone depletion ($40,41$) and to examine other acid ionizations, e.g. that of HF, in water.

Acknowledgments

This work was supported in part by an NIH Shannon Award, and by grants NSF CHE88-07852 and NSF CHE93-12267. K.A. was partly supported by a Fellowship of the Japan Society for the Promotion of Science for Japanese Junior Scientists. We thank Dr. A. Staib for fruitful discussions.

Literature Cited

(1) Bell, R. P. *The Proton in Chemistry*; Chapman and Hall: London, 1973.
(2) *Proton Transfer Reactions*; Caldin, E. F.; Gold, V., Eds.; Chapman and Hall: London, 1975.
(3) Ratajczak, H. In *Electron and Proton Transfer Processes in Chemistry and Biology*; Müller, A.; Ratajeczak, H.; Junge, W.; Diemann, E., Eds.; Elsevier: Amsterdam, 1992.
(4) van Duijnen, P. Th. *Enzyme* **1986**, *36*, 93.
(5) Jencks, W. P. *Catalysis in Chemistry and Enzymology*; McGraw-Hill: New York, 1969.
(6) Eigen, M. *Angew. Chem. (Int. Ed. Engl.)* **1964**, *3*, 1.
(7) Albery, W. J. *Progr. React. Kinet.* **1967**, *4*, 353.
(8) Legon, A. C.; Willoughby, L. G. *Chem. Phys. Lett.* **1983**, *95*, 449.
(9) Bacskay, G. B.; *Mol. Phys.* **1992**, *77*, 61.
(10) Bacskay, G. B.; Kerdraon, D. I.; Hush, N. S. *Chem. Phys.*, **1990**, *144*, 53.
(11) Borgis, D.; Lee, S; Hynes, J. T. *Chem. Phys. Lett.* **1989**, *162*, 19.
(12) Borgis, D.; Hynes, J. T. *J. Chem. Phys.* **1991**, *94*, 3619.
(13) Warshel, A.; Weiss, R. M. *J. Phys. Chem.* **1980**, *102*, 6218.
(14) Warshel, A.; Russel, S. *J. Amer. Chem. Soc.* **1986**, *108*, 6569.
(15) Staib, A.; Borgis, D.; Hynes, J. T. "Proton Transfer in Hydrogen-Bonded Acid-Base Complexes in Polar Solvents", submitted to *J. Chem. Phys.*
(16) Marcus, R. A. *Ann. Rev. Phys. Chem.* **1964**, *15*, 155.
(17) Ulstrup, J. *Charge Transfer Processes in Condensed Media*; Springer-Verlag: Berlin, 1979.
(18) Timoneda, J. J. i.; Hynes, J. T. *J. Phys. Chem.* **1991**, *95*, 10431.
(19) Halle, B.; Karlstrom, G. *J. Chem. Soc., Faraday Trans.2* **1983**, *79*, 1047.
(20) Newton, M. D. *J. Chem. Phys.* **1977**, *67*, 5535.
(21) Ando, K.; Hynes, J. T. to be submitted.
(22) Dupuis, M.; Watts, J. D.; Villar, H. O.; Hurst, G. J. B. HONDO Ver. 7.0, *QCPE* **1987**, *544.*
(23) Spangler, D.; Williams, I. H.; Maggiora, G. *J. Comp. Chem.* **1983**, *4*, 524.
(24) Binkley, J. S.; Pople, J. A.; Hehre, W. J. *J. Amer. Chem. Soc.* **1980**, *102*, 939.
(25) Hehre, W. J.; Ditchfield, R.; Pople, J. A. *J. Chem. Phys.* **1972**, *56*, 2252.
(26) Allen, M. P.; Tidesley, D. J. *Computer Simulation of Liquids*; Clarendon: Oxford, 1987.
(27) Møller, C.; Plesset, M. S. *Phys. Rev.* **1934**, *45*, 618.
(28) Krishnan, R.; Frisch, M. J.; Pople, J. A. *J. Chem. Phys.* **1980**, *72*, 4244.

(29) Dunning, T. H.; Hay, P. J. In *Methods of Electronic Structure Theory*, Schaefer III, H. F., Ed.; Plenum: New York, 1977.

(30) Urban, M.; Cernusak, I.; Kellö, V.; Noga, J. In *Methods in Computational Chemistry*, Wilson, S., Ed.; Plenum: New York, 1987.

(31) Jorgensen, W. L.; Chandrasekhar, J.; Madura, J. D.; Impey, R. W.; Klein, M. L. *J. Chem. Phys.* **1983**, *79*, 926.

(32) King, G.; Warshel, A. *J. Chem. Phys.* **1990**, *93*, 8682.

(33) Ando, K.; Kato, S. *J. Chem. Phys.* **1991**, *95*, 5966.

(34) Ebert, L. *Naturwiss* **1925**, *13*, 393.

(35) Robinson, R. A. *Trans. Faraday Soc.* **1936**, *32*, 743.

(36) Hynes, J. T. In *The Theory of Chemical Reaction Dynamics*; Baer, M., Ed.; CRC: Boca Raton, 1985, Vol. 4.

(37) Truhlar, D. G.; Hase, W. L.; Hynes, J. T. *J. Phys. Chem.* **1983**, *87*, 2664.

(38) Smith, B. B.; Staib, A.; Hynes, J. T. *Chem. Phys.* **1993**, *176*, 521.

(39) Fonseca, T.; Ladanyi, B. M. In *Ultrafast Reaction Dynamics and Solvent Effects: Experimental and Theoretical Aspects*; Gauduel, Y.; Rossky, P. J., Eds.; AIP: New York, 1994.

(40) Hanson, D. R.; Ravishankara, A. R. *J. Phys. Chem.* **1992**, *96*, 2682.

(41) Kroes, G. -J.; Clary, D. C. *J. Phys. Chem.* **1992**, *96*, 7079.

RECEIVED May 2, 1994

Chapter 11

Water-Assisted Reactions in Aqueous Solution

Jean-Louis Rivail, Serge Antonczak, Christophe Chipot,
Manuel F. Ruiz-López, and Leonid G. Gorb[1]

Laboratoire de Chimie Théorique, Unité de Recherche Associée
au Centre National de la Recherche Scientifique 510, Université
de Nancy 1, B.P. 239, 54506 Vandoeuvre-lès-Nancy Cedex, France

The results of ab initio computations including MP2 intramolecular
electron correlation corrections on two chemical processes, amide
hydolysis and ion pair formation in hydrochloric acid, are
reported. The influence of water is simulated by considering a
supermolecular system with one and two water molecules and the
effect of the bulk solvent is modeled by a *continuum* surrounding
the solute.
A mechanism involving a water dimer in the amide hydrolysis
appears to be energetically favored in the case of both the neutral
and the acid-catalyzed reaction. This mechanism can be interpreted
as a water-assisted reaction in which the second water molecule is
considered as an ancillary molecule (or equivalently as a
bifunctional catalytic process).
Using the same model, the ionic dissociation of HCl appears to be
impossible when a single water molecule is involved to solvate the
proton which is expected to be part of the process. Conversely,
solvating the proton, by a water dimer leads to an easy dissociation
in a medium having a dielectric permittivity greater than 15.
The particular role of the water dimer in an aqueous solution is
discussed in the present contribution, and some general
conclusions are proposed.

Aqueous solutions, which are probably the most widely studied chemical systems,
constitutes a fascinating subject for chemists. The reason for this special interest is
partly due to the fact that liquid water is an essential component of the physical world
and that it plays a crucial role in life, but it also comes from the unique properties of the
water molecule.
Among the amazing properties of water, one notices that, in the molecule, three atoms
have very marked and complementary properties: the two hydrogen atoms are electron
deficient to a non-negligible extent, which enables them to form strong hydrogen bonds
with atoms bearing electron lone pairs. This is actually the case of the oxygen atom
possessing two lone pairs which are very efficient in hydrogen bond formation and
cation solvation *(1)*, in particular, the H^+ cation coming from the ionic dissociation of
another water molecule. Through such interactions, a proton can easily jump from one

[1]Current address: Institute of Colloid and Water Chemistry, Ukrainian Academy of
Sciences of Ukraine, 252142 Kiev, 142 Ukraine

water molecule to another one, which in turn, can give one proton to another neighbor. A water molecule can, therefore, act as a relay in proton transfer phenomena. Finally, the large dipole moment of the molecule, which is even increased by hydrogen bond formation, gives rise to a high dielectric constant modifying to a large extent the structure and energetics of the chemical species present in the solution, in comparison to what can be expected for these species in a gaseous state.

All these remarks lead us to the conclusion that it is not wise to analyze the chemical process in solution by only considering the interaction of the solute with one water molecule. The other molecules of the solution may play a key-role either directly, by acting specifically as an intermediate in the reaction, or indirectly through the modifications of the electrostatic interactions due to the dielectric permittivity of the solvent.

Specific chemical interactions may be observed in chemical processes in which the presence of an extra (or ancillary) water molecule that does not enter the stoichiometry of the process, but modifies the reaction path. Such processes have been analyzed in terms of bifunctional catalysis *(2)*. Typically, these phenomena can be simulated by means of quantum chemical computations on clusters consisting of the solute and a given number of water molecules (usually one or two). The role of the water molecules in the bulk can be investigated by introducing the electrostatic interactions between the cluster and the solvent in the quantum chemical computation.

Methodology

The quantum chemical computations have been carried out using the conventional procedures and the standard basis sets available in the Gaussian92 *(3)* ab initio package. In most cases, the energy values discussed in this paper were obtained employing the 6-31G** basis set ; the second-(MP2) and occasionally third-(MP3), order Møller-Plesset theory was used to assess the influence of intramolecular electron correlation. In the case of small systems, a full geometry optimization was performed with these energy data. However, for larger systems, the optimization was carried out at the Hartree Fock level of approximation using the 3-21G basis set, and the energy was computed on the optimized structure by means of the Self-Consistent Reaction Field (SCRF) approach in which the solvent is represented by a *continuum* characterized by its dielectric permittivity ε ($\varepsilon=78.5$ for water at 25°C) and the solute is embedded in a deformable cavity adapted to its shape. The solute-solvent interactions are evaluated by means of the reaction field factors *(4-6)*, and the charge distribution of the solute is expanded in a multipole series up to order 6. This method allows us to carry out all the usual computations of quantum chemistry on a solvated molecule, including Møller-Plesset perturbation calculations *(7-8)*, with a cavity having a shape adapted to the geometry of the solute. Nevertheless, in order to reduce the computer time, we sometimes used the ellipsoidal approximation *(9)* for the cavity, in particular for geometry optimization.

Such an approach provides us with the equilibrium structure of the molecule in the solvent. The computed energy corresponds to the sum of the molecular energy and the electrostatic and induction contributions to the solute-solvent free energy of interaction *(10-11)*. In order to evaluate the total free energy of interaction, it is necessary to include two additional terms. The first term, namely the cavitation free energy, represents the energy spent to create the cavity in the solvent. This term can be obtained from several empirical formulae. In this work, we have adapted the formula proposed by Tuñon *et al. (12)*. The second term corresponds to the dispersion energy. It has been shown that when it is incorporated in the SCF computation, this term does not modify the equilibrium properties of the solute substantially *(13)*. In particular, it does not vary much when the geometry of the solute changes. For this reason, we have decided to drop it in this study. This approximation should be kept in mind when one discusses the energy variations along a reaction path and thinks of the possible variations of the dispersion contribution.

The Water Dimer in Aqueous Solution.

The properties of each individual molecule in liquid water are expected to be different from those of single isolated molecules because of the molecular interactions with the rest of the liquid. As usual, these interactions may be classified into (i) short-range interactions, which are depicted as intermolecular hydrogen bonds and, (ii) long-range interactions, which are, in the present approach, assimilated to the effects of the polarizable *continuum*.

The effects of molecular association can be analyzed by comparing the properties of a water dimer and those of the isolated monomer. This basic problem has been studied thoroughly and, recently, some accurate quantum chemical results have been published *(14-15)*. The influence of the liquid surroundings on these species has also been investigated *(16)*. This study has been carried out using the 6-311++G(2d,2p) basis set, at both the HF and the MP2 levels of approximation, even for the geometry optimizations within the framework of SCRF computations in the ellipsoidal approximation.

In order to distinguish the two kinds of effects that a given water molecule experience from the other molecules pertaining to the liquid, we shall briefly summarize the results of the study, in which we successively analyze the influence of a single H-bond formation in a water dimer, and that of the surroundings on a single molecule, as well as on its dimer. The modifications of the chemical properties of each molecule are followed by the variations of the Mulliken net atomic charges *(17)*. Although the partitioning of the electron density proposed by Mulliken has been criticized *(18-19)*, especially for the hydrogen atoms, it is commonly admitted that the variations of these charges is chemically meaningful and may be used to anticipate a variation of a chemical property depending on local charges.

Influence of Hydrogen Bonding. In the isolated state, one linear dimer is found and will be entitled dimer I in this study. However, a second structure (or cis-dimer) has been recognized as an energy minimum in the liquid state. It corresponds to a transition state in the gas phase and will be considered here under the name of dimer II (Figure 1). The geometries of these structures, optimized at the MP2 level, are available in ref. *(16)* and the SCF Mulliken charges are reported in Table I

Table I. Mulliken charges in each water molecule under various association states

	H_1	O_1	H_2	O_2	H_3 H_4	Δq	$\mu(D)$
Monomer	0.2410	-0.4819	0.2410	-0.4819	0.2410	-	2.063
Dimer I	0.2408	-0.5608	0.3177	-0.5230	0.2626	0.0023	2.848
Dimer II	0.2337	-0.5593	0.3248	-0.5275	0.2641	0.0008	4.375

Apart from the variations of the properties of atoms H_2 and O_2 involved in the hydrogen-bond formation, the most noticeable result of this comparison is the strong variations of the charge borne by atom O_1, which becomes more negative, and by those of atoms H_3 and H_4, which become more positive. One is, therefore, led to consider that the result of the dimerization is an increase of the basic properties of molecule 1 and an increase of the acidic properties of molecule 2, regardless of the conformation of the dimer.

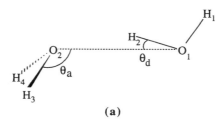

(a)

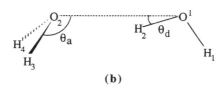

(b)

Figure 1: The water dimer
a) dimer I
b) dimer II

(Adapted from reference 16)

Influence of the Surroundings. The general effect of a liquid surroundings on a molecule is the polarization that usually increases the molecular electric moments. This effect can be analyzed through the differences in the Mulliken populations, which usually increase under the influence of the solvent. This is visible in the case of a single water molecule as shown on Table II.

Table II: Mulliken charges of oxygen (q_O) and hydrogen (q_H) and dipole moment (μ) of the water monomer in a *vacuum* ($\varepsilon=1$) and in water ($\varepsilon=78$)

	$\varepsilon=1$	$\varepsilon=78$	variation
q_O	-0.4819	-0.5860	-0.1041
q_H	0.2410	0.2930	+0.0520
μ(D)	2.063	2.382	0.319

The modifications of the electronic properties of the molecules are quite substantial. Qualitatively, the effects have many analogies with those intervening in H-bond formation, and one notices that the predicted variations of both the basic properties of oxygen and the acidic properties of hydrogen are even more pronounced than the largest variations anticipated from the analysis of the dimers.

Joined Effects of Hydrogen Bonding and Solvation. In order to have a global view of the possible modifications of the chemical properties of a water molecule under the joined influence of short- and long-range interactions, we now analyze the variations of the Mulliken charges between a single isolated water molecule and a solvated dimer. The results relative to the dimers I and II are given in Table III.

Table III: Variation of the Mulliken electron population due to the joined effects of hydrogen bonding and solvation

Atoms	H_1	O_1	H_2	O_2	H_3 H_4
Dimer I	+0.0701	-0.1832	+0.1145	-0.1449	+0.0776
Dimer II	+0.0714	-0.1955	+0.1136	-0.1472	+0.0787

This table shows that the combined effects increase further the contrast between the charges borne by the oxygen and the hydrogen atoms. If one discards the hydrogen bonded atoms O_2 and H_2 which are less available for reacting (although O_2 has a second lone pair), we notice that the basic or nucleophilic properties of the oxygen O_1 increases noticeably, as well as the acidic or electrophilic properties of the hydrogen atoms (especially H_3 and H_4, but also H_1 which is almost not modified in the isolated dimer).

The differences between dimer I and dimer II are negligible. Dimer II undergoes a stronger solvent effect due to its larger dipole moment which is also responsible for the fact that its conformation, in the liquid state, corresponds to an energy minimum which is only 0.02 kcal.mol^{-1} above dimer I, both at the HF or MP2 level.

The conclusion of this analysis is that the chemical properties of water may be very different in the liquid state and in the isolated unassociated state, which is often the only structure considered in quantum chemical computations.

These modifications are multiple. In addition to the enhancement of the nucleophilic and electrophilic properties of oxygen and hydrogen atoms, especially those which are not part of hydrogen bonds, one notices that the dimer appears to be quite deformable. In other words, the pair O_1-H_3 or H_4, is able to react with a pair of electrophilic and nucleophilic sites in a partner molecule, even if the relative position of these sites is not favorable for a reaction with a single water molecule. After the reaction, a water molecule may be regenerated from the dimer. This is the reason why one sometimes talks about ancillary molecules, or of bifunctional catalysis *(2)*.

Hydrolysis of Amides *(20)*

This reaction is a model of the chemical hydrolysis of peptides. In this case, it is known to be catalyzed by several proteolytic enzymes, but the reaction may also occur in aqueous solutions, especially in acidic media.

The overall stoichiometry of the reaction may be written as shown in Scheme 1.

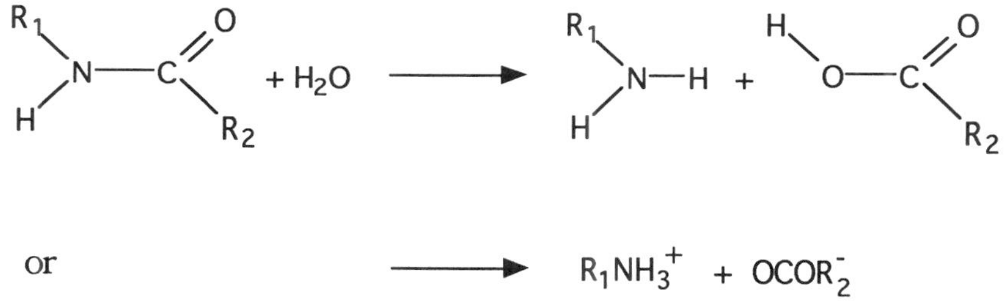

Scheme 1

Where only one water molecule enters the reaction stoichiometry.

In this study, we carefully analyze the simplest case in which $R_1=R_2=H$ (formamide), using the theoretical methodology summarized above.

The dissociative addition of a single water molecule to formamide (FA) has been studied previously in the isolated state by Bader *et al. (21)*. These authors found a transition state 42.0 kcal.mol^{-1} above the reactants.

In the present study, we consider an alternative process in which a water dimer is added to the amide molecule. These two processes are then investigated in the case of a reaction catalyzed by an acidic medium. Following the results of the earlier study by Bader *et al. (21)*, the presence of an excess of protons in the solution is simulated considering the complex formed by a hydronium H_3O^+ ion interacting with the oxygen atom of the amide. These reactions are first considered in the isolated state. The influence of the bulk water represented by a *continuum* will be examined afterwards.

The computations have been carried out at the MP2 or MP3/6-31G**//3-21G level of approximation. In addition to the usual energy calculations, the free energies have been estimated by computing the vibrational contribution to the partition function.

Are There One or Two Water Molecules Directly Involved in the Process? We report in Table IV the energy variations $\Delta E^{\neq}$ and the free energy variations $\Delta F^{\neq}$ between the transition state and the reactants for these two processes, in the case in which the molecular species are assumed to react *in vacuo*. We consider the neutral (uncatalyzed) reaction and the hydronium-catalyzed one. These results correspond to the effects due to intermolecular MP3 computation of electron correlation. They, therefore, slightly depart from the MP2 data reported earlier *(21)*.

Table IV: Energy barrier ($\Delta E^{\neq}$) and free energy barrier ($\Delta F^{\neq}$) (in kcal.mol^{-1}) of different processes of the hydration of formamide (FA). (Energies computed at the MP3/6-31G**//3-21G level)

	$\Delta E^{\neq}$		$\Delta F^{\neq}$	
reactants	H_2O + FA	$(H_2O)_2$ + FA	H_2O + FA	$(H_2O)_2$ + FA
neutral reaction	48.3	33.0	59.3	55.5
H_3O^+-catalyzed reaction	26.4	-3.8	37.2	20.1

This table shows that the assistance of a second water molecule lowers the energy barrier. The difference is less marked if one considers the free energy variations. This is due to the fact that the decrease of the entropy when passing from the reactants to the transition state is greater in the case of the water-assisted process.

If one considers the H_3O^+-catalyzed reaction, which is the most realistic one, it is clear that the assistance of a second water molecule is very favorable. One may then easily imagine that, in an aqueous solution, the probability for a pair of water molecule to react is quite large, and that the process is very likely to involve such a dimer. The structure of the transition state and of the reaction intermediates I_1, on the reactant side, and I_2, on the product side, are schematized on Figure 2.

Role of the Surroundings. The presence of a *continuum* does not greatly modify the conclusions emerging from the study carried out for the isolated species. For the water-assisted neutral reaction, the energy barrier is increased by an amount of 1.8 kcal.mol^{-1}. The most noticeable effect of the solvent appears on the reaction coordinate. This effect is visible on Table V, which indicates the main changes for the transition state of the neutral, water-assisted reaction in a gaseous state ($\varepsilon=1$), and in the solution ($\varepsilon=78.4$). The atoms are numbered according to Figure 2.

Table V: Main interatomic distances in the transition state for the water-assisted neutral molecule in a *vacuum* ($\varepsilon=1$) and in water ($\varepsilon=78.4$)

bondlengths	CN	CO_1	CO_2	O_2H_1	O_3H_1	O_3H_2	NH_2
$\varepsilon=1$	1.572	1.239	1.674	1.205	1.190	1.190	1.298
$\varepsilon=78.4$	1.530	1.273	1.595	1.293	1.120	1.098	1.460

The CN and CO_2 bondlengths are substantially shorter in solution, *i.e.*, the cleavage of the bond is retarded, although the nucleophilic attack on the carbon atom is favored. Conversely, the NH_2 bond is longer in solution. The same tendency has been pointed out in the case of the catalyzed reaction.

These modifications indicate a strong coupling between the reacting molecule and the surroundings which may influence the reaction kinetics to a larger extent that one would anticipate from the above energetics considerations.

One, therefore, concludes that the water-assisted process has to be considered as a good candidate to explain the hydrolysis of amides. In this process, the OH group that is bonded to the carbon atom to generate a carboxylic acid comes from one water molecule, and the hydrogen that is fixed to the nitrogen atom comes from another one. The two remaining moieties recombine to produce a water molecule, so that the reaction balance does not depend on the presence of the second water molecule. The reaction appears to proceed through an intermediate product which can be described as a cluster

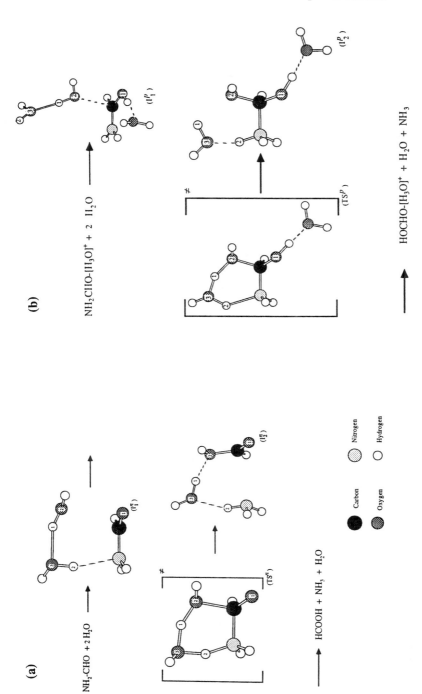

Figure 2: Reaction path of the water-assisted formamide hydrolysis reaction *in vacuo*
 a) neutral reaction
 b) H_3O^+ catalyzed reaction

(Reproduced from reference 20. Copyright 1994 American Chemical Society)

formed by an ammonia molecule and a formic acid bridged to the extra water molecule, in the case of the neutral reaction. In the case of the acid-catalysed reaction, the greater electrophilicity of the carbon atom, induced by the protonation of the carbonyl group, favors a structure which is closer to the tetrahedral intermediate, as shown on Figure 2.

This multicenter mechanism is not too surprising. Enzymatic hydrolysis of peptides, especially in serine proteases is also analyzed as a multicenter process in which the functionnal chains present in the active site play a role similar to the second water molecule *(22)*.

Ionic Dissociation of Hydrogen Chloride in Water *(23)*

Most of the common textbooks explain the ionic dissociation of hydracids in water by the high dielectric permittivity of the solvent. Specific solvation processes are considered too, and among them the solvation of the proton is usually assumed to produce the H_3O^+ hydronium ion. Nevertheless, on the basis of vibrational spectroscopy data, some authors have postulated that a second water molecule is directly involved in the process leading to the $(H_5O_2)^+$ protonic species *(24-26)*. In order to study the influence of various factors on the formation of the ion pair $H_3O^+Cl^-$ we report the results of computations performed at the MP2/6-31+G** level. This study analyzes the influence of one and two water molecules on the formation of the ion pair, in the isolated state and in a solvent. The structure of the species on which the computations have been performed is shown schematically on Figure 3.

HCl Interacting With One Water Molecule. In the gaseous state, the two molecules form a complex which can be depicted by considering a hydrogen bond between the hydrogen atom of HCl and a lone pair from H_2O.

The presence of a water-like *continuum* around this complex does not modify our conclusions substantially. The potential energy surface exhibits only one minimum (Figure 4) and the complex is closely related to the gas phase ($\varepsilon=1$) structure (Table VI).

Table VI: Geometric and energetic characteristics of the H_2O-HCl complex (atom numbers as on Figure 3)

	d_{Cl-H4} (Å)	d_{H4-O2} (Å)	< H4-O2-H1 (degree)	energy (a.u.)
$\varepsilon=1$	1.282	1.903	145.7	-536.46532
$\varepsilon=78.4$	1.289	1.847	134.5	-536.48072

Note: the ion pair $H_3O^+Cl^-$ has not been observed.

The System $(H_2O)_2$HCl. In the gaseous state, HCl again forms a hydrogen-bonded complex with the water dimer, but now, in a medium with a high dielectric constant ($\varepsilon>15$), our computations indicate a dissociation of the $H_5O_2^+Cl^-$ ion pair which is manifested by the fact that the Cl-H bondlength increases continuously, hence preventing a geometry optimization process to converge.

At the other end of the solvent polarity scale ($\varepsilon=2$), the hydrogen-bonded complex is again the only stable structure.

In order to study the influence of the surroundings, we have varied the dielectric permittivity of the *continuum*. For $\varepsilon=10$, two local minima appear on the potential

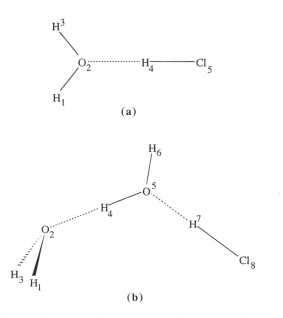

(a)

(b)

Figure 3: mono- (a) and dihydrated (b) complexes of hydrogen chloride.

(Reproduced from reference 23. Copyright 1994 American Chemical Society)

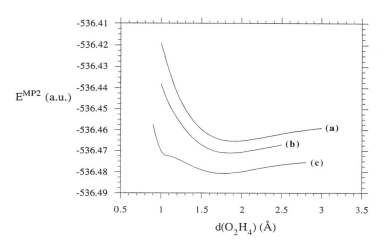

E^{MP2} (a.u.)

$d(O_2H_4)$ (Å)

Figure 4: Potential energy curve of the H_2O-HCl complex.

    a) *in vacuo*

    b) in a non polar medium ($\varepsilon=2$)

    c) in water ($\varepsilon=78.3$)

(Reproduced from reference 23. Copyright 1994 American Chemical Society)

energy surface: one corresponding to the hydrogen-bonded complex, and the other one to the expected ion pair, 1.2 kcal.mol^{-1} lower than the first one (see Table VII).

Table VII: Geometric and energetic features of the $(H_2O)_2HCl$ complex

	d_{Cl-H7} (Å)	d_{H7-O5} (Å)	d_{O2-H4} (Å)	energy (a.u.)
$\varepsilon=1$	1.291	1.809	1.860	-612.71491
$\varepsilon=10$	1.305	1.758	1.950	-612.74499
	2.085	0.994	1.408	-612.74685

These results clearly indicate that the modification of the basicity of the oxygen atom of the first water molecule, under the influence of dimerization, obviously plays an important role. This is visible on the results obtained in the gas phase, where HCl is bonded to a single water molecule; the Cl-H bondlength is shorter than when HCl interacts with the dimer (1.282 Å vs 1.291 Å). In addition, the length of the hydrogen bond decreases accordingly (1.903 Å vs 1.809 Å). These results are in agreement with a stronger interaction with the dimer than with the monomer. Nevertheless, this interaction per se is not strong enough to dissociate the HCl molecule. The influence of the dielectric constant then becomes crucial to allow this chemical modification.

When the dielectric constant of the solvent is increased, this ion pair is stabilized. From $\varepsilon=10$ to $\varepsilon=15$, the ionic complex is, as expected, further stabilized, leading to a very flat minimum which preludes the ionic dissociation. For this modification of the permittivity, the main structural change is an increase of the Cl-H bondlength from 2.085 Å to 2.178 Å. The H_7-O_5 bond is slightly shorter (0.991 Å) and the H_4-O_2 bond is almost unchanged (1.409 Å). The Mulliken net charge of the chlorine atom reaches -0.954, hence confirming the ionic nature of the system. This ion pair can be interpreted as a contact ion pair between Cl^- and a solvated H_3O^+ cation or a $H_5O_2^+$ polarized by the anion. The departure of the proton is expected to imply a second activated process which would be a jump of proton H_4 from the neighborhood of oxygen O_5 to the neighborhood of another oxygen atom like O_2, thus giving rise to a Cl^-/H_3O^+ ion pair separated by one water molecule.

In brief, the ionic dissociation of HCl appears to require at least one pair of water molecules, in accordance to Librovich's proposal (24-26). The resulting proton can then be present in solution in presumably various transient solvated forms which differ from each other by the number of water molecules (27).

Conclusion

We have focused our attention on two typical reactions occuring in an aqueous solution: hydrolysis and ionic dissociation. Both examples emphasize the prominent role of the water dimer which appears to be very important chemical species among the many possible multimers which may be considered as individual components of liquid water (28). The results reported here confirm the reasoning proposed in part III after having analyzed the properties of the liquid dimer in solution. In particular, the dimerization magnifies the acid-base properties of the water molecule and gives rise to a deformable species which reduces the possible geometric constraints occuring when the reaction requires a positively charged hydrogen atom and a nucleophilic oxygen atom simultaneously. This explains the efficiency of bifunctional catalytic (or water-assisted) hydrolysis. In the example that we have considered here, the role of the surroundings appears to be less decisive, which may probably be a general trend in such reactions which do not involve large charge transfers. One must, therefore, keep in mind that the observed effects, which are quite important when one considers energy variations, are less marked on free energy differences because of the low entropy associated to the transition state.

In contrast with what one have might expected in the case of the ionic dissociation of hydrogen chloride, the influence of the high dielectric permittivity of the medium, which is here very important, is not the only factor that influences the process. Again, the strong interaction between the oxygen atom of the water molecule and the hydrogen atom of hydrogen chloride plays such a role that a single water molecule appears not to be basic enough, and that a dimer is required for the dissociation to occur.

A question then arises immediately: what about higher multimers, and in particular, trimers which are expected to occur in liquid water? Two kinds of trimers can be considered: a linear trimer or a bifurcated one, as represented on Figure 5.

In the case of bifunctional catalysis, these species are expected not to be very important because the modifications of the atomic charges in the linear trimer are not greater than in the dimer, the charge transfer occuring mainly between the two terminal molecules (29). In addition the entropy variations are expected to be greater than for the dimer. As for the bifurcated trimer, the basicity or proton affinity of the central oxygen atom is known to be increased by an electron transfer from the two hydrogen-bonded molecules, but the resulting positive charge is shared between the four terminal hydrogen atoms.

This interesting property of the bifurcated trimer may become of greater importance when only the properties of oxygen are involved in the process. This is the case of the ionic dissociation of Brønsted acids. Indeed, the complex $(H_2O)_3HCl$ exhibits, in the case of the bifurcated trimer, all the features of a stronger interaction. At the same computational level as for the dimer, carried out in a *vacuum*, the optimized geometry

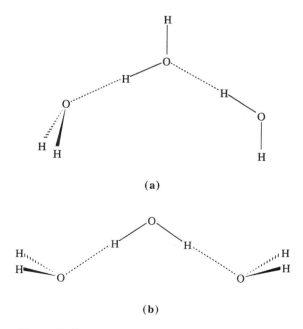

(a)

(b)

Figure 5 : linear (a) and bifurcated (b) water trimers.

of the complex between HCl and the bifurcated dimer shows a longer H-Cl bond: 1.300 Å instead of 1.291 Å and a shorter O-H bond: 1.732 Å instead of 1.809 Å in the case of the $(H_2O)_2HCl$ complex (Chipot, C. unpublished results). The assistance of two water molecules, or more, may, therefore, become crucial for the ionic dissociation of weaker acids (30-32). These examples show that, in aqueous solution, the number of water molecules chemically involved in a reaction (i.e. participating to the exchange of electrons) may be larger than one would anticipate, just by considering the stoichiometry of the process.

In addition, the quantum chemical computations have to take the electron correlation into account, since the Hartree-Fock approximation favors the ionic structures and may lead, as in the H_2O-HCl complex to spurious minima.

This, of course, may complicate to a large extent a realistic simulation of reactions in aqueous media. The approximation which consists of considering the solvent as a *continuum*, although it has some obvious limitations, allow us to undertake these sophisticated computations on a rather small subsystem. It is, therefore, still a very useful tool for exploring a rather complex reality.

Literature Cited

(1) Clementi, E.; *Determination of Liquid Water Structure*; Lecture notes in Chemistry 2,. Springer Verlag, Heidelberg **1976**.

(2) Lledós, A. and Bertrán, J. *Tetrahedron Lett.* **1981**, 22, 775; Ventura, N.; Coitiño E. L.; Lledós, A. and Bertrán, J. *J. Comp. Chem.* **1992**, 13, 1037.

(3) Frisch, M.J.; Trucks, G.W.; Head-Gordon, M.; Gill, P.M.W.; Wrong, M.W.; Foresman, J.B.; Johnson, B.G.; Schlegel, H.B.; Robb, M.A.; Reprogle, E.S.; Gomperts, R.; Andres, J.L.; Raghavachari, K.; Binkley, J.S.; Gonzales, C.; Martin, R.L.; Fox, D.J.; Defrees, D.J.; Baker, J.; Stewart, J.J.P.; Pople, J.A.; *Gaussian 92*, Carnegie-Mellon Quantum Chemistry Publishing Unit, Pittsburgh, PA, **1992**.

(4) Rinaldi, D.; Rivail, J.-L. *Thero. Chim. Acta* **1976**, 32, 57; Rinaldi, D.; Rivail, J.-L. Chem. Phys **1976**, 18, 233.

(5) Rivail, J.-L.; Terryn, B. *J. Chim. Phys.* **1982**, 79, 1; Rinaldi, D.; Ruiz-López, M. F.; Rivail, J.-L. *J. Chem. Phys.* **1983**, 78, 834.

(6) Dillet, V.; Angyán, J. G.; Rinaldi, D.; Rivail, J.-L. *Chem. Phys. Lett.* **1993**, 202, 18; Dillet, V.; Rinaldi, D.; Rivail, J.-L. *J. Phys. Chem.* **1994**, 98, 5034.

(7) Rivail, J.-L. *C.R. Acad. Sci. Paris*, **1990**, 311, 307.

(8) Chipot, C.; Rinaldi, D.; Rivail, J.-L. *Chem. Phys Lett.* **1992**, 191, 287.

(9) Rinaldi, D.; Rivail, J.-L.; Rguini, N. *J. Comp. Chem.* **1992**, 13, 675.

(10) Buckingham, A. D. In *Intermolecular Interactions, from Diatomics to Biopolymers*, Pullman, B, .Ed.; Wiley, J.: Chichester, **1978**; p1.

(11) Claverie, P. In *Intermolecular Interactions, from Diatomics to Biopolymers*, Pullman, B, .Ed.; Wiley, J.: Chichester, **1978**; p. 69.

(12) Tuñon, I.; Silla, E.; Pascual-Ahuir, J. L. *Chem. Phys. Lett.* **1993** 203, 289.

(13) Costa-Cabral, B. J.; Rinaldi, D.; Rivail, J.-L. *C.R. Acad. Sci. Paris 16*, **1984**, p675; Rinaldi, D.; Costa-Cabral, B. J.; Rivail, J.-L. *Chem Phys. Lett.* 125.

(14) Frisch, M. J.; Del Bene, J. E.; Binkley, J. S.; Schaefer III, H. J. *J. Chem. Phys.* **1986**, 84, 2279.

(15) Chakravorty, S. J.; Davidson, E. J. *J. Phys. Chem.* **1993**, 97, 6373.

(16) Bertrán, J.; Ruiz-López, M. F.; Rivail, J.-L. *Theor. Chim. Acta.* **1992**, 84, 181.

(17) Mulliken, R.S., *J. Chem. Phys.* **1955**, 23, 1833

(18) Williams, D.E., Yan, J. M., *Adv. At. Mol. Phys.* **1988**, 23, 87.

(19) Chipot, C.; Maigret, B.; Rivail, J.-L.; Scheraga, H. A., *J. Phys. Chem.* **1992**, 96, 10276.

(20) Antonczak, S.; Ruiz-López, M. F.; and Rivail, J.-L. *J. Am. Chem. Soc.* **1994**, 116, 3912.

(21) Krug, J.P.; Popelier, P.L.A.; Bader, R.F.W. *J. Phys. Chem.* **1992**, 96, 7604.

(22) Warshel, A. In *Computer Modeling of Chemical Reactions in Enzymes and Solutions*; Wiley, J. and sons: New-York, **1991**, p. 170.

(23) Chipot, C.; Gorb, L. G. and Rivail, J.-L. *J. Phys. Chem.* **1994**, *98*, 1601.

(24) Librovich, N. B.; Maiorov, V. D.; Vinnik, M. I., *Zh. Strukt. Khim.* **1973**, *14*, 17.

(25) Librovich, N. B.; Maiorov, V. D.; Savel'ev, V. A., *Dokl. Akad. Nauk.* SSSR **1975**, *225*, 1358.

(26) Maiorov, V. D.;Librovich, N. B., Russ. *J. Phys. Chem.* **1977**, *51*, 141.

(27) Tuñon, I.; Silla, E.; Bertrán, J., *J. Phys. Chem.* **1993**, *97*, 5547.

(28) Rao, C.N.R. In *Water, a comprehensive treatise*; Franks, F. Ed., Plenum Press, New York; **1972**, p. 93, vol. 1.

(29) Rinaldi, D. Thesis, Nancy **1969**.

(30) Legon, A. C.; Millen, D. J., *Chem. Soc. Rev.* **1992**, *21*, 71.

(31) Amirand, C.; Maillard, D., *J. Mol. Struct.* **1988**, *176*, 181.

(32) Hannachi, Y.; Schriver, A.; Perchard, J. P., *Chem. Phys* **1989**, *135*, 285.

RECEIVED June 8, 1994

Chapter 12

Transition-State Structures
From Gas Phase to Solution

J. Bertrán, J. M. Lluch, A. González-Lafont, V. Dillet, and V. Pérez

Departament de Química, Universitat Autònoma de Barcelona, 08193
Bellaterra (Barcelona), Spain

Although the concept of a transition state structure is well defined in
the gas phase, it needs clarification for reactions in solution. One
aspect discussed in this paper is whether the separation between the
chemical system and the solvent can be introduced or not. In this
paper, both possibilities will be analyzed. To illustrate this point
several reactions have been studied. In particular, the Menshutkin,
S_N1, Friedel-Crafts, proton transfer, and dissociative electron transfer
reactions have been treated.

For reactions in the gas phase at low pressure the transition state structures are well
defined in conventional transition state theory. McIver and Komornicki *(1)* define the
transition state structure as the point that fulfils the following four conditions: a) it
is a stationary point, that is, it has zero gradient; b) the force constant matrix at the
point must have only one negative eigenvalue; c) it must be the highest energy point
on a continuous line connecting reactants and products; d) it must be the lowest
energy point which satisfies the above three conditions. Because of the two first
mathematical conditions a conventional transition state structure is a saddle point.
As a matter of fact it has been recognized that the conventional transition state *(2)*
is actually a configuration-space hypersurface that divides reactants from products
(the so-called dividing surface) and is centered at the saddle point. This dividing
surface is generally constructed perpendicular to the minimum energy path (MEP),
which is the path of steepest-descent from the saddle point into the reactant and
product valleys. Hereafter in this paper we will use the term transition state as
equivalent to the entire dividing surface, in contrast with the term transition state
structure that is just a particular point in the dividing surface.

The rate constant depends on the difference between the free energy of the
ensemble of configurations corresponding to the dividing surface and the free energy
of reactants. The fundamental assumption of conventional transition state theory is
that the net rate of forward reaction at equilibrium is given by the flux of trajectories
across the dividing surface in the product direction, and this assumption would be
true if trajectories that pass across this surface never return. As this non-recrossing
assumption is not always valid, in variational transition state theory a set of dividing
surfaces is constructed, and a search is conducted for the one that maximizes the free
energy barrier, which is equivalent to minimizing the net forward rate. As a

0097–6156/94/0568–0168$08.00/0

consequence the variational transition state is a dividing surface that does not necessarily pass through the saddle point *(3)*.

As stated above, the location of transition state structures is the first step in calculating reaction rate constants from variationally determined free energy barriers. A strategy that has been widely used in recent years to search the transition state structures is the so-called reaction coordinate methodology. Following this methodology one or a few degrees of freedom are chosen as independent variables, and for each set of values of these independent variables, an optimization of the other degrees of freedom is carried out in order to minimize the potential energy of the system *(4)*. In this way one can build up reduced hypersurfaces where it is easier to locate the transition state structure and the MEP. However an ill-advised selection of the independent variables used to define the reaction coordinate leads to unphysical reaction paths, as has been shown in several examples in the bibliography *(4c, 5)*.

The goal of this paper is to discuss the meaning and the usefulness of the concept of transition state structure when reactions in solution are considered. It is well known *(6)* that reactions in clusters often fill the gap between processes in gas phase and in solution. In particular, reactions in clusters are well suited for studying the participation of solvent degrees of freedom in the reaction coordinate. In the next section we consider the question of transition state structure for processes in clusters, and in the following section we consider it for reactions in solution.

Reactions in Clusters

Reactions between species solvated by a few solvent molecules can be observed experimentally*(7)*. For that reason the theoretical study of such reactions is interesting in its own right. However, as we have already pointed out, our interest on this paper on studying reactions in clusters is focused on analyzing the similarities between those processes taking place in clusters and the same reactions in solution.

One way to treat solvation effects is to assume that solvent always remains in equilibrium with the solute. This equilibrium hypothesis supposes that for fixed solute coordinates all the solvent coordinates are at a minimum energy point, with all the solvent gradient components being zero. However, in several previous papers *(8)*, we have shown that in an S_N2 reaction with reactants solvated by a small number of water molecules, the solvent coordinates are significant components of the transition vector. Due to this participation of solvent coordinates in the reaction coordinate, the hypothesis of an equilibrium between the solvent and the chemical system is no longer strictly valid. As a consequence, it is not possible to separate internal solute coordinates from external solvent coordinates.

Nevertheless, in a nice paper *(9)* on the effect of nonequilibrium solvation on chemical reaction rates using variational transition state theory in the microsolvated reaction $Cl^-(H_2O)+CH_3Cl$, Truhlar and Tucker have shown that the equilibrium hypothesis is a good approximation for this reaction. The extent of nonequilibrium solvation is treated by comparing calculations in which the water molecule degrees of freedom participate in the reaction coordinate to those in which they do not.

In order to further discuss the validity of the equilibrium approximation we will next present some of our results for the Menshutkin reaction, $NH_3(H_2O)$ + $CH_3Br(H_2O)$ *(10)*. The solvent is treated by discrete coordinates (as opposed to a continuum model) using the supermolecule approach, with one water molecule solvating bromine and another solvating ammonia in a symmetric fashion, in such a way that the oxygens of both waters are aligned along the molecular N-C-Br axis. This geometrical arrangement prevents artificially breaking the symmetry of the reaction.

Ab initio calculations have been carried out at SCF level with the 3-21G basis set. In Figure 1 is depicted the transition state structure of the microsolvated reaction, with the symmetry restrictions mentioned above, together with the main components of the transition vector. The figure clearly shows significant participation of the solvent coordinates in the reaction coordinate. With the advance of the reaction, two charges of opposite sign are created; therefore the two solvation shells around the charged fragments are contracted.

In the first two rows of Table I are presented the main internuclear distances for the transition state structure in the gas phase and in the microsolvated cluster, along with the corresponding values of a parameter defined as $R_c = d_{C-Br} - d_{C-N}$, which measures the advance of the reaction. The relative energies of the transition state structures referenced to the corresponding reactants are also included. Comparing the transition state structure in gas phase (first row) and in the microsolvated cluster (second row), we observe a dramatic change in the geometry. The values of the R_c parameter indicate that the transition state structure is found earlier when two water molecules are attached. Furthermore, the energy barrier decreases noticeably in going from the gas phase to the microsolvated cluster.

Table I Main internuclear distances (in Å), advance parameter (in Å), and energetic barriers (in kcal/mol) at the transition states structures in gas phase and in the microsolvated clusters for the Mentshukin reaction

	d_{C-Br}	d_{C-N}	R_c	d_{N-O}	d_{Br-H}	$\Delta E^{\neq}$
Gas Phase	2.605	1.883	0.722	-	-	23.3
Solvated[a]	2.480	2.014	0.466	2.875	2.594	10.8
Solvated[b]	2.492	2.028	0.464	2.905	2.597	11.6

(a) Solvent molecules not equilibrated with the solute. (b) Solvent molecules equilibrated with the solute.

Up to now, we have made no system-bath separation, that is, we have treated all of the coordinates of the cluster system explicitly. A reduction of the dimensionality of the potential energy hypersurface can be achieved if such a system-bath separation is made. To explore this we have effected such a separation by first calculating the MEP in the gas phase and then introducing, along this MEP, two water molecules equilibrated with the chemical system. The corresponding results are presented in the last row of Table I. The comparison between the geometries and the barriers of the transition state structures found in both cluster calculations only shows slight differences, which indicates that the hypothesis of equilibrium between the solute and solvent, although not exact, is a good approximation for this reaction. As we have already mentioned, in a reaction like the Menshutkin process, where charge is being created, the solvation shells around both charged fragments contract. The fact that the movement of the solvent is strongly coupled to the separation of charge, and consequently to the motion of the chemical system along its MEP, makes the equilibrium approximation acceptable.

The discrete representation of the solvent with a reduced number of solvent molecules is especially well suited to show the participation of solvent coordinates in the reaction coordinate, but, as Jorgensen has emphasized, the cluster reactions do not include the significant contribution of bulk reorganization to the activation barrier in solution and the effect of statistical averaging (11).

Reactions in solution

In order to model a reaction in solution, we need to include a great number of solvent molecules in our system. The number of degrees of freedom involved in such a calculation makes it computationally impossible (with current capabilities) to locate the transition state structure of the reaction on the complete potential energy hypersurface.

A first strategy to bypass this problem is to carry out a separation between the solute and the solvent treated as a thermal bath. There are three possible options to implement this strategy. The first one, which is the traditional assumption, considers that the transition state structure is the same in the gas phase as in solution. The second option incorporates the solvent effect along the gas phase MEP, in such a way that usually in solution a shift of the transition state structure along the monodimensionally gas phase MEP curve is produced(*10,12*). A variant of this methodology consists in choosing two independent variables from the chemical system to construct a reduced hypersurface, first in gas phase and second in solution, in order to evaluate the modification of the transition state structure location. The third option consists in a relocation of the transition state structure taking into account all the coordinates of the chemical system and a large number of solvent molecules, which may be done easily, in the present state of the art, by using a continuum model. This third option needs to be adopted if the interaction between the solvent and the chemical system is so important that it changes the topology of the gas phase potential hypersurface.

The use of a continuum model to relocate the transition state structure in solution means that the coordinates of the chemical system are considered as independent variables, and for each set of values of those variables the solvent is taken to be in equilibrium with the chemical system. This is similar to what is done in order to construct reduced hypersurfaces for gas-phase reactions as described in the previous section. The difference between the gas-phase and solution treatment is that in the gas-phase methodology one minimizes the energy with respect to the coordinates not taken as independent variables while in the continuum model a statistical average is carried out over the solvent variables. Note that when a continuum model is used, potential energy and free energy terms are mixed.

The second strategy consists in not separating the solvent parameters from the chemical system. Not assuming separability will be important in those systems where the internal variables and the solvent coordinates participate with similar weight in the reaction coordinate, as happens, for instance, in some electron transfer processes. A practical way to treat such electron transfer reactions is to define a nongeometrical coordinate in such a way that we can represent the free energy of the whole system versus this generalized coordinate.

In the following section, several examples will be given in order to illustrate and discuss the different options.

Treatments involving separability

First option. The first option, mentioned above, is to take the same transition state structure in the gas phase as in solution. This option is a valid approximation for those reactions where the chemical system-solvent interactions are very weak. It is also valid for some reactions, like the S_N2 process between chloride anion and methyl chloride, where those interactions are strong. This option is valid for such reactions because of the symmetry of the process and the minimum interaction with the solvent at the transition state structure, due to the delocalization of charge. As a matter of fact, Jorgensen and Buckner carried out condensed-phase simulations of the S_N2 process and observed just a modest elongation of the C-Cl bond length, by ca. 0.05 Å, from the transition state structure in gas-phase to the transition state in solution *(13)*.

By means of the continuum model developed by Tomasi and coworkers *(14)*, calculations were carried out on the S_N2 process between fluoride anion and methyl fluoride. These calculations were based on the same transition state structure in solution as in gas phase *(8a)*. The results are very similar to those obtained by Jorgensen and coworkers using a Monte Carlo methodology on chloride exchange *(11,15)*.

Second option. In the second option, the solvent effect is introduced along the gas-phase MEP. Examples of processes that can be treated accurately following this option are S_N1 and Menshutkin reactions. In both processes there is a creation and separation of charge along the MEP, in such a way that the dipole moment of the system increases along the MEP. Consequently, the stabilization of the chemical system by the solvent also increases along the MEP, which produces an advance in the transition state structure location.

Using the continuum model of Tomasi and coworkers *(14)*, we studied the CH_3Cl dissociation process *(16)* and the Menshutkin reaction $NH_3 + CH_3Br$ *(10)*. In the CH_3Cl dissociation process there is a transition state structure in solution, while in gas phase we obtained an energy profile increasing monotonically from reactants to the final products. This result is similar to the one obtained by Kim and Hynes in studying the t-BuCl dissociation in different solvents *(17)*. This last study is carried out by means of two diabatic hypersurfaces corresponding to the ionic and the covalent state. This procedure allows to account for the mutual influence of the solute electronic structure and solvent polarization, both in equilibrium and nonequilibrium conditions. This last aspect of nonequilibrium solvent polarization leading to dynamic effects is totally absent in our CH_3Cl dissociation study.

The main results of the study of the Menshutkin reaction in several solvents with different polarities, turned out to be, on one hand, that the energy barrier decreased upon increase in solvent polarity, and, on the other hand, that the transition state structure was found earlier along the MEP. The polarization of the solute by the reaction field created by the solvent polarization is one of the most important aspects of the coupling between the solvent and the chemical system. This translates into an increase of the weight of the charge-transfer configuration versus its weight in gas phase.

It is interesting to compare here the results above mentioned for this reaction with the explicit microsolvation with those obtained with the continuum model. In both models an advance of the transition state structure along the reaction coordinate and a decrease in the potential energy barrier is observed, those two effects being more pronounced when the continuum model is used (R_c=0.157 Å, $\Delta E^{\neq}$=8.1 kcal/mol).

Third option. In the third option, the transition state structure needs to be relocated in solution taking into account all the parameters of the chemical system. The first example we will discuss is the Friedel-Crafts alkylation reaction where the region of the transition state structure is the region of maximum polarity. For that reason the solvent effect will be especially important at the transition state structure and we might suspect that in solution it will be far away from the gas phase MEP *(18)*. In order to reduce the computational time we modelled the Friedel-Crafts process by selecting a molecule of CH_4 as substrate and a molecule of HF as the alkylating agent. We studied both the uncatalyzed and the catalyzed process by BF_3 (Tuñon, I.; Silla, E.; Bertrán, J. *J. Chem. Soc. Faraday Trans.*, in press). All calculations were carried out at the HF/6-31+G level, using the continuum model developed by the group of Rivail *(19)*. In this model especial attention is devoted to the electrostatic term, although, in the calculation of the cavitation energy hydrophobic effects are also considered to some extent. We directly located the stationary points on the potential energy hypersurfaces in solution by using the recently developed analytical derivatives of the electrostatic term *(20)*. This reaction is a good example where the solute-solvent equilibrium hypothesis turns out to be very adequate. This fact was illustrated with calculations using a discrete model with a reduced number of water molecules, where an analysis of the transition vector showed a very small participation of the solvent parameters. Given the maximum polarity of the system at the transition state structure, the solvent stabilization reaches a maximum that is, the transition state structure is precisely the point where the motion of the first shell changes from the contraction phase to the expansion one.

In Table II the main intermolecular distances of the transition state structure, both in the gas phase and solution, are presented for the uncatalyzed and the two catalyzed reactions. The meaning of all these parameters is illustrated by the schemes shown in Figure 2.

Table II Main internuclear distances (in Å) of the transition state structure, both in gas phase ($\varepsilon=1$) and solution ($\varepsilon=78.4$), for the uncatalyzed and the two catalyzed reactions

Reaction	$\varepsilon^{(a)}$	d_1	d_2	d_3	d_4	d_5	d_6
Uncatalyzed	1	1.246	1.253	1.470	1.459	-	-
	78.4	1.212	1.217	1.574	1.558	-	-
	$78.4^{(b)}$	1.236	1.275	1.485	1.442	-	_
Bifunctional	1	1.198	1.217	1.614	1.555	1.478	1.487
catalyzed	78.4	1.224	1.188	1.645	1.817	1.478	1.467
Acid	1	1.198	1.202	1.675	1.656	1.531	-
catalyzed	78.4	1.189	1.189	1.815	1.787	1.492	-

(a) Dielectric constant. (b) Solvated MEP.

The transition state structures for the uncatalyzed reaction, in the gas phase and in solution, are nearly symmetrical with regard to F-H and C-H bonds. The main difference is that in solution the F-H distances lengthen while the C-H distances shorten. In passing from the gas phase to solution the negative charge on the fluorine atom evolves from -0.68 to -0.81. Consequently, the weight of the ionic pair state of the wave function is significantly increased in solution. Taking into account both the nuclear and the electronic relaxation, the transition state structure in solution is clearly an ionic pair species. It is not surprising that the potential barrier in solution decreases 15 kcal/mol from its value in the gas phase. This transition state structure in solution is also very different from the one obtained by solvating the points of the gas-phase MEP without any condensed-phase reoptimization. The maximum of the energy profile of the solvated MEP is only slightly advanced on the reaction coordinate. From the internuclear distances shown in Table II, one can observe that the geometry of this last transition state structure is clearly asymmetric. The charge on the fluorine atom has changed from -0.81 for the case corresponding to relocation in solution to -0.75 for the solvated MEP case, while the barrier has increased 2.7 kcal/mol.

For the catalyzed reactions by BF$_3$ in gas phase, it has been shown that the Friedel-Crafts process takes place by two different mechanisms *(21)*. While one of them corresponds to a traditional acid catalysis, in the other one the BF$_3$ acts as a bifunctional catalyst. This last mechanism becomes to be the most efficient mechanism. The solvent effect in both catalyzed reactions has been introduced by means of the continuum model. From the internuclear distances given in Table II, one observes that the transition state structure for both catalyzed processes, in gas phase as well as in solution, present a nearly symmetric cyclic structure. In going from gas phase to solution, the B-F bonds shorten while the F-H bonds lengthen in both catalyzed processes, although those changes are smaller in the bifunctional catalyzed reaction. Also in both catalyzed mechanisms the formation of the BF_4^- and CH_5^+ ions is more advanced in solution than in gas phase. The negative charge on the BF_4^- group changes from -0.75 to -0.84 in the bifunctional catalysis and from -0.84 to -0.90 in the acid catalysis. The energy barrier of both mechanisms diminishes due to the solvent effect; however, this decrease is more dramatic in the acid catalysis.

While in gas phase the bifunctional mechanism was the most efficient, as we have already stated, in solution the transition state structure for the acid catalysis is 7.4 kcal/mol more stable and the acid mechanism becomes more efficient than the bifunctional one. In conclusion the solvent has changed the mechanism by which the catalyzed reaction proceeds as a consequence of the modification on the topology of the potential energy hypersurface.

A similar result to the one, just discussed, for the Friedel-Crafts reaction, was obtained for the proton transfer through a water molecule between two oxygens atoms of an organic acid, in particular, formic acid. This chemical system reaches a maximum of polarity at the transition state structure. For that reason, when the solvent is introduced, the distances corresponding to the solute-solvent interactions attain a minimum and these parameters have no participation at all in the transition vector. This hypothesis was confirmed by a set of calculations carried out with a discrete representation of the solvent that consisted in adding just one other water molecule solvating the nascent hydronium ion (Andrés, J.L; Lledós, A.; Bertrán, J. submitted for publication), so both water molecules play a very different role. The first one is involved in the chemical system and the second one belongs to the solvation shell. Consequently, this is again a good example where it is valid to adopt the equilibrium hypothesis, and the continuum model can be used to introduce solvent effects. In order to understand this solvent effect, we present schematically in Figure 3, the transition state structure for the chemical system, both in gas phase and in solution where the solvent molecules are represented by the continuum model. From the internuclear distances shown in Figure 3, the greater separation between the H-COO and H_3O fragments in solution structure than in gas phase one, can be clearly observed. Moreover, the decrease of the O-H distances from 1.07 Å to 1.01 Å indicates that the hydronium ion is more definitely formed in solution than in gas phase. This is also corroborated by the electronic charge on the H_3O group, which increases from +0.77 for the gas phase chemical system to +0.89 for the one in solution. At the same time, the dipole moment changes from 2.82 D to 5.45 D, which clearly shows that in solution the transition state structure is an ionic pair. One could think that the structure obtained in solution is a minimum on the potential energy hypersurface. However, it was characterized as a true transition state structure with one and only one imaginary frequency of 328 cm^{-1}. Nevertheless, the value of this frequency has dramatically decreased from its value in the gas phase (1186 cm^{-1}).

Treatments that do not assume separability

From a strict point of view, the hypothesis of separability, which implies a conceptual partition of the global system into two sets of degrees of freedom (often the solute plus the solvent surrounding it), is always false for a chemical reaction in solution. Although in some cases, such as mentioned above, it can be employed as a practical assumption that provides a good enough approximation, for others is not

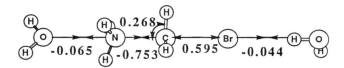

Figure 1. Transition state structure of the microsolvated cluster located without performing system-bath separation for the Menshutkin reaction, together with the main components of the transition vector.

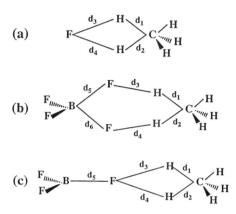

Figure 2. Transition state structures of the Friedel-Crafts reaction: (a) for the uncatalyzed reaction, (b) for the bifunctional catalyzed reaction, and (c) for the acid catalyzed reaction.

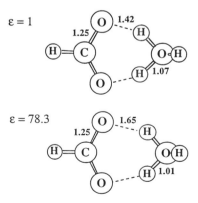

Figure 3. Transition state structures (distances in Å) for the proton transfer through a water molecule between two oxygen atoms in the formic acid, both in gas phase ($\varepsilon=1$) and in solution ($\varepsilon=78.3$) where the solvent molecules are represented by the continuum model.

acceptable at all. Many electron transfer reactions in solution are a typical example of chemical processes that belong to this last group.

Electron transfer reactions are one of the most important reaction processes, and the elucidation of the molecular mechanism that regulates their rate is one of the most important problems in chemistry and biology. The key role of the solvent fluctuations in these reactions was recognized in the pioneering work of Marcus (22), who showed the importance of using free energy rather potential energy in order to understand the kinetics of electron transfers. His seminal theory, based on a macroscopic solvent continuum model, was originally developed for outer-sphere electron transfer reactions, in which no chemical bonds are broken or formed. One of the most fundamental achievements of Marcus' theory is the nowadays well known, widely used quadratic driving force-activation free energy Marcus' relationship (22a,b,e,23). The scenario is clearly different (and more complicated) for the inner-sphere electron transfer reactions, where both the solute and the solvent coordinates have a very significant weight in the definition of the reaction coordinate. On the basis of a Morse curve description of the solute internal potential energies and a dielectric continuum approximation for the solvent fluctuational reorganization, the separability between both degrees of freedom being assumed, Savéant (24) has recently devised a simple model for dissociative electron transfer reactions in polar solvents, in an attempt to extend the applicability of Marcus' relationship. These are reactions in which the transfer of the electron and the breaking of a bond are concerted processes, being a special kind of inner-sphere electron transfer reactions. With the aim to show that the separability hypothesis is not correct for dissociative electron transfer reactions, we have performed a microscopic Monte Carlo simulation of the electrochemical reduction of methyl chloride in water, to give methyl radical and chloride anion (Pérez, V.; Lluch, J.M.; Bertrán, J. submitted for publication).

To describe the electron transfer we have employed a diabatic two-state model consisting of the methyl chloride surrounded by the solvent plus an electron inside an electrode (precursor complex), and the methyl chloride anion immersed in the solvent, once the electron has already shifted from the electrode (successor complex). Assuming a classical frame (22d,25), the radiationless electron transfer must take place at the S^* intersection region of the diabatic potential energy hypersurfaces corresponding, respectively, to the precursor complex (H_{pp}) and the successor complex (H_{ss}). Random thermal fluctuations in the nuclear configurations of the precursor complex, involving the nuclear coordinates of both the solute and the solvent, occur until this S^* region is reached, then the energies of both diabatic states becoming equal and the electron jump happening. The appearance of the proper fluctuations costs free energy. It is this free energy that determines the rate of the reaction. The electronic coupling integral between both diabatic states is supposed to be large enough for the reactants to be converted into products with unit probability in the intersection region, but small enough to be neglected in calculating the amount of internal energy required to reach S^*.

The energies of the H_{pp} and H_{ss} diabatic hypersurfaces have been obtained by adding three kinds of pairwise additive potential functions: the solute internal potential energy, the solvent-solvent interaction, and the solute-solvent interaction. The first one is the potential corresponding to the gas phase reaction. In the absence of the solvent, the precursor complex consists of the methyl chloride plus an electron inside an electrode. Its energy is calculated as the sum of the methyl chloride energy and a constant value that represents the Fermi level energy of the electrode. This Fermi level energy has been chosen as the value that makes the reaction energy equal to zero in the gas phase. On the other hand, the successor complex in the gas phase is considered to be the methyl chloride anion in the electronic state that leads to the diabatic dissociation in chloride anion and methyl radical. From a set of ab initio electronic structure calculations, analytic functions to describe the solute internal potential energy of the precursor and the successor complexes versus the d_{C-Cl} interatomic distance corresponding to the bond that is broken, are obtained. For the sake of simplicity, the methyl group has been modeled by an unique interaction center, in such a way that this d_{C-Cl} parameter is enough to specify the solute geometry.

For the water-water interactions the well-known TIP4P potential *(26)* has been used. The water-solute potential was described by means of coulombic interactions plus Lennard-Jones terms calculated between pairs of sites.

Thermal fluctuations have been generated by means of the Monte Carlo method *(27)* in a system that includes the solute and 200 water molecules at T=298K. For each generated configuration the value $\Delta E = H_{ss} - H_{pp}$ has been calculated. Then, as is known, the parameter ΔE can be used as the reaction coordinate for the reaction *(28)*. The configuration space was partitioned into subsets S, each one being associated with a particular value ΔE_S of the reaction coordinate ΔE. For practical purposes, the criterion $|\Delta E - \Delta E_S| \leq 5$ kJ/mol has been adopted to classify a given configuration as belonging to a subset S. We have identified the reactants' region (S_R) with the most populated interval when the H_{pp} potential is used. The intersection region S^* corresponds to the interval centered at the value of $\Delta E_S = 0$ kJ/mol. As a matter of fact the S^* is the dividing surface that appears in the conventional transition state theory. So, the free energy barrier $\Delta F^{\neq}$ corresponds to the evolution from S_R to S^*, in such a way that the factor exp ($-\Delta F^{\neq}/kT$) expresses the probability that the reaction system will be at the transition state S^* relative to the probability of being at S_R. Because of S^* is characterized by the value $\Delta E_S = 0$ kJ/mol, and this value is the result of the balance between the energy terms associated to all the degrees of freedom, it is evident that no separation can be performed for this kind of reactions. To achieve a complete sampling of the configuration space, we have used a strategy developed by Warshel and coworkers (29) based on a mapping potential and statistical perturbation theory.

As a direct result of the Monte Carlo simulation a free energy barrier of $\Delta F^{\neq} = 82.2$ kJ/mol was obtained. Note that this value comes from the ensemble of configurations belonging to the transition state, that is, to the dividing surface S^*. An analysis of the geometrical features of the configurations corresponding to the transition state S^* can be achieved by doing a scanning of this ensemble versus the d_{C-Cl} distance. The corresponding histogram bars are shown in Figure 4. In the gas phase just the d_{C-Cl} distance is required to specify the geometry of the system. Then, the gas phase transition state structure corresponds to the crossing point between the two diabatic solute potential energy curves, appearing at 2.28 Å. In solution, the successor diabatic solute potential energy curve (that implies charged species) is noticeably stabilizated due to the interaction with the polar solvent. Then the spread of C-Cl distances associated to the configurations of the transition state clearly appears at lower d_{C-Cl} values than the gas phase transition state structure. On the other hand, it has to be remarked that the existence in the present case of many configurations of S^* with different geometry, but the same value of the reaction coordinate ($\Delta E_S = 0$) is completely analogous to the appearance of many structures of the dividing surface arising from vibrations orthogonal to the MEP and centered at the saddle point for a normal chemical reaction in gas phase. The transition state structure for our dissociative electron transfer reaction could be defined as the lowest energy structure among all that belong to the transition state S^*. However the definition of this structure has no relevance at all, because the numerical simulation provides directly the free energy barrier, the location of the transition state structure requiring additional work and being unnecessary.

On the other hand, until now we have assumed that the transition state coincides with the S^* region, where the electron transfer itself occurs. However, this is an assumption that corresponds to conventional transition state theory: that is, the transition state is supposed to appear for the value of the reaction coordinate ($\Delta E_S = 0$)

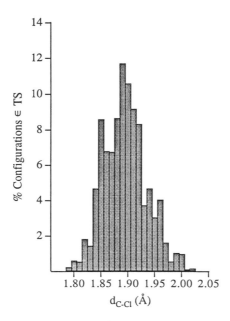

Figure 4. Histogram bars corresponding to the scanning of the configurations belonging to transition state (S^* region) in respect the d_{C-Cl} distance, for the electrochemical reduction of CH_3Cl.

that implies the maximum internal energy along it. Really, it has to be pointed out that, according the variational transition state theory, the bottleneck of the reaction can be located at a dividing surface different from S^*, i.e., at a dividing surface that imposes a maximum of free energy (3). Unfortunately, in our electrochemical reduction of methyl chloride the statistical noise is somewhat too much large for slight displacement of the transition state along the reaction coordinate ΔE being detected.

At this point it should be remarked that if we had incorporated the solvent effect by means of the continuum model along the methyl chloride and the methyl chloride anion gas phase diabatic curves (that is, introducing the solute-solvent separation in the way devised in the second option above), a transition state structure with just an unique d_{C-Cl} value would have appeared, due to the fact that the separation of coordinates had been assumed. Conversely, when the nongeometrical parameter ΔE is chosen to define the reaction coordinate within the Monte Carlo simulation, a transition state dividing surface involving a wide range of d_{C-Cl} values is obtained. The choice of this internal energy parameter to represent the reaction coordinate is particularly effective for studying reactions in condensed phases, since it permits to project easily a problem involving a huge dimensionality into a single dimension.

To summarize, when the solvent coordinates intervene significantly in the reaction coordinate, the separation assumption not being valid, a nongeometrical parameter like ΔE can be a good and practical choice to define the reaction coordinate.

Conclusions

In this paper we have shown, by means of the study of cluster reactions, the participation of solvent coordinates in the reaction coordinate. For that reason, it is questionable the separation between the internal solute coordinates and the external solvent coordinates. When the interaction between the chemical system and the solvent is weak, this separation is, clearly, a good approximation. However, in some cases, even if the solute-solvent interactions are strong the separation between internal solute coordinates and external solvent coordinates is still a valid approximation under certain conditions.

Firstly, the chemical system-solvent separation can be introduced in the processes where the region of the transition state structure is the region of maximum polarity. Then at the transition state structure, the solvent stabilization reaches a maximum and the transition vector shows a very small participation of the solvent parameters. As an example of those kinds of processes, we have presented the study of the Friedels-Crafts reaction and the results of the proton transfer through a water molecule between two oxygens in formic acid.

In the case of reactions involving charge transfer, as S_N2 processes, or involving charge separation or creation, as Menshutkin reaction or S_N1 processes, the separation can also be introduced. This fact is due to the strong coupling between the internal solute coordinates and the external solvent coordinates that takes place by means of the charge motion of the process.

Finally, in the processes where both external solvent coordinates and internal solute coordinates have an important weight in the reaction coordinate, the separation assumption should not be used as we have shown in the dissociative electron transfer reaction of CH_3Cl.

Literature Cited

[1] McIver, J.W.; Komornicki, A. *J. Am. Chem. Soc.* **1974**, *94*, 2625.
[2] Truhlar, D.G.; Hase, W.L.; Hynes, J.T. *J. Phys. Chem.* **1983**, *87*, 2664.
[3] Truhlar, D.G.; Isaacson, A.D.; Garrett, B.C. In *Theory of Chemical Reaction Dynamics;* Baer, M., Ed.; CRC Press: Boca Raton, FL,1985, Vol. IV, Chap. 2, p. 65.
[4] (a) Empedocles, P. *Theoret. Chim. Acta* **1969**, *13*, 139. (b) Empedocles, P. *Int. J. Quantum Chem.* **1969**, *35*, 47. (c) Dewar, M.J.S; Kirschner, S. *J. Am. Chem. Soc.* **1971**, *93*, 4290. (d) Ermer, O. *Struct. Bonding (Berlin)*, **1976**, *27*, 161, 202.
[5] (a) Hayes, D.M.; Hoffmann, R. *J. Phys. Chem.* **1972**, *76*, 656. (b) Baskin, C.; Bender, C.F.; Bauschlicher, C.W.; Schaefer III, H.F. *J. Am. Chem. Soc.* **1974**, *96*, 2709.
[6] (a) Riveros, J.M.; José, S.M.; Takashima, K. *Adv. Phys. Org. Chem.* **1985**, *21*, 197. (b) Castleman Jr., A.W.; Keesee, R.G. *Acc. Chem. Res.* **1986**, *19*, 413.
[7] (a) Hierl, P.M.; Ahrens, A.F.; Henchman, M.J. Viggiano, A.A.; Paulson, J.F.; Clary; D.C. *Faraday Discuss. Chem. Soc.* **1988**, *85*, 37. (b) Bernstein, E.R. *J. Phys. Chem.* **1992**, *96*, 10105.
[8] (a) Alemany, C.; Maseras, F.; Lledós, A.; Duran, M.; Bertrán, J. *J. Phys. Org. Chem.* **1989**, *2*, 611. (b) Jaume, J.; Lluch, J.M.; Oliva, A.; Bertrán, J. *J. Chem. Phys. Lett.* **1984**, *106*, 232. (c) Bertrán, J. In *New Theoretical Concepts for Understanding Organic Reactions;* Csizmadia, I.G., Ed.; Kluwer Academic Press: Dordretch, 1989, p. 231.
[9] Tucker, S.C.; Truhlar, D.G. *J. Am. Chem. Soc.* **1990**, *112*, 3347.
[10] Solà, M.; Lledós, A.; Duran, M.; Bertrán, J.; Abboud, J.L.M. *J. Am. Chem. Soc.* **1991**, *113*, 2873.

[11]   Chandrasekhar, J.; Smith, S.F.; Jorgensen, W.L. *J. Am. Chem. Soc.* **1985**, *107*, 154.
[12]   (a) Blaque, J.F.; Jorgensen, W.L. *J. Am. Chem. Soc.* **1991**, *113*, 7430. (b) Jorgensen, W.L.; Linn, D.; Blaque, J.F. *J. Am. Chem. Soc.* **1993**, *115*, 2936.
[13]   Jorgensen, W.L.; Buckner, J.K. *J. Phys. Chem.* **1986**, *90*, 4651.
[14]   (a) Miertus, S.; Scroco, E.; Tomasi, J. *Chem. Phys.* **1981**, *55*, 117. (b) Pascual-Ahuir, J.L.; Silla, E.; Tomasi, J.; Bonnacorsi, R. *J. Comput. Chem.* **1987**, *8*, 778. (c) Floris, F.; Tomasi, J. *J. Comput. Chem.* **1989**, *10*, 616.
[15]   Chandrasekhar, J.; Smith, S.F.; Jorgensen, W.L. *J. Am. Chem. Soc.* **1984**, *106*, 3049.
[16]   Solà, M.; Carbonell, E.; Lledós, A.; Duran, M.; Bertrán, J. *J. Mol. Struct. Theochem* **1992**, *255*, 283.
[17]   Kim, H.J.; Hynes, J.T. *J. Am. Chem. Soc.* **1992**, *114*, 10508, 10528.
[18]   Mestres, J.; Duran, M. *Int. J. Quantum Chem.* **1993**, *47*, 307.
[19]   (a) Rivail, J.L.; Rinaldi, D. *Chem. Phys.* **1976**, *18*, 233. (b) Rinaldi, D.; Ruiz-López, M.F.; Rivail, J.L. *J. Chem. Phys.* **1983**, *78*, 834. (c) Rivail, J.L.; Terryn, B.; Rinaldi, D.; Ruiz-López, M.F. *J. Mol. Struct. Theochem* **1985**, *120*, 387.
[20]   Rinaldi, D.; Rivail, J.L.; Rgnini, N. *J. Comput. Chem.* **1992**, *13*, 675.
[21]   Branchadell, V.; Oliva, A.; Bertrán, J. *J. Mol. Catal.* **1988**, *44*, 285.
[22]   (a) Marcus, R.A. *J. Chem. Phys.* **1956**, *24*, 966. (b) Marcus, R.A. *J. Chem. Phys.* **1956**, *24*, 979. (c) Marcus, R.A. *Annu. Rev. Phys. Chem.* **1964**, *15*, 155. (d) Marcus, R.A. *J. Chem. Phys.* **1965**, *43*, 679. (e) Marcus, R.A.; Sutin, N. *Biochim. Biophys. Acta.* **1985**, *811*, 265. (f) Marcus, R.A. *J. Phys. Chem.* **1986**, *90*, 3460.
[23]   Sutin, N. *Acc. Chem. Res.* **1982**, *15*, 275.
[24]   Savéant, J.M. *J. Am. Chem. Soc.* **1987**, *109*, 6788.
[25]   (a) Sutin, N. *Annu. Rev. Nucl. Sci.* **1962**, *12*, 285. (b) Hush, N.S. *Trans. Faraday Soc.* **1965**, *57*, 155.
[26]   (a) Jorgensen, W.J. *J. Am. Chem. Soc.* **1981**, *103*, 335. (b) Jorgensen, W.J.; Chandrasekhar, J.; Madura, J.D.; Impey, R.W.; Klein, M.L. *J. Chem. Phys.* **1983**, *79*, 926.
[27]   (a) Wood, W.W. In *Fundamental Problems in Statistical Mechanics III*; Cohen, E.G.D, Ed.; North-Holland: Amsterdam, 1975; pp 331-388. (b) Valleau, J.P.; Whittington, S.G. In *Statistical Mechanics. Part A: Equilibrium Techniques*; Berne, B.J., Ed.; Plenum Press: New York, 1977; pp 137-168. (c) Metropolis, N.; Rosenbluth, A.W.; Rosenbluth, M.N.; Teller, A.H.; Teller, E. *J. Chem. Phys.* **1953**, *21*, 1087.
[28]   (a) Warshel, A. *J. Phys. Chem.* **1982**, *86*, 2218. (b) Churg, A.K.; Weiss, R.M.; Warshel, A.; Takano, T. *J. Phys. Chem.* **1983**, *87*, 1683. (c) Zichi, D.A.; Ciccotti, G.; Hynes, J.T.; Ferrario, M. *J. Phys. Chem.* **1989**, *93*, 6261. (d) Tachiya, M. *J. Phys. Chem.* **1989**, *93*, 7050. (e) Carter, E.A.; Hynes, J.T. *J. Phys. Chem.* **1989**, *93*, 2184. (e) Yoshimori, A.; Kakitani, T.; Enomoto, Y.; Mataga, N. *J. Phys. Chem.* **1989**, *93*, 8316.
[29]   (a) Hwang, J.K.; Warshel, A. *J. Am. Chem. Soc.* **1987**, *109*, 715. (b) Hwang, J.K.; King, G.; Creighton, S.; Warshel, A., *J. Am. Chem. Soc.* **1988**, *110*, 5297. (c) King, G.M; Warshel, A. *J. Chem. Phys.* **1990**, *93*, 8682.

RECEIVED April 5, 1994

ORGANIC REACTIONS

Chapter 13

Probing Solvation by Alcohols and Water with 7-Azaindole

F. Gai, R. L. Rich, Y. Chen, and J. W. Petrich[1]

Department of Chemistry, Iowa State University, Ames, IA 50011

The nonradiative pathways of 7-azaindole are extremely sensitive to solvent. In alcohols, 7-azaindole executes an excited-state double-proton transfer. In water, this tautomerization is frustrated. Proton inventory experiments suggest a concerted double-proton transfer in the alcohols and point to another nonradiative process in water. We propose the following idealized picture. Whereas at room temperature 7-azaindole can form a cyclic hydrogen-bonded intermediate with a single alcohol molecule facilitating tautomerization, in water more than one solvent molecule coordinates to the solute and thus prohibits the concerted process. More detailed measurements, however, indicate that water and alcohols do not solvate 7-azaindole in fundamentally different ways, but rather that they represent two extremes of the same phenomenon.

7-Azaindole (Figure 1) is the chromophoric moiety of the nonnatural amino acid, 7-azatryptophan. Recently, we have proposed 7-azatryptophan as an alternative to tryptophan as an optical probe of protein structure and dynamics (*1-8*, Gai, F.; Rich, R. L.; Petrich, J. W. *J. Am. Soc. Chem.* in press). 7-Azatryptophan can be incorporated into synthetic peptides and bacterial protein (*1,2*, Smirnov, A. V.; Rich, R. L.; Petrich, J. W. *Biochem. Biophys. Res. Commun.*, in press; Rich, R. L.; Gai, F.; Lane, J. W.; Petrich, J. W.; Schwabacher, A. W. *J. Am. Soc. Chem.*, in press). Its steady-state absorption and fluorescence spectra are sufficiently different from those of tryptophan that selective excitation and detection may be effected. Most important for its use as an optical probe, however, is that the fluorescence decay for 7-azatryptophan over most of the pH range, when emission is collected over the entire band, is single exponential. For tryptophan, on the other hand, a nonexponential fluorescence decay is observed (Chen, Y.; Gai, F.; Petrich, J. W. *J. Phys. Chem.*, in press). The potential utility of 7-azatryptophan as an optical probe suggests a thorough investigation of the photophysics of its chromophore, 7-azaindole, in

---

[1]Corresponding author

0097–6156/94/0568–0182$08.00/0

order to characterize its fluorescence properties and to elucidate its pathways of nonradiative decay.

7-Azaindole has undergone considerable study in nonpolar solvents (*6,9-13*). Kasha and coworkers (*9*) discovered that 7-azaindole can form dimers that undergo excited-state tautomerization (Figure 1a). It has also been demonstrated that excited-state tautomerization occurs for 7-azaindole in alcohols (*6,11-13*). In alcohols the fluorescence spectrum of 7-azaindole is bimodal. In methanol, for example, the maximum of the higher energy band is at 374 nm and that of the lower-energy band is at 505 nm. The former band arises from the so-called "normal" species that decays into the latter band by double-proton transfer. In alcohols, the tautomerization or double-proton transfer reaction has been traditionally depicted (Figure 1b) as being mediated by one solvent molecule, which forms a cyclic complex with the solute. In water, on the other hand, significantly different behavior is observed as illustrated by the fluorescence emission with a single maximum at 386 nm and the single-exponential fluorescence decay when emission is collected with a wide band-pass, 910 ps (*4-6*).

The aims of this article are to investigate the apparent difference between alcohols and water on the excited-state reactivity of 7-azaindole and to obtain more detailed information on the nature of the tautomerization process.

Solvation of 7-Azaindole in Water

7-Azaindole exhibits a single-exponential fluorescence decay of 910 ± 10 ps in water at neutral pH and 20°C if emission from the entire band ($\lambda_{em} \geq 320$ nm) is collected (*4-6*). The fluorescence decay, however, deviates from single exponential if emission is collected with a limited bandpass. For $\lambda_{em} \leq 450$ nm, a single exponential does not provide a satisfactory fit. An acceptable fit is obtained using two exponentially decaying components and indicates that about 20% of the fluorescent emission decays with a time constant between 40 to 100 ps (depending on the full-scale time base chosen for the experiment). A component with a 70-ps decay time is also detected in the transient absorbance of 7-azaindole in water (*3*). There is no such rapid component in the fluorescence decay or the transient absorption of the 7-methyl- and 1-methyl-derivatives of 7-azaindole (*3,4*). We have thus attributed this rapid component to a *small* population of 7-azaindole molecules that undergo excited-state tautomerization. For the duration of the discussion, we shall refer to this transient as the 70-ps component because it is more clearly resolved in the transient absorption measurements (*3*).

The 910 ps component that is resolved for $\lambda_{em} \leq 450$ nm or when emission is collected over the entire band is attributed to the majority of the 7-azaindole molecules that are not capable of excited-state tautomerization because they exist in a "blocked" state of solvation (Figure 1c). This assignment will be described in more detail below.

When $\lambda_{em} \geq 505$ nm, the fluorescence decay can be fit to the form F(t) = -0.69 exp(-t/70 ps) + 1.69 exp(-t/980 ps). The long-lived component is observed to lengthen from 910 to 980 ps. This lengthening of the lifetime at long emission wavelengths was reported earlier (*6*), but no significance was drawn to it. If the rise time of the fluorescence emission can be attributed to the

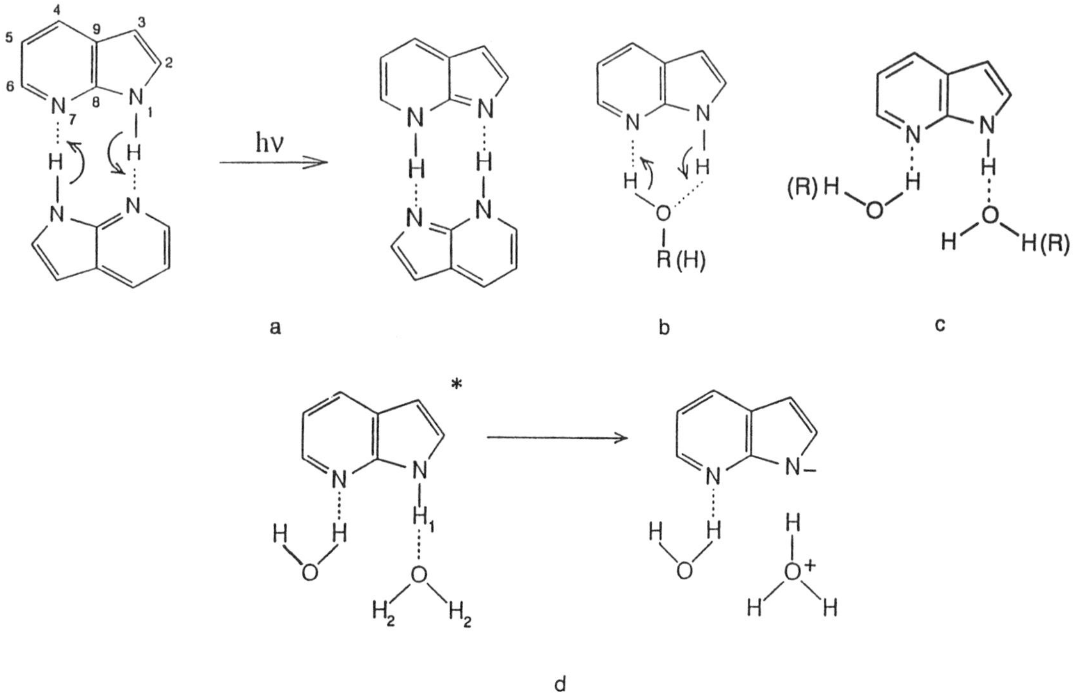

Figure 1. Idealized structures for excited-state tautomerization in (a) dimers of 7-azaindole and in (b) complexes of 7-azaindole with linear alcohols. We have argued that water (6), and to a certain extent, alcohols (13) can solvate 7-azaindole in such a fashion (c) that excited-state tautomerization is frustrated. We suggest, however, that abstraction (d) of the N_1 proton by the coordinated water molecule is an important nonradiative pathway.

appearance of tautomer, then for $\lambda_{em} \geq 505$ nm $|0.69/1.69| \sim 0.40$ is the fraction of tautomer present. The rest of the emission arises from 7-azaindole molecules incapable of tautomerization and characterized by a 910 ps lifetime. Thus, 980 ps represents the weighted average of 910 ps and a longer lifetime, namely ~ 1100 ps. This decay time is *identical* to that of protonated (pH < 3) 7-azaindole (*4*).

Comparison of 7-Azaindole in Water and Alcohols

At ambient temperature, the fluorescence decay of the normal band of 7-azaindole (commercially-available or purified) in alcohols can always be fit well by a single exponential plus a small amount of longer-lived component. Figure 2 illustrates the increase in the magnitude of this long component with decreasing temperature for 1-butanol. The amplitude of the longer-lived component increases from about 5% at 20°C to about 44% at -6°C. This result renders the assignment of this component in alcohols to an impurity untenable. Assuming that the extinction coefficient and the radiative rate of a putative impurity are relatively insensitive to temperature, such a large change in the amplitude is unlikely. The longer-lived component is *predominant* in polyalcohols even at 20°C: ethylene glycol, $F(t) = 0.31\exp(-t/141ps) + 0.69\exp(-t/461ps)$; and propylene glycol, $F(t) = 0.31\exp(-t/197ps) + 0.69\exp(-t/816ps)$. This longer-lived component is taken as evidence for the presence of a blocked state of solvation in alcohols such as has been already discussed for water (Chen, Y.; Gai, F.; Petrich, J. W. *Chem. Phys. Lett.*, submitted).

Application of the Proton Inventory to the Nonradiative Process in 7-Azaindole

7-Azaindole in Methanol. The isotope effect on proton transfer reactions is rarely a linear function of solvent deuterium content. Gross and Butler explained this phenomenon by noting that either the H/D composition in the proton site can be different with respect to the solvent or more than one proton is in flight during the rate-limiting step (*14*). The Gross-Butler equation (below) relates the rate of the process in the protiated solvent, k_o, to the rate in a solution of mole fraction n of the deuterated solvent and to all the protons in the reactant and transition states involved:

$$k_n = k_o \, \frac{\displaystyle\prod_{i}^{\nu} (1-n+n\phi_i^T)}{\displaystyle\prod_{i}^{\nu} (1-n+n\phi_i^R)} \sim k_o \, \prod_{i}^{\nu} (1-n+n\phi_i^T)$$

where ν is the total number of protons involved. The $\phi^{T,R}$ are the fractionation factors in the transition and the reactant states, respectively. ϕ is the ratio of the preference in a site in a molecule for deuterium over protium relative to the preference for deuterium over protium in a solvent molecule (*14*). In other words, ϕ is the equilibrium constant for the generalized reaction: XH + ROD $\rightleftharpoons$ XD + ROH. It is customary in most analyses to take $\phi^R = 1$ for an NH or an OH site, as indicated above. (These ϕ^R are for *the ground state*. In order to

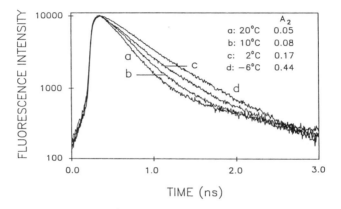

Figure 2. Fluorescence decay of the normal band of 7-azaindole in 1-butanol as a function of temperature (320 nm < λ_{em} < 460 nm): (a) F(t) = 0.95exp(-t/234 ps) + 0.05exp(-t/1818 ps); (b) F(t) = 0.92exp(-t/280 ps) + 0.08exp(-t/1406 ps); (c) F(t) = 0.83exp(-t/329 ps) + 0.17exp(-t/916 ps); (d) F(t) = 0.56exp(-t/360 ps) + 0.44exp(-t/760 ps). Because the full-scale time base for the photon-counting measurement is only three nanoseconds, an accurate determination of the duration of the longer-lived component is difficult.

apply them directly to our problem, we must assume that the ϕ^R are identical in the excited state. We have considered this possibility elsewhere (*8*).)

The downward bulging curve for 7-azaindole in methanol (and in ethanol (*8*)) (Figure 3) is such that a plot of $(k_n/k_o)^{1/2}$ vs n yields a straight line. This result suggests that only two protons are involved in the excited-state tautomerization of 7-azaindole in alcohols. This result is also consistent with the "cyclic complex" of 7-azaindole and alcohol (Figure 1b) that has been traditionally assumed to be required for the tautomerization to proceed.

7-Azaindole in Water. The downward bulging of the curve obtained for 7-azaindole in H_2O/D_2O mixtures suggests that more than one proton is involved in the transition state of the nonradiative deactivation process. Fitting k_n/k_o vs n to a quadratic model (i.e., a two-proton process) gives imaginary ϕ^T for the data in water ($\phi^T = 0.43 \pm i0.30$). Imaginary ϕ^T can be obtained when there are two or more competing parallel pathways and if at least one of the transition states involves *at least* two protons (*14*). We have, however, argued elsewhere (*3-5,8*) that not more than 20% of the 7-azaindole population in water is capable of executing double proton transfer and that this process can be observed only under conditions of sufficient wavelength and time resolution (*3,4*). In fact, double proton transfer of 7-azaindole in water is a minor nonradiative pathway compared to monophotonic ionization (*3,7*, Chen, Y.; Gai, F.; Petrich, J. W. *J. Phys. Chem.*, in press). The failure of the quadratic model to fit the proton inventory data coupled with the previous evidence against the importance of excited-state tautomerization in water argue against a concerted two-proton process in this solvent. (If two protons are being transferred by 7-azaindole in water, they are not transferred concertedly between N_1 and N_7).

For the sake of simplicity and because of experimental precedent with another system, we discuss the proton inventory data of 7-azaindole in water as a three-proton process. This three-proton process involves the abstraction of hydrogen from N_1 by a coordinated water molecule.

For the three-proton process shown in Figure 1d, the data in Figure 4b yield an excellent fit to the equation $k_n/k_o = (1-n+0.48n)(1-n+0.69n)^2$. Furthermore, a plot of $(k_n/k_o)^{1/2}$ vs n does not yield a straight line, which is inconsistent with a concerted two-proton process as observed in the alcohols.

Wang et al. observed essentially identical behavior in ribonuclease (*14*). In this enzyme there is an isomerization between two of its conformations that are characterized by $pK_a > 8$ and $pK_a = 6.1$. These workers measured a solvent isotope effect of 4.7 ± 0.4. Their proton inventory measurements were best described by the relation $k_n/k_o = (1-n+0.46n)(1-n+0.69n)^2$. They assigned the rate-limiting step in this isomerization to proton transfer to a water molecule from the protonated imidazole group of a histidine (*14*). Within experimental error, the proton inventory rate parameters for 7-azaindole in water are identical to those for the isomerization of ribonuclease. In both cases, the shuttling of a proton from nitrogen to a water molecule is proposed to be the rate-determining step.

A fundamental assumption made in deriving the Gross-Butler equation is that the rate of H/D exchange between the solute and the solvent is significantly greater than the rate of proton transfer being investigated. In other words, *the decay of the entire reactant population must be characterized by a rate*

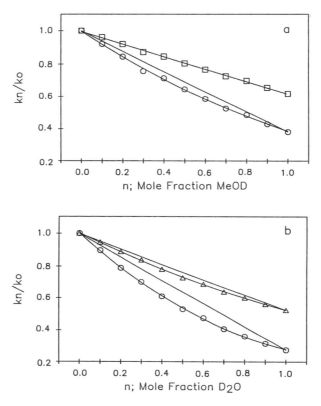

Figure 3. (a) Ratio of tautomerization rate of 7-azaindole in MeOH, k_o, to that in a mixture of protiated and deuterated methanol that is mole fraction, n, in MeOD, k_n. The open circles represent k_n/k_o vs n. The solid line through the data represents the fit assuming a two-proton process with $\phi^T = 0.62$. Directly above is plotted the straight line that would result from a one proton process, i.e., the average of k_o and k_1 weighted by the respective mole fractions of protiated and deuterated solvents (19). The open squares represent $(k_n/k_o)^{1/2}$ vs n. The linearity of this plot verifies the two-proton process in methanol, assuming the validity of the Gross-Butler equation. (b) Proton inventory data for 7-azaindole in H_2O and D_2O at 20°C. The open circles represent k_n/k_o vs n. The pH at n=0 is 6.8. The solid line through these data represents the fit assuming a three proton process: $k_n/k_o = (1 - n + 0.48 n)(1 - n + 0.69 n)^2$. The straight line plotted directly above is the result expected for a one-proton process. The open triangles represent $(k_n/k_o)^{1/2}$ vs n. The solid line through the open triangles is only meant to guide the eye. This plot deviates significantly from the straight line just above it. Hence, the proton inventory data in water are different from those in the alcohols. Assuming the validity of the Gross-Butler equation, the water data are inconsistent with a two-proton process. In all cases, the error bars lie within the symbols.

constant that does not change with time; that is, first-order decay kinetics must be obtained.   If solvent exchange is not rapid, then the observed decay is a superposition of the decays of the individual isotopically substituted species.   For the case of 7-azaindole, at least four individual rate constants may be involved (see below).   In practice, it is often very difficult to distinguish experimentally between genuine first-order kinetics, which are characterized by a single exponential decay time, and the superposition of several single-exponential decays characterized by different time constants.   For this reason, we have presented (8) several different methods of analyzing the 7-azaindole data (Table I).

In order for application of the Gross-Butler equation to the excited-state process of 7-azaindole to be valid, we require that 7-azaindole exchange its N_1 ligand with solvent protium or deuterium much faster than the actual tautomerization reaction depicted in Figure 1b.   Since the fluorescence lifetime of 7-azaindole in the solvents used here ranges from 140 to 900 ps, an appropriate time constant for ligand exchange with the solvent would be a few picoseconds.   Such a rapid exchange seems unlikely.   NMR measurements of ground-state indoles indicate that N_1 exchanges its proton on a time scale of seconds with the solvent (20).   The strong likelihood of slow exchange in the excited state requires us to consider the kinetics in more detail.

The Criteria for a Concerted Reaction.   Figure 4 presents the four cases that may arise if two protons are involved in the deactivation of excited-state 7-azaindole.   In Figure 4, the reactants and products are denoted A and D, respectively.   B and C denote *intermediates* that would exist if the excited-state tautomerization of 7-azaindole proceeded by either the stepwise pathway ABD or ACD involving first the breaking of the N_1-H bond and then the formation of the N_7-H bond, and vice versa.   Given such a reaction scheme, in order to demonstrate that the tautomerization is a concerted process, it is necessary, but not sufficient, to show that $k^{HD} = k^{DH}$ and that $k^{HD} = (k^{HH}k^{DD})^{1/2}$.   This latter criterion is referred to as "the rule of the geometric mean."   Use of the Gross-Butler equation assumes the applicability of the rule of the geometric mean. This relationship is very restrictive and demands that many requirements be satisfied (8,16).   For the examples illustrated in Figure 4, one of the most important of these requirements is that for the concerted double-proton transfer, the secondary isotope effect at the N_7 (or N_1) site is *equal* to the primary isotope effect at the N_1 (or N_7) site.   We shall also see that in order for this relationship to be satisfied, the reaction must be "symmetric"; that is, the rate constants for the decay of the intermediate B (or C) to A and D must be equal.

The significance of the rule of the geometric mean is that if there is a concerted reaction, both protons must be "in flight" in the transition state. Under these circumstances and in the absence of other effects such as tunneling (16), one thus expects the multiple sites in a single transition state to behave independently with respect to isotopic substitution.

The Nonradiative Process of 7-Azaindole in Water.   Glasser and Lami (17) and Wallace and coworkers (18) have discussed the importance of fission of the NH bond as a nonradiative process in gas phase indole.   Barkley and

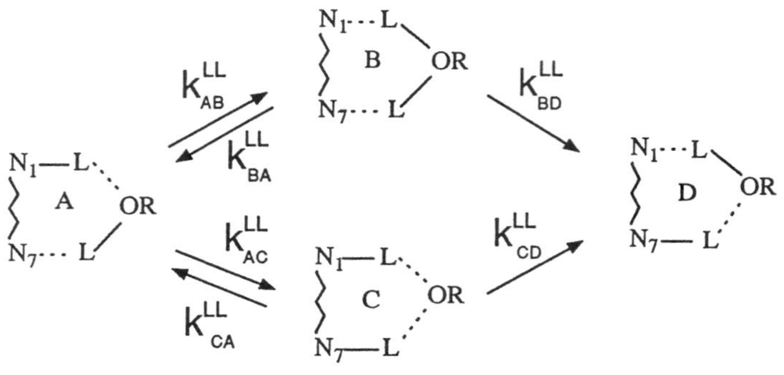

Figure 4. Excited-state tautomerization reactions for each of the four cases of isotopic substitution considered in the text. L = H or D. The paths ABD and ACD represent *stepwise* processes where B and C are distinct intermediates.

Table I
Rate Constants for Proton Transfer Steps[a]

rate constant $(s^{-1} \times 10^{-9})$	MeOH/MeOD	EtOH/EtOD	H_2O/D_2O
k^{HH}	7.19 ± 0.10	5.43 ± 0.08	1.13 ± 0.02
k^{DD}	2.74 ± 0.04	1.93 ± 0.02	0.31 ± 0.01
$(k^{HH}k^{DD})^{1/2}$	4.43 ± 0.05	3.24 ± 0.03	0.59 ± 0.01
$k^{HD,b}$	4.42 ± 0.06	3.25 ± 0.02	
$k^{DH,b}$	4.59 ± 0.04	3.27 ± 0.02	
$k^{HD,c}$	4.29 ± 0.11	3.24 ± 0.09	0.48 ± 0.02

[a] Fluorescence lifetime measurements from which the rate constants were obtained were performed at 20 °C.
[b] Obtained from equations 18 and 19 of reference 12. Because this method requires fitting the data to a double-exponential fluorescence decay, the corresponding rate constants could not be determined for water, where a single exponential is sufficient to describe the decay curves.
[c] Obtained from equation 22 of reference 12. This method of analysis assumes that $k^{HD} = k^{DH}$. The values cited are for n = 0.5. If n = 0.2, then for MeOH/MeOD and EtOH/EtOD, k^{HD} is 4.14 × 10^9 s^{-1} and 3.18 × 10^9 s^{-1}, respectively. If n = 0.7, then for MeOH/MeOD and EtOH/EtOD, k^{HD} is 4.18 × 10^9 s^{-1} and 3.37 × 10^9 s^{-1}, respectively.

coworkers (*19*) have performed detailed investigations of the deuterium isotope effect on the photophysics of tryptophan, indole, and some of their derivatives. They have proposed at least six different mechanisms to explain the isotope effect ranging from photoionization, hydride transfer from the NH, proton transfer from the solvent to the ring, solvent mediated NH exchange, tautomerization resulting in NH abstraction, and exciplex formation.

We propose that the isotope effect observed in indole derivatives can be rationalized by the same mechanism that we illustrate for 7-azaindole in Figure 1d. We suggest that in indole this process is much less efficient because there is no N_7 nitrogen coordinated with a solvent proton. Such an interaction could establish a partial positive charge on N_7 that would help to stabilize the negative charge generated on N_1.

The Origin of the Isotope Effect. Finally we must comment on the origin of the isotope effect. In large part because of the rapid (1.4 ps) tautomerization observed in dimers of 7-azaindole (*10*), the tautomerization of dilute solutions of 7-azaindole in alcohols has been discussed in terms of a two-step process (*11-13*). The first step involves obtaining the correct solvation of the solute by the alcohol; the second step, double-proton transfer. The interpretation of our isotopic substitution experiments depends on whether the two-step model is appropriate and, if it is, whether the solvation step is slow, fast, or comparable to tautomerization. If the rate-limiting step in the double-proton transfer reaction is the formation of the cyclic complex, then the isotope effects we discuss above require reinterpretation. Additional experimental and theoretical work is necessary in order to answer this question definitively. For the moment, we suggest that if solvation were the rate-limiting step in the excited-state tautomerization of 7-azaindole in alcohols, it would be extremely fortuitous that the rule of the geometric mean holds (Table I). In addition, dimers of 7-azaindole may not be an appropriate paradigm for the tautomerization of the 7-azaindole-alcohol complex. For example, Fuke and Kaya (*20*) observe that in supersonic jets the rate of excited-state double-proton transfer of 7-azaindole dimers is 10^{12} s^{-1}, while in dimers of 1-azacarbazole and in complexes of 7-azaindole with 1-azacarbazole the rate is 10^9 s^{-1}. The reduction in rate by a factor of 10^3 is initially surprising given the very similar hydrogen bonding in the three types of complexes. It is therefore most likely premature to assume that tautomerization in a 7-azaindole complex occurs as rapidly as in a 7-azaindole dimer. Fuke and Kaya suggest that detailed considerations of the coupling of proton motion with intermolecular vibrational motion are required in order to predict the rate of such tautomerization reactions (*20*).

Figure 5 presents a plot of the time constant for excited-state proton transfer in 7-azaindole at 20°C against pK_{auto} for a wide range of solvents. K_{auto} is the equilibrium constant for autoprotolysis and characterizes both the proton accepting and proton donating abilities of a solvent (S) for the reaction: $2SH \rightarrow SH_2^+ + S^-$ (*21*). The correlation is exceptionally good, especially when one considers that previous correlations attempted between the proton transfer times and viscosity or polarity ($E_T(30)$) *are strongly dependent upon the molecular structure of the solvent* (e.g., primary as opposed to secondary alcohols or

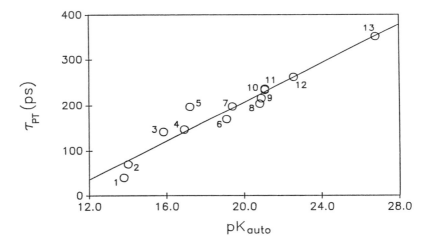

Figure 5. Correlation of the time constant for excited-state tautomerization of 7-azaindole in various solvents with pK_{auto}. (1) 2,2,2-trifluoroethanol; (2) water; (3) ethylene glycol; (4) methanol; (5) propylene glycol; (6) ethanol; (7) 1-propanol; (8) 1-pentanol; (9) 1-butanol; (10) 2-propanol; (11) 2-methyl-1-propanol; (12) 2-butanol; (13) 2-methyl-2-propanol. For 2,2,2-trifluoroethanol, the pK_{auto} is estimated from the pK_a.

polyalcohols or water) and in general are quite scattered (*13,22*). The linear free energy relation presented in Figure 5, however, comprises very disparate kinds of solvents. Even water fits well into this relationship. This correlation is consistent with the requirement of a cyclic solute-solvent complex for excited-state tautomerization and with the proton-transfer event being the rate-limiting step. The larger the autoprotolysis constant (the smaller the pK_{auto}), the easier it is for the solvent to accept a proton from N_1 and to donate a proton to N_7.

Summary and Conclusions

Recently two related studies of 7-azaindole in water have been performed. Chou et al. (*23*) investigated 7-azaindole in mixtures of water and aprotic solvents. Small additions of water to polar aprotic solvents produced tautomer-like emission. They proposed that excited-state tautomerization is possible only when there are significant concentrations of 1:1 complexes of 7-azaindole and water. They further proposed that in pure water the formation of higher-order aggregates inhibits tautomerization during the excited-state lifetime.

Chapman and Maroncelli have studied 7-azaindole fluorescence in water and in mixtures of water and diethyl ether (*22*). They too observe long-wavelength, tautomer-like emission at low water concentrations. In pure water they also observe a rapid rise time at long wavelengths. They, however, take a different point of view, namely that excited-state tautomerization occurs for the entire 7-azaindole population in pure water and that the 7-azaindole fluorescence lifetime is dominated by this reaction. Using a two-state kinetic model in conjunction with steady-state spectral data they conclude that the rapid rise time is associated with the nonradiative decay rate of the tautomer. They propose that the longer, ~ 900 ps, decay time of the entire emission band is a measure of the tautomerization rate. Their scheme requires that the nonradiative decay rate of the tautomer is greater than the rate of tautomerization. They estimate that the rate of tautomerization is $1.2 \times 10^9 \text{ s}^{-1}$.

Our observations and conclusions more nearly approach those of Chou et al., although there is a small population of 7-azaindole molecules that do tautomerize in addition to the majority of the population in which this reaction is frustrated. That the fluorescence lifetime of 7-azaindole is not dominated by excited-state tautomerization is demonstrated by the observation of three distinct fluorescence lifetimes: ~ 70 ps, the normal decay time; ~ 980 ps (i.e., 1100 ps (*4*)), the tautomer decay time; and 910 ps, the decay time of the blocked solute. Further evidence is provided by the spectral inhomogeneity of the emission band (*4*). Our major conclusions concerning water can be summarized as follows:

Only a small fraction ($\lesssim$ 20%) of 7-azaindole molecules in pure water are capable of excited-state tautomerization on a 1-ns time scale.

The majority of the 7-azaindole molecules are solvated in such a fashion that tautomerization is blocked. More than 10 ns (*4*) are required to achieve a state of solvation that facilitates tautomerization, that is, to convert the "blocked" species into a "normal" species.

No significant emission intensity is observed for 7-azaindole in water at 510 nm because so little tautomer is produced and because the tautomer that is produced is rapidly protonated and has an emission maximum at ~ 440 nm.

Most importantly these results clarify the photophysics of 7-azaindole for use as the intrinsic chromophore of the probe molecule, 7-azatryptophan. In particular, the minor amount of tautomerization will contribute to the decay kinetics only if emission is collected at wavelengths red of 505 nm or with a relatively narrow spectral bandpass (with adequate temporal resolution). This is not a serious restriction since experiments are not likely to be performed with such spectral resolution owing to the low fluorescence intensity. When emission is collected over a large spectral region and on a full-scale time base coarser than 3 ns, the tautomerization reaction is imperceptible. On the other hand, the appearance of long-wavelength emission of a protein containing 7-azatryptophan in water would definitely signal a change of environment that facilitates tautomerization.

By analogy with the types of solvation possible in water, we propose that the long-lived fluorescence decay component observed for 7-azaindole in alcohols can be understood by attributing it to a "blocked" form of solvation. *In other words, alcohols and water represent different extremes of solvation; but in neither case is excited-state tautomerization completely permitted or completely prohibited.* Similar blocked states of solvation have been observed in argon matrices at 10 K for the much studied model of excited-state proton transfer, 3-hydroxyflavone (24). The groups of Barbara (25), Kasha (26), and Harris (27) have discussed the importance of intermolecular hydrogen bonding, cyclic hydrogen-bonded complexes with one solvent molecule, and doubly solvated hydrogen bonded complexes.

We have performed the first application of the proton inventory technique to an excited-state process. The data suggest that the excited-state tautomerization of 7-azaindole in alcohols proceeds by a concerted, two-proton process that is consistent with the structure of the cyclic solute solvent complex presented in Figures 1a,b. (The data, however, do not prove the existence of a cyclic complex of one solvent molecule with the solvent. There is the possibility that the double proton transfer involving N_1 and N_7 occurs via two different alcohol molecules that interact with each other sufficiently strongly to effect the concerted reaction.) These proton inventory experiments provide further evidence to support the model (Figure 1c) of 7-azaindole being solvated by water in such a way that double-proton transfer—as it occurs in alcohols at room temperature—is negligible.

Acknowledgments

F. G. and R. L. R. are recipients of Fellowships from Phillips Petroleum and Amoco, respectively. J. W. P. is an Office of Naval Research Young Investigator. This work was partially supported by the ISU Biotechnology Council, University Research Grants, and a Carver Grant.

Literature Cited

1. Négrerie, M.; Bellefeuille, S. M.; Whitham, S.; Petrich, J. W.; Thornburg, R. W. *J. Am. Chem. Soc.* **1990**, *112*, 7419.
2. Rich, R. L.; Négrerie, M.; Li, J.; Elliott, S.; Thornburg, R. W.; Petrich, J. W. *Photochem. Photobiol.* **1993**, *58*, 28.
3. Gai, F.; Chen, Y.; Petrich, J. W. *J. Am. Chem. Soc.* **1992**, *114*, 8343.
4. Chen, Y.; Rich, R. L.; Gai, F.; Petrich, J. W. *J. Phys. Chem.* **1993**, 97, 1770.
5. Rich, R. L.; Chen, Y.; Neven, D.; Négrerie, M.; Gai, F.; Petrich, J. W. *J. Phys. Chem.* **1993**, *97*, 1781.
6. Négrerie, M.; Gai, F.; Bellefeuille, S. M.; Petrich, J. W. *J. Phys. Chem.* **1991**, *95*, 8663.
7. Négrerie, M.; Gai, F.; Lambry, J.-C.; Martin, J.-L.; Petrich, J. W. *J. Phys. Chem.* **1993**, *97*, 5046.
8. Chen, Y.; Gai, F.; Petrich, J. W. *J. Am. Chem. Soc.* **1993**, *115*, 10158.
9. Taylor, C. A.; El-Bayoumi, M. A.; Kasha, M. *Proc. Natl. Acad. Sci. U.S.A.* **1969**, *63*, 253.
10. Share, P. E.; Sarisky, M. J.; Pereira, M. A.; Repinec, S. T.; Hochstrasser, R. M. *J. Lumin.* **1991**, *48/49*, 204.
11. McMorrow, D.; Aartsma, T. J. *Chem. Phys. Lett.* **1986**, *125*, 581.
12. Konijnenberg, J.; Huizer, A. H.; Varma, C. A. G. O. *J. Chem. Soc., Faraday Trans. 2* **1988**, *84*, 1163.
13. Moog, R. S.; Maroncelli, M. *J. Phys. Chem.* **1991**, *95*, 10359.
14. Schowen, K. B. J. In *Transition States of Biochemical Processes*; Gandour, R. D., Schowen, R. L., Eds.; Plenum: New York, NY, 1978; 225-.
15. Wüthrich, K. *NMR of Proteins and Nucleic Acids*; Wiley: New York, NY, 1986; Chapter 2.
16. Limbach, H.-H.; Henning, J.; Gerritzen, D.; Rumpel, H. *Faraday Discuss. Chem. Soc.* **1982**, *74*, 229.
17. Glasser, N.; Lami, H. *J. Chem. Phys.* **1981**, *74*, 6526.
18. Demmer, D. R.; Leach, G. W.; Outhouse, E. A.; Hagar, J. W.; Wallace, S. C. *J. Phys. Chem.* **1990**, *94*, 582.
19. McMahon, L. P.; Colucci, W. J.; McLaughlin, M. L.; Barkley, M. D. *J. Am. Chem. Soc.* **1992**, *114*, 8442.
20. Fuke, K.; Kaya, K. *J. Phys. Chem.* **1989**, *93*, 614.
21. Reichardt, C. *Solvents and Solvent Effects in Organic Chemistry*; VCH: Weinheim, 1988; .
22. Chapman, C.; Maroncelli, M. *J. Phys. Chem.* **1992**, *96*, 8430.
23. Chou, P.-T.; Martinez, M. L.; Cooper, W. C.; Collins, S. T.; McMorrow, D. P.; Kasha, M. *J. Phys. Chem.* **1992**, *96*, 5203.
24. Brucker, G. A.; Kelley, D. F. *J. Phys. Chem.* **1987**, *91*, 2856.
25. Barbara, P. F.; Walsh, P. K.; Brus, L. E. **J. Phys. Chem. 1989**, *93*, 29.
26. McMorrow, D.; Kasha, M. *J. Phys. Chem.* **1984**, *88*, 2235.
27. Schwartz, B. J.; Peteanu, L. A.; Harris, C. B. *J. Phys. Chem.* **1992**, *96*, 3591.

RECEIVED April 5, 1994

Chapter 14

Theoretical Models of Anisole Hydrolysis in Supercritical Water

Understanding the Effects of Pressure on Reactivity

Susan C. Tucker and Erin M. Gibbons

Department of Chemistry, University of California, Davis, CA 95616

Electrostatic contributions to the free energies of activation for the title reaction in supercritical water are calculated as a function of pressure. The calculations model the solvent as an incompressible fluid having a pressure dependent dielectric constant. A number of implementations of this model, both with and without solute polarizability, are considered and contrasted. The calculated equilibrium solvation effects are found to adequately explain the experimentally observed pressure dependence of this reaction. However, an effective local dielectric which differs significantly from the bulk value for supercritical water must be invoked to obtain agreement between theory and experiment, indicating that there is a large degree of solvent-solute clustering for this reaction. Additionally, the free energies of solvation, and hence the solvent-solute interactions, are found to depend strongly on reaction coordinate position. These results raise questions about the importance of reaction path dependent clustering on the free energy of activation and solute reactivity.

Supercritical water provides a unique solvent environment because significant changes in its solvating properties may be effected by modest changes in thermodynamic conditions. Supercritical water (SCW) thus provides the basis for a number of industrial processes, from SCW extraction (*1*) to the oxidative destruction of hazardous wastes (*2*). More recently, SCW has been proposed as a medium for selective synthetic chemistry (*3,4*). From a more fundamental point of view, SCW is important because it enables one to study the effect of changing solvent properties while the underlying chemical interactions remain fixed.

Recent experiments in the vicinity of, but not at, the critical point have shown that in SCW heterolytic reactions are favored at higher pressures and lower temperatures, while homolytic free radical processes are favored at lower pressures and higher temperatures (*3-6*). It has been proposed that the increased predominance of heterolytic reaction products with increasing pressure is correlated with the increase in the SCW ion product K_w with pressure (*3*). It is reasonable to attribute these observations to an increase in the microscopic kinetic rate constants for heterolytic reactions (and a corresponding decrease in those for homolytic reactions) with the increasingly ionic character of the solvent. In fact, it has been proposed that changes in electrostatic interactions due to known variations in SCW's static

0097–6156/94/0568–0196$08.00/0

dielectric constant with pressure may be an important cause of these observed rate variations (*7,8*). Here we provide a test of these ideas.

The rate of the (heterolytic) hydrolysis of anisole in SCW has been studied experimentally and found to increase with pressure, as suggested above (*4,9*). In the present study, we examine the effect of changing electrostatic interactions on the rate of this reaction in order to determine whether these effects can explain the experimentally determined pressure dependence of the rate. In particular, we examine the electrostatic contribution to the free energy of solvation along the reaction coordinate for this reaction, as the free energy of solvation determines the free energy of activation profile for the solvated reaction system. From the free energy of activation, we determine the equilibrium solvent effects on the rate of this reaction. Because the hydrolysis of anisole involves two neutrals evolving into two ions in a polar medium, it is reasonable to expect that electrostatic interactions will dominate the solvent effects. Additionally, in this initial study we neglect nonequilibrium solvent effects due to the frequency dependence of the orientational polarization, as these effects are expected to produce a linear, rather than an exponential, variation in the rate (*10*).

Here we model SCW as a continuum fluid characterized solely by its dielectric constant; yet, it is well know that reactivity in SCW differs from reactivity in organic solvents of similar polarity under standard conditions. While it may appear that this simple model must therefore miss the essential physics of reactivity in SCW, this is not true. First, perhaps the most important, but least exotic, reason for the appearance of novel reaction mechanisms in SCW is accounted for within this simple model—the elevated temperature. An energetic barrier to reaction which is prohibitively high at room temperature, say ~ 100 kT, will become effectively half as high, ~ 46 kT, at the critical temperature of water, allowing the reaction to occur. Second, SCW differs from typical low polarity organic solvents in that it is a hydroxylic solvent. Since the simple dielectric model does not account for hydrogen bonding effects, it is not an appropriate model for comparing reactivity in SCW (or any other hydroxylic solvent) with reactivity in aprotic solvents. In contrast, this simple model does provide a useful tool for comparing reactivity in SCW at different pressures/polarities, since SCW is hydroxylic at all pressures, and hence the neglected molecular effects should approximately cancel. Third, unusual reactivities are observed in SCW in part because reactant solubilities in SCW are different than they are in more conventional solvents. For the hydrolysis reaction considered here, SCW provides one of the reactants at high concentration—something not possible in many organic solvents.

In addition to the factors just given, SCW solutions exhibit two unique behaviors which may be important in determining the pressure dependence of the reaction rate. First, there is growing evidence that solvent-solute clustering occurs in SCW and other supercritical fluid solvents, due to the large isothermal compressibility of these solvents (*8,11,12*). Solvent-solute clustering means that the solvent density in the vicinity of the solute will be, on average, higher than the bulk density, and as a result the solute will be surrounded by a region having a higher dielectric constant than that of the bulk solvent. As a result, the variation in the electrostatic solvent effects with pressure may track a local static dielectric constant which is larger than the bulk dielectric constant (*8,13,14*). In addition, the degree of this clustering may vary along the reaction path, especially in the charge separation reaction considered here. Such a variation would cause the local dielectric constant, and hence the solvent-solute electrostatic interactions, to also vary along the reaction path. Free energy differences along the reaction path would thus be further altered from those for the reacting solute in an incompressible dielectric. Entropic effects, which make a contribution to the electrostatic free energy of solvation in the work required to reorient the solvent dipole density (*15*), would also be altered by variation in clustering along the reaction path. Specifically, there would be an additional loss of entropy along the reaction path if the solvent clustering increases along the reaction path. This entropic effect is not incorporated into standard electrostatic free energy calculations which assume an incompressible continuum fluid. The second important behavior which has been proposed to occur to an unusually large degree in SC fluids is solute-solute clustering (16). Hence it is possible that as the ion product K_w of SCW increases, protons from dissociated water molecules may be found near the reacting solute more frequently than random statistics would predict, thus allowing for an acid catalyzed mechanism. Although such a mechanism has been proposed to explain the pressure dependence of the hydrolysis of 2-methoxynapthalene (17), we do not consider this catalyzed mechanism in the present work. Also, recent mixed quantum mechanical/molecular mechanical simulations of the benzene dimer in SCW show no evidence of enhanced solute-solute clustering (18).

We first discuss the experimental data for anisole hydrolysis in SCW. Next we describe the reaction path energetics calculated for the gas phase reaction. We then consider the solution phase energetics, calculated via incompressible, continuum solvation models. We compare the free energies of solvation from various continuum models in order to gauge the reliability of our results and to explore the importance of solute polarization. The results from our most reliable model are then used to determine the free energy of activation curve for the title reaction as a function of the solvent dielectric constant. We then estimate the dependence of the solute rate constant on solvent dielectric constant and, by comparison with experiment, ascertain whether solvent clustering is an important factor in determining the rate.

Experimental Data

Klein and coworkers have studied the rates of hydrolysis of substituted anisoles in SCW (4). The experiments were performed at 380°C, which corresponds to a reduced temperature of $T_r = T/T_c = 1.01$, where $T_c = 374.0°C$ is the critic temperature of water. The reactions were studied in constant volume batch reactors at four reduced densities, ranging from $\rho_r = \rho/\rho_c = 0.8$ to $\rho_r = 2.0$, where the critical density of water is $\rho_c = 0.32$ g/cm^3. These densities correspond to a range of pressures from 23.2 MPa to 49 MPa (19,20). The products were analyzed by GC/MS. Klein and coworkers have also determined the mechanism for the hydrolysis of various phenyl ethers, and found that it involves S_N2 attack by a water molecule (9). Assuming this same mechanism for anisole, one has

$$PhOCH_3 + H_2O \rightarrow PhO^- + CH_3OH_2^+. \tag{1}$$

In fact, the experimentally measured products are PhOH and MeOH, but we assume that the proton transfers occur as rapid subsequent steps. The intrinsic bimolecular rate constant k_{exp} (L/mol s) was found to increase with pressure, i.e. $k_{exp} = 3.3 \times 10^5$ at $P = 23.2$ MPa, 6.7×10^5 at 24.0 MPa, 11×10^5 at 27.3 MPa and 18×10^5 at 49 MPa (4).

Gas Phase Energetics

The anisole hydrolysis reaction is a charge separation S_N2 reaction, and in the gas phase it is expected to be highly endothermic. We performed HF/SCF ab initio calculations on this reaction with both the 3-21G* and the 6-31G** basis sets using Gaussian 92 (21). The computed overall endoergicity at the HF/6-31G** level is 178.95 kcal/mol. The conversion to endothermicity can be approximated from the reactant and product zero point energies. Thus, we calculated frequencies at the HF/3-21G* level and, by comparison with experimental frequencies

Table I. HF/6-31G** energies
along the reaction path

Point	r_c (Å)	E (kcal/mol)
Reac	$-\infty$	0.00
DD	-1.83804	-1.76
x1	-0.06955	57.92
x2	0.23609	73.77
x3	0.58367	84.08
x4	0.98367	90.55
Prod	∞	178.95

for anisole (*22,23*), found that the calculated frequencies should be reduced by 11.29%. The calculated 6-31G** endothermicity, based on the scaled 3-21G* frequencies, is 172.60 kcal/mol, in excellent agreement with the experimental value of 172 kcal/mol. The experimental value was computed from heats of formation taken from References *24* (H_2O), *25* ($PhOCH_3$), and *26* (PhO^-, $CH_3OH_2^+$).

Recent studies on other charge separation S_N2 reactions indicate that they may be expected to show the classic double-well reaction profile found in charge transfer S_N2 reactions (*27,28*). However, we did not find a double well profile for this reaction with either basis set at the Hartree-Fock level. Instead, we found only three stationary points along the reaction path: the reactants (Reac), the products (Prod) and the water-anisole dipole-dipole complex (DD). No saddle point and no ion-ion complex were found along the S_N2 backside attack reaction path ($PhO...CH_3...OH_2$). While it is likely that an ion-ion complex exists at another geometry (*e.g.* $PhO...H_2O...CH_3$), the calculations indicate that there is a barrier separating the reaction path from any such complex, so we did not search further for this complex. The lack of a saddle point makes it infeasible to follow the minimum energy path at the Hartree-Fock level. Instead, we follow an assumed reaction coordinate, r_c, which is defined as the methyl carbon to phenyl oxygen distance, $r1$, minus the methyl carbon to water oxygen distance, $r2$. A shoulder was found along this reaction path with the 3-21G* basis. We chose four points in the vicinity of this shoulder to characterize the reaction path and evaluated the energies of these points optimized with respect to everything but r_c at the 6-31G** level. The HF/6-31G** relative energies along this reaction path are given in Table I and illustrated in Figure 1.

Energetics in Supercritical Water

The bulk static dielectric constant of SCW at 380° C ranges from $\sim$ 2 at 20 MPa to $\sim$ 14 at 50 MPa (*19,20*). Hence, to determine the electrostatic effects of SCW on the energetics of this reaction, we model SCW as a continuous dielectric with a pressure dependent dielectric constant. We considered a range of dielectrics from 2 to 80, rather than only from 2 to 14, because clustering may raise the local dielectric felt by the solute above the bulk value of the dielectric constant. To determine the free energy of activation curves for this reaction in SCW as a function of pressure, we calculate the free energy of solvation at 7 points along the reaction path for each of a series of dielectric constants.

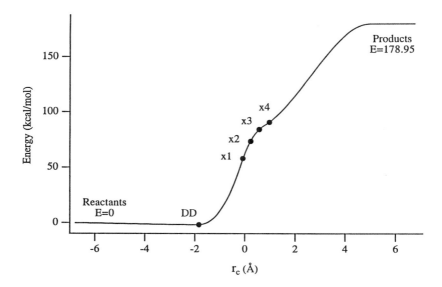

Figure 1. Calculated gas phase energies along the reaction path. The solid line is a spline fit to the data.

Comparison of Methods. A variety of methods have been proposed for calculating solvation energies within the continuum solvent approximation, as discussed in a recent review (*29*). Methods which include semiempirical parameters to account for cavitation and other nonelectrostatic effects, such as the AM1-SM2 model of Cramer and Truhlar (*30*) or the recent models of Hehre and coworkers (*31*), are not appropriate for studies of SCW because the semiempirical corrections are fit to data at room temperature and pressure. Thus, we restrict ourselves to models which incorporate only electrostatic effects. Note that all of these continuum solvation methods assume an incompressible fluid, and hence ignore the effects of electrostriction. This is liable to be a poor approximation in SCW since SCW has a large isothermal compressibility. In the present study we assume that this compressibility need not be taken into account explicitly with a position dependent dielectric, but that it may be accounted for by an effective local dielectric constant which differs from the bulk value. This idea has been used previously to explain spectroscopic data (*8,13,14*). We will address the incorporation of electrostriction into the continuum model in future work.

In the present study we consider two models of solute-incompressible solvent electrostatic free energies. First we consider a numerical grid based algorithm for solving Poisson's equation for an arbitrary, fixed charge distribution (*32,33*). We use the implementation available in the program Delphi, distributed as part of Biosym (*34*). Solvation energies evaluated by this method are generally in good agreement with experiment for ions (*30,33,35*). For neutrals, where the electrostatic interactions often do not dominate the solvation energy, the Delphi energies show a significantly weaker correlation with experiment, although they are frequently within a few kcal/mol of the experimental results (*29*). Also, Delphi results are somewhat dependent upon the choice of input parameters, including the atomic radii, point charge distribution and solute dielectric constant (*29,35,36*).

In the present calculations, the gas phase HF/6-31G** geometry and Mulliken charges are used to define the solute charge distribution. A study by Alkorta, *et al.* compares Delphi solvation energies using atomic charges determined from AM1 wave functions by two different methods: the Mulliken population analysis and a fit to the AM1 electrostatic potential (ESP) (*36*). The Mulliken charges fairly consistently yield a slightly smaller solvation energy than do the fitted charges. The maximum difference was 6 kcal/mol, with the average difference being significantly less, at 1.3 kcal/mol. Indeed, as the difference in solvation energy is fairly consistently in the same direction, the differences in relative solvation energies between the two methods is expected to be less than the difference in the absolute solvation energies. Thus, while the fitted charges provide a more accurate representation of the gas phase wave function, the extra effort of obtaining fitted charges was not considered warranted. In particular, the Delphi method neglects solute polarization in the presence of the solvent, which would alter the solute charge distribution from its gas phase value. Hence, within this method, a more accurate gas phase charge distribution will not necessarily yield a better estimate of the solvation energy.

Another parameter which is required for Delphi calculations is the dielectric constant of the solute molecule, ϵ_m. We consider two values of ϵ_m: $\epsilon_m = 1$, corresponding to a vacuum (Delphi1) and $\epsilon_m = 2$, corresponding to the usual value of the optical dielectric constant due to electronic polarizability (Delphi2). We discuss the relative merits of these two choices below. For the atomic radii, we simply use the default Van der Waals radii (C= 1.55 Å, O= 1.35 Å and H= 1.10 Å) (*34*). Finally, we checked the Delphi results for convergence with respect to the size and spacing of the numerical grid. We found that a border space of only 8 Å was required to converge the results to 0.1 kcal/mol. Convergence with respect to grid spacing was more erratic. Even by a grid spacing of 0.115, smooth convergence was not observed. The production calculations were all performed with a grid spacing of 0.2 Å, as suggested recently (*35*), as this value provides a reasonable compromise between convergence and memory requirements. Based on these convergence studies, we attribute an error of ±1 kcal/mol to the finite grid spacing. Note that coulombic boundary conditions were used with no focusing. Solvation energies were computed from the total electrostatic energy, rather than from the reaction field energy (*37*).

The second method we consider is the self consistent Born-Onsager Reaction Field approximation (BO) implemented in Gaussian 92 (*21,38*). This method models the molecule as a multipole expansion in a spherical cavity, but includes only the monopole and dipole terms. Since an ideal dipole in a spherical cavity will not go smoothly to the limit of two monopoles, this method is expected to fail at some point along the reaction path as the complex separates into two charged species. The advantage of this method is that it self-consistently solves the electronic structure problem in the presence of the solvent reaction field, allowing for polarization of the solute wave function. Such a self-consistent treatment is tractable by virtue of the computational simplicity of evaluating the reaction field in the BO approximation. Knowledge of the solvent polarization effects for this reaction will be important for the development of model potentials for use in explicit simulation studies of solvent clustering effects. Specifically, the importance of solute polarization will determine whether it is reasonable to use a molecular mechanics potential to study this system, as such potentials do not generally allow for solute polarization.

The BO method has been shown to give good relative solvation energies for a few isomerization reactions (*29*). We implement the BO method at the HF/6-31G** level. Like the numerical grid methods, the BO method is sensitive to cavity size. The cavity radii used here are determined from an isodensity surface of the gas phase wave function as implemented in Gaussian 92 (*21,39,40*).

It is appropriate to mention a third method which combines the advantages of the two methods considered here, although at additional computational expense. Specifically, the Polarized Continuum Model (PCM) developed by Tomasi and coworkers self-consistently solves the electronic structure in the presence of a reaction field (apparent surface charge) determined by numerical solution of Laplace's equation (*41*). A similar method has also recently been developed by Tannor, et al. (*42*). While still suffering from the same parameter sensitivity as Delphi, these methods will allow for a more accurate study of the solute polarization effects than the BO method (see below). The PCM method has recently been used to study a similar reaction, the Menshutkin reaction of ammonia with methyl bromide (*28*). In that study, solute polarization effects were found to be significant. These results, along with those of the present study, suggest that a detailed description of the solute polarization will be required to accurately model the dynamics of the anisole hydrolysis reaction.

Comparison of Reactant and Product Solvation Energies. In order to provide an estimate of the reliability of the methods used here, we compare the calculated solvation energies of the reactants and products with solvent dielectric $\epsilon_s = 80$ to experimental values of the solvation energies in water at room temperature and pressure (*29,43,44*). The results are tabulated in Table II. In addition to the methods we use here, BO, Delphi1 and Delphi2, we list values calculated by other authors using related methods (*29,36*). Two of these are also Delphi calculations, but they are implemented with AM1 optimized geometries and charges; AM1ED uses the ESP fitted charges while AM1MD uses the Mulliken charges (*36*). Both of these implementations use a solute dielectric constant of $\epsilon_m = 1$. The last method listed is the AM1-SM2 semiempirical model (*30*).

Looking first at the BO method, one sees that the magnitudes of all of the solvation energies are severely underestimated, especially for the ions. While the difference in solvation energies between the two ions and between the two neutrals

Table II. Solvation energies in kcal/mol for Reactants and Products ($\epsilon_s = 80$)

Solute	Expt.	BO	Delphi1	Delphi2	AM1ED	AM1MD	AM1 − SM2
Water	−6.3	−2.1	−8.7	−6.6	−3.0	−2.1	−6.3
Anisole	−2.4	−0.2	−15.7	−11.2	...	...	−2.3
PhO$^-$	−72.0	−44.9	−79.5	−74.9	−60.7	−58.8	−65.6
CH$_3$OH$_2^+$	−83.	−55.9	−82.8	−81.0	−70.8	−70.0	−84.2

is reasonable, to count on such a large cancellation of error is very risky. Also, the ion to neutral difference is quite poor. Clearly, the cavity shape (*e.g.* for PhO$^-$) and higher order electric moments are indeed important, as would be expected. Thus, the BO method only reproduces the qualitative trends.

Overall, the 4 implementations of Delphi perform much better, as compared to experiment, than does the BO method. Anisole provides an exception; its solvation energy is significantly overestimated by Delphi. Delphi is known to overestimate the solvation of related compounds, such as benzene, toluene and analine (*29*). Comparing AM1ED and AM1MD with Delphi1 (since these methods all use $\epsilon_m = 1$), it is clear that the difference which results from basing the calculation upon AM1 rather than HF/6-31G** wave functions (AM1MD vs. Delphi1) is much greater than the difference which results from using ESP rather than Mulliken charges from the same wave function (AM1ED vs. AM1MD). In fact, choice of wave function alters the solvation energy by as much as 21 kcal/mol, whereas the maximum change due to choice of charge model is 1.9 kcal/mol. This supports the present choice of using the Mulliken charges. Also, the HF/6-31G** wave functions provide better agreement with experiment than do the AM1 wave functions.

The Delphi2 results, which use $\epsilon_m = 2$, are in the best agreement with experiment, in accord with previous results (*35*). The usual justification for this choice of solute dielectric constant is that because a frozen charge model has been used for the solute, an optical dielectric constant should be included to account for the neglected solute polarizability (*45,46*). There is a fault in this logic when it is applied to small molecule calculations (but not necessarily when it is applied to proteins), as follows. A dielectric constant is a macroscopic quantity which accounts for the polarizability of electrons, dipoles, etc. which *are not included explicitly* in the model. Charges which are included explicitly (microscopic treatment) cannot simultaneously be represented by a dielectric constant (macroscopic treatment). In small molecules, then, when all of the charges are included explicitly, *albeit* represented by a fixed distribution of point charges, there is no neglected charge distribution for the solute dielectric of $\epsilon_m = 2$ to be accounting for, and a value of $\epsilon_m = 1$ is appropriate.

Additionally, using a solute dielectric of $\epsilon_m = 2$ will *decrease* the magnitude of the solvation energy relative to that for $\epsilon_m = 1$ (as illustrated in Table II and Reference *35*). In contrast, the allowance of explicit solute charge polarization will always *increase* the magnitude of the solvation energy relative to the fixed charge distribution value, otherwise the solute charges would not readjust. It follows that setting the solute dielectric to 2 cannot account for the neglect of solute polarizability incurred by using a fixed set of point charges. Hence, we consider the use of $\epsilon_m = 2$ in small molecule calculations to be an empirical correction for other errors in the model, such as the choice of wave function, etc. In fact, while using $\epsilon_m = 2$ in the Delphi HF/6-31G** calculations improves the comparison with

Table III. Solvation Energy Relative
to Reactants in kcal/mol ($\epsilon_s = 40$)

Point	Delphi1	Delphi2	BO
Reac	0.0	0.0	0.0
DD	2.4	1.7	0.8
x1	−7.7	−6.8	−6.1
x2	−15.8	−13.8	−13.0
x3	−26.6	−23.0	−23.6
x4	−37.8	−33.0	−34.5
Prod	−136.2	−136.4	−97.3

experiment, using this choice of ϵ_m in the AM1ED or AM1MD calculations would further reduce the agreement of these calculations with experiment. Thus, despite the agreement of Delphi2 with experiment under ambient conditions, Delphi2 is not considered to be the most reliable method for the present studies.

The AM1-SM2 semiempirical model, included for comparison, gives noticeably better agreement with experiment than any of the Delphi implementations. This model even provides a good value for the solvation energy of anisole. As suggested above, this model was not considered because of its semiempirical character. The parameters accounting for cavitation, dispersion and solvent rearrangement are fit to room temperature and pressure data. It is not known how these effects are altered in the supercritical regime; hence the inclusion of these terms in a semiempirical way would simply complicate the analysis of our results.

Comparison of Reaction Path Solvation Energies. The focus of the present study is to determine relative changes in the solvation energy along the reaction path. This is a less demanding test than is computation of the absolute solvation energies, so we also compare the relative solvation energies along the reaction path for the three methods, Delphi1, Delphi2 and BO. The results for $\epsilon_s = 40$ are presented in Table III and their negatives are pictured in Figure 2.

The relative solvation energies are in better agreement than were the absolute solvation energies. Most striking is the agreement of the BO method with the Delphi methods at all points except the products. In fact, the relative energies suggest that the ideal dipole approximation does not break down as rapidly as expected with increasing dipole separation. However, a more detailed look at the BO results (see below) indicates that the observed agreement is somewhat fortuitous.

The differences in the relative energies of the two Delphi implementations, though small, vary along the reaction path; so we studied this difference in more detail. The difference in the relative solvation energies as computed by Delphi1 and Delphi2 at points along the reaction path is shown in Figure 3 for six different values of the solvent dielectric constant. The magnitude of the relative solvation energy increases more rapidly with increasing charge separation along the reaction path for the calculations with the solute dielectric $\epsilon_m = 1$ than for those with $\epsilon_m = 2$. Additionally, the difference in the results for these two choices of ϵ_m is greater the greater the solute dielectric constant. Hence this choice will affect the trends calculated for the equilibrium solvent effects for this reaction. Because the use of $\epsilon_m = 1$ makes intuitively more sense for this calculation, as discussed above, the Delphi1 calculations should more accurately reflect the relative solvation energies under different conditions than do the Delphi2 calculations. We therefore take

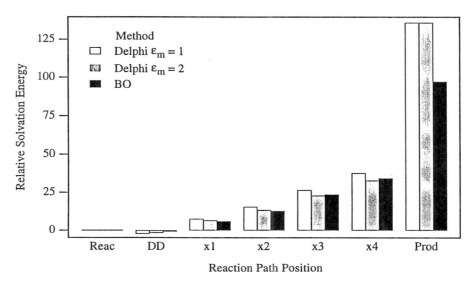

Figure 2. Negative of the relative free energy of solvation along the reaction path for a solvent dielectric constant of 40.

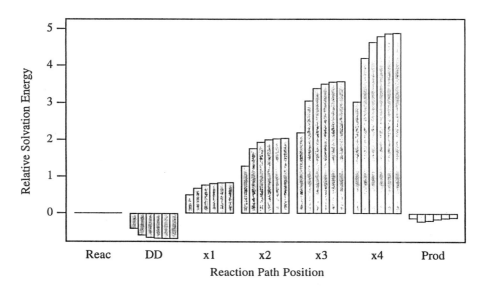

Figure 3. Difference in the calculated values of the relative solvation energy for the Delphi2 and Delphi1 methods. For each reaction path position, the bars from left to right correspond to ϵ_s values of 2, 4, 8, 15, 40 and 80, respectively.

the Delphi1 calculations as the most reliable, despite the better comparison of the Delphi2 absolute solvation energies with experiment.

Solute Polarization. Since the BO method allows for self-consistent solution of the electronic structure problem in the presence of the solvent, we used the BO method to study the importance of solute polarization for this reaction. The BO method predicts quite large induced dipoles for points intermediate along the reaction path. In the most extreme example, point x4 at $\epsilon_s = 80$, the BO method predicts an induced dipole of 7.5 D, on top of the computed gas phase value of 17.7 D. This dipole moment change is due almost entirely to charge redistribution; the change in the distance $r1 + r2$—which is directly related to the oxygen-oxygen distance—from its gas phase value to its value upon solvation at $\epsilon_s = 80$ is less than 0.03 Å.

The energy cost of the solute distortion at this point, x4 with $\epsilon_s = 80$, is 14.0 kcal/mol. To evaluate the additional polarization energy gained by this change in dipole moment, we evaluated a frozen-BO polarization energy at this point, *i.e.* we evaluated the BO solvation energy for the unpolarized gas phase wave function and geometry. The frozen-BO polarization energy is −25.4 kcal/mol as compared to the relaxed-BO polarization energy of −51.7 kcal/mol. Hence, a gain in polarization energy of 26.3 kcal/mol offsets the solute distortion cost of 14.0 kcal/mol to yield a net gain in solvation energy upon solute polarization of 12.3 kcal/mol.

The large gain in polarization energy upon solvation of the polarized solute predicted by the BO approximation is in contrast to what is found with Delphi calculations of the same charge distributions. Specifically, Delphi1 calculations using the gas phase geometry and charges for this point yield a polarization energy of −62.8 kcal/mol. When the relaxed-BO geometry and charges are used instead, the Delphi1 calculations predict a polarization energy of −69.8 kcal/mol, only 7 kcal/mol more favorable than for the unrelaxed molecule. The gain in polarization energy predicted by Delphi is not even sufficient to offset the solute distortion cost, indicating that if the more accurate Delphi solvation model were used in the self-consistent calculations the BO-predicted charge redistribution would not occur.

This observation can be explained by the fact that the BO approximation neglects moments higher than the dipole moment. At the point x4, the gas phase Mulliken charge distribution grouped by moiety is −0.03 on the phenyl ring, −0.89 on the phenyl oxygen, +0.61 on the methyl group, and +0.30 on the water. This charge distribution can be loosely thought of as being comprised of an overall dipole plus one dipole on each of the PhO and MeOH$_2$ fragments which give rise to the quadrupole and higher order terms. Since the BO approximation neglects the quadrupole moment, it will gain an increase in solvation energy if the dipole moment is increased at the expense of the quadrupole moment. Indeed, this is what happens. The relaxed-BO charge distribution increases the charge separation between the PhO and MeOH$_2$ groups by only 0.05. The large induced dipole arises primarily from a shift of negative charge *away* from the phenyl oxygen and onto the most extended hydrogen of the phenyl ring, thus increasing the dipole moment but decreasing the quadrupole moment. Specifically, the relaxed-BO charge distribution is −0.17 on the phenyl ring, −0.80 on the phenyl oxygen, +0.63 on the methyl group and +0.34 on the water. We conclude that the BO model compensates for the neglect of higher moments by inducing an artificially large dipole. By virtue of this compensation, the BO approximation attains a reasonable estimate of the relative solvation energy of this compound; yet, the physics of the induced polarization is incorrect. Thus, we do not consider the BO results sufficiently meaningful to tabulate them here. We note that, despite this nonphysical behavior, the magnitude of the BO induced dipole exhibits solute polarization trends similar to those observed by Solà, *et al.* (*28*) for a Menshutkin reaction using the method of Tomasi and coworkers (*41*). These results indicate that, while the BO model treats the solute polarization incorrectly, solute polarization effects may still be important for the anisole hydrolysis reaction.

Free Energy of Activation Profile. The free energy of activation profile for the anisole hydrolysis reaction in SCW is determined by the solvation energies along the reaction path. The magnitude of these solvation energies, calculated by the most reliable method considered, Delphi1, are shown for 6 values of the solute dielectric constant ($\epsilon_s = 2, 4, 8, 15, 40, 80$) in Figure 4. Note first that the solvation energies are asymmetric, *i.e.* they increase in magnitude as the reaction proceeds. This behavior is expected for a charge separation reaction, because as the reaction proceeds the solute becomes more ionic and is thus more highly stabilized by a polar solvent. Such asymmetric solvation has been discussed recently for Menshutkin reactions (*27,28*). Second, the solvation energies at a dielectric of $\epsilon_s = 2$, the bulk dielectric of SCW at 380°C and ~ 20 MPa, are already at 40 to 50% of their "maximum" values at $\epsilon_s = 80$. At $\epsilon_s = 15$, the bulk dielectric constant of SCW at 380°C and ~ 57 MPa, the solvation energies are at $\sim 90\%$ of their $\epsilon_s = 80$ values.

Combining the calculated free energies of solvation with the gas phase reaction profile yields the free energy of activation curve in solution. These curves are presented schematically as a function of dielectric constant in Figure 5. The reaction path value (r_c) of each point shown is given in Table I. The zero of energy is in all cases taken to be the energy of the reactants at a dielectric constant of $\epsilon_s = 80$, so that all energies will be positive. From Figure 5 it is clear that the asymmetrical nature of the solvation energies causes the qualitative shape of the reaction path to be altered as a function of dielectric constant. The products remain the most energetically unfavorable configuration for all $\epsilon_s \leq 4$. A saddle point is observed to grow in at $\epsilon_s \geq 4$ and to move to earlier locations as the dielectric constant is increased, in agreement with the Hammond postulate and previous results on Menshutkin reactions (*27,28*). Note that in contrast to the Menshutkin reaction results, an ion-ion complex is only observed at $\epsilon_s \approx 4$. For lower dielectric solvents the ion-ion configuration is less stable than earlier points on the reaction path, while for higher dielectric solvents the ion-ion configuration is less stable than the products. Also, the dipole-dipole complex disappears somewhere between $\epsilon_s = 4$ and $\epsilon_s = 8$.

The potential barrier for this reaction changes from the gas phase endoergicity of 179 kcal/mol in vacuum, through the endoergicity value of 74 kcal/mol at $\epsilon_s = 4$, to the saddle point value of ~ 61 at $\epsilon_s = 8$ and of ~ 58 at $\epsilon_s = 80$. Thus, the continuum solvation model predicts that the energetics of this reaction change dramatically over the range of dielectrics explored by SCW under the conditions of Klein's experiments.

Discussion and Conclusions

The theoretical treatment presented here is restricted to a one dimensional reaction coordinate picture, and hence is not expected to generate good estimates of the absolute reaction rates. Instead, we concentrate on the variation in the rates with pressure. In fact, it is more informative to compare the pressure dependence of the theoretical and experimental activation energies than it is to compare the pressure dependence of the resulting reaction rates. Thus, we extract differences in the free energy of activation as a function of pressure, $\Delta(\Delta G^\ddagger)$, from the experimental rate constants (*4,47*). To do so, we assume a standard transition state theory expression for the rate constant, *i.e.*

$$k \propto e^{-\Delta G^\ddagger/RT} \qquad (2)$$

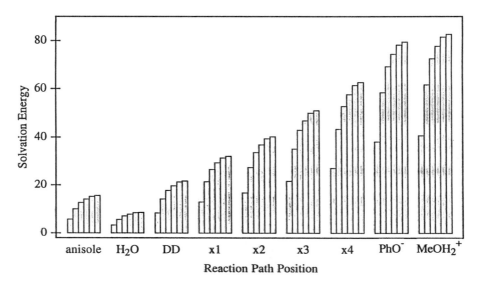

Figure 4. The magnitude of the absolute solvation energy. For each reaction path position, the bars from left to right correspond to ϵ_s values of 2, 4, 8, 15, 40 and 80, respectively.

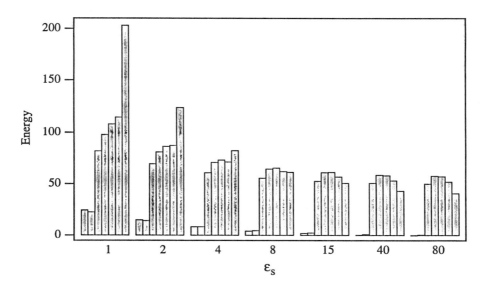

Figure 5. The calculated free energy of activation curves. For each value of ϵ_s, the bars from left to right correspond to the reaction path positions Reac, DD, x1, x2, x3, x4 and Prod, respectively.

Table IV. Calculated vs. experimental free energy of activation differences $\Delta(\Delta G^{\ddagger})$ in kcal/mol (*20*)

ρ_r	$\rho(g/cm^3)$	P(MPa)	ϵ^a		$\Delta G^{\ddagger}$ calc	$\Delta(\Delta G^{\ddagger})$ calc	$\Delta(\Delta G^{\ddagger})$ exp
0.8	0.25	23	4		74.2		
						−13.3	−1.5
1.6	0.50	27	10	(8)	60.9		
						−1.9	−0.7
2.0	0.63	49	14	(15)	59.0		
						−1.0	...
...	...	...		(40)	58.0		

^{a}The value in parentheses is the value used in the theoretical calculations.

where R is the ideal gas constant and T is temperature. To correlate the experimental free energy differences with the calculated values, we evaluate the bulk dielectric constant of SCW at the experimental densities(*20*). The results are presented in Table IV. Note that the calculated free energies of activation $\Delta G^{\ddagger}$ are not exact, even for the method used, because the position of the maximum on the (assumed) reaction path was determined from a very coarse grid of points.

The comparison of free energy differences $\Delta(\Delta G^{\ddagger})$ given in Table IV equates the experimental and theoretical values based on the bulk dielectric constant of the solvent. This correlation is extremely poor. The calculations predict a significantly larger pressure effect on the rate than is observed experimentally. However, if solvent clustering is assumed to occur, then the local dielectric constant, which will be greater than the bulk dielectric constant, should be used in the calculations. Flarsheim, *et al.* have calculated the local density of SCW around I⁻ at 385° C as a function of pressure using an electrostriction model (*11*). For a pressure of ~ 25 MPa they find the SCW density at 2.2 Å from the I⁻ ion to be ~ 0.8 (g/cm³), as compared to a bulk density of ~ 0.3 (g/cm³). This corresponds approximately to a local dielectric constant (at 2.2 Å) of ~ 20, as compared to a bulk dielectric of ~ 5 (*20*). For the present reaction, we make a simple estimate of the degree of solvent clustering and the associated local dielectric constants based on the calculated $\Delta(\Delta G^{\ddagger})$ results. Specifically, we let the local dielectric constant be approximately double the bulk value, *i.e.* $\epsilon_{bulk} = 4 \rightarrow \epsilon_{local} = 8$, $8 \rightarrow 15$ and $15 \rightarrow 40$. This estimate of the local dielectric constant is similar in magnitude to the clustering effects observed by Flarsheim, *et al.*, and yields a much better correlation between the calculated and experimental $\Delta(\Delta G^{\ddagger})$'s than was found using the bulk dielectric values. In Table IV, this transformation corresponds to comparing the calculated $\Delta(\Delta G^{\ddagger})$ value of −1.9 with the experimental value of −1.5 and the calculated value of −1.0 with the experimental value of −0.7. Given the small free energy changes observed in the experiments, the accuracy of the present calculations is insufficient to determine reliably the effective local dielectric constant as a function of pressure. However, the results indicate quite clearly that clustering is an important factor in determining the pressure dependence on the rate of this reaction. Also, it seems likely that electrostatic effects indeed dominate the pressure dependence observed for this reaction.

Since we have determined clustering to be significant for the anisole hydrolysis reaction, it will be important to determine how large are the neglected entropy of activation effects due to variable clustering along the reaction path. For the charge separation reaction studied here, the clustering is expected to be stronger at the saddle point than at reactants, thus causing an additional negative contribution to the entropy of activation. This effect, like the local dielectric effect of clustering, would reduce the degree to which the reaction rate constant increases with pressure. It seems likely that these variable clustering effects will have to be taken into account if we are to make more definitive conclusions about the degree of solvent clustering for this reaction. We are thus pursuing the explicit inclusion of electrostriction effects into the free energy calculations as a function of the reaction coordinate. Also, experimental results which explore lower values of the effective dielectric constant where much larger changes in the rate are predicted would provide a significantly more challenging and reliable test of these ideas.

Acknowledgments. We thank Martin Head-Gordon for showing us a very useful Z-Matrix trick. This work has been supported by grant no. CHE-9307679 from the National Science Foundation and by the San Diego Supercomputer Center.

Literature Cited

1. Brennecke, J. F. and Eckert, C. A., *AIChE J.* **1989**, *35*, 1409.
2. Subramaniam, B. and McHugh, M. A., *Ind. Eng. Chem. Res.* **1986**, *25*, 1.
3. Antal, Jr., M. J., Brittain, A., DeAleida, C., Ramayya, S., and Roy, J. C., in *Supercritical Fluids, ACS Symposium Series 329*, Squires, T. G. and Paulaitis, M. E., Eds., ACS, Washington **1987**.
4. Klein, M. T., Mentha, Y. G., and Torry, L. A., *Ind. Eng. Chem. Res.* **1992**, *31*, 182.
5. Narayan, R. and Antal, Jr., M. J., *J. Amer. Chem. Soc.* **1990**, *112*, 1927.
6. Abraham, M. A. and Klein, M. T., in *Supercritical Fluids, ACS Symposium Series 329*, Squires, T. G. and Paulaitis, M. E., Eds., ACS, Washington **1987**.
7. Wu, B. C., Klein, M. T., and Sandler, S. I., *Ind. Eng. Chem. Res.* **1991**, *30*, 822.
8. Johnston, K. P. and Haynes, C., *AIChE J.* **1987**, *33*, 2017.
9. Klein, M. T., Torry, L. A., Wu, B. C., Townsend, S. H., and Paspek, S. C., *J. Supercritical Fluids* **1990**, *3*, 222.
10. Hänggi, P., Talkner, P., and Borkovec, M., *Rev. Mod. Phys.* **1990**, *62*, 250.
11. Flarsheim, W. M., Bard, A. J., and Johnston, K. P., *J. Phys. Chem.* **1989**, *93*, 4234.
12. Cochran, H. D., Cummings, P. T., and Karaborni, S., *Fluid Phase Eq.* **1992**, *71*, 1.
13. Blitz, J. P., Yonker, C. R., and Smith, R. D., *J. Phys. Chem.* **1989**, *93*, 6661.
14. Ikushima, Y., Saito, N., and Arai, M., *J. Phys. Chem.* **1992**, *96*, 2293.
15. Laidler, K. J., *Chemical Kinetics, 2nd ed.*, Mc-Graw Hill, New York **1965**.
16. Chialvo, A. A. and Debendetti, P. G., *Ind. Eng. Chem. Res.* **1992**, *31*, 1391.
17. Baur, H. L. C., Master's Thesis, Eindhoven, 1991.
18. Gao, J., *J. Amer. Chem. Soc.* **1993**, *115*, 6893.
19. Haar, L., Gallagher, J. S., and Kell, G. S., *NBS/NRC Steam Tables*, Hemisphere, Washington,D. C. **1984**.
20. Pressures were determined from densities by simple linear extrapolation of the steam tables, Ref. *19*. Static dielectric constants were determined using eq. C.4 of Ref. *19*.
21. Frisch, M. J., Trucks, G. W., Head-Gordon, M., Gill, P. M. W., Wong, M. W., Foresman, J. B., Johnson, B. G., Schlegel, H. B., Robb, M. A., Replogle, E. S., Gomperts, R., Andres, J. L., Raghavachari, K., Binkley, J. S., Gonzalez, C., Martin, R. L., Fox, D. J., Defrees, D. J., Baker, J., Stewart, J. J. P., and Pople,

J. A., Gaussian, Inc., Pittsburgh PA, 1992; also, Gaussian 92, Revision C.2 and Gaussian 86, Revision C.

22. Fewster, S., Ph.D. Thesis, University of Manchester, UK, 1970.
23. Vincent, M. A. and Hillier, I. H., *Chem. Phys.* **1990**, *140*, 35.
24. Wagman, D. D., Evans, W. H., Parker, V. B., Schumm, R. H., Halow, I., Bailey, S. M., Churney, K. L., and Nuttall, R. L., "The NBS Tables of Chemical Thermodynamics Properties", *J. Phys. Chem. Ref. Data* **1982**, *11*, supp. No.2.
25. Cox, J. D. and Pilcher, G., *Thermodynamics of Organic and Organometallic Compounds*, Academic Press, New York **1970**.
26. Lias, S. G., Bartmess, J. E., Liebman, J. F., Holmes, J. L., Levin, R. D., and Mallard, W. G., "Gas Phase Ion and Neutral Chemistry", *J. Phys. Chem. Ref. Data* **1988**, *17*, Supp. No. 1.
27. Gao, J. and Xia, X., *J. Amer. Chem. Soc.* **1993**, *115*, 9667.
28. Solà, M., Lledos, A., Duran, M., Bertrán, J., and Abboud, J.-L. M., *J. Amer. Chem. Soc.* **1991**, *113*, 2873.
29. Cramer, C. J. and Truhlar, D. G., "Continuum Solvation Models: Classical and Quantum Mechanical Implementations", *Rev. Comp. Chem.*, *6* , in press.
30. Cramer, C. J. and Truhlar, D. G., *Science* **1992**, *256*, 213.
31. Dixon, R. W., Leonard, J. M., and Hehre, W. J., *Spartan 3.0*, Wavefunction Inc., Irvine, CA **1993**.
32. Davis, M. E. and McCammon, J. A., *Chem. Rev.* **1990**, *90*, 509.
33. Gilson, M. K. and Honig, B. H., *Proteins* **1988**, *4*, 7.
34. Biosym Technologies of San Diego, Version 2.4, 1992.
35. Mohan, V., Davis, M. E., McCammon, J. A., and Pettit, B. M., *J. Phys. Chem.* **1992**, *96*, 6428.
36. Alkorta, I., Villar, H. O., and Perez, J. J., *J. Comp. Chem.* **1993**, *14*, 620.
37. *Delphi User Guide, version 2.4*, Biosym Technologies, San Diego **1993**.
38. Wong, M. W., Frisch, M. J., and Wiberg, K. B., *J. Amer. Chem. Soc.* **1991**, *113*, 4776.
39. Wong, M. W., Wiberg, K. B., and Frisch, M. J., *J. Amer. Chem. Soc.* **1992**, *114*, 1645.
40. Radius values used in the BKO calculations were 4.11 Å (anisole), 2.53 Å (water), 4.40 Å (DD), 4.43 Å (x1), 4.41 Å (x2), 4.41 Å (x3), 4.43 Å (x4), 3.93 Å (PhO$^-$) and 2.99 Å (CH$_3$OH$_2^+$). All volumes were calculated at the HF/6-31G** level except that for (CH$_3$OH$_2^+$), which was evaluated at the HF/3-21G*//6-31G** level.
41. Miertuš, S., Scrocco, E., and Tomasi, J., *Chem. Phys.* **1981**, *55*, 117.
42. Tannor, D. J., Marten, B., Murphy, R., Friesner, R. A., Nicholls, A., Honig, B., and Ringnalda, M., "Accurate First Principle Calculation of Molecular Charge Distributions and Solvation Energies from Ab Initio Quantum Mechanics and Continuum Dielectric Theory", preprint.
43. Kuyper, L. E., Hunter, R. N., Ashton, D., K. M. Merz, J., and Kollman, P. A., *J. Phys. Chem.* **1991**, *95*, 6661.
44. Pearson, R. G., *J. Amer. Chem. Soc.* **1986**, *108*, 6109.
45. Tanford, C. and Kirkwood, J. G., *J. Amer. Chem. Soc.* **1957**, *79*, 5333.
46. Gilson, M. K., Rashin, A., Fine, R., and Honig, B. H., *J. Mol. Biol.* **1985**, *183*, 503.
47. The activated complex is a transient species, so it is considered to be in infinite dilution. In this limit no *PV* work is done, so it is reasonable to use quasi-Gibbs free energies for the activated complex, even though the experiments were performed at constant volume.

RECEIVED May 2, 1994

Chapter 15

Simulating Solvent Effects on Reactivity and Interactions in Ambient and Supercritical Water

Jiali Gao and Xinfu Xia

Department of Chemistry, State University of New York at Buffalo, Buffalo, NY 14214

The effects of hydration on the rate acceleration of the Claisen rearrangement of allyl vinyl ether and the Menshutkin reaction of ammonia and methyl chloride were investigated by a hybrid quantum mechanical and classical Monte Carlo simulation method. In addition, the potentials of mean force for the ion pair Na^+Cl^- in ambient and supercritical water were determined. The results provided valuable insights on intermolecular interactions for these processes in solution.

Study of chemical transformations in solution is important because of the connection to biological processes in life. Of great challenge is to gain an atomic level understanding of structure and reactivity in aqueous solution. However, the difficult task of describing gas-phase reactions is further complicated by the need to consider the solvent effects on the reaction dynamics and potential surface (1,2). Significant progress has been made in the past decade through computer simulations and a number of methods are being developed to investigate chemical reactions in solution and enzymes (3-7). In short, the computational procedure, as summarized by Jorgensen (6), typically involves three major steps: (1) determination of the minimum energy reaction path (MERP) in the gas phase as a function of a single geometrical variable, (2) development of empirical potential functions for the reaction profile as well as for solute-solvent interactions along the entire MERP, and (3) free energy simulations to estimate the solvent effects. Valuable insights have been obtained for organic reactions in solution. Nevertheless, a major difficulty in these studies is the requirement for an accurate, analytical description, i.e., empirical potential functions, of solute-solvent interactions along the whole reaction path. The parametrization process was laborious and difficult due to a lack of experimental data, while the empirical molecular mechanics-type potentials are generally not appropriate for treatment of bond formation and

0097–6156/94/0568–0212$08.00/0

breaking processes (*4*), which involve electronic structure reorganizations. This has limited the application to only a few well-defined systems (*6-12*). The problem is further escalated by specific consideration of solvent polarization effects (*13*), which have been treated in an average sense in the past.

An alternative approach is to use a combined quantum mechanical and classical approach, in which the reacting system is treated explicitly by a quantum mechanical (QM) method, while the environmental solvent which is the most time consuming part in the computation is approximated by a standard molecular-mechanics (MM) force field (*5,14-17*). Since the reactant electronic structure and solute-solvent interactions are determined quantum-mechanically, the procedure is appropriate for studying chemical reactions, and importantly, there is no need to develop empirical potential functions for new systems. Furthermore, it has the advantage of taking into account the solvent polarization effects (*18*). Details of such a combined QM/MM potential and contributions by other groups are available in several recent reviews (*16-19*). In this paper, the focus will be on results from our group on organic reactions in aqueous solution. In addition, we describe a Monte Carlo simulation of the ion pair Na^+Cl^- in ambient and supercritical fluid water.

Methodology

The Combined QM/MM Potential. We employ a combine quantum mechanical and molecular mechanical (QM/MM) model with the semiempirical AM1 and TIP3P interface to describe solute-solvent interactions in solution (*16,17*). The method has been reviewed previously (*16-19*). Thus, only a brief summary is presented here. In this approach, the condensed-phase system is partitioned into (1) a QM region consisting of the reacting solute molecules, which are represented by electrons and nuclei and described by Hartree-Fock molecular orbital theory, and (2) an MM region containing the surrounding solvent, which is approximated by an empirical force field. Consequently, the total effective Hamiltonian for the system is

$$\hat{H}_{eff} = \hat{H}_{qm}^o + \hat{H}_{mm} + \hat{H}_{qm/mm} \qquad (1)$$

where $\hat{H}^o$ is the Hamiltonian for the QM solute, $\hat{H}_{mm}$ is the solvent-solvent interaction energy, and $\hat{H}_{qm/mm}$ is the solute-solvent interaction Hamiltonian. The total energy of the system is given by equation 2.

$$E_{tot} = <\Psi_{aq}|\hat{H}_{eff}|\Psi_{aq}> = E_{qm} + E_{mm} + E_{qm/mm} \qquad (2)$$

It should be pointed out that $\hat{H}_{qm/mm}$ depends on the partial charges and positions of the solvent interaction sites (atoms). As a result, only the one-electron integral part in the Fock matrix needs to be modified and standard molecular orbital computation methods can be directly used. The wave function obtained through this procedure, Ψ_{aq}, however, includes the solvent effects, based upon which the computed properties are further averaged in Monte Carlo simulations (*18*). Of particular interest is the solvent polarization energy. Given the wave functions for the solute in aqueous solution, Ψ_{aq}, and in the gas phase, Ψ^o, solvent polarization contributions to the total solvation free energy can be determined via equation 3.

$$-\Delta G_{pol} \;=\; -kT \; \ln <e^{\,[E_{qm/mm}(\Psi^o)-E_{qm/mm}(\Psi_{aq})]/kT}>_{\Psi_{aq}} \tag{3}$$

where k is Boltzmann's constant, T is absolute temperature, and $< \;>_{\Psi aq}$ indicates ensemble average using the aqueous wave function.

Ab initio molecular orbital method or density functional theory would be perfectly suited for this combined QM/MM approach since it has been well-tested and is convenient to use (*20,21*). However, to compute the energies of the QM solute molecule throughout fluid simulations, a computationally efficient method must be employed. Therefore, the semiempirical Austin Model 1 (AM1) theory developed by Dewar and coworkers is adopted in our calculation (*22*), which is coupled with Jorgensen's TIP3P model for water (*23*). Although good results for many organic systems have been obtained using the AM1, a major disadvantage in use of the semiempirical method is its empirical, parametrized nature. For systems in which poor results are produced, there is no systematic procedure to improve the computation other than systematic re-parametrization. The present approach is a compromised consideration to take into account both the theoretical capacity that a combined QM/MM approach can offer and its practical applicability in organic and biological systems due to computer limitations.

Monte Carlo Simulation of Free Energy Profiles.   Over the last few years, methods for determining free energies of solvation have been maturing, and have been reviewed recently by several authors (*24-26*). The focus here will be on our approach to obtain free energy surfaces for organic reactions in solution with Monte Carlo QM/MM calculations. The principal objective has been determining the minimum energy reaction path in the gas phase and then evaluating the solvent effects along this MERP. Here, we describe a simple two-step procedure which will allow experimental chemists to use conveniently without the need of empirical parametrization.

The initial step, which is similar to the computational procedure proposed by Jorgensen (*6*), involves determination of the reaction path and energetics using ab initio molecular orbital calculations. The transition state (TS) is first located, from which an intrinsic reaction path (IRC) will be followed via standard method available in Gaussian 92, leading to the starting materials and products, respectively (*21-29*). In the past, such a one dimensional reaction coordinate is typically used in order to reduce the computational costs (*6-12*). This would be adequate for simple reactions having "balanced" solvent effects on both sides of the TS as demonstrated in the S_N2 reaction between Cl$^-$ and CH_3Cl (*30-32*). However, for reactions involving heterolytic bond cleavage such as the S_N2 Menshutkin reaction, both the position of the TS and the reaction path are significantly influenced by the effects of solvation (*10,33,34*). Consequently, a multidimensional surface should be employed. This can conveniently be done with the combined QM/MM potential, whereas it is difficult to obtain high quality parameters for multi-dimension empirical potentials. It should be noted that the use of the combined QM/MM potential, in particular, the present AM1/TIP3P model, is primarily for evaluating the effects of solvation. Although the semiempirical energy and geometry may have deviation from high level ab initio results in certain cases (35), the intermolecular interaction between the AM1 solute and MM solvent can still be adequately determined. Thus, as have been suggested by others (36), the energetics for a chemical process in the gas phase should come from high level ab initio calculations unless the AM1 results are in good accord with the experimental data (37). This will then be supplemented by the solvation free energies obtained using the combined QM/MM-AM1/TIP3P potential.

Having defined the reaction path, the next step is to evaluate the solvent effects on the energetics as well as the path itself through statistical perturbation theory in molecular dynamics or Monte Carlo simulations (17,38). The procedure effectively steps along the reaction path to yield differences in free energy of hydration according to equation 4 (6).

$$\Delta G_h(R_j) - \Delta G_h(R_i) = -kT \ln <e^{-[E(R_j)-E(R_i)]/kT}>_{E(R_i)} \qquad (4)$$

where R_i and R_j are geometries generated from the reaction path following calculation for frames i and j. The average in equation 4 is for sampling based on the energy of R_i. Therefore, R_j may be regarded as a perturbation to R_i along the reaction coordinate. Typically, the difference between R_i and R_j is about 0.1 to 0.15 Å for distances. A procedure termed "double-wide" sampling is often used (39), which allows the perturbation calculation to be performed for transformations from R_i to R_{i-1} and to R_{i+1}. Finally, the potential of mean force (pmf) as a function of the reaction coordinate is constructed by summing up the gas phase and solvation free energies (equation 5).

$$\Delta G_{aq}(R) = \Delta G_{gas}(R) + \Delta G_h(R) \qquad (5)$$

The procedure outlined above appears to be similar to that used by Jorgensen (6). However, the major difference is on their second, most crucial step regarding the empirical parameter fitting. Although improvements have been made (8,9), it is still difficult to obtain the partial charges needed in fluid simulations. Since solute-solvent interactions are directly enumerated by quantum mechanical calculations during the fluid simulation here, this step is entirely eliminated. Furthermore, the combined QM/MM method has the advantage of both allowing solute electronic structure relaxation in solution (18), which is not included in the empirical potential approach, and considering explicit solute-solvent interactions, which is treated as a continuum medium in the self-consistent reaction field (SCRF) method (40,41).

We have developed a computer program, MCQUB (42), which combines the semiempirical AM1 theory and molecular mechanics force field (22). The

Monte Carlo sampling is now carried out using Jorgensen's BOSS program (43), while the QM calculations are performed with Stewart's MOPAC (44). For the organic systems described in this paper, the reacting molecules are typically solvated by 250-500 solvent molecules in tetragonal periodic boxes. Intermolecular interactions are feathered to zero from 9 to 10 Å between roughly the center of mass for the solvent molecule (water oxygen) and any solute atoms. All simulations, unless specifically identified, are carried out in the isothermal-isobaric ensemble (NPT) at 25 °C and 1 atm.

Applications

The Claisen Rearrangement of Allyl Vinyl Ether. The potential of mean force for the Claisen rearrangement of allyl vinyl ether (AVE) in aqueous solution was computed as a function of the IRC determined previously for the gas phase process using the 6-31G(d) basis set by Severance and Jorgensen (9,45). The IRC was obtained through the reaction path following procedure in Gaussian 92, starting from the transition state toward the forward and backward directions (9). A total of 143 frames of structures along the path were generated, of which 69 were used here in the aqueous simulation. This required a total of 34 simulations using double-wide sampling. Each simulation involved 0.5-1.0×10^6 configurations for equilibration followed by 1.5×10^6 configurations of averaging.

The hydration effects are found to lower the activation free energy of the gas phase by -3.5 ± 0.1 kcal/mol (Figure 1), which translates to a predicted rate acceleration by a factor of 368. For comparison, the experimental rate acceleration was believed to be about 1000 for AVE at 75 °C taking into account various available experimental results (9,46-49). Considering the difference in temperature between experiments and the simulation, our results seem to be in accord with experimental findings. Another indication of the good performance of the AM1/TIP3P Monte Carlo simulation is recorded by the computed difference in free energy of hydration $\Delta\Delta G_h$ between 4-pentenal and AVE (-1.9 ± 0.2 kcal/mol, Figure 1). It is known experimentally that aldehydes are generally better hydrated than ethers by 1-2 kcal/mol (9,50).

There have been two recent theoretical studies of the hydration effects on the reaction rate of Claisen rearrangements among other analyses (52,53), one using an SCRF model by Cramer and Truhlar (51) and the other employing Monte Carlo simulations with the OPLS potential by Severance and Jorgensen (9). The SCRF approach, which relies on classical theories of Onsager and Kirkwood, incorporates the solvent effects via a continuum dielectric medium into molecular orbital calculations, and has been parametrized to yield solvation free energies of organic compounds with a generalized Born theory (40). Using their solvation model (AMSOL), Cramer and Truhlar (51) reported a rate acceleration of 16, or a $\Delta\Delta G_h^{\ddagger}$ of -1.6 kcal/mol, for AVE at 25 °C at the 6-31G(d) TS geometry (a rate acceleration of 3.5 was observed with the AM1 TS in water). More importantly, they concluded, based on electronic structure analyses, that solute electric polarization in addition to the hydrophobic effects in the first solvation layer is responsible for the rate acceleration of Claisen

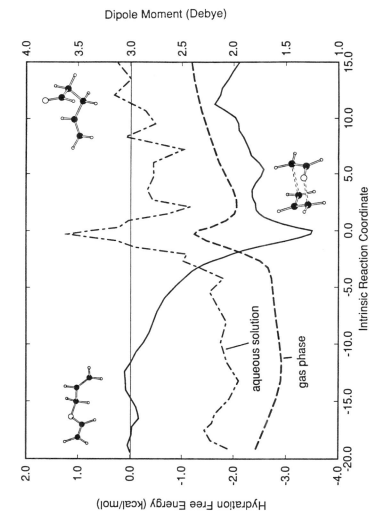

Figure 1. Computed free energy of hydration (solid curve) and the AM1 dipole moment in the gas phase (dashed curve) and in water (dash-dotted curve) along the 6-31G(d) IRC path (units are in amu$^{1/2}$bohr).

rearrangements in aqueous solution.   On the other hand, Severance and Jorgensen derived a set of atomic charges based on 6-31G(d) molecular electrostatic potentials along the reaction coordinate, and carried out a Monte Carlo calculation (9). The solvent effect from this OPLS potential is predicted to be -3.85 ± 0.16 kcal/mol, in good accord with our results. These authors have attributed the hydration effects to an increase in both the number and strength of solvent hydrogen bonding to the oxygen.

To characterize the origin of the solvent effects, we have computed the solute electric polarization contribution insinuated by Cramer and Truhlar, using equation 3 (45). The computed ΔG_{pol} is -0.55 ± 0.04 kcal/mol for AVE and -1.77 ± 0.08 kcal/mol for the TS. Therefore, the difference in solute electric polarization between the ground state and the TS is 1.2 kcal/mol, which amounts to 35% of the total $\Delta G_h^{\ddagger}$. These results are nicely mirrored by the computed induced dipole moments (Figure 1), which show a much greater increase at the TS ($\Delta\mu_{ind}$ = 1.0 ± 0.1 D) than in the ground state ($\Delta\mu_{ind}$ = 0.3 ± 0.1 D). In addition, Mulliken population analyses suggest that it is important to differentiate between the solvent-induced polarization and solvent-induced heterolytic bond cleavage in the Claisen rearrangement (47,48). The computed difference in charge transfer between the allyl and $H_2C=CHO$ groups on going from AVE to the TS only increases by 0.03 e as a result of hydration. However, the charge distribution within the $H_2C=CHO$ fragment has a substantial alteration. In water, the partial charges on oxygen are -0.25 e in AVE and -0.42 e in the TS, an increase by 0.17 e. On the other hand, the change is only 0.10 e in the gas phase.

The Menshutkin Reaction of H_3N + CH_3Cl. The next reaction to be studied was the S_N2 Menshutkin reaction of H_3N + CH_3Cl → $CH_3NH_3^+$ + Cl^-. A major challenge in theoretical treatment of this reaction is due to the large solvent effects involved in the charge separation process in water (10,33,34). Hydration brings this highly unfavored process into an exothermal reaction in water, which is accompanied by changes in the position of the TS and in the reaction path.   It would be difficult to investigate these changes if empirical potential functions are used in liquid simulation because the solvent influence on the structure of the transition state and the reaction path is not known a priori, and these potentials are typically derived based on results for gas-phase hydrogen bonding complexes.   Consequently, it is necessary to carry out simulations of the Menshutkin reaction with a quantum mechanical treatment, which should include solvent polarization effects on the solute electronic structures, and cover the bond formation and breaking processes independently in order to obtain the reaction path in aqueous solution.

Ab initio 6-31+G(d) calculations provided the gas phase energies and geometries for the reactants, TS, and products of the reaction (10), which are then compared with predictions using the AM1 theory (54).   The AM1 calculations yield an excellent agreement with the 6-31+G(d) structures for the reactants and products (Figure 2). The largest deviations are only 0.04 Å for bond lengths and 1.3° for bond angles. However, the AM1 TS is much tighter than the ab initio results, which are 0.24 and 0.23 Å longer for the C-N and

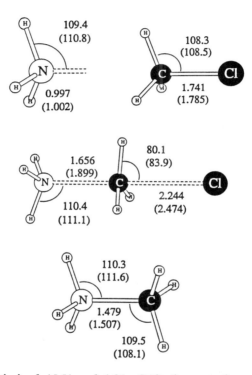

Figure 2. Optimized AM1 and 6-31+G(d) (in parentheses) geometries: bond lengths in angstroms and angles in degrees.

C-Cl distance (Figure 2). Although this is of concern to use the AM1 geometry for the Menshutkin reaction, it seems to be reasonable in the present study since our primary interest is the solvent effects on the change in the TS structure. To ensure the validity of the combined QM/MM AM1/TIP3P model to describe solute-water interactions for the Menshutkin reaction, partial geometry optimizations were carried out at the 6-31+G(d) level for a total of 24 complexes between $[H_3N-CH_3-Cl]$ and water at different stages of the reaction; the results are compared with the AM1/TIP3P optimization (54). In the latter calculation, $[H_3N-CH_3-Cl]$ is treated quantum-mechanically by the AM1 theory, while water is represented by the TIP3P model. The accord is good for an energy range of -1 to -12 kcal/mol. The overall root-mean-square (RMS) deviation is 0.5 kcal/mol. The agreement supports the use of the combined AM1/TIP3P potential for the Menshutkin reaction.

The effects of hydration on the TS structure are assessed by mapping out a two-dimensional potential surface in water, treating the C-N and C-Cl bond variations independently. A linear approach by the nucleophile, NH_3, to CH_3Cl is assumed, which appears to be reasonable; however, a full consideration of the reaction surface should also include the angular average of the nucleophilic attack. Extension beyond the two-dimensional map becomes computationally prohibited. To limit the computational costs, the Monte Carlo simulation is restricted to the region near the TS (Figure 3). The free energy surface shown in Figure 3 was constructed via a grid search method (54). First, a series of potential of mean forces as a function of R_{C-N} were determined at given distances of R_{C-Cl} using statistical perturbation theory. Then, the relative height of two such neighboring profiles was computed by another perturbation calculation with respect to R_{C-Cl}, this time at a fixed value of R_{C-N}. Finally, the potential surface was anchored relative to the free energy at a reaction coordinate of -2 Å (see below). In all, a total of 87 simulations were executed, each involved at least 5×10^5 configurations for equilibration and 10^6 configurations of data collection.

The most striking finding from this study in Figure 3 is the change in the position of the TS structure on going from the gas phase (shown by an O) into aqueous solution (indicated by an X). Nevertheless, the result is in good accord with the expectation according to Hammond postulate (55). The structural change features a lengthening of R_{C-N} of 0.30 Å from its gas phase value of 1.66 Å, and a decrease in R_{C-Cl} by 0.15 Å (1.94 Å in the gas phase). Consequently, the TS in the Menshutkin reaction occurs much earlier in water than in the gas phase. A separate study of a similar reaction involving H_3N and CH_3Br was reported by Sola et al. (34), using a continuum SCRF method in ab initio MO calculations. They obtained similar qualitative features for the TS when a dielectric constant of 78 was used to represent water.

The familiar potential of mean force was located as a function of the reaction coordinate defined by equation 6. Thus, RC is 0 at the TS.

$$RC = R_{C-Cl} - R_{C-N} - [R_{C-Cl}(TS) - R_{C-N}(TS)] \qquad (6)$$

Addition calculations along this path were carried out to extend the path leading to the reactants and products, which is depicted in Figure 4 along with the gas phase energy. In Figure 4, the pmf is virtually flat for $|RC| \geq 1.0$ Å and was zeroed at RC = -2.0 Å. The computed activation free energy is 26.3 ± 0.3 kcal/mol in water, which is in agreement with the experimental activation energy (23.5 kcal/mol) for H_3N + CH_3I in water. This represents a lowering of the gas phase barrier height by 23.7 kcal/mol thanks to hydration. The computation also yielded the free energy change for the overall reaction, ΔG_{rxn}, in aqueous solution (-18 ± 2 kcal/mol), which translates to a solvent stabilization of 155 kcal/mol relative to the gas phase process. The experimental estimate of ΔG_{rxn} based on standard free energies of formation and hydration is about -34 ± 10 kcal/mol. The difference here is primarily due to poor performance of the AM1 theory for Cl^-, which overestimates its heat of formation by 18 kcal/mol (22). If the experimental value was used, the computed ΔG_{rxn} would be -36 kcal/mol. The calculation by Sola et al. gave values of -27 to -44 kcal/mol using various basis sets (34).

The present calculation also yields detailed structural features concerning the stabilization of the TS by the solvent. Integrating the hydrogen bonding peaks of the radial distribution functions and solute-solvent energy pair distributions (Figure 5) indicates that the total number of hydrogen bonds between the reacting system $[H_3N\text{-}CH_3\text{-}Cl]$ and water increases from zero for the reactants at RC = -2.0 Å to about 9 (three bound to $CH_3NH_3^+$ and six to Cl^-) for the product ion pair at RC = +2.0 Å. Accompanying the number increase is an increase in hydrogen bond strength. It is noteworthy to contrast the present result to the previous finding for the type I S_N2 reaction of Cl^- + CH_3Cl in water, where the number of hydrogen bonds is roughly equal along the whole reaction coordinate (30). The differential solvation is due to variation in hydrogen bond strength. Both are critical in the Menshutkin reaction (54). It is also interesting to notice that Figure 4 shows a unimodal energy profile in aqueous solution for the Menshutkin reaction, a feature consistent with traditional assumption (56).

Finally, the solvent effects on the electronic charge separation is characterized by a comparison of the computed partial charges on chlorine in the gas phase and in water along the reaction coordinate (equation 6). This gives a good indication of the charge development in the course of the reaction. Charge population analyses have been performed by Bash et al. (57) and by Hwang et al. (58) in their molecular dynamics studies of the chloride-methyl chloride exchange reaction. In contrast to findings for the type I reaction by Bash et al. (57), in which charge transfer in water lags behind the process in the gas phase, the present type II S_N2 reaction shows clearly a solvent-induced charge separation. At the TS, a charge separation of more than 65% in water is predicted by the AM1/TIP3P Monte Carlo simulation. This may be compared with the gas phase charge transfer of about 50%.

Na^+Cl^- Ion Pair in Ambient and Supercritical Water. Ionic interactions are of fundamental importance in chemistry and biological science. Potential of mean force is usually used to characterize ion pair formation in solution (59). The

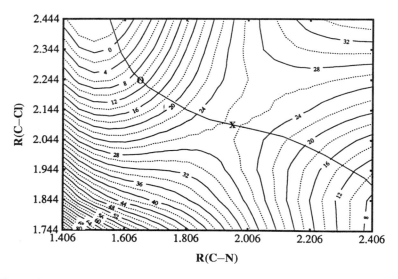

Figure 3. Compute free energy surface for the type II S_N2 reaction of NH_3 + CH_3Cl in water. The transition states in the gas phase and in aqueous solution are marked by an O and an X, respectively. The reaction path defined by equation 6 is indicated by the curve across the diagram. Energies are given in kilocalories per mole and distances are in angstroms.

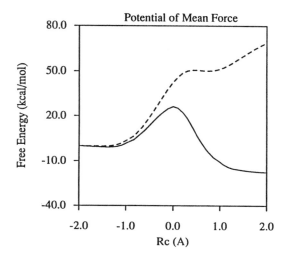

Figure 4. Potential of mean force for the Menshutkin reaction in water (solid curve) and in the gas phase (dashed curve).

first molecular dynamics simulation of pmf was reported by Berkowitz et al. for Na^+Cl^- in water using importance sampling (60). Since then, there have been several theoretical studies of ion pair in water (59-64). The key features from these investigations are the contact and solvent-separated free energy minima. Importantly, the energy barrier dividing the contact and solvent-separated species is only 1-3 kcal/mol, allowing the ions to dissociate into the bulk.

The interest in ion association at high temperature and pressure conditions, i.e., supercritical and subsupercritical water, comes from a number of disciplines, including geochemistry and hazardous waste destruction (65-68). In the latter application, organic materials and oxygen become completely miscible with supercritical water (SCW) and are destroyed to form water, CO_2 and inorganic acids and salts, which rapidly precipitate out. The low solubility of electrolytes in SCW is not surprising in view of the fact that the fluid dielectric constant is only about 2-10 at 450 °C and 200-1000 atm (69). Although numerous experimental studies have been carried out to refine the optimal conditions for organic waste oxidation in SCW (65-68) and to measure solubilities of inorganic salts at elevated temperature and pressure (70), little information is available on the structural and energetic details of ion-ion and ion-solvent interactions at the atomic level. An important issue in supercritical solvation is the formation of solvent clusters in the vicinity of a solute molecule, or a local density enhancement (molecular charisma) (71).

In a previous study (69), we found that the TIP4P model for water performs remarkably well to describe supercritical fluid water, despite the fact that the model was developed for ambient water at 25 °C (23). The computed fluid density and dielectric constant for SCW at 400 °C were in good accord with the experimental data obtained at 450 °C for pressure ranging from 350 to 2000 atm. The need to compare the computed results with experimental data at a slightly different temperature was attributed to an underestimate of the critical temperature by 30-50 °C by the TIP4P model (69). Thus, the computed and experimental reduced temperatures (T_r = T/T_c) are comparable. Similar observations were obtained for the SPC model (72), another widely used potential for ambient water. Thus, simulations of supercritical aqueous solution using the TIP4P or SPC model at 400 °C may be regarded as comparable to experimental conditions at 450 °C.

The potential of mean force for Na^+Cl^- ion pair in SCW was computed at 400 °C and 350 atm corresponding to bulk conditions of ρ_{calc} = 0.23 g/cm^3 and ϵ_{calc} = 3.6 (Figure 6). The system consisted of 390 TIP4P water molecules plus the ion pair, described by the OPLS potential (73), in a rectangular periodic cell. Perturbations of ± 0.1 or 0.15 Å were used in a total of 56 Monte Carlo simulations to cover R_{Na-Cl} from 2.20 to 9.95 Å. Each calculation was first subjected to 2 x 10^6 configurations of equilibration followed by 2 x 10^6 configurations of averaging. To compare the ion pairing interaction in ambient water, the simulations were repeated for a system containing 740 waters plus Na^+Cl^- at 25 °C and 1 atm. These pmf's are anchored at the longest separation distances to the results from the "primitive model" obtained by dividing the Coulombic energy by the computed bulk dielectric constant (61).

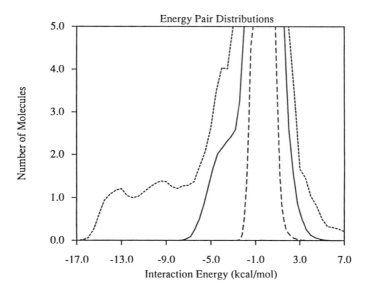

Figure 5. Computed solute-water energy pair distribution for the reactant (dashed curve), transition state (solid curve), and product (dotted curve). The ordinate gives the number of water molecules bound to the solute, with energy shown on the abscissa.

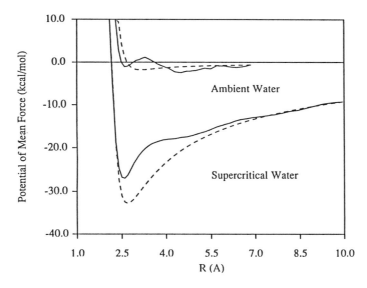

Figure 6. Calculated potential of mean force for separating the ion pair Na^+Cl^- in ambient water at 25 °C and 1 atm, and in supercritical water at 400 °C and 350 atm.

As shown in Figure 6, the minima for the ion pair in ambient water were found at 2.6 and 4.5 Å, corresponding to the contact and solvent-separated ion pairs. There are no specific structural features beyond the solvent-separated species. This is in good accord with previous studies (*60*). However, the present results indicate that the solvent-separated ion pair is ca. 1.3 kcal/mol lower in energy than the contact pair, while the barrier for the ion pair to escape from the contact cage is about 2 kcal/mol. The calculation reported previously yielded a contact minimum 1.3 kcal/mol deeper in free energy than the solvent separated minimum (*60*). Note that the TIPS2 model for water was used in that study (*60*).

In contrast, the pmf for the Na^+Cl^- ion pair in SCW shows remarkable differences than that in ambient water (Figure 6). First, the Coulombic interaction between the two ions is much stronger in SCW than in ambient water; the minimum energy is -27 kcal/mol in SCW as opposed to ca. -1 kcal/mol in ambient water relative to infinite separation. This is, of course, expected in view of the much smaller fluid dielectric constant at supercritical conditions (*69*). However, such a strong interaction between Na^+ and Cl^- in SCW shall prevent them from dissociating. Thus, inorganic salts have low solubilities of about 10-100 wt% ppm in SCW (*65-68,70*). Secondly, there is no solvent-separated minimum in the pmf in SCW, although a leveling effect is evident for ion separations between 4 and 5 Å. This suggests that ion-water interactions are still significant; however, the balance between the favorable ion-ion attraction and the costs to remove water molecules separating the ions is not enough to result in an observable minimum on the potential surface. This might be an indication of an active dynamic feature for ion-water interactions in SCW (*66*). Finally, a comparison between the computed pmf and the primitive model is informative. The importance of explicit ion-water interaction is evident in the case of ambient water. In SCW, the agreement between the two models at large ion separations beyond 7 Å is nearly exact, which indicate that the computed bulk dielectric constant is quite reasonable (*69*). The difference between the computed pmf and the primitive model at short ion separations is also significant. The weaker interaction (by about 6 kcal/mol) predicted from the Monte Carlo simulation suggests that the local dielectric constant is somewhat higher than the value in the bulk, an indication of enhanced local solvent clustering or charisma (*69,71,74*). In SCW, there are on average 5.3 nearest neighbors around both Na^+ and Cl^-. This may be compared with coordination numbers of 6.0 for Na^+ and 7.5 for Cl^- in ambient conditions (*73*). Clearly, the local density for the ions in SCW within the first solvation layer is similar to that in ambient water and much higher than in the fluid bulk. Note that at longer Na^+Cl^- separations, the solvent-clustering surrounding each ion will not affect the bulk dielectric behavior, leading to an agreement between the simulation and continuum model.

Concluding Remarks

The ability to model organic reactions in solution at the atomic level has greatly increased in the past decade. The method presented here, combining quantum mechanical and molecular mechanics (QM/MM) Monte Carlo simulation techniques, can be applied to a wide range of problems in chemistry and biochemistry. These computations have already yielded a wealth of information on the nature of intermolecular interactions in solution, and the knowledge gained in these studies shall provide a better understanding and control of chemical reactivity and dynamics. Undoubtedly, there is a great potential of growth in methodology development. We anticipate that a new generation of computational tools for modeling chemical processes in solution will become available as we welcome the arrival of the twenty-first century.

Acknowledgments. Gratitude is expressed to the National Science Foundation and National Institutes of Health for support of this research.

Literature Cited

1.   Reichardt, C. *Solvents and Solvent Effects in Organic Chemistry*; VCH: New York, NY, 1990.
2.   Truhlar, D. G.; Hase, W. L.; Hynes, J. T. *J. Phys. Chem.* **1983**, *87*, 2664.
3.   Karplus, M.; Petsko, G. A. *Nature*, **1990**, *347*, 631.
4.   McCammon, J. A.; Harvey, S. C. *Dynamics of Proteins and Nucleic Acids*; Cambridge University Press: Cambridge, 1987.
5.   Warshel, A. *Computer Modeling of Chemical Reactions in Enzymes and Solutions*; Wiley: New York, NY, 1991.
6.   For a review, see: Jorgensen, W. L. *Adv. Chem. Phys.* **1988**, *70*, 469.
7.   Kim, H. J.; Hynes, J. T. *J. Am. Chem. Soc.* **1992**, *114*, 10508.
8.   Blake, J. F.; Jorgensen, W. L. *J. Am. Chem. Soc.* **1991**, *113*, 7430.
9.   Severance, D. L.; Jorgensen, W. L. *J. Am. Chem. Soc.* **1992**, *114*, 10966.
10. Gao, J. *J. Am. Chem. Soc.* **1991**, *113*, 7796.
11. Peng, Z.; Merz, K. M., Jr. *J. Am. Chem. Soc.* **1992**, *113*, 2733.
12. Weiner, S. J.; Singh, C.; Kollman, P. A. *J. Am. Chem. Soc.* **1985**, *107*, 2974.
13. Dang, L. X.; Rice, J. E.; Caldwell, J.; Kollman, P. A. *J. Am. Chem. Soc.* **1991**, *113*, 2481.
14. Warshel, A.; Levitt, M. *J. Mol. Biol.* **1976**, *103*, 227.
15. Weiner, S. J.; Seibel, G. L.; Kollman, P. A. *Proc. Natl. Acad. Sci.* **1986**, *83*, 649.
16. Field, M. J.; Bash, P. A.; Karplus, M. *J. Comput. Chem.* **1990**, *11*, 700.
17. Gao, J. *J. Phys. Chem.* **1992**, *96*, 537.
18. Gao, J.; Xia, X. *Science* **1992**, *258*, 631.
19. Warshel, A. *Chem. Rev.* **1993**, *93*, 2523.
20. Hehre, W. J.; Radom, L.; Schleyer, P. v. R.; Pople, J. A. *Ab Initio Molecular Orbital Theory*; Wiley: New York, NY, 1986.
21. Parr, R. G.; Yang, W. *Density-Functional Theory of Atoms and Molecules*; Oxford University Press: New York, NY, 1989.
22. Dewar, M. J. S.; Zoebisch, E. G.; Healy, E. F.; Stewart, J. J. P. *J. Am. Chem. Soc.* **1985**, *107*, 3902.

23. Jorgensen, W. L.; Chandrasekhar, J.; Madura, J. D.; Impey, R. W.; Klein, M. L. *J. Chem. Phys.* **1983**, *79*, 926.
24. Beveridge, D. L.; DiCapua, F. M. *Ann. Rev. Biophys. Chem.* **1989**, *18*, 431.
25. Jorgensen, W. L. *Acc. Chem. Res.* **1989**, *22*, 184.
26. Kollman, P. A.; Merz, K. M., Jr. *Acc. Chem. Res.* **1990**, *23*, 246.
27. Frisch, M. J.; Head-Gordon, M.; Trucks, G. W.; Foresman, J. B.; Schlegel, H. B.; Raghavachari, K.; Robb, M.; Binkley, J. S.; Gonzalez, C.; Defrees, D. J.; Fox, D. J.; Whiteside, R. A.; Seeger, R.; Melius, C. F.; Baker, J.; Marin, R. L.; Kahn, L. R.; Stewart, J. J. P.; Topiol, S.; Pople, J. A. *Gaussian 90*, Gaussian Inc.: Pittsburgh, PA, 1990.
28. Fukui, K. *Acc. Chem. Res.* **1981**, *14*, 363.
29. Gonzalez, C.; Schlegel, H. B. *J. Phys. Chem.* **1990**, *94*, 5523.
30. Chandrasekhar, J.; Smith, S. F.; Jorgensen, W. L. *J. Am. Chem. Soc.* **1985**, *107*, 154.
31. Jorgensen, W. L.; Buckner, J. K. *J. Phys. Chem.* **1986**, *90*, 4651.
32. Tucker, S. C.; Truhlar, D. G. *J. Am. Chem. Soc.* **1990**, *112*, 3347.
33. Abboud, J. M.; Notario, R.; Bertran, J.; Sola, M. *Prog. Phys. Org. Chem.* **1993**, *19*, 1.
34. Sola, M.; Lledos, A.; Duran, M.; Bertran, J.; Abboud, J. M. *J. Am. Chem. Soc.* **1991**, *113*, 2873.
35. Gao, J. *J. Am. Chem. Soc.* **1993**, *115*, 2930.
36. Cramer, C. J.; Truhlar, D. G. *J. Am. Chem. Soc.* **1993**, *115*, 8810.
37. Gao, J.; Pavelites, J. J. *J. Am. Chem. Soc.* **1992**, *114*, 1912.
38. Zwanzig, R. W. *J. Chem. Phys.* **1954**, *22*, 1420.
39. Jorgensen, W. J.; Ravimohan, C. *J. Chem. Phys.* **1985**, *83*, 3050.
40. Cramer, C. J.; Truhlar, D. G. *Science*, **1992**, *256*, 213.
41. Karelson, M. M.; Zerner, M. C. *J. Phys. Chem.* **1992**, *96*, 6949.
42. Gao, J. *MCQUB (Monte Carlo QM/MM at the University at Buffalo)*; SUNY at Buffalo, 1993.
43. Jorgensen, W. L. *BOSS 2.9*; Yale University, 1990.
44. Stewart, J. J. P. *MOPAC, Version 5.0*: QCPE 455, 1986.
45. Gao, J. *J. Am. Chem. Soc.* **1994**, *116*, 1563.
46. White, W. N.; Wolfarth, E. F. *J. Org. Chem.* **1970**, *35*, 2196, 3585.
47. Coates, R. M.; Rogers, B. D.; Hobbs, S. J.; Peck, D. R.; Curran, D. P. *J. Am. Chem. Soc.* **1987**, *109*, 1060.
48. Gajewski, J. J.; Jurayj, J.; Kimbrough, D. R.; Gande, M. E.; Ganem, B.; Carpenter, B. K. *J. Am. Chem. Soc.* **1987**, *109*, 1170.
49. Brandes, E.; Grieco, P. A.; Gajewski, J. J. *J. Org. Chem.* **1989**, *54*, 515.
50. Ben-Naim, A.; Marcus, Y. *J. Chem. Phys.* **1984**, *81*, 2016.
51. Cramer, C. J.; Truhlar, D. G. *J. Am. Chem. Soc.* **1992**, *114*, 8794.
52. Gajewski, J. J. *J. Org. Chem.* **1992**, *57*, 5500.
53. Dewar, M. J. S.; Jie, C. *J. Am. Chem. Soc.* **1989**, *111*, 511.
54. Gao, J.; Xia, X. *J. Am. Chem. Soc.* **1993**, *115*, 9667.
55. Issacs, N. S. *Physical Organic Chemistry*; Wiley: New York, NY, 1987.
56. Ingold, C. K. *Structure and Mechanism in Organic Chemistry*; 2nd ed.; Cornell University Press: Ithaca, NY, 1969.

57. Bash, P. A.; Field, M. J.; Karplus, M. *J. Am. Chem. Soc.* **1987**, *109*, 8092.
58. Hwang, J.; King, G.; Greighton, S.; Warshel, A. *J. Am. Chem. Soc.* **1988**, *110*, 5297.
59. Jorgensen, W. L.; Buckner, J. K.; Gao, J. In *Chemical Reactivity in Liquids*; Moreau, M.; Turq, P., Eds.; Plenum: New York, NY, 1988; p 253.
60. Berkowitz, M.; Karim, O. A.; McCammon, J. A.; Rossky, P. J. *Chem. Phys. Lett.* **1984**, *105*, 577.
61. Jorgensen, W. L.; Buckner, J. K.; Huston, S. E.; Rossky, P. J. *J. Am. Chem. Soc.* **1987**, *109*, 1891.
62. Dang, L. X.; Pettitt, B. M. *J. Am. Chem. Soc.* **1987**, *109*, 5531.
63. Hoston, S. E.; Rossky, P. J. *J. Phys. Chem.* **1989**, *93*, 7888.
64. Kusalik, P.; Patey, G. *J. Chem. Phys.* **1988**, *89*, 5843.
65. Thomason, T. B.; Modell, M. *Hazardous Waste*, **1984**, *1*, 453.
66. Shaw, R. W.; Brill, T. B.; Clifford, A. A.; Eckert, C. A.; Franck, E. U. "Supercritical Water, a Medium for Chemistry", *Chem. Eng. News* **1991**, *69(51)*, 26.
67. Armellini, F. J.; Tester, J. W. *J. Supercrit. Fluid* **1991**, *4*, 254.
68. Barner, J. E.; Huang, C. Y.; Johnson, T.; Jacobs, G.; Martch, M. A.; Killilea, W. R. *J. Hazardous Materials* **1992**, *31*, 1.
69. Gao, J. *J. Am. Chem. Soc.* **1993**, *115*, 6893.
70. Armellini, F. J.; Tester, J. W. *Fluid Ph. Equil.* **1993**, *84*, 123.
71. Eckert, C. A.; Ziger, D. H.; Johnston, K. P.; Ellison, T. K. *Fluid Ph. Equil.* **1983**, *14*, 167.
72. Cummings, P. T.; Cochran, H. D.; Simmonson, J. M.; Mesmer, R. E. *J. Chem. Phys.* **1991**, *94*, 5606.
73. Chandrasekhar, J.; Spellmeyer, D. C.; Jorgensen, W. L. *J. Am. Chem. Soc.* **1984**, *106*, 903.
74. Zagrobelny, J.; Bright, F. V. *J. Am. Chem. Soc.* **1992**, *114*, 7821.

RECEIVED May 9, 1994

Chapter 16

Factor Analysis of Solvent Effects on Reactions

Application to the Claisen Rearrangement

Joseph J. Gajewski and Nancy L. Brichford

Department of Chemistry, Indiana University, Bloomington, IN 47405

Solvent effects on reactions have been successfully subjected to a factor analysis using as parameters the Kirkwood-Onsager dielectric function, the Hildebrand cohesive energy density, and the chloride ion and potassium ion solvent transference data after subtraction of the K-O and CED contributions. The coefficients of the parameters provide information about the change in dipole moment, the volume change, and the hydrogen bond donor and acceptor requirements in the reaction. Application to the rates of solvolysis of *tert*-butyl chloride reveals not only the increased dipole moment in the transition state and increased hydrogen bonding (presumably to the leaving chloride ion), but to the overwhelming contribution of ground state destabilization in solvents with high CED. The thermally induced 3,3-shift of allyl vinyl ether (the Claisen rearrangement) responds only to the solvent CED and hydrogen bond donor ability and not to the K-O function. Consistent with this analysis are secondary deuterium kinetic isotope effects at the bond breaking and making sites that reveal little change in transition state structure in more aqueous media.

The understanding of solvent effects on rates and equilibrium of chemical reactions has been a long-standing goal of chemistry. Understanding can take many forms, but for an organic chemist interested in rates and stereochemistry, understanding must first occur at the level of the macroscopic behavior of the solvent. Thus, correlation of rate and equilibrium data with physical and chemical properties of the solvent should provide not only mechanistic insight but prediction. One popular approach is the correlation with solvent effects on uv transitions of dye molecules. A good example of such a quantity is the solvent E_T whose values and applications are given by Reichardt (*1*). However, if a correlation is obtained with some physical phenomenon that has been theoretically related to free energy changes, then a more fundamental understanding of solvent effects is possible. For instance, solvent dielectric constant, when cast in terms of the Kirkwood function, $(\varepsilon - 1) / (2\varepsilon + 1)$, has

0097–6156/94/0568–0229$08.00/0

been moderately successful in correlating the rate response of some reactions over a narrow range of solvents.   The importance of this approach is the fact that the coefficient of the Kirkwood function is proportional to the square of the change in dipole moment from initial to transition (or final) state and inversely proportional to the cube of the cavity radius associated with the reactants...assuming that the reactants are spherical (2,3).   The fact that the Kirkwood function is not generally successful has led others to examine multiparameter approaches.   Palm was the first to do so (circa 1970) using the Kirkwood function, a polarizability function derived from solvent refractive indices, and terms that might characterize solvent electrophilicity and nucleophilicity (4).   Subsequently, Makitra and Pirig(5) added a cohesive energy density term, namely Hildebrand's δ^2 values, $(\Delta H_{vap}\text{-}RT)/V_{molar}$, to characterize the energy necessary to generate a cavity in the solvent (6,7).   The origin of the solvent electro- and nucleophilicity parameters is of some concern, as pointed out by Abraham, Taft, and coworkers (8,9)   Taft and coworkers provided their own multiparameter approach to solvent effects using a solvatochromically derived parameter to represent dipole-dipole interactions called π^*, and solvent hydrogen bond donating, α, and accepting, β, parameters derived from nmr measurements, and the Hildebrand δ^2 parameter (although early papers used the square root of δ^2), see equation 1.:

$$\ln k = p\pi^* + a\alpha + b\beta + v\delta^2 + c \qquad (1)$$

More sophisticated variations on the Taft equation include a polarizability correction in π^*.   The range of successive applications of the Taft equation is remarkable.   If there are any criticisms of the Taft approach, they focus on the use of a solvatochromically derived parameter, π^*, which leads to difficulty in interpreting the coefficients of the parameters without reference to other correlations.

      Our interest in solvent effects on reactions began with the revelation by Carpenter (10) and by Grieco and Brandes (11) that addition of water to the Claisen rearrangement reaction medium increased the rate of reaction. Carpenter attempted to correlate this behavior with solvent E_T values but found that highly aqueous media provided rate accelerations larger than predicted on the basis of correlations with less polar media.   The increases in rate observed with addition of water are not enormous. Typically, factors of 3-20 are obtained over a limited range of solvents which rarely includes pure water since neutral allyl vinyl ethers are not soluble in water.   Yet, mechanistic hypotheses invoking ion-pairs was promoted by Carpenter with a stimulus by Arigoni. This is in a paper jointly authored by the Cornell group and the Indiana group at the request of the editors despite the fact that each group had different hypotheses and conclusions (10).

      In order to quantify solvent effects over the entire range of solvent possibilities, Brandes determined the relative rates of rearrangement with an allyl vinyl ether having a small hydrocarbon chain ending in an ester function or a carboxylic acid function (see Table I).   The ester is soluble in less polar media while the carboxylate is soluble in

Table I. Relative Rate constants for the Claisen Rearrangement

SOLVENT	k_{rel} (R = Na)	k_{rel} (R = CH$_3$)
Water	214.	
Trifluoroethanol	31.	56.
Methanol	9.4	8.6
Ethanol		6.1
Isopropyl alcohol		5.0
Dimethyl sulfoxide		3.2
Acetonitrile		3.1
Acetone		2.1
Benzene		2.0
Cyclohexane		1.0

more hydroxylic media. In solvents where both materials are soluble, the rates differ by no more than a factor of two.

The rate response of the rearrangement to different solvents cannot be understood by recourse to a single solvent parameter. For instance, the correlation with E_T values gives:

$$\ln k = -.141 \, E_T/RT \; - 4.87 \quad r = 0.893 \quad (SD = 0.725, \; Range = 5.37) \tag{2}$$

If only the hydroxylic solvent data (using, in addition, mixed solvent data provided in ref 11) is analyzed, correlation is obtained with Grunwald-Winstein Y values (*12,13*) where Y is defined as log k (solvolysis *tert*- butyl chloride in the solvent relative to log k of solvolysis in 80% aqueous ethanol (1:4 water, ethanol v/v):

$$\log k(\text{Claisen}) = .33 * Y + c \tag{3}$$

However, despite the correlation with ionization reactions, the sensitivity of the response to Y is much smaller than any yet observed for a solvolysis reaction including that for allylic rearrangements via ion-pairs studied by Goering (*14*). The conclusion was that there is not much polar character in the Claisen rearrangement transition state. When the Taft factors are applied to the Brandes data (10 pure solvent points) an excellent correlation is obtained:

$$\ln k = 1.273\pi^* + 1.69\alpha - 0.77\beta + 2.6\delta^2 - 0.344 \quad (r = 0.994) \tag{4}$$

Besides the concern over what the coefficients mean in terms of difference between ground and transition state, there is the concern over the extent to which the parameters are independent and are complete.   Completeness is inferred by the goodness of fit, but independence is more problematic. The π^* parameter may have some contributions from the other parameters since the π^* parameters were largely unchanged after addition of the cohesive energy density term.   But again, the coefficients have meaning only in comparison to other correlations. For instance, application of the Taft equation to the solvolysis of *tert*-butyl chloride in a wide variety of solvents gives:

$$\ln k = 11.9\pi^* + 9.93\alpha + 3.32\beta + 6.75\delta^2 - 15.6 \ (r = 0.981) \tag{5}$$

Clearly, the Brandes data has a much smaller response to all parameters than the solvolysis data, but the α and δ^2 coefficients are relatively larger than the π^* and β coefficients relative to the solvolysis data suggesting greater importance of hydrogen bonding and solvent-solvent interaction in the Claisen rearrangement.   But a question of quantification remains.

In an effort to provide more insight into solvent effects and stimulated by the work of Beak (*15,16*), we utilized a four parameter equation using the Kirkwood-Onsager function (*2,3*), the δ^2 parameter (cast in units to provide an energy when multiplied by a volume in ml/mole), (*6,7*) and α' and β' values derived from single ion transference data of A. J. Parker (*17*) and Y. Marcus (*18*) from water to a variety of solvents using chloride ion and potassium ion, respectively, to represent the specific donation of a hydrogen bond from solvent and specific hydrogen bond accepting ability of the solvent, respectively.   This equation was called the KOMPH equation (equation 7) after Kirkwood, Onsager, Marcus, Parker, and Hildebrand (*19*).   These parameters would appear to be independent and would have coefficients which by themselves would provide insight into the ground and transition states for reactions.

$$\ln k = p \ (\varepsilon-1)/(2\varepsilon+1) \ + a \ \alpha' + b \ \beta' + v \ \delta^2 + \text{const} \tag{6}$$

These parameters were chosen after examination of the *tert*-butyl chloride solvolysis rate data provided by Abraham (*20*). Here, the free energy of partitioning of *tert*-butyl chloride between various solvents and dimethylformamide (DMF) was determined revealing the *tert*-butyl chloride was less stable in water than in methanol by roughly 5 kcal/mol.   Further, we found that the Hildebrand cohesive energy density alone correlated the data directly.   Thus, the solvents with great intermolecular attractions, like water, ethylene glycol, and formamide, are destabilized by addition of a hydrocarbon-like material.   From the solvent effect data on the ground state and the effect of solvent on the rate of solvolysis of *tert*-butyl chloride, Abraham could determine the effect of solvent on the transition state.   This information is remarkable in that the transition state is stabilized to only a small extent by the dielectric constant of the solvent provided that the dielectric constant is greater than 10. For instance, in water the solvolysis transition state is only 1kcal/mol more stable than it is in methanol.   The effect of dielectric constant is seen only in the examination of solvents with low dielectric constant. Indeed, the data would seem to roughly correlate with the Kirkwood function except for solvents like trifluoroethanol and hexafluoroisopropyl alcohol in which the transition state for solvolysis is much more stable than expected on the basis of dielectric constant.   These two solvents are unique in their ability to donate hydrogen bonds, and this is exactly what is necessary to stabilize the departing chloride ion in the solvolysis transition state.   It therefore seemed appropriate that a multiparameter correlation equation should include the Kirkwood function, the Hildebrand cohesive energy density and some measure of hydrogen bond donating ability, and, because of the work of pioneers in the area, a term involving hydrogen bond accepting ability.   Thus, the chloride and potassium ion transference data seemed appropriate starting points for these parameters.

In published work (*19*), the free energies of transference were used directly without correction for solvent dielectric constant or δ^2.   The dielectric constants for all solvents listed were sufficiently high that no effect should be noted, but these led to an arbitrary choice of zero for the gas phase and very non-polar solvents.   For the solvents not listed, interpolations from Taft's data were used.   It was also assumed that differential cohesive energy density effects were small, an assumption which in retrospect (see below) was unnecessary, but fortunately not disastrous.   Despite the assumptions, this equation provided great insight not only into the effect of solvents on reactions but into important attributes of the reaction itself.   Thus, the *tert*-butyl chloride solvolysis reaction was characterized by:

$$\ln k = 18.9\ (\varepsilon\text{-}1)/(2\varepsilon\text{+}1)+ 52.8\ \alpha' - 0.76\ \beta' + 10.1\ \delta^2 - 15.4\ (r = 0.989) \qquad (7)$$

Here the coefficient of the Kirkwood-Onsager function is related to the square of dipole moment change from ground state to transition state divided by the cube of the solvent cavity (assuming it doesn't change much in the reaction).   For a cavity of 4 Angstroms, a dipole moment change calculated is 7 D.   Hydrogen bonding is important, and the coefficient of the α' term is about 90% of the value for chloride ion transfer between solvents.   The β' value is negligible consistent with little nucleophilic participation by the solvent.   The coefficient of the δ^2 term is the negative of a volume change from ground state to transition state, and it should be compared to - 20 cc/mole, the Activation Volume for the solvolysis reaction determined from external pressure studied in a variety of reactions.   When the ground state relative energies and transition state relative energies are examined separately, it is clear that the *tert*-butyl chloride solvolysis reaction is governed by ground state destabilization, which is correlated with the δ^2 term in more polar media, and by transition state stabilization, which is correlated with the KO function in less polar media.   Solvent hydrogen bonding to chloride leaving group in the transition state is always important.

In addition to the successful application of the KOPMH equation to the solvolysis reaction its application to Brandes' Claisen rearrangement data lead to important insights:

$$\ln k = 9.55\ \alpha' - 2.44\ \beta' + 3.77\ \delta^2 -0.071\ (r = 0.968) \qquad (8)$$

Here the solvent dielectric constant is unimportant and the alpha' and δ^2 terms are most important. This suggests no change in dipole moment from ground state to transition state, but a smaller transition state volume, which should be the case, and increased hydrogen bonding in the transition state...presumably because the oxygen is more negatively charged than in the vinyl ether ground state. Compensating changes in charges lead to the same dipole moment as in the ground state. The conclusion of increased hydrogen bonding in the transition state was reinforced by solvent Monte Carlo calculations of Jorgensen who also found little change in dipole moment in the transition state in an *ab initio* calculation (*21*). This contrasts with suggestions by Carpenter (*10*) and by Cramer and Truhlar (*22*) of polarization in the transition state.

KOMPH2: Updated Hydrogen Bond Donating and Accepting Parameters
    Recently, we have made efforts to more appropriately parameterize the hydrogen bonding α' and β' values by correcting the single ion transference data for the effect of dielectric constant in the form of the Kirkwood-Onsager parameter and for δ^2 and placed the values on a scale that appropriately gives the energy of the reference ions, chloride and potassium ion, in the gas phase. Thus, for the solvents HMPA, acetone, acetonitrile, nitromethane, nitrobenzene, DMF, formamide, DMSO, methanol, ethanol, trifluoroethanol, water, ethylene glycol, and sulfolane, the α' and β' values were obtained by adjustment in the correlation the transference data of chloride ion and potassium ion, respectively, with the free energies of aquation of the gas phase ions as part of the data set to anchor the relative values. When this was done the parameters led to the following correlation of the ions with the KOPMH2 equation (each with r = 0.999):

$$\ln K \ (Cl^-) = 219.2 \ (\varepsilon-1)/(2\varepsilon+1) + 99.95 \ \alpha` - 0.08 \ \beta` - 6.74 \ \delta^2 - 127.09 \qquad (9)$$

$$\ln K \ (K^+) = 254.1 \ (\varepsilon-1)/(2\varepsilon+1) + 0.36 \ \alpha` + 100.3 \ \beta` - 5.31 \ \delta^2 - 136.5 \qquad (10)$$

The values of the hydrogen bond donating and accepting ability were adjusted to give 100. for the coefficient in the above correlations so that a measure of the importance of these parameters in any correlation could be compared with that for interaction with the reference ions.
    Unfortunately, ion transference data is not available for all of solvents used in organic chemistry so we attempted to generate α' and β' values from Taft's α and β parameters by examining the correlation of the values already determined with the Taft values. The following coefficients were obtained:

$$\alpha = \ 5.28 \ \alpha' -0.15 \ (r = 0.982, \ SD = 0.11, \ range = 1.51\text{-removed nitrobenzene}) \qquad (11)$$

$$\beta = 4.15 \ \beta' - 0.45 \ \delta^2 +0.03 \ (r = 0.902, \ SD = 0.13, \ range = 1.05\text{-removed EtOH}) \ (12)$$

Since the alpha values were reasonably correlated with one another, the α' values for other solvents were adjusted to provide a reasonable correlation with Taft's. The same was done with the beta values where β' for the other solvents were adjusted to give the same coefficients as obtained with the solvents from which the K^+ transference data was obtained. The validity of this approach can be questioned, particularly with solvents whose alpha and beta values approach 0.0 where the intercepts for the Taft correlations differ from those from the ion transference data. Thus the α' and β' parameters for the weakly donating solvents or those not included in the transference data set can only be regarded as temporary. Nonetheless, their values are small in any event, so they cannot be substantially in error.

Taft π^* with KOMPH2

As a sidelight here, it is interesting to subject the Taft π^* parameters for the solvents for which transference data is available to the KOMPH2 equation. Here is was found that all terms but the β' parameter are important, but overall the correlation is relatively poor.

$$\pi^* = 1.65 \ (\varepsilon\text{-}1)/(2\varepsilon+1) - 0.90 \ \alpha' + 0.54 \ \delta^2 - 0.01 \ (r = 0.870, SD = 0.13,$$
$$\text{range} = 1.09) \tag{13}$$

Itt is clear that π^* has additional contributing factors that must be responsible for the success of the Taft equation. Over the broader range of solvents, the correlation of π^* with the KO function gave a correlation coefficient of 0.790.   Only the addition of the rest of the parameters of the KOMPH2 equation provided a better correlation:

$$\pi^* = 2.10 \ (\varepsilon\text{-}1)/(2\varepsilon+1) - .86 \ \alpha' - 0.94 \ \beta' + .71 \ \delta^2 - .18 \ (r = 0.842, SD = 0.159,$$
$$\text{range} = 1.09) \tag{14}$$

The variances in all the coefficients except the KO parameter are unacceptably large, so this reinforces concern over other factors that contribute to π^*.   Given the great success of the Taft approach, some effort might be spent understanding the physical origin of this parameter.

E_T Values with KOMPH2

It is revealing to correlate another solvatochromically derived parameter that is in popular use, namely solvent E_T values by using just the solvents for which transference data is available.   Clearly, E_T is less dependent on solvent dielectric constant than on hydrogen bond donating and accepting ability, and addition of the KO parameter and the δ^2 parameter only slightly improve the correlation:

$$E_T *1000/RT = 158.8 \ \alpha' + 69.1 \ \beta' + 52.5 \ (r = 0.987, SD = 2.65, \text{range} = 54.3) \tag{15}$$

When the entire range of solvents is included in the correlation, the cohesive energy density term assumes some importance:

$$E_T*1000/RT = 116.2 \ \alpha' + 34.5 \ \beta' + 21.3 \ \delta^2 + 53.8 \ (r = 0.958, SD = 4.6,$$
$$\text{range} = 54.3) \tag{16}$$

The solvents that deviate the most are acetic acid, tert-butyl alcohol, and benzonitrile. The former two may have overestimated α' values and the latter may have an underestimated β' value by comparison to correlations of the solvolysis of *tert*-butyl chloride.   Indeed, if these are removed from the correlation, the fit improves although with less emphasis on the cohesive energy density term which can be removed with only a moderate degradation in quality of fit:

$$E_T*1000/RT = 142.8 \; \alpha' + 63.8 \; \beta' +53.5 \; (r = 0.966, SD = 4.40, range = 54.1) \quad (17)$$

The importance of the specific solvation terms in E_T values and the success of the many correlations with E_T stresses the need to include specific solvation in any theoretical treatment.

tert-Butyl Chloride Solvolysis with KOMPH2

Application of the KOMPH2 equation to the solvolysis of tert-butyl chloride in only the solvents used to derive ion transference data provides important insights:

$$\ln k = 46.1 \; \alpha' + 11.4 \; \delta^2 - 6.23 \; (r = 0.979, SD = 1.22, range = 19.0) \quad (18)$$

The data has little contribution from solvent dielectric or from solvent basicity. The latter is expected since cation solvation is difficult due to steric effects, but the former is quite unexpected until there is recognition of the near invariance of the KO parameter in this solvent regime. The KO function approaches its maximum rapidly and assumptotically with increasing dielectric constant. What is remarkable is that in this solvent regime, the hydrogen bond donation stabilizes the transition state, and the cohesive energy density destabilizes the ground state. In water both effects contribute nearly equally. In TFE, hydrogen bond donation is much more important. Secondarily, with the new hydrogen bond donating parameters, the extent of stabilization of chloride ion in the transition state by this effect is only half of that expected for transfer of chloride ion; this contrasts with nearly 90% transfer as suggested from the previous parameters.

If a much larger range of solvents is examined, the dielectric effect is found to contribute:

$$\ln k = 36.0 \; (\varepsilon-1)/(2\varepsilon+1) + 39.9 \; \alpha' + 11.7 \; \delta^2 -22.8 \; (r = 0.985, SD = 1.12,$$
$$range=25.4) \quad (19)$$

The dielectric term suggests a change in dipole moment of roughly 10 D with a cavity radius of 4 A. Some concern should be expressed here since the mechanism of an ionization process in media with very poor dielectric properties may be different from that in the solvents in which ion transfer data was obtained.

Claisen Rearrangement with KOMPH2

With the new parameters, the correlation of the Claisen rearrangement rate data gathered by Brandes is slightly different from the previous correlation in that the solvent basicity term is no longer significant:

$$\ln k = 9.43 \; \alpha' + 3.27 \; \delta^2 -0.29 \; (r = 0.969, \; SD = 0.40, range = 5.37) \quad (20)$$

The magnitudes of the contributions from hydrogen bonding and from the cohesive energy density are comparable in water while in TFE the hydrogen bonding is more important.   Perhaps significant is the fact that the contribution of cohesive energy density is important in solvents such as cyclohexane and benzene.   This may provide an explanation for the larger rate of Claisen rearrangements in solution than in the gas phase.

Again, the correlation analysis is consistent with Jorgensen's recent calculations on the Claisen rearrangement in aqueous solution where increased hydrogen bonding in the transition state was found to be most important and changes in dipole moment from ground to transition state were much less important.   Unclear is whether any insight on the effect of cohesive energy density arises from Jorgensen's calculations.

New Experimental Data on the Claisen Rearrangement

In an effort to characterize the transition state of the Claisen rearrangement in more aqueous media, we determined the secondary deuterium kinetic isotope effects at C-4 and C-6 of the parent allyl vinyl ether in xylene and aqueous methanol solvent systems at 100°C.   The results (see Table II), when converted to i values using the equilibrium isotope effects (i values are the ratio of the logarithms of the kinetic isotope effect to the equilibrium effect), reveal a structure in all media that is similar to that in the gas phase (*23*).   That is, there is little evidence for a change in transition state structure upon submersion of allyl vinyl ether in water despite a 15-30 fold rate enhancement relative to xylene solvent.

TABLE II. Secondary Deuterium Kinetic Isotope Effects for the 3,3-Shift of Allyl Vinyl Ether in *m*-xylene, and in 28% and 75% aqueous methanol (% refers to volume percent methanol; Standard Deviations are in parentheses)

Cpd.	Temp °C.	Solvent	k^H / k^D_2	ln kie/ln EIE	$k^H (10^6 sec^{-1})$
4-D_2	100.	*m*-xylene	1.119 (.019)	0.37 (.04)	4.7
6-D_2	100.	*m*-xylene	0.953 (.015)	0.155 (.05)	
4-D_2	100.	75% MeOH	1.059 (.007)	0.191 (.02)	46.1
6-D_2	100.	75% MeOH	0.981 (.018)	0.06 (.06)	
4-D_2	100.	28% MeOH	1.145 (.04)	0.44 (.09)	180.3
6-D_2	100.	28% MeOH	0.958 (.04)	0.14 (.15)	
4-D_2	160.3	gas phase	1.092 (.005)	0.33 (.02)	0.19(extra-polated)
6-D_2	160.3	gas phase	0.98 (.005)	0.16 (.05)	

Concern as to what to expect for ionization to an ion pair lead to examination of the solvolysis of allyl mesylate in aqueous methanol. Remarkably, with the 1,1-dideuterio material, no rearrangement of the label in either starting material or product was observed while the reaction was being monitored by proton NMR using deuterated solvents at room temperature. Unsubstituted allyl systems have long been suspected to undergo direct substitution by solvent, and this is further proof of that hypothesis (24).

This, however, raises still another concern about mechanisms involving ion pairs. If a good leaving group attached to an allyl moiety does not induce heterolysis, how should a substantially poorer leaving group? To pursue this more quantitatively, if the rate constant for reaction of allyl mesylate in aqueous methanol is a maximum value for the rate constant for heterolysis, then replacement of the mesylate by an enol ether should reduce the heterolysis rate constant by roughly the K_a difference of the corresponding conjugate acids. This is roughly a factor of ten powers of ten. Since the Claisen rearrangement is conducted at 100°C., the heterolysis rate of allyl vinyl ether at that temperature might be a thousand times faster which is still a factor of roughly 10 million times slower than the actual rearrangement. We conclude that there is no evidence for heterolysis, even to a small extent, with unsubstituted allyl vinyl ether.

$$\text{CD}_2\text{OMs} \xrightarrow[\text{CD}_3\text{OD/D}_2\text{O}]{25^\circ \text{ C}} \text{CD}_2\text{OMe (OD)}$$

Claisen Rearrangements with Increased Polarity in the Transition State.
The Claisen rearrangement reaction that appears to have the most polar character is that of C-4 alkoxy-substituted allyl vinyl ethers studied by Coates and Curran (25,26). Thus, 4-methoxy allyl vinyl ether rearranges roughly 100 times faster than the parent allyl vinyl ether in benzene solution. Further, the rearrangement is 18 times faster in methanol than in benzene compared with a factor of 1.7 for the parent system in these two solvents.

$$\text{(MeO-allyl vinyl ether)} \xrightarrow{80^\circ\text{C}} \text{(MeO product)}$$

k (benzene) = 96 x k parent (benzene)
k (MeOH) = 18 x k (benzene)

In a more dramatic comparison, the rate of rearrangement of a 1'-ethoxy substituted O- allyl-1-cyclopentenyl ether in 80% aqueous ethanol is 70 times faster than the rearrangement of the same ether in benzene. However, under these conditions the rate difference for *tert*-butyl chloride ionization can be estimated to be 10,000,000. Thus the differential solvent stabilization of the Claisen rearrangement transition state is only about a fifth of the free energy of the differential stabilization of the transition state for ionization.

k (80% aq. EtOH) = 70 x k (benzene)

Diels-Alder Reactions

Blokzijl, Engberts, and Blandamer studied the reaction of cyclopentadiene with 5-methoxy-1,4-naphthoquinone and provided the most extensive list of solvent rate effects on the Diels-Alder reaction yet published (*26*). With the current set of parameters the KOMPH2 correlation equation provides:

$$\ln k = 12.90\ \alpha' - 6.28\ \delta^2 - 7.63 \quad (r = 0.979,\ SD = 0.486,\ range = 8.7) \tag{21}$$

If the methanol point is removed the correlation improves to:

$$\ln k = 13.98\ \alpha' - 6.12\ \delta^2 - 7.61 \quad (r = 0.989,\ SD = 0.362,\ range = 8.7) \tag{22}$$

This correlation suggests the importance of hydrogen bonding, perhaps in the form of general acid catalysis of the reaction by protonation of the carbonyl groups of the dienophile, and the importance of cohesive energy density destabilizing the ground state relative to the transition state recognizing there is a reduced volume in the transition state. Most Diels-Alder reactions have activation volumes of -20 to -35 ml/mole so the coefficient of the cohesive energy density is not tracking activation volumes directly.

Blokzijl, *et. al.* found that in dilute aqueous solvents the Diels-Alder rate effect arises primarily from changes in the potential energy of the initial state and that the activated complex is only moderately sensitive towards solvent effects. This initial state effect was called the "enforced hydrophobic interaction" and is clearly evident in the correlations provided here (*27,28*).   Whether or not hydrogen bonding plays an important role in the Blokzijl *et. al.* analysis is not clear.   It is interesting that Jorgensen's calculations on simple Diels-Alder reactions reveal the importance of increased hydrogen bonding in the transition state (*29*).

Summary

It is possible to model, with reasonable accuracy, the effect of solvents on solvolysis reactions, Claisen rearrangements, and Diels-Alder reactions using a multiparameter approach employing the Kirkwood-Onsager function, the Hildebrand cohesive energy density, and hydrogen bond donor and acceptor parameters based on chloride and potassium ion transference data.   These parameters even provide insight into the origin of the success of the single parameter solvent characterization scheme

APPENDIX 1. Solvent Parameters for the KOMPH2 equation

#	SOLVENT	$\varepsilon-1/2\varepsilon+1$	Alpha'	Beta'	CED
34	CS2	.2611	.0000	.0000	.1000
33	Sulfolan*	.4829	.0445	.1680	.2800
32	Pyridine	.4441	.0000	.1500	.1880
31	Et3N	.2432	.0000	.2031	.1000
30	EtGlyc*	.4804	.2065	.1703	.3580
29	HCOOH	.4873	.2973	.0000	.3600
28	HOAc	.3864	.2493	.1250	.3435
27	HFIP	.4993	.4273	.0000	.1507
26	H2O*	.4905	.2582	.1664	.9270
25	TFE*	.4724	.2902	.0203	.2314
24	MeOH*	.4773	.1958	.1305	.3461
23	EtOH*	.4698	.1789	.1203	.2736
22	iPrOH	.4632	.1424	.1953	.2250
21	tBuOH	.4423	.1246	.2344	.1890
20	DMSO*	.4841	.0685	.2016	.2850
19	HCONH2*	.4933	.1763	.1594	.6105
18	DMF*	.4798	.0431	.1984	.2344
17	PhNO2*	.4788	.0961	.0672	.2060
16	CH3NO2*	.4794	.0913	.0852	.2675
15	PhCN	.4708	.0000	.0953	.2074
14	CH3CN*	.4803	.0662	.1241	.2326
13	CH3COCH3*	.4646	.0356	.1758	.1530
12	EtOAc	.3850	.0000	.1000	.1338
11	CH2Cl2	.4217	.0623	.0000	.1650
10	CHCl3	.3587	.0890	.0000	.1500
9	HMPA*	.4751	.0071	.2266	.1240
8	THF	.4072	.0000	.1234	.1458
7	Dioxane	.2232	.0000	.0938	.1688
6	Et2O	.3447	.0000	.0984	.0950
5	nBu2O	.2917	.0000	.0969	.1000
4	Ph-H	.2302	.0089	.0313	.1414
3	CCl4	.2263	.0000	.0000	.1246
2	cyc-C6	.2024	.0000	.0000	.1134
1	gas*	.0000	.0000	.0000	.0000

*alpha' and beta' data are derived from transference data and free energies of aquation of chloride and potassium ion, respectively.   The solvent parameters are in units appropriate for the equation:

   $\ln Keq(\text{or } k) = dmu*(e-1/2e+1) + a*\alpha' + b*\beta' + V*CED + C$

where: $dmu = 24.2*((mu**2/r**3)ts - (mu**2/r**3)gs)$; mu is in D; r is in Angstroms;
        e is the dielectric constant;
        a is the specific H bonding percent relative to that of Cl⁻;
        b is the specific cation solvation percent relative to that of K^+;
        V is in ml/mol;
        C is the gas phase value of ln Keq or ln k.

in the form of solvent E_T values which are based on uv wavelength shifts.   An important take-home lesson is that the most important effect of solvent on these organic reactions is hydrogen bonding and response to solvent cohesive energy density.   Dielectric effects, generally are unimportant unless the solvent regime capable of being studied includes very low and moderately high dielectric constants, and even then, the reaction must involve large changes in ionic character.   The Claisen rearrangement and the Diels-Alder reaction are not among the reactions in this latter category.

Acknowledgment
We thank the National Science Foundation and the Department of Energy for support of this work.

References
1. Reichardt, C. *Solvents and Solvent Effects in Organic Chemistry*, 2nd ed.; VCH; Weinheim, 1988.
2. Kirkwood, J.G. *J. Chem. Phys.* **1934** *2*, 351.
3. Onsager, L. *J. Am. Chem. Soc.* **1936**, *58*, 1486.
4. Koppel, I.A.; Palm, V.A. *Org. React. USSR* **1971**, *8*, 291 (*CA* **1972** *76*, 28418k).
5. Makitra, R.G.; Pirig, Y. N.; Zeliznyi, A.M.; Daniel de Aguar, M.A.; Mikolajev, V.A.; Ramanov, V.A. *Org. React. USSR*, **1977**, *14*, 421.
6. Hildebrand, J.H.; Prausnitz, J.M.; Scot, R.L. *Regular and Related Solutions*; Van Nostrand-Reinhold; Princeton, 1970.
7. Hildebrand, J.H.; Scott, R.L. *Regular Solutions*; Prentice-Hall; Englewood Cliffs, NJ, 1962.
8. Kamlet, M.J.; Abboud, J.-L.M.; Taft, R.W. *Prog. Phys. Org. Chem.* **1981**, *13*, 485.
9. A recent parameter list: Abraham, M.H.; Grellier, P.L.; Abboud, J.-L.M.; Doherty, R.M.; Taft, R.W. *Can. J. Chem.* **1988**, *66*, 2673.
10. Gajewski, J.J.;Jurayj, J.; Kimbrough, D.R.; Gande, M.E.; Ganem, B.; Carpenter, B.K. *J. Am. Chem. Soc.* **1987**, *109*, 1170.
11. Brandes, E.; Grieco, P.A.; Gajewski, J.J. *J. Am. Chem. Soc.* **1989**, *54*, 515.
12. Grunwald, E.; Winstein, S. *J. Am. Chem. Soc.* **1948**, *70*, 846.
13. Smith, S.G.; Fainberg. A.H.; Winstein, S. *J. Am. Chem. Soc.* **1961**, *83*, 618.
14. Goering, H.L.; Anderson, R.P. *J. Am. Chem. Soc.* **1978**, *100*, 6469.
15. Mills, S.G.; Beak, P. *J. Org. Chem.* **1985**, *50*, 1216.
16. Beak, P.; Covington, J.B.; White, J.M. *ibid* **1980**, *45*, 1347.
17. Cox, B.G.; Hedwig, G.R.; Parker, A.J. *Aust. J. Chem.* **1974**, *27*, 477.
18. Marcus, Y. *Pure Appl. Chem.* **1983**, *55*, 977.
19. Gajewski, J.J. *J. Org. Chem.* **1992**, *57*, 5500; see Appendix for Data. Note that previous formulations used a negative sign for the coefficient of the δ^2 term which would be the appropriate if the coefficient were an activation volume.   To avoid confusion, the sign of all coefficients are now exactly what results from the regression analyses.
20. Abraham, M.H.; Grellier, P.L.; Nashzadeh, A.; Walker, R.A.C. *J. Chem. Soc.*, **1988**, 1717.
21. Severance, D.L.;Jorgensen, W.L. *J. Am. Chem. Soc.* **1992**, *114*, 10966.

22. Cramer, C.J.; Truhlar, D.G. *J. Am. Chem. Soc.* **1992**, *114*, 8794.

23. Gajewski, J.J.; Conrad, N.D. *J. Am. Chem. Soc.* **1979**, *101*, 6693.

24. DeWolfe, R.H.; Young, W.G. *Chem. Rev.* **1956**, *56*, 753.

25. Coates, R.M.; Rogers, B.D.; Hobbs, S.J.; Peck, D.R.; Curran, D.P. *J. Am. Chem. Soc.* **1987**, *109*, 1160.

26. Blokzijl, W.; Engberts, J.B.F.N.; Blandamer, M.J. *J. Am. Chem. Soc.* **1991**, *113*, 4241.

27. Blokzijl, W.; Engberts, J.B.F.N. *J. Am. Chem. Soc.* **1992**, *114*, 5440.

28. For an extensive review of hydrophobic effects see Blokzijl, W.; Engberts, J.B.F.N. *Angew. Chem. Int. Ed. Engl.* **1993**, *32*, 1545-1579.

29. Blake, J.F.; Jorgensen, W.L. *J. Am. Chem. Soc.* **1991**, *113*, 7430.

RECEIVED July 5, 1994

Chapter 17

Claisen Rearrangement of Allyl Vinyl Ether

Computer Simulations of Effects of Hydration and Multiple-Reactant Conformers

Daniel L. Severance and William L. Jorgensen[1]

Department of Chemistry, Yale University, New Haven, CT 06511-8118

A free energy of hydration profile for the Claisen rearrangement of allyl vinyl ether (AVE) has been obtained from Monte Carlo statistical mechanics simulations at 25 °C. The gas-phase minimum energy reaction path through the chair transition state was determined from ab initio 6-31G(d) calculations and was followed in a periodic cell containing 838 water molecules. The transition state is computed to be 3.85 ± 0.16 kcal/mol better hydrated than the reactant, which corresponds to a rate increase by a factor of 664 over the gas-phase reaction. This large effect is shown to be consistent with available experimental data. The origin of the effect is analyzed. The energetic impact of multiple conformational states for the reactant has also been considered.

For chemical reactions, changes in polarity between the reactants and transition state lead to rate variations in different solvents (1). The effects can be profound with rate ratios of 10^8 or more for comparisons of alternate solvents and up to 10^{20} when the gas phase is included (2). However, some classes of reactions are relatively immune to solvent effects since they have "isopolar transition states" (1). With the exception of dipolar cycloadditions, many pericyclic reactions have been considered to be in this category (1). However, the generality of this notion was strikingly challenged by observations that simple Diels-Alder reactions could show rate accelerations by factors of 10^2 - 10^4 in aqueous solution over hydrocarbon solvents (3,4). Scrutiny of the literature reveals comparable solvent dependence for the rates of Claisen rearrangements as well (5). The data of White and Wolfrath from 1970 are particularly notable (5a). They found a rate increase by over a factor of 300 for the rearrangement of allyl p-tolyl ether in going from the gas phase to progressively more polar solvents; the greatest rate was in p-chlorophenol, though they did not obtain data in pure water.

Efforts in our laboratory have been directed at understanding the origin of such solvent effects at the molecular level through computational modeling. This is not only technically challenging, but also affords the opportunity to better characterize the electronic structure of transition states and to reveal details about their solvation, which

[1]Corresponding author

0097–6156/94/0568–0243$08.00/0

can provide insights for catalyst design. The present work focuses on the parent Claisen rearrangement of allyl vinyl ether (AVE) in the gas phase and aqueous solution. In addition to the effect of hydration on the free energy of activation, which was the subject of a prior communication (6), consideration is also given here to the existence of multiple conformational states for the reactant. The population of this manifold of states is not normally considered in computational studies on reaction energetics.

Computational Procedure

The computational approach is an updated version of our efforts on S_N2, addition, and association reactions (7). Ab initio molecular orbital calculations are used to locate the transition state(s) for the reaction and to obtain a minimum energy reaction path (MERP) in the gas phase. The ab initio calculations are also used to provide partial charges for the reactants along the reaction path, which are needed for the potential functions that describe the intermolecular interactions between the reacting system and solvent molecules. The reacting system is then placed in a periodic cell with hundreds of solvent molecules and the changes in free energies of solvation along the reaction path are computed in Monte Carlo simulations with statistical perturbation theory. These calculations provide the key thermodynamic results and great detail on the variations in solvent-solute interactions along the reaction path.

Ab Initio Calculations.   The present ab initio calculations were performed with the GAUSSIAN 92 program using the 6-31G(d) basis set, which includes a set of d-orbitals on carbon and oxygen (8). Such RHF/6-31G(d) calculations were previously carried out by Vance et al. to locate structures for the reactant, chair transition state, and product (9). Reaction path following (10) was then performed (8) to generate a 143 frame "movie" along the minimum energy reaction path from one conformer of AVE through the transition state to 4-pentenal. 6- 31G(d)//6-31G(d) optimizations were also executed to locate eight additional low-energy minima for AVE. All stationary points were confirmed via calculations of the vibrational frequencies. This further permitted computation of the relative free energies of the stationary points in the gas phase using standard statistical mechanical procedures (11). For the latter purpose, the computed vibrational frequencies were scaled by 0.91 to be more consistent with experimental values, and scaled frequencies below 500 cm^{-1} were treated as classical rotations with $E_v = RT/2$ (12). The nine conformers for AVE are shown in Figure 1 along with their relative 6-31G(d)//6-31G(d) energies and dipole moments.

Fluid Simulations.   Monte Carlo calculations were performed along the gas-phase MERP using statistical perturbation theory to compute the changes in free energies of hydration. The intermolecular interactions are described by potential functions with the potential energy, ΔE_{ab}, consisting of Coulomb and Lennard-Jones terms between the atoms i in molecule a and the atoms j in molecule b, which are separated by a distance r_{ij} (equation 1) (13). All atoms are explicit and

$$\Delta E_{ab} = \sum_i \sum_j \left\{ q_i q_j e^2 / r_{ij} + 4\varepsilon_{ij} \left[\left(\sigma_{ij} / r_{ij} \right)^{12} - \left(\sigma_{ij} / r_{ij} \right)^6 \right] \right\} \tag{1}$$

the TIP4P model was used for water (13a). In all, 59 frames (structures) along the MERP were utilized. Partial charges (q) were obtained for each of these from the 6-31G(d) ab initio calculations using the CHELPG procedure for fitting to the electrostatic potential surface (*14*). Mulliken charges were also considered; however, they showed more variation in going to the 6-31+G(d,p) basis set, computed dipole moments with the Mulliken charges deviated significantly (0.5 - 1.4 D) from the 6-31G(d) and CHELPG values, and the general superiority of 6-31G(d) CHELPG charges for computing free energies of hydration was previously demonstrated (*15*). The CHELPG charges are given in Table I for the reactant, transition state, and product along the MERP with the atom numbering illustrated below. The conformation of AVE on the MERP corresponds to structure **4** in Figure 1. Standard

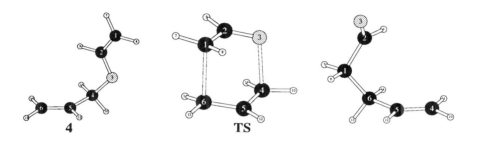

all-atom Lennard-Jones parameters (σ, ε) (*13*) were adopted with geometric combining rules for the reactant and product and scaled as hybridization changed along the MERP. The variations in σ and ε are modest, as summarized in Table I.

The simulations were executed in the NPT ensemble at 25 °C and 1 atm with Metropolis and preferential sampling (*16*). The system consisted of 838 water molecules and the solute in a periodic cell ca. 30 Å on a side. The BOSS program (*17*) perturbed the system between frames and computed the change in ΔG_{hyd} via statistical perturbation theory (*16,18*). Utilization of the 59 frames required 29 separate Monte Carlo simulations with double-wide sampling (*19*). The perturbations are performed from one frame to the two adjacent frames on either side; the three images of the reacting system are overlaid maximally in an rms sense for the atomic positions, and a common dummy center point is used as the origin for the rigid-body rotations so that the three images do not separate upon rotation. Each simulation entailed 1×10^6 (1M) configurations of equilibration followed by 4M configurations of averaging. Solute-water interactions were included for waters with an oxygen within 10.5 Å of any solute atom, and the water-water cutoff was at 10.5 Å based on the O-O distance. The interactions were feathered to zero quadratically between 10.0 and 10.5 Å. The individual molecules were kept rigid in the simulations; only translations and rigid-body rotations were sampled.

SCRF Calculations for AVE.   Estimates of the relative free energies of hydration for the 9 conformers of AVE in Figure 1 were obtained from four self-consistent reaction field (SCRF) procedures. These are needed to obtain the energetic effects of including all 9 conformers as opposed to just **4** on the free energy of activation for the reaction in water. Only the chair conformer of the transition state required consideration since the alternative boat form is 4-7 kcal/mol higher in energy (*9*). In each case the 6-31G(d) optimized geometries were used without reoptimization in the reaction field; for a given SCRF method, reoptimization for neutral molecules normally has negligible energetic effects (*20*), and for the present

Table I. Potential Function Parameters for the Claisen Rearrangement of AVE [a]

Atom	Reactant (4)			Transition State			Product		
	q	σ	ε	q	σ	ε	q	σ	ε
C1	-0.574	3.550	0.070	-0.416	3.545	0.070	-0.071	3.500	0.066
C2	0.216	3.550	0.070	0.296	3.549	0.070	0.542	3.550	0.070
O3	-0.413	3.000	0.170	-0.485	2.995	0.175	-0.533	2.960	0.210
C4	0.407	3.500	0.066	0.161	3.506	0.067	-0.429	3.550	0.070
C5	-0.122	3.550	0.070	-0.479	3.546	0.070	-0.101	3.550	0.070
C6	-0.425	3.550	0.070	0.014	3.545	0.070	0.145	3.500	0.066
H7	0.179	2.420	0.030	0.105	2.428	0.030	0.014	2.500	0.030
H8	0.218	2.420	0.030	0.183	2.428	0.030	0.037	2.500	0.030
H9	0.088	2.420	0.030	0.045	2.422	0.030	-0.032	2.420	0.030
H10	-0.019	2.500	0.030	0.090	2.490	0.030	0.172	2.420	0.030
H11	-0.019	2.500	0.030	0.082	2.490	0.030	0.175	2.420	0.030
H12	0.120	2.420	0.030	0.207	2.427	0.030	0.108	2.420	0.030
H13	0.173	2.420	0.030	0.092	2.428	0.030	-0.013	2.500	0.030
H14	0.173	2.420	0.030	0.105	2.428	0.030	-0.013	2.500	0.030

[a]Charges in electrons, σ in Å, ε in kcal/mol.

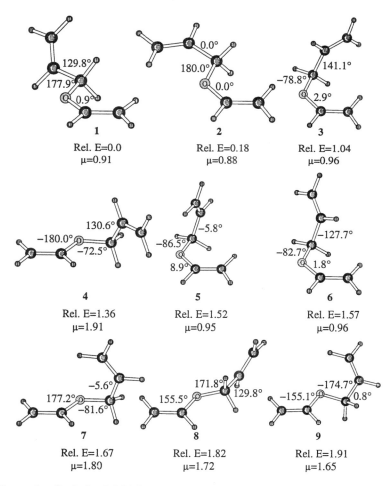

Figure 1. Optimized 6-31G(d)//6-31G(d) structures for the nine conformers of allyl vinyl ether. Relative energies are given in kcal/mol and dipole moments are in Debyes. The optimized values for the three central dihedral angles are shown.

comparative purposes, it is desirable to keep the geometries the same for all methods. Two of the four procedures involved an AM1 calculation with the SM2 method (20) and a PM3 calculation with the SM3 method (21) using the AMSOL program (22). The MST method (23) was also used with AM1 and 6-31G(d) wavefunctions in a modified version of GAUSSIAN 92 (8,23c). The relative free energies of hydration for the 9 conformers could be obtained from Monte Carlo simulations with the explicit TIP4P water model. This was not done in view of the associated large computational effort; the SCRF results should be adequate to illustrate the magnitude of the energetic effects of including the additional conformers.

Results and Discussion

Gas-Phase Energetics.   The key variables for the conformational search for AVE were the three dihedral angles of the C2-O-C4-C5 fragment since the double bonds restrict rotation about C1-C2 and C5-C6. Starting structures for optimizations were obtained by systematic variation of the three central dihedral angles. This resulted in the 9 structures in Figure 1, whose status as true minima was verified by the vibrational frequency calculations at the 6-31G(d)//6-31G(d) level. Basically, the dihedral angle about C2-O prefers to be near 0° or 180° to maintain conjugation for the vinyl ether fragment, while dihedral angles near 0°, 120°, and 240° are preferred about C4-C5 to provide the usual eclipsing of a vinyl group to the bonds attached to an adjacent sp^3 carbon (C4). The normal preference for staggered geometries about an sp^3 C - O ether bond is then expected and yields preferred dihedral angles near 180° and ±60° for the central C4-O bond. Of the 2 x 3 x 3 = 18 likely conformers, symmetry equivalence and steric clashes reduce the number of resultant unique minima to the 9 shown in Figure 1. Remarkably, their relative energies fall in a less than 2 kcal/mol range at the 6-31G(d)//6-31G(d) level, so significant populations of more than one conformer are expected near room temperature. It may also be noted that the computed dipole moments for the structures fall into two groups with values near 1.0 D and 1.8 D. The difference is primarily related to the C1-C2-O-C4 dihedral angle, which is near 0° and 180° for the two groups, respectively.

    The computed structures and vibrational frequencies were then used to compute the relative free energies for **1 - 9**, as summarized in Table II. Though conformer **1** remains the lowest in free energy, there is substantial reshuffling of the order of the other conformers between energy and free energy. It may be noted that **2**, the 0° - 180° - 0° conformer, which is only 0.18 kcal/mol above **1** in energy is 1.23 kcal/mol higher in free energy. Part of the change results from an RT ln 2 (0.41 kcal/mol) symmetry correction disfavoring **2** relative to the other conformers since it is the only achiral conformer. **2** was taken as the geometry for AVE in the study by Vance et al. (9). On the other hand, **4**, which is 1.36 kcal/mol higher in energy than **1**, becomes the second most favorable conformer in free energy, only 0.11 kcal/mol above **1**.

    Nine structures along the computed MERP are illustrated in Figure 2. Details on the 6-31G(d)//6-31G(d) geometries for the transition state are reported elsewhere (9,24) The length of the breaking bond, O-C4, is 1.918 Å and the length of the forming bond, C1-C6, is 2.266 Å at the 6-31G(d)//6-31G(d) level. These values increase to 2.100 and 2.564 Å, respectively, in the (6/6)CASSCF/6-31G(d) optimized geometry, which yields improved agreement with observed kinetic isotope effects (24). However, both transition state structures correspond to a concerted process, which also emerges from the experimental isotope effects study (25). As noted previously, the 6-31G(d)//6-31G(d) activation energy and free energy are too high, ca. 49 kcal/mol in Table II; MP2 corrections lower the figures by about 24 kcal/mol to a reasonable range (9).

    Given the results for the nine conformers of AVE in Table II, the effect of

Table II. Computed Relative Energies, Relative Free Energies, and Absolute Free Energies of Hydration (kcal/mol) for Conformers **1 - 9**, the Transition State, and Product (4-Pentenal) [a]

Conformer	ΔE^0	ΔG^{298}	$\Delta G_{hyd}(298)$			
			AM1/SM2	PM3/SM3	AM1/MST	6-31G(d)/MST
1	0.00	0.00	-0.74	-0.63	-1.43	-1.26
2	0.18	1.23	-0.44	-0.39	-0.95	-0.92
3	1.04	1.09	-0.72	-0.74	-1.09	-1.06
4	1.36	0.11	-0.84	-0.76	-1.80	-1.99
5	1.52	2.22	-0.58	-0.62	-1.06	-1.16
6	1.57	1.67	-0.79	-0.73	-1.20	-1.27
7	1.67	1.00	-0.69	-0.61	-1.63	-1.96
8	1.82	0.78	-0.90	-0.65	-2.29	-2.05
9	1.91	0.90	-0.53	-0.39	-1.42	-1.36
TS	48.93	48.48	-2.50	-3.00	-3.95	-4.26
Pdct.	-21.28	-19.88	-1.80	-3.87	-2.40	-3.70
$\Delta\Delta G^{\ddagger b}$		0.73	0.66	0.63	0.61	0.44
$\Delta\Delta G^{\ddagger}(g{\to}w)^c$			-0.08	-0.10	-0.12	-0.29

[a] ΔE^0 and ΔG^{298} from 6-31G(d)//6-31G(d) calculations. [b] Increase in the free energy of activation from including the population of all conformers, not just **4**. [c] Decrease in the free energy of activation in water relative to the gas phase for population of all conformers.

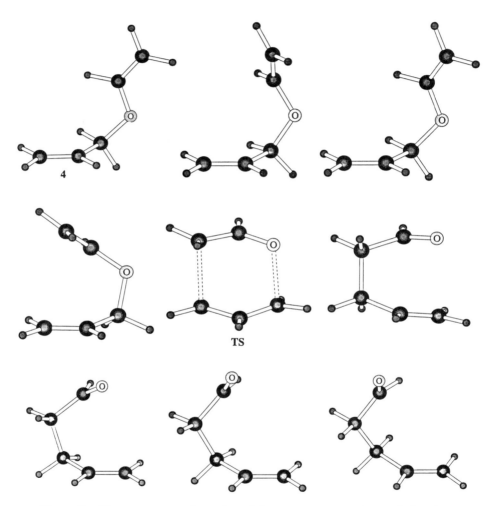

Figure 2.   Nine structures along the minimum energy reaction path for the Claisen rearrangement at the 6-31G(d)//6-31G(d) level.

including all of them rather than just one on the free energy of activation can be addressed. The free energy difference between populating just one conformer k and the full manifold of conformers is given by equation 2 where Q_k and Q are the corresponding partition functions and the sum is over all conformers i.

$$\Delta G = -RT\ln(Q/Q_k) = -RT\ln\left[\sum_i \exp\left(-(G_i - G_k)/RT\right)\right] \qquad (2)$$

Of course, population of the other conformers makes ΔG negative, though it is the least negative for choosing k as the conformer with lowest free energy. In the present case, with conformer 1 as k, $\Delta G = -0.62$ kcal/mol. Thus, the computed free energy of activation in the gas phase is raised by 0.62 kcal/mol by considering all of the AVE conformers. However, if conformers 2 or 4 were taken as the reference point for calculation of the free energy of activation, then considering all conformers would add 1.85 and 0.73 kcal/mol, respectively. For accurate work, such corrections may be significant.

Effects of Hydration.   The Monte Carlo simulations in water yielded the change in the free energy of hydration, ΔG_{hyd}, along the MERP. The results starting from conformer 4 at frame 1 are illustrated in Figure 3 by the 4 curves, which correspond to averaging for 1M to 4M configurations. It is apparent that the results are well-converged after 2M configurations. The TS at frame 83 has the lowest ΔG_{hyd}, 3.85 ± 0.16 kcal/mol (26), below the reactant. The curvature of the ΔG_{hyd} curve is less than for the gas-phase MERP, so the transition state is not predicted to shift along the MERP upon transfer to water. Without adjusting for the population of other AVE conformers in water, which will not have great effect because of the low free energy for 4 (Table II), the implied rate increase is a factor of 644 over the gas phase. This assumes no solvent dynamical effects on the barrier crossing, as in classical TS theory (27). Perusal of the experimental data starts with the measured relative rates for AVE of 1:4:58 in di-n-butyl ether, ethanol, and 2:1 methanol/water at 75 °C (5c). These give a Grunwald-Winstein m value of 0.4, which leads to a predicted relative rate of 876 in water; however, kinetic data on 4-substituted AVE's in solvents from cyclohexane to water suggest a relative rate in water closer to 150 (5d). Furthermore, the gas-phase kinetics indicate a rate of 0.031×10^{-6} sec^{-1} at 75°C (28), while the rate in di-n-butyl ether extrapolated to 75 °C is 0.276×10^{-6} sec^{-1} (5e), a factor of 9 higher. Thus, when the range of media is extended from the gas phase to water, a substantial rate acceleration for AVE indeed emerges, ca. 10^3 at 75 °C. Another potential comparison is for $\Delta\Delta G_{hyd}$ between 4 and the product, -2.88 kcal/mol in Figure 3. Experimental data are not available for these compounds; however, aldehydes are well-established to have lower ΔG_{hyd}'s than analogous ethers by 1-2 kcal/mol (29). The SCRF values in Table II are 1-3 kcal/mol.

Turning to the SCRF results for ΔG_{hyd} of 4 versus the transition state, the computed changes for the free energy of activation in water as compared to the gas phase are summarized in Table III. Results from Still's generalized Born/surface area (GB/SA) method are also included using the 6-31G(d) structures and CHELPG charges (30). The SCRF results are all qualitatively correct; however, the predicted rate accelerations are all too small. AM1/SM2 results were reported by Cramer and Truhlar as $\Delta\Delta G_{hyd} = 0.75$ kcal/mol and $k_{rel} = 4$ (31). The differences from the results in Table III stem from the use of AM1 optimized structures in their study and a reactant structure corresponding to 3 rather than 4; the results are improved with use of the 6-31G(d) structures and in going to the SM3 or MST models (Table III). In

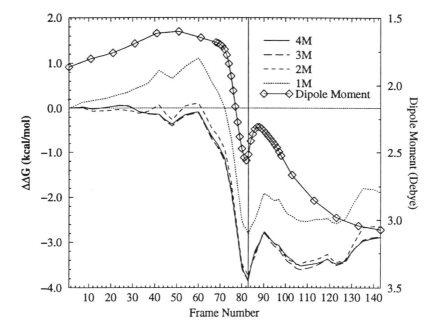

Figure 3. Computed changes in ΔG_{hyd} by averaging for 1M to 4M configurations from the Monte Carlo simulations and the variation in the 6-31G(d) dipole moment for the reacting system. The x-axis gives the frame number along the MERP; the transition state is at frame 83.

view of the number of features and terms in the SCRF treatments, it is difficult to trace the origin of their underestimates of the hydration effect. Cramer and Truhlar provide some discussion on this and suspect that the fault lies in the AM1 charge distributions for the AM1/SM2 results (*31*). However, the results from the newer PM3/SM3 model are better, while switching to the 6-31G(d) wavefunction with the MST procedure and the 6-31G(d) CHELPG charges with the GB/SA continuum model does not lead to substantial improvement. With the Monte Carlo approach and the associated two-

Table III. Computed Changes in the Free Energy of Activation (kcal/mol) in Water versus the Gas Phase and the Corresponding Rate Enhancements, k_{rel}, at 25 °C

Method	$-\Delta\Delta G_{hyd}$[a]	k_{rel}
AM1/SM2	1.66	16
PM3/SM3	2.24	44
AM1/MST	2.15	38
6-31G(d)/MST	2.27	46
GB/SA	2.10	35
Monte Carlo	3.85	664
Experiment[b]	(4.1)	(1000)

[a]The difference in free energy of hydration of conformer **4** and the chair transition state. [b]Estimated - see text.

body potential functions with 6-31G(d) CHELPG charges, the reliability of the charges is undoubtedly the dominant element in getting correct relative free energies of hydration (*15*). This approach can be criticized for ignoring solute polarization by the solvent since the partial charges are fixed from the CHELPG calculations. It appears that the polarization is included to some extent in an average way owing to overestimation of the polarity of molecules at the 6-31G(d) level (*15,31*). Another recent study of the effect of hydration on the rearrangement of AVE should be noted; Gao obtained a $-\Delta\Delta G_{hyd}$ of 3.5 kcal/mol ($k_{rel} = 368$) with his combined AM1/Monte Carlo approach (*32*).

The effect of the population of other conformers besides **4** can be addressed for the free energy of activation in water. Equation 2 is again applied where the relative free energies of the conformers in water are given by the sum of the gas-phase value, ΔG^{298}, and the SCRF $\Delta G_{hyd}(298)$ results in Table II. The results for the gas phase and the four hydration models are given in Table II as $\Delta\Delta G^{\ddagger}$. The effects in water, 0.4-0.7 kcal/mol, are seen to be very similar to those in the gas phase, 0.7 kcal/mol. This results from the limited variation in the calculated free energies of hydration of the 9 conformers. Consequently, there is little net solvent effect from this source; $\Delta G^{\ddagger}$ is lowered in water relative to the gas phase by 0.04 - 0.29 kcal/mol from population of the manifold of conformers rather than just **4**. These corrections could be added to the $\Delta\Delta G_{hyd}$ values in Table III and slightly improve the agreement with the experimental estimate. Overall, for this reaction, the conclusion is that starting from a low-energy conformer of the reactant, the effects of including all conformers raises the computed free energy of activation by ca. 1 kcal/mol and the differential effect between the gas phase and aqueous solution is close to negligible.

Origin of the Rate Acceleration in Water.   A fundamental issue for Claisen rearrangements is the extent of dipolarity in the transition state (TS) (5). Substantial rate increases are observed in protic solvents for substituted cases that enhance putative enolate/allyl cation character. However, the origin of the computed acceleration from the Monte Carlo simulations is not partial ionization. The net CHELPG charges on the $H_2C=CHO$ unit in AVE and the TS are -0.287 and -0.273 e (Table I), though solvent polarization of the solute was not included, as noted above. Even less charge separation is found in the MCSCF/6-31G(d) results (24).
      An alternative possibility emerged from the solute-water energy pair distributions, which were computed for the reactant (**4**), transition state, and product during the Monte Carlo simulations. These are given in Figure 4, which shows the number of water molecules on the y-axis that interact with the solute with the potential energy given on the x-axis. The bimodal nature of the curves reflects the hydrogen-bonded water molecules in the bands at low energy and the interactions with the many distant water molecules in the spike between -2 and +2 kcal/mol. The number and average strength of the hydrogen bonds are clearly in the order reactant < product < transition state. Integration of the curves to -3.0 kcal/mol, where the break to the central spike occurs, yields estimates of 0.5, 1.7, and 1.8 hydrogen bonds for the reactant, product, and transition state. The -3.0 kcal/mol limit is probably too severe for the ether reactant; graphical examination of configurations from the simulation normally shows one water molecule hydrogen bonded to the ether oxygen, which is typical for ethers (15).

Hydrogen Bonding Analysis.   A full hydrogen-bonding analysis was then performed on configurations saved during the Monte Carlo simulations. A less stringent energetic cutoff for hydrogen bonding, -2.25 kcal/mol, was supplemented by a geometric limit of 2.5 Å for an O ⋯ H hydrogen-bond length, which coincides with the first minimum in OH radial distribution functions. The analysis gave average numbers of hydrogen bonds of 0.9, 1.7, and 1.9 for the reactant, product, and TS with average strengths of -3.4, -4.5, and -4.7 kcal/mol. Typical examples of the water structure near the solutes are shown in Figures 5-7. The 1 and 2 hydrogen bonds with water for ether and carbonyl oxygens are normal (15). Clearly, the rate acceleration in water can largely be attributed to the increased number and strength of the hydrogen bonds to the ether oxygen in progressing from the reactant to the transition state. Also, the oxygen in the transition state is behaving in a hydrogen-bonding sense as carbonyl-like rather than ether-like. There is enhanced polarization of the HC2-O unit in proceeding from AVE to the TS that promotes hydrogen bond acceptance; the O becomes more negative by 0.07 e, which more than offsets the +0.04e change for HC2 (Table I). Continuing to the product, the O gains an additional 0.05 e; however, HC2 loses 0.17 e. In addition, the lengthening of the C4-O bond to 1.92 Å in the TS from 1.41 Å in AVE increases the solvent-accessibility of the oxygen. It should be noted that hydrogen bonding also emerged as the dominant

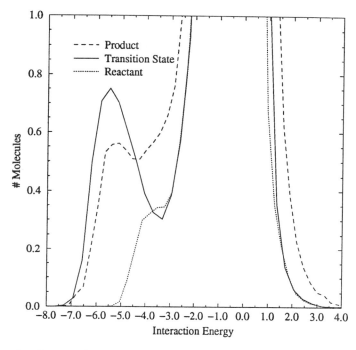

Figure 4. Computed solute-water energy pair distributions for the reactant (**4**), transition state, and product from the Monte Carlo simulations. The y-axis records the number of water molecules that interact with the reacting system with the potential energy in kcal/mol shown on the x-axis. The units for the y-axis are number of molecules per kcal/mol.

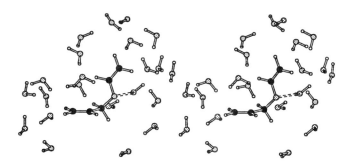

Figure 5. Stereoplot of a random configuration from a Monte Carlo simulation of the reactant (**4**) in water. Only the first shell of water molecules is shown.

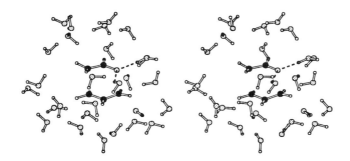

Figure 6.   Stereoplot for the transition state in water as in Figure 5.

Figure 7.   Stereoplot for the product, 4-pentenal, in water as in Figure 5.

element in a multiparameter analysis of solvent effects on the kinetics of Claisen rearrangements (*33*).

The present results give some insights into desirable features for catalyst design. The hydrogen bonding results demonstrate the advantage of having two hydrogen-bond donating groups positioned to interact with the oxygen atom in the chair transition state. An illustrative structure extracted from the Monte Carlo simulations is shown below. A related, important recent event is the determination of the crystal structure of a chorismate mutase with a bound endo-oxabicyclic transition state analog (*34*). Chorismate mutases catalyze the Claisen rearrangement of chorismate to prephenate. The ether oxygen of the inhibitor makes a single hydrogen bond with a side-chain NH of arginine-90 (*34*). This is consistent with the normal hydrogen-bonding characteristics of ether oxygens, as noted above (*15*). The position of the ether oxygen is at the edge of a solvent-exposed region of the binding cleft.

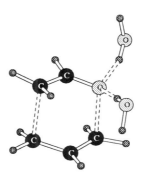

Based on the present results, it seems likely that in the transition state for rearrangement of a substrate, a second hydrogen bond to the oxygen would be present with the Arg-90 side chain or with a water molecule.

The charge and structural reorganization is also reflected in changes in the dipole moment for the reacting system. As shown in Figure 3, the ΔG_{hyd} and dipole moment curves mirror each other until a little past the transition state, then the ΔG_{hyd} profile levels off, while the dipole moment increases to the product. The 6-31G(d) dipole moments are 1.9 D for **4**, 2.5 D for the transition state, and 3.1 D for the product.

Conclusion

A combined quantum and statistical mechanical approach has provided insights into the acceleration of the Claisen rearrangement of allyl vinyl ether in aqueous solution. Rather than partial ionization, an increase in the number and strength of the hydrogen bonds with the ether oxygen in going to the transition state has been identified as the chief contributor to the catalysis. The observed magnitude of the lowering of the free energy of activation in water is well-reproduced by the Monte Carlo simulations in an all-atom format with explicit representation of the solvent molecules. Four SCRF treatments with continuum solvent models were found to uniformly underestimate the rate acceleration in water. The other principal issue to be addressed was the effect of including the populations of all conformers of the flexible reactant on the free energies of activation in water. Starting from a low-energy conformer such as **4**, the effects are less than 1 kcal/mol and similar in the gas phase and in water. This does not mean that such contributions should be ignored, particularly since the effects could be much greater if a higher-energy conformer is taken as the reference point in the absence of a full conformational search.

Acknowledgments

Gratitude is expressed to the National Science Foundation for support of this work and to Professor Modesto Orozco for computational assistance and advise on the SCRF calculations.

Literature Cited

(1)   Reichardt, C., *Solvents and Solvent Effects in Organic Chemistry*; VCH; Weinheim, 2nd edn, 1988, Chap 5.
(2)   See, for example: Kemp, D. S. and Paul, K. G. *J. Am. Chem. Soc.* **1975**, 97, 7305. Olmstead, W. N.; Brauman, J. I. *J. Am. Chem. Soc.* **1977**, 99, 4219.
(3)   (a) Rideout, D. C.; Breslow, R. *J. Am. Chem. Soc.* **1980**, 102, 7816. (b) Breslow, R. *Acc. Chem. Res.* **1991**, 24, 159.
(4)   Blokzijl, W.; Blandamer, M. J.; Engberts, J. B. F. N. *J. Am. Chem. Soc.* **1991**, 113, 4241.
(5)   (a) White, W. N.; Wolfarth, E. F. *J. Org. Chem.* **1970**, 35, 2196. (b) Coates, R. M.; Rogers, B. D.; Hobbs, S. J.; Peck, D. R.; Curran, D. P. *J. Am. Chem.Soc.* **1987**, 109, 1160. (c) Gajewski, J. J.; Jurazj, J.; Kimbrough, D. R.; Grande, M. E.; Ganem, B.; Carpenter, B. K. *J. Am. Chem. Soc.* **1987**, 109, 1170. (d) Brandes, E.; Grieco, P. A.; Gajewski, J. J. *J. Org. Chem.* **1989**, 54, 515. (e) Burrows, C. J.; Carpenter, B. K. *J. Am. Chem. Soc.* **1981**, 103, 6983.
(6)   Severance, D. L.; Jorgensen, W. L. *J. Am. Chem. Soc.* **1992**, 114, 10966.
(7)   For reviews, see: (a) Jorgensen, W. L. *Adv. Chem. Phys.*, Part II **1988**, 70, 469. (b) Jorgensen, W. L. *Acc. Chem. Res.* **1989**, 22, 184.
(8)   Frisch, M. J.; Trucks, G. W.; Head-Gordon, M.; Gill, P. M. W.; Wong, M. W.; Foresman, J. B.; Johnson, B. G.; Schlegel, H. B.; Robb, M. A.; Replogle, E. S.; Gomperts, R.; Andres, J. L.; Raghavachari, K.; Binkley, J. S.; Gonzalez, C.; Martin, R. L.; Fox, D. J.; Defrees, D. J.; Baker, J.; Stewart, J. J. P.; Pople, J. A. GAUSSIAN 92, Revision A; Gaussian, Inc., Pittsburgh, PA, 1992.
(9)   Vance, R. L.; Rondan, N. G.; Houk, K. N.; Jensen, F.; Borden, W. T.; Komornicki, A.; Wimmer, E. *J. Am. Chem. Soc.* **1988**, 110, 2314.
(10)   Gonzalez, C.; Schlegel, H. B. *J. Phys. Chem.* **1990**, 94, 5523.
(11)   Hehre, W. J.; Radom, L.; Schleyer, P. v. R.; Pople, J. A. *Ab Initio Molecular Orbital Theory*; Wiley: New York, 1986.
(12)   Grev, R. S.; Janssen, C. L.; Schaefer, H. F., III *J. Chem. Phys.* **1991**, 95, 5128.
(13)   (a) Jorgensen, W. L.; Chandrasekhar, J.; Madura, J. D.; Impey, R. W.; Klein, M. L. *J. Chem. Phys.* **1983**, 79, 926. (b) Jorgensen, W. L.; Tirado-Rives, J. *J. Am. Chem. Soc.* **1988**, 110, 1657. (c) Jorgensen, W. L.; Severance, D. L. *J. Am. Chem. Soc.* **1990**, 112, 4768.
(14)   Breneman, C. M.; Wiberg, K. B. *J. Comp. Chem.* **1990**, 11, 361.

(15)   Carlson, H. A.; Nguyen, T. B.; Orozco, M.; Jorgensen, W. L. *J. Comp. Chem.* **1993**, 14, 1240.

(16)   For reviews, see: Beveridge, D. L.; DiCapua, F. M. *Annu. Rev. Biophys. Biophys. Chem.* **1989**, 18, 431. Allen, M. P.; Tildesley, D. J. *Computer Simulation of Liquids*; Clarendon Press: Oxford, England, 1987.

(17)   Jorgensen, W. L. BOSS, Version 3.2 **1992**, Yale University, New Haven, CT.

(18)   Zwanzig, R. W. *J. Chem. Phys.* **1954**, 22, 1420.

(19)   Jorgensen, W. L.; Ravimohan, C. *J. Chem. Phys.* **1985**, 83, 3050.

(20)   Cramer, C. J.; Truhlar, D. G. *Science* **1992**, 256, 213.

(21)   Cramer, C. J.; Truhlar, D. G. *J. Comput. Chem.* **1992**, 13, 1089.

(22)   Cramer, C. J.; Lynch, G. C.; Truhlar, D. G. AMSOL 3.0, Univ. of Minnesota, Minneapolis, Minn., 1992.

(23)   (a) Miertus, S.; Scrocco, E.; Tomasi, J. *Chem. Phys.* **1981**, 55, 117. (b) Miertus, S.; Tomasi, J. *Chem. Phys.* **1982**, 62, 539. (c) Bachs, M.; Luque, F. J.; Orozco, M. *J. Comp. Chem.* **1994**, 15, 0000. In press. (d) Negre, M.; Orozco, M.; Luque, F. J. *Chem. Phys. Lett.* **1992**, 196, 27. (e) Luque, F. J.; Negre, M.; Orozco, M. *J. Phys. Chem.* **1993**, 37, 4386.

(24)   Yoo, H. Y.; Houk, K. N., in press.

(25)   Kupczyk-Subotkowska, L.; Saunders, W. H., Jr.; Shine, H. J.; Subotkowski, W. *J. Am. Chem. Soc.* **1993**, 115, 5957.

(26)   The reported uncertainty is ±1s and was obtained from the cummulative fluctuations in the free energy changes starting from the reactant. The fluctuations are computed by the batch means procedure from separate averages for each block of 200,000 configurations in the runs of 4,000,000 configurations.

(27)   (a) Truhlar, D. G.; Hase, W. L.; Hynes, J. T. *J. Phys. Chem.* **1983**, 87, 2664. (b) Gertner, B. J.; Wilson, K. R.; Hynes, J. T. *J. Chem. Phys.* **1989**, 90, 3537.

(28)   Schuler, F. W.; Murphy, G. W. *J. Am. Chem. Soc.* **1950**, 72, 3155.

(29)   Hine, J.; Mookerjee, P. K. *J. Org. Chem.* **1975**, 40, 292.

(30)   Still, W. C.; Tempczyk, A.; Hawley, R. C.; Hendrickson, T. *J. Am. Chem. Soc.* **1990**, 112, 6127. Hollinger, F. P.; Still, W. C., personal communication.

(31)   Cramer, C. J.; Truhlar, D. G. *J. Am. Chem. Soc.* **1992**, 114, 8794.

(32)   Gao, J. *J. Am . Chem. Soc.* **1994,** 114, 1563.

(33)   Gajewski, J. J. *J. Org Chem.* **1992**, 57, 5500.

(34)   Chook, Y. M.; Ke, H.; Lipscomb, W. N. *Proc. Natl. Acad. Sci. USA* **1993**, 90, 8600.

RECEIVED May 9, 1994

Chapter 18

Case Studies in Solvation of Bioactive Molecules

Amiloride, a Sodium Channel Blocker, and β-Cyclodextrin, an Enzyme Mimic

Carol A. Venanzi[1], Ronald A. Buono[1], Victor B. Luzhkov[1,4], Randy J. Zauhar[2], and Thomas J. Venanzi[3]

[1]Department of Chemical Engineering, Chemistry, and Environmental Science, New Jersey Institute of Technology, Newark, NJ 07102
[2]Biotechnology Institute and Department of Molecular and Cell Biology, 519 Wartik Laboratory, Pennsylvania State University, University Park, PA 16802
[3]Department of Chemistry, College of New Rochelle, New Rochelle, NY 10805

Molecular dynamics and the Lagevin Dipole, Induced Polarization Charge Boundary Element, and Self-Consistent Reaction Field static solvation models are shown to agree in their prediction of the relative solvation energy of conformers of amiloride. This is important since NMR experiments were unable to distinguish between the conformers in solution. Molecular dynamics simulation also showed that the protonated form of amiloride tends to remain planar on the average due to a high OCCN torsional barrier. This lends support to the hypothesis that amiloride most likely binds to DNA in a planar conformation. The Langevin Dipole model is also used with the AM1 semiempirical quantum mechanical method to compare the reaction paths for acetate phenol cleavage by hydroxide ion and by cyclodextrin. The results show that acylation at the 3'-hydroxyl of cyclodextrin is favored over the 2'-position by about 10 kcal/mol and that cyclodextrin lowers the activation energy barrier for acylation by about 10 kcal/mol compared to attack by hydroxide ion.

Amiloride, A Potent Sodium Channel Blocker

Amiloride, **1**, is a novel acylguanidine diuretic which has been shown to be a potent inhibitor of sodium transport in a variety of cellular and epithelial systems (1-4). Amiloride also binds to other systems such as the a- and ß-adrenergic receptors (5,6) and DNA (7). Studies in this laboratory (8-11) have been directed towards the interpretation of structure-activity data (12,13) that relates the binding of amiloride analogs to inhibition of sodium transport through the epithelial sodium channel in the frog skin. Smith, et al (14) carried out the first computational and NMR analysis of conformers of the free base (**1a**) and protonated (**1b**) forms of amiloride. However,

[4]Current address: Institute of Chemical Physics, Chernogolovka, Moscow Region, Russia 142432

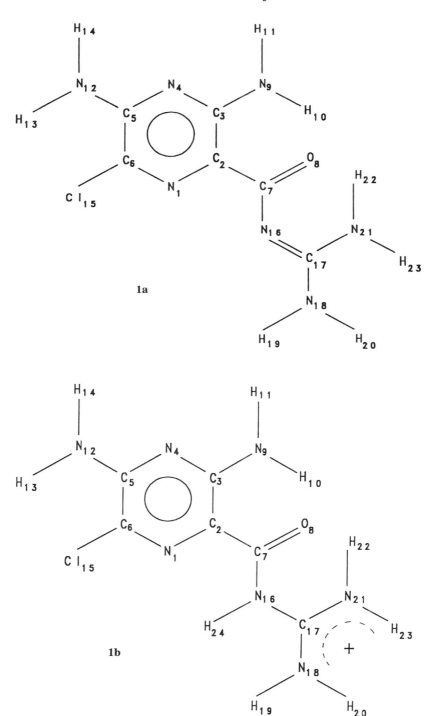

1a

1b

the NMR data were not able to distinguish between the A1 (OCCN=180°) and A4 (OCCN=0°) conformers in solution. The calculation of the barrier to rotation around the $-C_2C_7-$ bond was carried out in the CNDO/2 method without geometry optimization. So, although the calculations found A1 to be more stable than A4, the level of approximation in the calculations was such as to leave some uncertainty in the resolution of the question. Recently (8), using the 3-21G* basis set and full geometry optimization, we found the barrier to A1/A4 conversion to be 19 kcal/mol with A1 2.50 kcal/mol more stable than A4. A4 was found to have a much larger dipole moment than A1: 7.08 D compared to 2.68 D. So, although the gas phase calculations found A1 to be more stable than A4, the large difference in dipole moments suggests that A4 might be more stable in solution. In order to settle this question, as described below, we calculated the relative solvation energy of the free base conformers using both static and dynamic microscopic solvation models. The results provide not only a clarification of the NMR results, but also a valuable comparison of several solvation methods.

The work of Smith, et al also identified the F1 conformer (OCCN=180°), shown in **1b**, as the protonated form of lowest energy. The F1 conformer shows significant intramolecular hydrogen bonding which would also be found in analogs with substitutions at the 5- and 6-position of the pyrazine ring. In order to interpret the structure-activity data of Li, et al (12,13), we carried out a molecular electrostatic potential analysis of amiloride analogs with pyrazine ring substitutions, assuming each analog to be in an F1-like orientation with OCCN=180°. From this study (9), we were able to determine important electrostatic features which may influence the formation of a stable analog-channel blocking complex. Since it is the protonated form of amiloride which binds to the sodium channel and is, therefore, active as a sodium channel blocker, it is important to identify other conformers of low energy which might be involved in interaction with the ion channel. We carried out an extensive conformational analysis of the protonated species in the 3-21G* basis set (11). We found that the energy of the protonated species rises slowly from 0 to 4 kcal/mol as the OCCN torsional angle changes from 180° to 160°. Beyond this region, the energy increases steeply and monotonically, reaching a maximum of 33 kcal/mol at the F4 (OCCN=0°) conformation. This indicates that while the planar conformation is the one of lowest energy, other conformers within ±20° of OCCN=180° may be accessed with little expenditure of energy. It is possible that interactions with the ion channel protein could stabilize some slightly nonplanar conformations of amiloride. Since the structure of the ion channel is not known, it is impossible to model exactly the amiloride-channel interaction. Instead we have chosen to obtain information on the energetics of potential hydrogen bonding interactions with the channel by considering the hydrogen bonding interactions of amiloride with discrete water molecules. To this end we carried out the molecular dynamics simulation described below to determine if interaction with discrete solvent molecules tends to break up the intramolecular hydrogen bonding pattern and stabilize nonplanar conformers.

Methodology: Molecular Dynamics and Static Solvation Models. The methodology has been described in detail in a recent publication (11). Only a summary of relevant information is given here. Calculations were carried out with the Langevin Dipole method (15-18), the Induced Polarization Charge Boundary Element (IPCBE) method (19-23), and the Self-Consistent Reaction Field (SCRF) method (24-29) and compared to the results of molecular dynamics (MD) simulations. The MD, LD, IPCBE, and SCRF techniques differ in the manner in which they treat the solvent, the solvent accessible surface and internal energy of the solute, and the effect of the solvent on the energy of the solute. Only the MD method treats the solvent explicitly, which is important for studying hydrogen bonding between water and polar solutes like amiloride. Only the quantum mechanical SCRF and LD/AMPAC (17) methods take into account the effect of the solvent on the electronic structure of the solute by incorporating the electrostatic solvent effect as an additional term in the Hamiltonian.

The LD method and the IPCBE method treat the solute charge distribution as a collection of fixed point charges, although a more exact treatment of solute charge distribution for the IPCBE method has been recently developed within the framework of the INDO and INDO/S-CI formalisms (23).

The IPCBE and SCRF methods are reaction field techniques in which the solvent is treated as a continuous medium around a cavity which is defined by the molecular surface of the solute.   The IPCBE method defines the cavity by triangulation (21) of the solvent accessible surface, whereas the SCRF method uses a simple sphere.   In the IPCBE approach, the solute is modeled as a cavity of arbitrary shape carved out of a continuum of high dielectric constant representing the solvent. The solute charge distribution is embedded in the cavity, resulting in an electric field consisting of a contribution arising from the solute charges alone and a contribution due to the polarization of the solvent (i.e. the reaction field), which depends on both the arrangement of solute charges and the specific shape of the molecule surface (dielectric boundary).   In the SCRF method, the reaction field is taken to be proportional to the molecular dipole moment of the solute, with a constant of proportionality which depends on the dielectric constant of the medium. In this formalism, the reaction field, dipole moment, and correlated energy of the solute in the reaction field are determined by an iterative, self-consistent formalism.

In the LD method, the solvent system around the solute is divided into two regions.   The first region is confined by a large sphere around the center of mass of the solute.   The solvent molecules within this sphere are treated in a microscopic manner as polarizable point dipoles.   The average polarization of the solvent dipoles is related to the corresponding local electric field in a self-consistent iterative way using the Langevin-type equation.   This formula approximates the average orientational energy of the solvent molecules by an energy function that represents the average orientation of the solvent dipoles in the field of the solute charges.   In the second region, the bulk solvent is treated as a continuum using the Born (30) reaction field formulas.

Although all the techniques allow for geometry optimization of the solute in the presence of the solvent, calculations (other than the MD simulations) were carried out on the same, fixed, 3-21G*-optimized geometries so that the SCRF, LD, and IPCBE methods could be compared independently of differences in basis sets or force fields.   It should be noted that the LD, IPCBE, and SCRF methods estimate only the electrostatic contribution to the relative free energy of hydration, while ignoring contributions such as dispersion effects.   While it has been shown that this is a good approximation to the electrostatic contribution to the relative enthalpy of hydration for ions (15,31) and tautomeric systems (32), we use the IPCBE method to show that this is also true for amiloride.

Molecular Dynamics Simulation.   The GROMOS (33,34) molecular simulation program was used for the calculation of the molecular dynamics trajectories for the A1, A4, and F1 conformers of amiloride.   Since the GROMOS force field was developed to treat only proteins, nucleic acids, and cyclodextrins, new parameters needed to be added for amiloride.   We derived Lennard-Jones parameters for neutral chlorine and used the information from our 3-21G* calculations of the barrier to rotation around the $-C_2C_7-$ bond in **1a** and **1b** to derive appropriate torsional potential functions.   The 3-21G* atomic point charges were adjusted to conform to the GROMOS charge group concept.   Molecular dynamics simulations of 30 ps (including 5 ps of equilibration) were carried out on the A1, A4, and F1 conformers, using the 3-21G* optimized geometry as the starting structure in each case.   Each solute was placed in a box of approximately 400 water molecules.   The simulations were carried out at constant temperature and pressure by weakly coupling the system to a thermal bath at 300 K and a pressure bath at 1 atm.

The Langevin Dipole Model.   The LD method was used in two ways.   For direct comparison with the IPCBE method and for investigation of the sensitiviy of the relative hydration free energy to the point charge set used, the electrostatic contribution to the hydration free energy, E_{elec}, was calculated for seven free base and eleven protonated conformers using three different point charge sets (3-21G* Mulliken, 3-21G* potential-derived, and the 3-21G* charges adjusted to the GROMOS charge group concept). Only the A1, A4, F1, and F4 results are reported here. In addition, for direct comparison with the SCRF method, the LD/AM1 method (in the LD/AMPAC module of the POLARIS program (35)) was used to calculate the hydration free energy of the A1 and A4 conformers.

The Induced Polarization Charge Boundary Element Model. The IPCBE method was used to calculate the electrostatic contribution to the hydration free energy for the same conformers and charge sets described above. In addition, the enthalpies of the A1, A4, F1, and F4 conformers were calculated by the method of Rashin and Namboodiri (36) using all three point charge sets.

Self-Consistent Reaction Field Method. The SCRF method was used with the 3-21G* basis set to calculate the electrostatic contribution to the relative hydration free energies of the A1 and A4 conformers in water. The cavity radius was defined in two ways: by scaling the electron density envelope to obtain the molecular volume (27) and by derivation from the molecular greatest dimension (25). The first method gave what seemed to be an unphysically small radius, so only the results using the second radius, 6.2 Å are reported here. Full details are given in ref. (11). The GAUSSIAN92 program (37) was used for the calculations.

Results.

Molecular Dynamics Simulation. The results of the molecular dynamics simulation are given in Table I. The internal energy term, E_{int}, shows the A1 conformer to be more stable than the A4 by 3.4 kcal/mol, in agreement with the results of the 3-21G* gas phase calculation (2.50 kcal/mol). The amiloride-water interaction energy term, ΔE_{A-W}, gives the A4-water interaction term as 1.4 kcal/mol more stable than A1. The sum of these two terms, ΔE_{tot}, indicates that A1 is more stable in solution than A4 by 2.0 kcal/mol. Analysis of the OCCN torsional angle indicates that its average value tends to vary by less than $\pm 10°$ from planarity for the A1, A4, and F1 trajectories. Analysis of the intramolecular hydrogen bonding patterns from the three trajectories shows that the median $O_8...H_{22}$ and $O_8...H_{10}$ distances are somewhat larger than the value of about 1.9 Å found in the 3-21G*-optimized geometries of the A1, A4, and F1 conformers. This is due to fluctuations in the OCCN torsional angle from the planar value.

The Langevin Dipole Method. Table II gives the results of the LD and LD/AM1 calculations. Inspection of the electrostatic contribution to the relative solvation energy difference, ΔE_{elec}, for both approaches shows that the A4-water interaction term is more favorable than the A1-water term in agreement with the MD results. This is related to the fact that the A4 conformer has the higher dipole moment, independent of the charge set used in the LD model. However, when the difference in gas phase energy is taken into acount in the calculation of $\Delta\Delta G_{solv}$, both approaches (except for a slightly negative value for the LD/3-21G* Mulliken model) predict the A1 conformer to be more stable than A4.

Induced Polarization Charge Boundary Element Model. The results of the IPBCE method are given in Table III. The values of ΔE_{elec} compare well to the LD values for each of the charge sets for both the free base and protonated conformers of amiloride and are consistent with the MD results. Again, when the difference in relative gas phase internal energies is taken into account, the IPCBE method predicts the A1 conformer to be more stable in solution. Table IIIB shows that, independent of point charge set, the electrostatic contribution to the relative hydration enthalpy,

Table I.   Average Inter- and Intramolecular Energy[a], Molecular Dynamics Simulations

	A1	A4	ΔE_{int}[f]	$\Delta E_{A\text{-}W}$[g]	ΔE_{elec}[h]	ΔE_{LJ}[i]	ΔE_{tot}[j]
E_{int}[b]	-66.8±3.0	-63.4±2.7	3.4				
$E_{A\text{-}W}$[c]	-34.3±3.4	-35.7±4.2		-1.4			
E_{elec}[d]	-15.9±3.5	-18.5±4.3			-2.6		
E_{LJ}[e]	-18.4±2.0	-17.2±2.2				1.2	
							2.0

	F1
E_{int}	-53.4±3.5
$E_{A\text{-}W}$	-97.3±7.7
E_{elec}	-83.8±8.5
E_{LJ}	-13.5±3.0

SOURCE: Adapted from ref. (*11*)
[a]In kcal/mol.
[b]Internal energy.
[c]Amiloride-water interaction energy.
[d]Electrostatic component of the amiloride-water interaction energy.
[e]Lennard-Jones component of the amiloride-water interaction energy.
[f]$\Delta E_{int} = E_{int}(A4) - E_{int}(A1)$.
[g]$\Delta E_{A\text{-}W} = E_{A\text{-}W}(A4) - E_{A\text{-}W}(A1)$.
[h]$\Delta E_{elec} = E_{elec}(A4) - E_{elec}(A1)$.
[i]$\Delta E_{LJ} = E_{LJ}(A4) - E_{LJ}(A1)$.
[j]$\Delta E_{tot} = \Delta E_{int} + \Delta E_{A\text{-}W}$.

Table II. Hydration Free Energy[a], Langevin Dipole Method

	A1	A4	ΔE_{elec}[b]	$\Delta \Delta G_{solv}$
E_{elec}[c]	-5.9	-7.0	-1.1	1.4[f]
E_{elec}[d]	-13.5	-16.1	-2.6	-0.1[f]
E_{elec}[e]	-18.2	-20.3	-2.1	0.4[f]
E_{elec}[g]	-20.0	-23.1	-3.1	
$\Delta H_g{}^S$ [h]	57.4	62.6		
				2.1[i]
Dipole[j]	2.3	10.7		

	F1	F4	ΔE_{elec}[b]	$\Delta \Delta G_{solv}$
E_{elec}[c]	-65.7	-69.0	-3.3	30.1[k]
E_{elec}[d]	-63.6	-70.2	-6.7	26.7[k]
E_{elec}[e]	-64.6	-79.4	-14.7	18.7[k]

SOURCE: Adapted from ref. (11).
[a]In kcal/mol.
[b]$\Delta E_{elec} = E_{elec}$ (A4) - E_{elec} (A1), or E_{elec} (F4) - E_{elec} (F1).
[c]Electrostatic contribution to the hydration free energy calculated with GROMOS charges.
[d]Electrostatic contribution to the hydration free energy calculated with 3-21G* Mulliken charges.
[e]Electrostatic contribution to the hydration free energy calculated with 3-21G* potential-derived charges.
[f]$\Delta \Delta G_{solv} = \Delta E_{elec} + E_{gas}{}^{3-21G*}$ (A4) - $E_{gas}{}^{3-21G*}$ (A1) = ΔE_{elec} + 2.50 kcal/mol. Gas phase relative energy difference taken from ref. (8).
[g]Electrostatic contribution to the hydration free energy calculated with the LD/AM1 method.
[h]Enthalpy of formation of the solute in the gas phase calculated with the LD/AM1 method.
[i]In the LD/AM1 method, $\Delta \Delta G_{solv} = \Delta E_{elec} + \Delta H_g{}^S$ (A4) - $\Delta H_g{}^S$ (A1).
[j]Dipole moment in the presence of solvent, in Debye.
[k]$\Delta \Delta G_{solv} = \Delta E_{elec} + E_g{}^{3-21G*}$ (F4) - $E_g{}^{3-21G*}$ (F1) = ΔE_{elec} + 33.4 kcal./mol. Gas phase relative energy difference from ref. (11).

Table III. Induced Polarization Charge Boundary Element Method

A. Hydration Free Energy[a]

	A1	A4	ΔE_{elec}[b]	$\Delta \Delta G_{solv}$
E_{elec}[c]	-5.7	-6.4	-0.7	1.8[f]
E_{elec}[d]	-17.3	-18.6	-1.3	1.2[f]
E_{elec}[e]	-23.9	-25.4	-1.5	1.0[f]

	F1	F4	ΔE_{elec}[b]	$\Delta \Delta G_{solv}$
E_{elec}[c]	-53.7	-58.5	-4.8	28.6[g]
E_{elec}[d]	-59.7	-66.4	-6.7	26.7[g]
E_{elec}[e]	-61.1	-78.7	-17.6	15.8[g]

B. Hydration Enthalpy[a]

	A1	A4	$\Delta E'_{elec}$[b]	$\Delta \Delta G_{solv}$
E'_{elec}[h]	-5.8	-6.6	-0.8	1.7[k]
E'_{elec}[i]	-17.8	-19.2	-1.4	1.1[k]
E'_{elec}[j]	-24.6	-26.1	-1.5	1.0[k]

	F1	F4	$\Delta E'_{elec}$[b]	$\Delta \Delta G_{solv}$
E'_{elec}[h]	-54.8	-59.8	-5.0	28.4[l]
E'_{elec}[i]	-60.9	-67.9	-7.0	26.4[l]
E'_{elec}[j]	-62.4	-80.4	-18.0	15.4[l]

SOURCE: Adapted from ref. (*11*).

[a]In kcal/mol.

[b]$\Delta E_{elec} = E_{elec} (A4) - E_{elec} (A1)$, or $E_{elec} (F4) - E_{elec} (F1)$.

[c]Electrostatic contribution to the hydration free energy calculated with GROMOS charges.

[d]Electrostatic contribution to the hydration free energy calculated with 3-21G* Mulliken charges.

[e]Electrostatic contribution to the hydration free energy calculated with 3-21G* potential-derived charges.

[f]$\Delta \Delta G_{solv} = \Delta E_{elec} + E_g^{3-21G*} (A4) - E_g^{3-21G*} (A1) = \Delta E_{elec} + 2.50$ kcal/mol. Gas phase relative energy difference from ref. (*8*).

[g]$\Delta \Delta G_{solv} = \Delta E_{elec} + E_g^{3-21G*} (F4) - E_g^{3-21G*} (F1) = \Delta E_{elec} + 33.4$ kcal/mol. Gas phase relative energy difference from ref. (*11*).

[h]Electrostatic contribution to the hydration enthalpy calculated with GROMOS charges.

[i]Electrostatic contribution to the hydration enthalpy calculated with 3-21G* Mulliken charges.

[j]Electrostatic contribution to the hydration enthalpy calculated with 3-21G* potential-derived charges.

[k]$\Delta \Delta H_{solv} = \Delta E'_{elec} + E_g^{3-21G*} (A4) - E_g^{3-21G*} (A1) = \Delta E'_{elec} + 2.50$ kcal/mol. Gas phase relative energy difference from ref. (*8*).

[l]$\Delta \Delta H_{solv} = \Delta E'_{elec} + E_g^{3-21G*} (F4) - E_g^{3-21G*} (F1) = \Delta E'_{elec} + 33.4$ kcal/mol. Gas phase relative energy difference from ref. (*11*).

$\Delta E'_{elec}$, is within 4% of the value of the electrostatic contribution to the relative hydration free energy, ΔE_{elec}.

Self Consistent Reaction Field Method. The results of the SCRF calculations are given in Table IV. The results agree with the LD, LD/AM1, IPCBE, and MD methods in that they all predict the A1 conformer to be more stable than A4 in solution by about 1-2 kcal/mol. Comparison to Table II and to the 3-21G* gas phase values for the dipole moment (2.7 D, A1 and 7.1 D, A4) shows that the LD/AM1 method predicts a larger change in the dipole moment upon solvation than does the SCRF method.

Conclusions. The results show close agreement between the LD and IPCBE methods, as well as between the LD/AM1 and SCRF methods. All agree with the MD results in predicting the A1 conformer to be more stable in solution and, therefore, shed some light on the NMR results of Smith and coworkers, who were unable to distinguish between these two conformers in solution. The agreement of the SCRF results is surprising since this technique uses a simple sphere to define the cavity and is, therefore, more appropriately applied to small, compact molecules rather than planar molecules like amiloride. However, the SCRF technique was also shown to successfully predict the solvent effect on the conformational equilibrium of furfural (25), a small planar molecule. It is also surprising that the average MD nonbonded solute-solvent interaction energies, which neglect solvent reorganization energy, agree closely with the results of the other methods. This approximation may prove to be a useful one in studies of other pharmacologically active molecules. The agreement of the IPCBE, LD, and SCRF methods with the results of the MD simulation indicate that these methods may be used to provide estimates of the hydration free energy in cases where specific hydrogen bonding information is not required. Calculation of the free energy of solvation and the enthalpy of solvation with the IPCBE method showed that these two quantities are nearly equal for the amiloride system, in agreement with similar results for ions (15,31) and tautomeric systems (32).

The implications for amiloride-channel binding are also significant. The MD simulations show that the protonated conformer remains planar on the average. The torsional barrier for F1/F4 conversion is high and essentially relegates the protonated species to an orientation within $\pm 10°$ of the F1 conformer. This indicates that amiloride most likely binds to the ion channel in the planar, F1, conformer since it would be difficult for the channel binding site to provide sufficient interactions to stabilize the high-energy nonplanar conformations. Recent nuclease footprinting studies (7) have shown that the protonated form of amiloride binds selectively to sites on DNA that are rich in adenine and thymine residues. This is true as well of analogs that are monosubstituted (N^5-propylamiloride) and disubstituted (N^5-isopropyl-N^5-methylamiloride) on the amino group at position 5. The results support an intercalative mode of binding and the authors suggest that the studies are consistent with the planar, intramolecularly hydrogen bonded conformation of F1, **1b**. Our results support this hypothesis by showing the degree to which the planar F1 conformer is energetically favored.

ß-Cyclodextrin, an Enzyme Mimic

ß-cyclodextrin, **2**, is a cycloheptaamylose with many uses in research and industry (38-40). The structure of cyclodextrins has been studied by X-ray and neutron diffraction (41-43), by molecular dynamics simulation in the gas (44), aqueous (45-47), and crystalline (48-50) states, and by molecular mechanics calculations (51-54). Inclusion complexes of cyclodextrins with various guests have been studied by NMR (55-62), X-ray (63), and other (64-66) techniques. Of particular interest is the ability of ß-cyclodextrin to accelerate the rate of acylation of ester substrates (67-70) and of functionalized ß-cyclodextrin to mimic the action of chymotrypsin (71,72) and other enzymes(73). Molecular mechanics calculations (74,75) have been carried out to elucidate the host-guest steric relationships which may influence the increased rate

Table IV. Hydration Free Energy, SCRF Method

	A1	A4	A4 - A1
E_{pol}[a]	-0.000421	-0.002878	
E_{sol}[b]	-1141.474771	-1141.475473	
E_{tot}[c]	-1141.474350	-1141.472595	
Dipole[d]	3.0	7.8	
ΔE_{pol}[e]			-1.5
ΔE_{sol}[f]			-0.4
ΔE_{tot}[g]			1.1

SOURCE: Adapted from ref. (*11*).
[a]Electrostatic contribution to the hydration free energy, in hartrees.
[b]Total energy of solute, in hartrees.
[c]$E_{tot} = E_{sol} - E_{pol}$, in hartrees.
[d]Dipole moment in the presence of solvent, in Debye.
[e]$\Delta E_{pol} = E_{pol} (A4) - E_{pol} (A1)$, in kcal/mol.
[f]$\Delta E_{sol} = E_{sol} (A4) - E_{sol} (A1)$, in kcal/mol.
[g]$\Delta E_{tot} = E_{tot} (A4) - E_{tot} (A1)$, in kcal/mol.

2

acceleration noted with ferrocene substrates. Only recently has solvent been taken into account in any analysis of cyclodextrin-guest complexation. Furuki et. al (*76*) used reaction field theory with the MNDO method to study the decarboxylation of phenylcyanoacetate anion catalyzed by cyclodextrins. In this study, however, the cyclodextrin molecule was not treated explicitly, but rather was modeled as a cyclinder of dielectric constant 2. In contrast, we present below the first quantum mechanical analysis of the hydrolysis of acetate phenol by ß-cyclodextrin using the LD/AM1 method. The structures of both the host and the guest are treated exactly. The only approximations are those inherent in the LD/AM1 method. In order to assess the effect of cyclodextrin as an enzyme mimic, we compare the reaction paths for cleavage of acetate phenol by attack by a secondary hydroxide ion of cyclodextrin to cleavage by attack by hydroxide ion in the absence of cyclodextrin.

Methodology: The Langevin Dipole Method. Full geometry optimization of the ß-cyclodextrin host, acetate phenol guest, and tetrahedral intermediates and transition states for hydrolysis of the substrate by hydroxide ion (Scheme I) and acylation of the substrate by cyclodextrin (Scheme II) was carried out with the MOPAC-93 program (*77*) using the AM1 method. The starting geometry for the cyclodextrin optimization was taken from the neutron diffraction structure of ß-cyclodextrin undecahydrate (*41*). The solvation energy of all the structures was also calculated using the LD method that we incorporated into the MOPAC-93 program. The solvation energy was calculated with both conventional Coulomb (C) charges and with potential-derived (PD) charges. Since product esters equilibrate between the 2'- and 3'-positions (*74*), it is difficult to determine at which position acylation initially occurs. There is evidence for tosylation at both positions (*78,79*). For that reason, reaction profiles for interaction with alkoxide ions formed at both the 2'- and 3'-positions were studied. Full details of the calculations and results will be given in a forthcoming publication (Luzhkov, V.B. and Venanzi, C.A., manuscript in preparation).

Results.

Attack by Hydroxide Ion in the Absence of Cyclodextrin. Figure 1 gives the gas phase and solvation energies for the points along the reaction curve in Scheme I. The energies are given relative to the sum of the gas phase enthalpies of formation of **1** and **2**. The gas phase curve (open boxes) moves downhill from the separated reactants **1** and **2** to the product **4**. The shallow minimum around 3.5 Å can be attributed to the formation of an ion-dipole complex. The small peak at 2.65 Å corresponds to the transition state of the reaction. The gas phase calculation predicts the relative energy for the pair (**5, 6**) to be more stable that that of the pair (**7, 8**) by about 6 kcal/mol. The experimental value, estimated from the relative enthalpies of solvation (*80,81*) for the molecule/ion pairs (**7, 5**) and (**6, 8**) is -8.3 kcal/mol. In contrast, the solvation energy curves show a very different reaction profile. The activation energy for the formation of the tetrahedral intermediate **3** is 26.0 kcal/mol in the C charge set (open circles) and 17.8 kcal/mol in the PD charge set (open diamonds). This is much closer than the gas phase result to the experimental value of 11.9 kcal/mol (*82*). The calculated heat of reaction for the sum of steps 1 and 2 is -35.5 kcal/mol in the C charge set and -37.3 kcal/mol in the PD charge set. These values are close to the average value of -40 kcal/mol observed for nucleophilic substitution at carbonyl groups (*83*). In addition, the relative energy for the pair (**5, 6**) is predicted to be less stable that that of the pair (**7, 8**) by -0.3 kcal/mol in the C charge set and -5.7 kcal/mol in the PD charge set, in agreement with the experimental value of -8.3 kcal/mol.

Acylation by Cyclodextrin. Figure 2 gives the gas phase and solvation energies for the points along the reaction curve in Scheme II. Figure 2a shows the gas phase and solvation energy reaction profile for interaction with the alkoxide ion formed at the 2' secondary hydroxyl; Figure 2b, at the 3' hydroxyl. Figure 2a shows that the activation

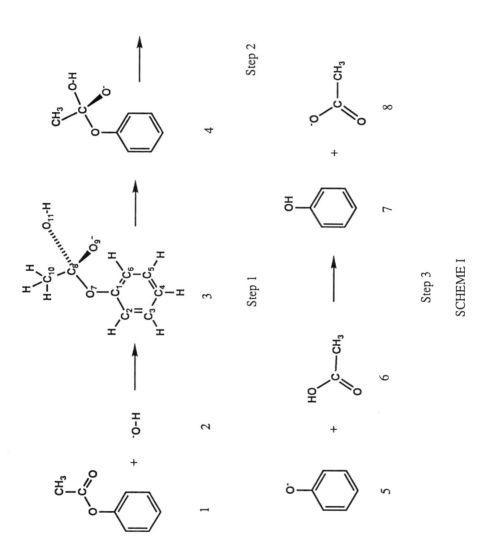

SCHEME I

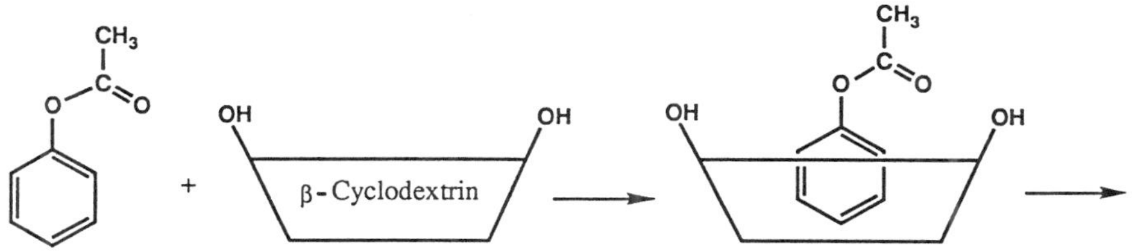

1 9 10

Step 1 Step 2

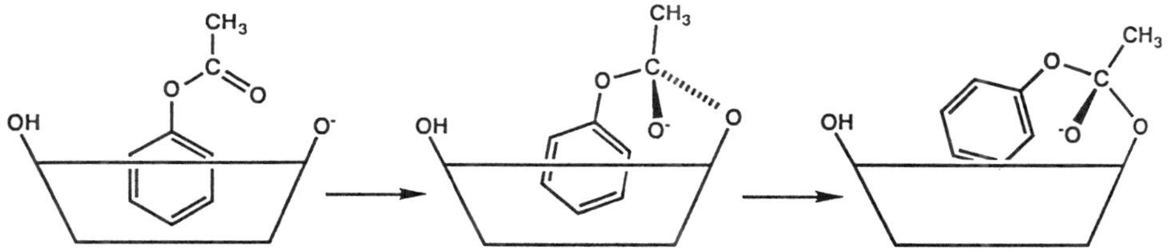

11 12 13

Step 3

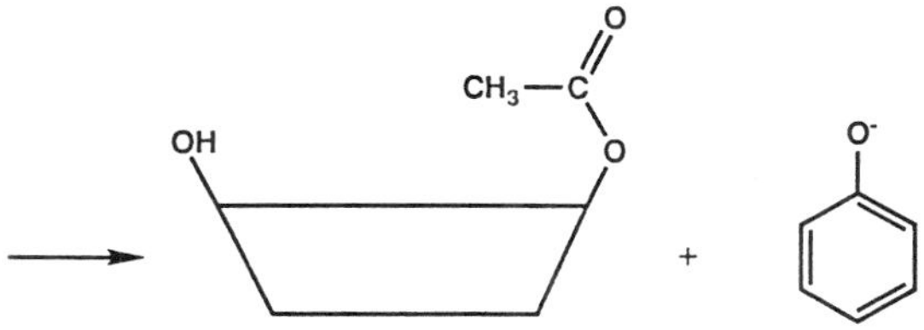

14 5

Step 4

SCHEME II

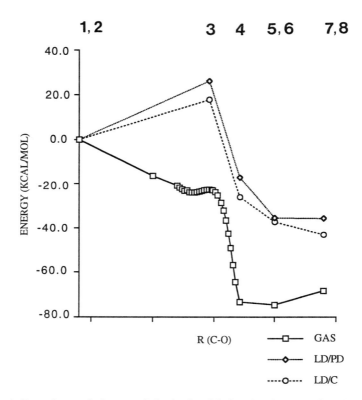

Figure 1. Reaction path for attack by hydroxide ion in absence of cyclodextrin (Scheme I). Points along the reaction curve are identified by the numbered structures along the top of the figure. Energies are given relative to ΔH_{gas} (**1**) + ΔH_{gas} (**2**). Gas phase AM1 calculation - open boxes; Langevin Dipole/AM1 potential-derived charge set calculation - open diamonds; Langevin Dipole/AM1 Coulomb charge set calculation - open circles.

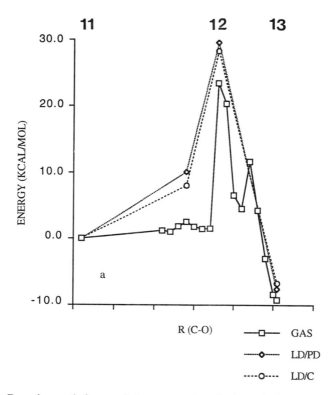

Figure 2. Reaction path for attack by a secondary hydroxyl of cyclodextrin (Step 3, Scheme II). Points along the reaction curve are identified by the numbered structures along the top of the figure. Energies are given relative to ΔH_{gas} (**11**). Symbols are the same as in Figure 1. (a) Reaction path for attack by the 2' hydroxyl of cyclodextrin. (b) Reaction path for attack by the 3' hydroxyl of cyclodextrin.

Continued on next page

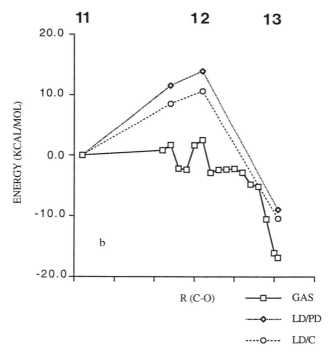

Figure 2.   Continued.

energy for the formation of the transition state **12** is very high for both the gas phase and solution phase calculations. In contrast, Figure 2b shows that reaction at the 3' hydroxyl is more favorable. The gas phase activation energy is 2.50 kcal/mol, which is lower than the activation energy for the reaction of **1** and **2** in water. The activation energy for the solution phase calculation is about 10 kcal/mol, with little difference between the two charge sets. So, despite the fact that the 2' hydroxyls of cyclodextrin are more acidic than the 3' hydroxyls (*84*), a more kinetically and thermodynamically favored path is obtained by reaction at 3'.

Conclusions. Figures 1 and 2 show that both the C and PD charge sets give qualitatively similar pictures of the reaction profile in water for Schemes I and II, although the PD charge set gives better agreement with the known experimental data for hydrolysis of acetate phenol in water. Comparison of the gas phase and solution curves in each figure shows that inclusion of solvent has both qualitative and quantitative effects on the reaction profile. Assuming reaction takes place at the more favorable 3'-hydroxyl, the effectiveness of cyclodextrin as an enzyme mimic can be seen in its ability to lower the activation energy barrier for step 3 of Scheme II compared to step 1 of Scheme I by about 10 kcal/mol. The type of static, energy minimization calculation reported here provides a picture of the gross features of the effect of solvent on the reaction profile and of the function of cyclodextrin as an enzyme mimic. A more accurate picture of the reaction profile could be obtained through a potential of mean force calculation, and this is the subject of work in progress.

Summary.

The importance of solvation has been illustrated in the elucidation of the behavior of two very different bioactive molecules: amiloride, a potent sodium channel blocker, and ß-cyclodextrin, a chymotrypsin mimic. Solvation was seen to lower the relative energy difference of amiloride conformers and to influence the nature of the reaction profile for hydrolysis of acetate phenol in water and in cyclodextrin. Good agreement was found between predictions of static and dynamic solvation models.

Acknowledgments. This work was funded by grants to C.A.V. from the New Jersey Commission on Science and Technology, the Campbell Institute for Research and Technology, the National Academy of Sciences Cooperation in Applied Science and Technology program, and by generous grants of computer time from New Jersey Institute of Technology and the Pittsburgh Supercomputing Center.

Literature Cited.
(1)   Cragoe, E. J., Jr.; Kleyman, T. R.; Simchowitz, L. *Amiloride and Its Analogs: Unique Cation Transport Inhibitors*; VCH: New York, 1992.
(2)   Garty, H.; Benos, D. J. *Physiol. Rev.* **1988**, *68*, 309.
(3)   Kleyman, T. R.; Cragoe, E. J., Jr. *J. Membr. Biol.* **1988**, *105*, 1.
(4)   Benos, D. J. In *Na+/H+ Exchange*; S. Grinstein, Ed.; CRC: Boca Raton, 1988; pp 121.
(5)   Howard, M. J.; Mullen, M. D.; Insel, P. A. *Am. J. Physiol.* **1987**, *253 (Renal Fluid Electrolyte Physiol.)*, F21.
(6)   Howard, M. J.; Hughes, R. J.; Motulsky, H. J.; Mullen, M. D.; Insel, P. A. *Molecular Pharmacol.* **1987**, *32*, 53.
(7)   Bailly, C.; Cuthbert, A. W.; Gentle, D.; Knowles, M. R.; Waring, M. J. *Biochemistry* **1993**, *32*, 2514.
(8)   Venanzi, C. A.; Plant, C.; Venanzi, T. J. *J. Comput. Chem.* **1991**, *12*, 850.
(9)   Venanzi, C. A.; C.Plant; Venanzi, T. J. *J. Med. Chem.* **1992**, *35*, 1643.

(10)   Venanzi, C. A.; Venanzi, T. J. In *Mechanisms of Taste Transduction*; S. A. Simon and S. D. Roper, Ed.; CRC: Boca Raton, 1993; pp 428.
(11)   Buono, R. A.; Venanzi, T. J.; Zauhar, R. J.; Luzhkov, V. B.; C.A.Venanzi *J. Am. Chem. Soc.* **1994**, *116*, 1502.
(12)   Li, J. H.-Y.; Cragoe, E. J., Jr.; Lindemann, B. *J. Membr. Biol.* **1985**, *83*, 45.
(13)   Li, J. H.-Y.; Cragoe, E. J., Jr.; Lindemann, B. *J. Membr. Biol.* **1987**, *95*, 171.
(14)   Smith, R. L.; Cochran, D. W.; Gund, P.; Cragoe, E. J., Jr. *J. Am. Chem. Soc.* **1979**, *101*, 191.
(15)   Warshel, A.; Russel, S. T. *Q. Rev. Biophys.* **1984**, *17*, 283.
(16)   Russel, S. T.; Warshel, A. *J. Mol. Biol.* **1985**, *185*, 389.
(17)   Luzhkov, V.; Warshel, A. *J. Comput. Chem.* **1992**, *13*, 199.
(18)   Lee, F. S.; Chu, Z. T.; Warshel, A. *J. Comput. Chem.* **1993**, *14*, 161.
(19)   Zauhar, R. J.; Morgan, R. S. *J. Mol. Biol.* **1985**, *186*, 815.
(20)   Zauhar, R. J.; Morgan, R. S. *J. Comput. Chem.* **1988**, *9*, 171.
(21)   Zauhar, R. J.; Morgan, R. S. *J. Comput. Chem.* **1990**, *11*, 603.
(22)   Zauhar, R. J. *J. Comput. Chem.* **1991**, *12*, 575.
(23)   Fox, T.; Roesch, N.; Zauhar, R. J. *J. Comput. Chem.* **1993**, *14*, 253.
(24)   Wong, M. W.; Wiberg, K. B. *J. Chem. Phys.* **1991**, *95*, 8991.
(25)   Wong, M. W.; Frisch, M. J.; Wiberg, K. B. *J. Am. Chem. Soc.* **1991**, *113*, 4776.
(26)   Wong, M. W.; Wiberg, K. B.; Frisch, M. J. *J. Am. Chem. Soc.* **1992**, *114*, 523.
(27)   Wong, M. W.; Wiberg, K. B.; Frisch, M. J. *J. Am. Chem. Soc.* **1992**, *114*, 1645.
(28)   Wong, M. W.; Wiberg, K. B.; Frisch, M. J. *J. Am. Chem. Soc.* **1993**, *115*, 1078.
(29)   Cieplak, A. S.; Wiberg, K. B. *J. Am. Chem. Soc.* **1992**, *114*, 9226.
(30)   Born, M. *Z. Physik* **1920**, *1*, 45.
(31)   Warshel, A. *J. Phys. Chem.* **1979**, *83*, 1640.
(32)   Rashin, A. *J. Phys. Chem.* **1990**, *94*, 1725.
(33)   van Gunsteren, W. F.; Berendesen, H. J. C., GROMOS, University of Groningen, The Netherlands, 1987.
(34)   van Gunsteren, W. F.; Berendsen, H. J. C.; Hermans, J.; Hol, W. G. J.; Postma, J. P. M. *Proc. Natl. Acad. Sci. USA* **1983**, *80*, 4315.
(35)   POLARIS, available from Molecular Simulations, Inc., Burlington, MA.
(36)   Rashin, A.; Namboodiri, K. *J. Phys. Chem.* **1987**, *91*, 6003.
(37)   Frisch, M. J.; Trucks, G. W.; Head-Gordon, M.; Gill, P. M. W.; Wong, M. W.; Foresman, J. B.; Johnson, B. G.; Schlegel, H. B.; Robb, M. A.; Replogle, E. S.; Gompers, R.; Andres, J. L.; Raghavachari, K.; Binkley, J. S.; Gonzalez, C.; Martin, R. L.; Fox, D. J.; Defrees, D. J.; Baker, J.; Stewart, J. J. P.; Pople, J. A. , GAUSSIAN92, Release A, Gaussian, Inc.: Pittsburgh, PA, 1992.
(38)   Szejtli, J. *Cyclodextrin Technology*; Kluwer: Dordrecht, 1988.
(39)   Szejtli, J. In *Inclusion Compounds*; J. L. Atwood; J. E. D. Davies and D. D. MacNicol, Ed.; Academic Press: New York, 1984; Vol. 3; pp 331.
(40)   Saenger, W. *Angew. Chem. Int. Ed. Engl.* **1980**, *19*, 344.
(41)   Betzel, C.; Saenger, W.; Hingerty, B. E.; Brown, G. M. *J. Am. Chem. Soc.* **1984**, *106*, 7545.
(42)   Zabel, V.; Saenger, W.; Mason, S. A. *J. Am. Chem. Soc.* **1986**, *108*, 3664.
(43)   Steiner, T.; Mason, S. A.; Saenger, W. *J. Am. Chem. Soc.* **1991**, *113*, 5676.
(44)   Wertz, D. A.; Shi, C.-X.; Venanzi, C. A. *J. Comput. Chem.* **1992**, *13*, 41.
(45)   Koehler, J. E. H.; Saenger, W.; van Gunsteren, W. F. *J. Biomol. Struct. Dynam.* **1988**, *6*, 181.
(46)   Koehler, J. E. H.; Saenger, W.; van Gunsteren, W. F. *J. Mol. Biol.* **1988**, *203*, 241.
(47)   Koehler, J. In *Molecular Dynamics. Applications in Molecular Biology*; J. M. Goodfellow, Ed.; CRC Press: Boca Raton, 1990.
(48)   Koehler, J. E. H.; Saenger, W.; van Gunsteren, W. F. *Eur. Biophys. J.* **1987**, *15*, 197.

(49)   Koehler, J. E. H.; Saenger, W.; van Gunsteren, W. F. *Eur. Biophys. J.* **1987**, *15*, 211.
(50)   Koehler, J. E. H.; Saenger, W.; van Gunsteren, W. F. *Eur. Biophys. J.* **1988**, *16*, 153.
(51)   Lipkowitz, K. B. *J. Org. Chem.* **1991**, *56*, 6357.
(52)   Lipkowitz, K. B.; Green, K.; Yang, J. A. *Chirality* **1992**, *4*, 205.
(53)   Lipkowitz, K. B.; Green, K. M.; Yang, J. A.; Pearl, G.; Peterson, M. A. *Chirality* **1993**, *5*, 51.
(54)   Venanzi, C. A.; Canzius, P. M.; Zhang, Z.; Bunce, J. D. *J. Comput. Chem.* **1989**, *10*, 1038.
(55)   Inoue, Y.; Katono, Y.; Chujo, R. *Bull. Chem. Soc. Jpn.* **1979**, *52*, 1692.
(56)   Inoue, Y.; Okuda, T.; Miyata, Y.; Chujo, R. *Carbohydrate Res.* **1984**, *125*, 65.
(57)   Inoue, Y.; Hoshi, H.; Sakurai, M.; Chujo, R. *J. Am. Chem. Soc.* **1985**, *107*, 2319.
(58)   Inoue, Y.; Kuan, F.-H.; Chujo, R. *Bull. Chem. Soc. Jpn.* **1987**, *60*, 2539.
(59)   Takahashi, S.; Suzuki, E.; Nagashima, N. *Bull. Chem. Soc. Jpn.* **1986**, *59*, 1129.
(60)   Takahashi, S.; Suzuki, E.; Amino, Y.; Nagashima, N.; Nishimura, Y.; Tsuboi, M. *Bull. Chem. Soc. Jpn.* **1986**, *59*, 93.
(61)   Wood, D. J.; Hruska, F. E.; Saenger, W. *J. Am. Chem. Soc.* **1977**, *99*, 1735.
(62)   Lipkowitz, K. B.; Raghothama, S.; Yang, J. A. *J. Am. Chem. Soc.* **1992**, *114*, 1554.
(63)   Hamilton, J. A.; Chen, L. *J. Am. Chem. Soc.* **1988**, *110*, 5833.
(64)   Kano, K.; Mori, K.; Uno, B.; Goto, M.; Kubota, T. *J. Am. Chem. Soc.* **1990**, *112*, 8645.
(65)   Eftink, M. R.; Andy, M. L.; Bystrom, K.; Perlmutter, H. D.; Kristol, D. S. *J. Am. Chem. Soc.* **1989**, *111*, 6765.
(66)   Inoue, Y.; Liu, Y.; Tong, L.-H.; Shen, B.-J.; Jin, D.-S. *J. Am. Chem. Soc.* **1993**, *115*, 10637.
(67)   Breslow, R.; Czarniecki, M. F.; Emert, J.; Hamaguchi, H. *J. Am. Chem. Soc.* **1980**, *102*, 762.
(68)   Breslow, R. In *Biomimetic Chemistry*; D. Dolphin; C. McKenna; Y. Murakami and I. Tabushi, Ed.; American Chemical Society: Washington, D.C., 1980; Vol. 191; pp 1.
(69)   Breslow, R.; Trainor, G.; Ueno, A. *J. Am. Chem. Soc.* **1983**, *105*, 2739.
(70)   Trainor, G. L.; Breslow, R. *J. Am. Chem. Soc.* **1981**, *103*, 154.
(71)   Bender, M. L. *J. Inclusion Phenom.* **1984**, *2*, 433.
(72)   D'Souza, V. T.; Bender, M. L. *Acc. Chem. Res.* **1987**, *20*, 146.
(73)   Breslow, R. In *Inclusion Compunds*; J. L. Atwood and J. E. D. Davies, Ed.; Academic Press: New York, 1984; Vol. 3; pp 473.
(74)   Thiem, H.-J.; Brandl, M.; Breslow, R. *J. Am. Chem. Soc.* **1988**, *110*, 8612.
(75)   Menger, F. M.; Sherrod, M. J. *J. Am. Chem. Soc.* **1988**, *110*, 8606.
(76)   Furuki, T.; Hosokawa, F.; Sakurai, M.; Inoue, Y.; Chujo, R. *J. Am. Chem. Soc.* **1993**, *115*, 2903.
(77)   Stewart, J. J. P. , MOPAC93, Fujitsu: Tokyo, 1993.
(78)   Fujita, K.; Tahara, T.; Imoto, T.; Koga, T. *J. Am. Chem. Soc.* **1986**, *108*, 2030.
(79)   Ueno, A.; Breslow, R. *Tetrahedron Lett.* **1982**, *23*, 281.
(80)   Pearson, R. C. *J. Am. Chem. Soc.* **1986**, *108*, 6109.
(81)   Wilson, B.; Giorgiadis, R.; Bartmess, J. E. *J. Am. Chem. Soc.* **1991**, *113*, 1762.
(82)   Burton, C. A.; O'Connor, C.; Turney, T. A. *Chem. Ind. (London)* **1967**, 1835.
(83)   Madura, J. F.; Jorgensen, W. L. *J. Am. Chem. Soc.* **1986**, *108*, 2517.
(84)   Casu, B.; Reggiani, M.; Gallo, G. G.; Vigevani, A. *Tetrahedron* **1968**, *24*, 803.

RECEIVED June 13, 1994

THE HYDROPHOBIC EFFECT

Chapter 19

Designing Synthetic Receptors for Shape-Selective Hydrophobic Binding

Craig S. Wilcox, Neil M. Glagovich, and Thomas H. Webb

Department of Chemistry, University of Pittsburgh, Pittsburgh, PA 15260

It has been 20 years since the discovery that simple water soluble cyclophanes can be effective synthetic receptors for hydrophobic binding of lipophilic substrates. These synthetic receptors are small enough that quite accurate structural and thermodynamic binding data are obtainable and yet are large enough to offer the computational chemist most of the challenges found in biological receptors. The course of development of water soluble cyclophanes has been notably conservative. Here we review some data from our own work and from other laboratories and attempt to identify results and observations that will interest the computational chemist. More intense use of calculations and computer aided design methods will benefit the experimentalist and the theoretical chemist. Predictions of the behaviour of these synthetic systems can provide a valuable test for the validity of contemporary quantitative theories and as confidence in the calculations grows, rapid and more innovative progress in receptor design will follow.

In the mid-1970's Tabushi initiated work on macrocyclic polyammonium ions **1** composed of alternating aromatic rings and aliphatic spacers (*1-2*). These molecules are structurally related to important ansate cyclophanes that Murakami was investigating at that time (*3*), and to the smaller water *in*soluble cyclophanes first prepared (and christened) by Cram (*4,5*). Tabushi pointed out that the "water soluble cyclophanes" contain lipophilic cavities and would provide a valuable model of enzyme-substrate and receptor-effector complexation (*6*). Since that time numerous

0097–6156/94/0568–0282$08.00/0

descendants of Tabushi's synthetic receptor have appeared. Representative examples (Figure 1) include macrocyclic receptors reported by Koga (who provided the first crystallographic evidence of inclusion complexation, the first examples of chiral recognition, and the first asymmetric syntheses using these molecules), Diederich, Dougherty, and Wilcox (*7-10*).

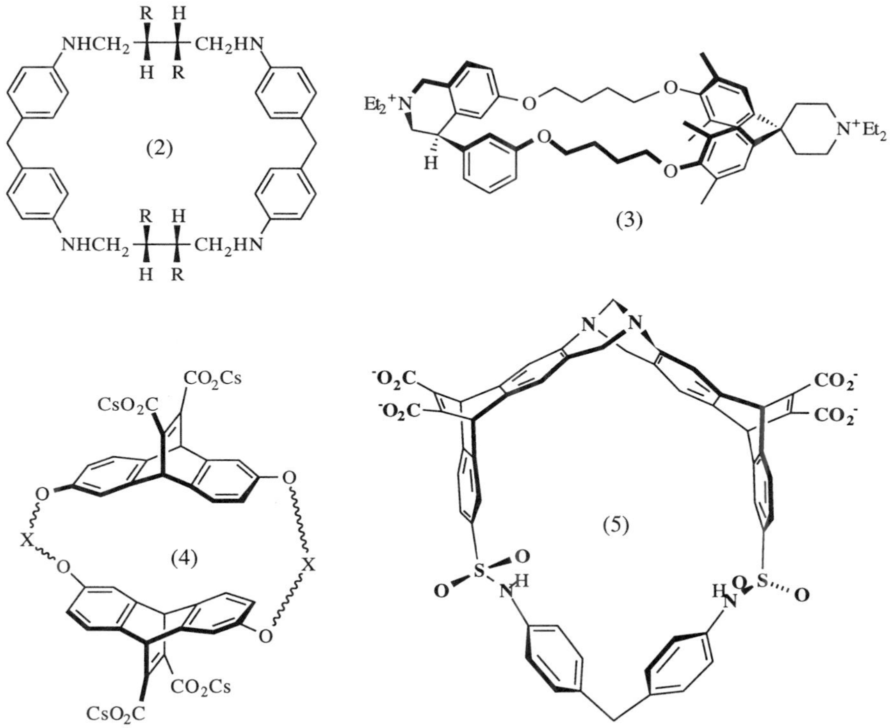

Figure 1. The descendants of Tabushi's receptor.

These four hosts illustrate how the field has evolved as chemists have pursued receptors that have greater substrate binding affinity and selectivity. Water soluble cyclophanes today are often chiral and enantiomerically pure. This allows binding affinities for enantiomeric substrates to be determined. In addition, in three of these hosts, the polar groups that are required for water solubility are placed at locations more remote from the interior pocket than the polar groups in Tabushi's host. This relocation was first pursued in the belief that binding would be increased because desolvation of the ammonium groups, proposed to occur upon guest binding in Koga's host, would be avoided. However, Schneider has recently shown that with electron rich aromatic guests, binding to a host related to **2** is in fact assisted by the positively charged group close to the guest (*11*). Finally, these hosts illustrate the trend toward increasingly rigid hosts.

Restricting Conformational Freedom.

Our initial goal in this area was to find synthetic methods for making hosts less flexible than the hosts reported by Tabushi and Koga (*12*). Such preorganization of the host was desirable on theoretical grounds: it should lead to enhanced binding affinities because the entropy change on binding (ΔS) would be less negative (*13, 14*).   Diminishing conformational flexibility would also offer a chance for greater selectivity in binding.

To achieve this goal we have used dibenzodiazocine substructures, and two early hosts we prepared incorporating this structural unit are illustrated in Figure 2 (*3,13,14*). Host **6** has more conformational freedom than host **7** and is the less effective receptor. One important difference between these receptors was revealed by studying the temperature dependence of binding. The entropy change for association of host (6) with 2,4,6-trimethylphenol is -15 cal mol^{-1}K^{-1}. Usually, hydrophobic association of non-polar solutes is characterized by positive entropy changes, but this negative ΔS is typical for the less organized synthetic receptors and suggests that the conformational freedom of the unbound receptor is substantially restricted when the guest arrives. In contrast, the entropy change for binding of host **7** to the same guest is +7 cal mol^{-1}K^{-1} (*15*). Better organization of the receptor prior to association leads to diminished loss of conformational freedom and stronger binding.

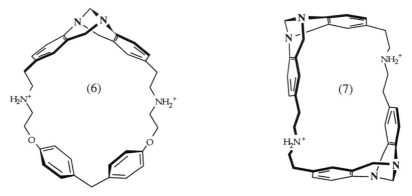

Figure 2.   The more rigid cyclophane **7** excludes aliphatic cyclohexanoid substrates.

An x-ray diffraction study verified that the shape and size of host **7** is perfectly suited to aromatic guests (Figure 3). Because the available space is complementary to aromatic guests but is too narrow for an aliphatic guest, cyclohexanoid substrates will not bind to host **7** and binding of aromatic substrates is not inhibited in the presence of menthol or cyclohexanol.

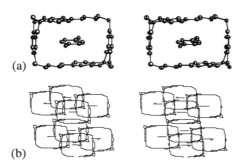

Figure 3. (a) Crystal structure of **7** bound to p-xylene. (b) Packing diagram.

To further pursue the rigidification and shape-optimization of these cyclophanes, we developed receptor **5** (Figure 1). It was created based upon a non-macrocyclic receptor that we first reported in 1988 (*16,17*). Receptor **5** binds cyclohexanoid substrates strongly in aqueous media and reverses the selectivity observed with receptor **7** in that it binds cyclohexanoid substrates better than aromatic substrates. Furthermore, this receptor is diastereoselective. Cyclohexane derivatives that have axial substituents are usually bound less well than diastereomeric derivatives with equatorial substituents. For example (+)-menthol (K_a = 3 900 M^{-1}) binds almost 3 times more strongly than (+)-isomenthol (K_a = 1 400 M^{-1}). Equatorial hydroxyl groups are in general better solvated than axial hydroxyl groups and therefore if relative K_a's depended only on guest solvation forces then *cis*-4-tert-butylcyclohexanol with an axial hydroxyl group would bind more strongly than its equatorial diastereomer. However the shape selectivity of receptor **5** overcomes this solvation factor and *trans*-4-tert-butylcyclohexanol binds seven times more strongly (K_a = 40 000 M^{-1}) than the cis isomer (K_a = 5 900 M^{-1}).

Molecular modeling using Amber and MM3 force fields allows the possible receptor-substrate interactions to be visualized (*18-20*). The qualitative results are consistent with the binding studies and indicate that the distance between the sides of receptor **5** is very well suited to forming a sandwich type complex with cyclohexane rings. These calculations also leave no doubt that an axial substituent will not be accommodated in this receptor (Figure 4).

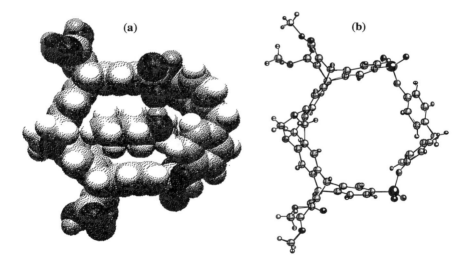

Figure 4. (a) Optimized geometry of menthol bound in receptor **5**. (b) Structure of the tetramethyl ester of receptor **5** as determined by x-ray diffraction methods.

The effects of temperature on binding of menthol to receptor **5** are particularly interesting because here, for the first time, a synthetic receptor exhibited temperature dependent association consistent with classical hydrophobic binding. As the temperature increased, there was a slight increase in association constant. A van't Hoff plot for the data indicates ΔH = +1.5 kcal mol^{-1} and ΔS = +22 cal mol^{-1}K^{-1}. The heats of hydration of alkanes such as menthol are negative while the entropy of hydration is also negative. Positive heats of association (an event leading to dehydration of some hydrophobic surface area) are typical of true hydrophobic binding.

Variations on Receptor 5.

We decided to vary the structure of receptor **5** to examine the effects of systematic structural variation. Three new receptors, **8, 9**, and **10**, were prepared by using structural components that were regioisomers of the components incorporated in receptor **5**. Our interest was in comparing binding affinity and selectivity among these four receptors and in developing computational methods to allow predictions of the relative binding affinities. (See Figure 5.)

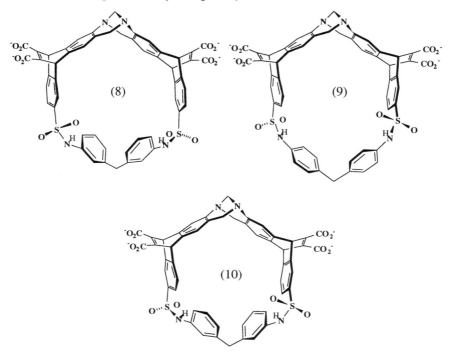

Figure 5. Structural isomers of receptor **5**.

Evaluation of binding. Receptors **8, 9, 10**, and **5** are structural isomers. NMR titration studies were used to evaluate binding (*21,22*). A summary of the reported association constants is given in Table I, where data for binding of all four hosts to menthol enantiomers can be compared.

Table I. Association constants for cyclophane hosts (8), (9), (10), and (5)

Substrate	Receptor			
	8	9	10	5
(-)-menthol	72 ± 7 M^{-1}	335 ± 6 M^{-1}	3 ± 1 M^{-1}	4300 ± 640 M^{-1}
(+)-menthol	*a*	293 ± 77 M^{-1}	*a*	3900 ± 480 M^{-1}

a Experiment not performed.

Receptors **8** and **10** exhibit extremely poor complexation with the enantiomers of menthol. In contrast, the association constants (K_a) of receptor **5** with menthol enantiomers are over 3500 M^{-1}. The data reveal receptor **9** is not much better, and binds only modestly with these guests ($K_a \approx 300$ M^{-1}). What is the reason for these failures? Are accessible computational methods capable of predicting such failures?

What went wrong. One important piece of evidence is provided by the ^{1}H-NMR data for receptors **8** and **10**. In aqueous buffer these receptors gave unusual spectra. The signals expected from the bridging methylene groups of the dianiline unit were shifted far upfield. Normally this signal occurs at about 3.5 ppm. But in aqueous solution the signals for **8** and **10** arose at approximately 1.8 and 0.87 ppm, respectively. This would happen if the protons of the methylene bridge were in the shielding region of one or more aromatic rings.

The reason receptors **8** and **10** are so ineffective as receptors in aqueous media is that they collapse ("deflate") to form crescent shaped molecules that have no binding site available. The methylene protons of the bridge are placed directly within the shielding region of the aryl rings of the receptor and large upfield shifts result. We say the hosts "sickle" because the shapes of the receptors are reminiscent of sickled erythrocytes.   Figure 6 show receptors **8** and **10** in two possible "sickled" conformations.

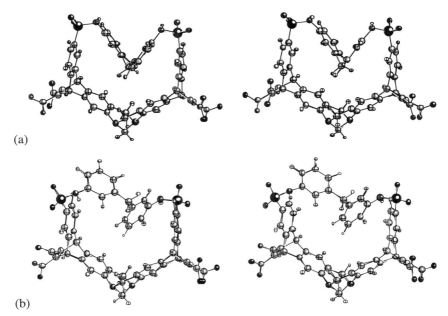

(a)

(b)

Figure 6. Stereoviews of receptors **8** (a) and **10** (b) in sickled or closed forms.

We find that sickling must be a solvation driven phenomenon because it is unique to aqueous solutions. The tetramethyl esters of the macrocycles do not sickle in chloroform. The sickling process involves motions that are too rapid to resolve by NMR. Only a weighted average of the sickled and unsickled receptor resonances could be observed. The barrier to sickling must therefore be less than 18 kcal mol^{-1}. Given a crude estimate that the upfield shift of a proton in this situation would be about 2.5 ppm, we can estimate that the host **10** is about 95% in the sickled form, while host **8** may be only about 60% in the sickled form. However, modeling results (discussed below) show that these hosts have alternative ways to eliminate molecular voids and can fold in a manner that would not place the bridging methylene in a strongly shielding environment (Figure 6b). So such very simple interpretations, though tempting, may be quite erroneous.

Receptor **9** is also inferior to **5**, but we propose this is not due to a propensity to invert like receptor **8**. NMR data indicate that the bridging diphenylmethane unit is exterior to the cavity and its protons resonate at unexceptional field strengths. We think that in **9**, where the 2,7-disubstituted ethenoanthracene unit was used, the distance across the cavity is too small to allow cyclohexanoid rings to easily enter. Conformational analysis of this receptor indicates the cleft is squeezed shut, and has a cleft more narrow than receptor **5** and too small to easily encapsulate a cyclohexane ring.

Are such results predictable? These data showed that solvent was an important factor in the sickling process. We carried out Monte Carlo conformational searches for each of the four macrocycles using Still's Generalized Born equation and surface area model, GB/SA (*23*). The results for receptor **5** calculations using a water model: Of 37 structures located, 14 were sickled, and they ranged in calculated energy up to 16 kcal mol^{-1} above the lowest energy structure. The remaining minima had open cavities, ranging in energy from 2.5-7 kcal mol^{-1} above the lowest energy conformer. In listing of all conformers in order of conformational energy, there was very little overlap between the domains of sickled and open conformers. In contrast, using a chloroform solvation model we located 49 conformational minima and the differences between sickled and open conformers were much diminished. The most stable open conformer was only 0.4 kcal mol^{-1} above the overall minimum, and although the 7 lowest energy conformers were sickled, the preference for sickled conformers was not as strong as it was in water. For receptor **8** the first open conformation was calculated to lie 5 kcal mol^{-1} above the overall minimum so these very crude calculations do confirm that receptor **8** is more prone to sickling than receptor **5**.

NMR data indicate, however, that receptor **5** is not so strongly sickled as these energy calculations suggest. The open form is probably favored for this receptor. Many improvements are required in these calculations. For example it is important to accurately assign atomic charges for all atoms in the receptor. The solvation calculations are very sensitive to atomic charge (*23*). Most importantly, we need a good way to use these energies to calculate relative free energies of the conformers - taking into account the effects of entropy in stabilizing open and sickled forms. The preference for the open form in receptor **5** might be due to the greater internal entropy available to that family of conformational substates.

Concluding Comments.

Systematic variations of this model receptor demonstrated that overall binding energies are very sensitive to small structural modifications. Molecular sub-units which are overly flexible, like the substituted diphenylmethane units, can lead to undesirable diminishment in association energy. They do so because they allow the receptors to deflate or "sickle" by folding in upon themselves. This folding process, like the binding event itself, is driven by hydrophobic forces.

Recently we prepared receptor **11** (Figure 7). This receptor eliminates the diphenylmethane unit and cannot sickle. Even though this receptor has rather poor complementarity with menthol, it binds strongly (K_a = 45 000 M^{-1}). Its free energy of association with menthol is thus about 1.4 kcal/mol greater than that measured for the more flexible receptor **5** - a 25% improvement. Complementarity without rigidity can give poor results (if sickling occurs) or good results (when sickling is not favored). Increasing rigidity, as in **11**, when achieved at the expense of complementary structure does not eliminate binding. On the contrary, receptor **11** is far better than receptor **5**. Our next goal will be to prepare a receptor that has good shape complementarity (Figure 4) but is more preorganized than receptor **5** and cannot sickle.

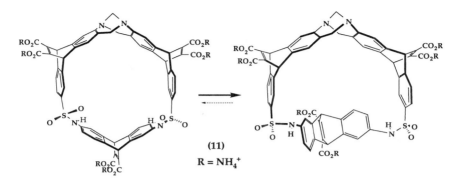

Figure 7. A receptor that cannot sickle, but does fold.

For confident planning and efficient progress in this field, the molecular engineer requires accurate and conveniently accessible computational tools.   The computational chemist requires precise data for testing new methods of calculation. The data obtained for these model receptors can provide a useful challenge to theorists and computational chemists. The uncertainties of measuring enthalpies and entropies of association between proteins and their substrates are large compared to the uncertainties that attend observations on model receptors. The computational chemist can quickly assess the success of an approach by testing it against binding affinities measured for synthetic receptors of the type described here. Predictions made on the basis of calculations can be quickly verified and additional data to test hypotheses arising from the calculations can be obtained at a cost below that required to prepare and characterize mutant proteins. Closer cooperation between computational chemists and synthetic chemists involved in preparing model receptors will accelerate the pace of progress in both fields.

Acknowledgment

Our work in this area has been supported by a grant from the National Institute of General Medical Sciences.

Literature Cited

1. Tabushi, I.; Kuroda, Y.; Kimura, Y. *Tetrahedron Lett.* **1976**, 8827.
2. Tabushi, I.; Kimura, Y.; Yamamura, K. *J. Am. Chem. Soc.* **1978**, *100*, 1304.
3. Murakami, Y. ; Aoyama, Y.; Ohno, K.; Dobashi, K.; Nakagawa, T.; Sunamoto, J. *J. Chem. Soc. Perkin Trans. I* **1976**, 1320-1326.
4. Cram, D. J.; Steinberg, H. *J. Am. Chem. Soc.* **1951**, *73*, 5691-5704.
5. Cram, D. J. in *Cyclophanes*; Kheen, P. M.; Rosenfeld, S. M., Eds.; Academic Press: New York, NY, 1983; pp 1-21.
6. Tabushi, I.; Yamamura, K. *Topics Curr. Chem.* **1983**, *113*, 145-182.
7. Odashima, K.; Itai, A.; Iitaka, Y.; Koga, K. *J. Am. Chem. Soc.* **1980**, *102*, 2504-2505.
8. Diederich, F. in *Cyclophanes*; Stoddart, J. F., Ed.; Royal Society: Cambridge, 1991.
9. Petti, M. A.; Sheppodd, T. J.; Barrans, R. E.; Dougherty, D. A. *J. Am. Chem. Soc.* **1988**, *110*, 6825-6840.
10. Cowart, M. D.; Sucholeiki, I.; Bukownik, R. R.; Wilcox, C. S. *J. Am. Chem. Soc.* **1988**, *110*, 6204-6210.

11. Schneider, H.-J.; Blatter, T.; Zimmerman, P. *Angew. Chem. Int. Ed. Engl.* **1990**, *29*, 1161-1162.
12. Wilcox, C. S. *Tetrahedron Lett.* **1985**, *26*, 5749-5752.
13. Wilcox, C. S.; Cowart, M. D. *Tetrahedron Lett.* **1986**, *27*, 5563-5566.
14. Wilcox, C. S. in *Inclusion Phenomena and Molecular Recognition*; Atwood, J. A., Ed.; Plenum: New York, NY, 1990.
15. Zawacki, F. PhD Dissertation, Univ. of Pittsburgh, 1993.
16. Wilcox, C. S.; Greer, L. M.; Lynch, V. *J. Am. Chem. Soc.* **1987**, *109*, 1865-1867.
17. Webb, T. H.; Suh, H.; Wilcox, C. S. *J. Am. Chem. Soc.* **1991**, *113*, 8554-8555.
18. Weiner, S. J.; Kollman, P. A.; Nguyen, D. T.; Case, D. A. *J. Comput. Chem.* **1986**, *7*, 230.
19. Allinger, N. L.; Zhi-qiang, S. Z.; Chen, K. *J. Am. Chem. Soc.* **1992**, *114* 6120-6133.
20. Mohamadi, F.; Richards, N. G. J.; Guida, W. C.; Liskamp, R.; Caufield, C.; Chang, G.; Hendrickson, T.; Still, W. C. *J. Comput. Chem.* **1990**, *11*, 440.
21. Wilcox, C. S. in *Frontiers in Supramolecular Organic Chemistry and Photochemistry*; Schneider, H.-J.; Dürr, H., Eds.; VCH:Weinheim, 1991.
22. Glagovich, N. M.; Webb, T. H.; Suh, H. Geib, S.; Wilcox, C. S. *Proc. Indian Acad. Sci.* **1994**, (in press).
23. Still, W. C.; Tempczyk, A.; Hawley, R. C.; Hendrickson, T. *J. Am. Chem. Soc.* **1990**, *112*, 6127-6129.

RECEIVED April 5, 1994

Chapter 20

Hydrophobic and Antihydrophobic Effects on Organic Reactions in Aqueous Solution

Ronald Breslow

Department of Chemistry, Columbia University, New York, NY 10027

The presence of hydrophobic effects on solubilities, binding constants, and reaction rates and selectivities can be diagnosed, and made quantitative, by the use of prohydrophobic and antihydrophobic additives. Antihydrophobic agents act mainly by bridging between solutes and water, not by disrupting water structure. "Water-like" solvents are quite different from water in their behavior. Examples of observed hydrophobic effects include the Diels-Alder reaction, the benzoin condensation, binding to monomeric and dimeric cyclodextrins, and DNA stacking.

The hydrophobic effect (*1,2*) is the lowering of the free energy of the system when non-polar molecules or segments cluster together in water so as to minimize the hydrocarbon-water interface. Sometimes a particular thermodynamic criterion for the presence of a hydrophobic effect is suggested, but this is not reliable. Hydrophobic association can be driven by either entropy or enthalpy; indeed, entropy-enthalpy compensation (*3*) is a common phenomenon in chemistry. Solvent molecules can bind or release to minimize free energy while trading enthalpy of solvation for the entropy associated with solvent freedom. Thus another criterion is needed to detect hydrophobic effects on binding constants or chemical reactions (rates and selectivities) in water solvent. We have used the contrasting effects of solutes that either increase or decrease the hydrophobic effect.

Prohydrophobic and Antihydrophobic Agents

Many common salts, such as sodium chloride, increase the hydrophobic effect (*4,5*). They promote association of apolar molecules or segments—and they also decrease the solubilities of apolar molecules—in water. The latter effect is called "salting out", and it is commonly used to promote laboratory extractions, for instance. Decreased solubility is related to the association of non-polar segments in solution, association simply being a microscopic analog of phase separation in which there may not be complete escape from contact with water.

Not all salts have this effect. For example, guanidinium chloride (GnCl) is a "salting in" agent. It increases the solubility of hydrocarbons in water solution (*6*), and it is a common denaturant of proteins and nucleic acids. In these macromolecules the

0097–6156/94/0568–0291$08.00/0

three-dimensional structure in water solution is produced in part by hydrophobic association of apolar segments, such as hydrocarbon sidechains in proteins or purines and pyrimidines in the nucleic acids. Decreasing the hydrophobic effect with GnCl disrupts the three-dimensional structure.

Antihydrophobic salts generally have large singly-charged cations or anions. They are more effective if both ions are large, with dispersed charges. For example, lithium perchlorate is also a salting-in material, while guanidinium perchlorate is more effective than is either guanidinium chloride or lithium perchlorate. The effectiveness of these ions follows (7) what is known as the Hofmeister series. It should be mentioned that not all denaturants are salts. Urea is a common denaturant, and it also increases the water solubility of hydrocarbons.

How do these Agents Work?

It has been realized for a long time that salting-out agents such as sodium chloride—that increase the hydrophobic effect—do so by electrostriction of the solvent (4,5). With the collapse of water around these ions, there is less space in which to insert an apolar solute. Putting it another way, it costs more energy to open a cavity in the water solvent to make the needed space. However, there have been two theories about the origin of the antihydrophobic effects of denaturants such as guanidinium ion.

One proposal was that guanidinium ion, urea, and similar materials break up the structure of water by providing alternate hydrogen bonding possibilities. This would make it easier to produce a cavity in which a non-polar material could be inserted. This theory was widely accepted, and the denaturants were often called "water structure breakers" because of it. The second possibility is that these agents directly assist the solvation of non-polar materials by water, acting as a bridge between the very polar water and the non-polar solute. Our work (8) makes it very likely that this is the correct explanation.

If a solubility modifier changes the energy needed for cavitation in a solvent, it should also change the surface tension of that solvent in a predictable way. This is true for salting out agents such as sodium chloride. Consistent with the proposal that they make cavitation more difficult, they also increase the surface tension of water. When a cavity is created in the solvent the surface must increase; the energy needed to create the cavity has a simple relationship to the energy of the new surface. If antihydrophobic agents such as guanidinium chloride act by disrupting water structure so as to make cavitation easier, they should also <u>decrease</u> the surface tension of water. Remarkably, this had not been looked into previously, although it had been reported (9) that urea increased the surface tension of water. The implication of this fact with respect to the two contrasting theories about antihydrophobic effects was not noticed.

We examined (8) the surface tensions of water solutions of guanidinium chloride and of lithium perchlorate. In contrast to the idea that they break up water structure, we found that both salts <u>increase</u> the surface tension of water. The effect is in the same direction as that for lithium chloride, but significantly smaller. If this is a guide to the ease of cavitation of the solvent, it is in the wrong direction to explain salting in effects. Thus the antihydrophobic effect of these salts must be to produce better solvation of the hydrocarbon solute, overcoming the small surface tension effect. We picture this as binding of the guanidinium ion, for instance, to a benzene molecule and thus bridging between the benzene and the water solvent. Recently Jorgensen (private communication) has calculated the energies of such solutions; his calculations indicate that indeed a guanidinium ion can bind to benzene in water to lower the total energy.

The only escape from our conclusion is the possibility that opening macroscopic cavities—which the surface tension measurement reflects—might be made more difficult, but that opening molecule-sized cavities might be made easier by antihydrophobic agents. There is no theoretical or experimental evidence for such a reversal of the effect as a function of cavity size. In the next section we will discuss further evidence for our conclusion.

Water-like Solvents

There are several other solvents that are considered "water-like", in the sense that they also have high cohesive energies and dissolve hydrocarbons poorly. We examined two of them: formamide and ethylene glycol. In these solvents the Diels-Alder reaction is accelerated by solvophobic effects related to hydrophobicity (*vide infra*) and apolar molecules bind into cyclodextrin cavities just as they do in water. What was the effect of the antihydrophobic agents guanidinium chloride, lithium perchlorate, and urea on hydrocarbon solubility in these solvents?

We saw (*10*) that the solubility of benzene in formamide or in ethylene glycol was <u>decreased</u> by added lithium perchlorate or guanidinium chloride, less than by lithium chloride but in the same direction. This is the reverse of the finding in water, where the antihydrophobic agents increase the solubility of benzene. At first we interpreted this to mean that indeed the antihydrophobic agents are interacting with the solvent water, as in the most popular theory, to make cavitation easier. However, with our more recent surface tension results we had to rethink the meaning of the contrast between water and "water-like" solvents. We believe that all the data can be accommodated by our preferred idea, that in water large ions like guanidinium act to solvate the benzene by bridging between it and water. The contrast with the behavior in water-like solvents has to do with relative polarities (*8*).

Guanidinium ion or perchlorate ion can bridge between benzene and water because they are of intermediate polarity; we propose that even with their charges they are less polar than water, which has a very high cohesive energy. However, they are more polar than formamide or ethylene glycol, so they cannot usefully bridge between these solvents and benzene. The result is that only the increase in surface tension that they cause (we found (*8*) that guanidinium chloride gave about the same surface-tension increase in formamide or ethylene glycol as did lithium chloride) affected solubility, by decreasing it. Even in water the increased surface tension contributed by the antihydrophobic agents would decrease solubility, but there the solvation bridging effect more than compensated for it. There was no such compensation in formamide or in ethylene glycol.

One type of antihydrophobic agent was seen to behave differently (*8*). Tetrabutylammonium chloride increases the solubility of benzene in water, and also in formamide and in ethylene glycol. Furthermore, it is known to <u>decrease</u> the surface tension of water, in common with detergents. Thus in principle one could propose that it simply makes cavitation easier, as it surely does. However, if relatively polar ions such as guanidinium can bridge between benzene and water, surely the less polar tetrabutylammonium ion can. We propose that it increases benzene solubility in water by <u>both</u> of these effects. It is sufficiently non-polar that it can even bridge between benzene and the solvent in formamide or in ethylene glycol, increasing benzene solubility in those solvents as well. However, for guanidinium ion or perchlorate ion in water, breaking of water structure to permit easier cavitation is no longer the preferred explanation of their denaturing antihydrophobic effects.

Hydrophobic Effects on Diels-Alder Reactions in Water

Some years ago we set out to examine the possibility that Diels-Alder reactions could be promoted by mutual binding of the diene and the dienophile into a cyclodextrin cavity. Molecular models suggested that both cyclopentadiene and acrylonitrile or methyl vinyl ketone could be bound together into the hydrophobic cavity of β-cyclodextrin, and in the correct geometry to undergo Diels-Alder reaction. We saw (*11*) that this was indeed the case: the addition of 10 mM β-cyclodextrin increased the rate of addition of cyclopentadiene and acrylonitrile (reaction 1) in water by 9-fold, and of cyclopentadiene and methyl vinyl ketone (reaction 2) by 2.5-fold.

Consistent with the idea that this reflected mutual binding of the reactants into the cavity, α-cyclodextrin had the opposite effect. With 5 mM α-cyclodextrin the acrylonitrile reaction (1) was decreased in rate by 20%, and with 10 mM α-cyclodextrin that (2) of methyl vinyl ketone by 40%. Models show that the smaller cavity of α-cyclodextrin can bind the diene, but not along with the dienophile, so the net result is inhibition. As additional evidence, the Diels-Alder reaction (3) between anthracene-9-carbinol and N-ethylmaleimide is inhibited even by β-cyclodextrin; they are so large that only one of them can fit into a β-cyclodextrin cavity.

1. X = CN
2. X = CO-CH₃

The reason the rate effects from binding of diene and dienophile into these cavities were so modest was the most important observation—water itself led to strong acceleration of the reactions by promoting association of the two reactants, so the cyclodextrin effects came on top of a strong independent binding mechanism in water (*11*). This was revealed both by the rate effects themselves and also by the effects of salting in and salting out agents on the rates.

In the reaction (1) of cyclopentadiene with acrylonitrile, the rate in methanol solvent was only twice that in isooctane, revealing a typical low sensitivity of the Diels-Alder reaction to solvent polarity. However, the rate in water was 15 times that in methanol. Such a large jump seemed unlikely to reflect only a polarity effect on the reaction; instead, hydrophobic packing of the reactants by water seemed to be involved. With methyl vinyl ketone (2) the polar solvent effects were larger, the rate showing a 13-fold increase in methanol relative to isooctane. However, in water the rate went up another 58-fold relative to that in methanol. This seemed a discontinuous jump in terms of solvent polarity alone.

The most striking argument from simple rate data had to do with the reaction (3) between anthracene-9-carbinol and N-ethylmaleimide. In this case the reaction rate in methanol was only 43% of that in isooctane, apparently because the methanol broke up a hydrogen-bonded association between diene and dienophile. However, the rate in water was 28 times that in isooctane, and 66 times that in methanol. Here the effect of solvent polarity on rate ran opposite the water effect, so hydrophobic binding seemed the most likely extra factor.

Additional evidence came from the effect of salting in and salting out agents. With 4.86 M LiCl reaction (2) was increased in rate by 2.5-fold, but 4.86 M GnCl decreased it by 2% (*11*). The antihydrophobic effect here was quite modest, but more striking with reaction (3). There 4.86 M LiCl increased the rate in water by 2.5 fold, but 4.86 M GnCl slowed the reaction by 3.0 fold (*12*). This is exactly as expected for a hydrophobic effect. In a more recent study (*13*) we saw that reaction (3) was accelerated 2.2 fold by 4.0 M LiCl, and slowed 2-fold by 2.0 M GnCl. It was also slowed 1.5 times by 4.0 M LiClO$_4$; with guanidinium perchlorate, in which both the cation and the anion are antihydrophobic, a 2.0 M solution slowed the rate of (3) by almost 3-fold. These last findings eliminate an alternative explanation (*13*) of the salt effects, and clearly establish the fact that the hydrophobic effect is contributing to the rate acceleration of this Diels-Alder reaction in water. Of course this does not eliminate the possibility that polar effects also contribute, as others have argued (*14*).

The hydrophobic packing of diene and dienophile in water can also affect the geometry of the product. We saw this (*12*) with the reactions of cyclopentadiene with methyl vinyl ketone (reaction 2) and with methyl acrylate, dimethyl maleate, and methyl methacrylate, and Grieco saw it with some other Diels-Alder reactions (*15*). The most striking effects (*16*) were in reaction (2). There the endo/exo ratio was 3.85 in neat cyclopentadiene, 8.5 in ethanol, but 25.0 in true water solution. There is almost no exo product. The addition of detergents decreased the ratio, as did the conversion of solvent from water to a microemulsion (*16*).

The endo/exo ratio was also somewhat altered by prohydrophobic and antihydrophobic agents (*16*). The 25.0 ratio went up to 28 ± 0.4 with 4.86 M LiCl, and down to 22 ± 0.8 with 4.86 M GnCl. Since the transition state leading to endo product is more compact, with less exposed hydrophobic surface, the trends are as expected even though the effects are small.

Diels-Alder Reactions in Water-like Solvents.   Schneider (*17*) examined the rate of the Diels-Alder reaction of cyclopentadiene with diethyl fumarate in various solvents including water, and showed that they correlated with a solvent solvophobicity parameter of Abraham (*18*) rather than with some simple polarity parameters. We also examined Diels-Alder reactions in nonaqueous polar solvents (*10*). We found that there was solvophobic binding of diene and dienophile in formamide or in ethylene glycol, and that β-cyclodextrin was also able to bind the two components together in these solvents and furnish further acceleration. However, the rate accelerations were not as fast as in water, and the endo/exo ratio was not increased strikingly, as it had been in water.

The most striking finding (*10*) was that in these solvents LiClO$_4$ and GnCl accelerated the reactions, in contrast to their effects in water. This is consistent with our observation—mentioned above—that they no longer increase the solubility of benzene, as they do in water. Again consistent with the results of our solubility studies, Bu$_4$NBr does decrease the rates in these solvents (*10*); it had also been antisolvophobic in the sense that it increased benzene solubility. As described above, we believe that the contrasting behavior of antihydrophobic agents in water and in water-like solvents reflects relative polarities that no longer favor bridging between the solute and the more organic water-like solvents, in contrast to the situation in water.

The Benzoin Condensation

Treatment of benzaldehyde with cyanide ion in various solvents produces benzoin. Under most conditions the addition of cyanide to benzaldehyde is rapid and reversible, and the rate determining step is the addition of the cyanohydrin anion to benzaldehyde (equation 4). Consideration of the likely geometry of this reaction suggested to us that the two phenyl rings in this transition state would probably pack face to face. If so, hydrophobic effects should favor this packing and speed the reaction.

We found (*19*) that the reaction was 200 times faster in water than in ethanol, but of course this does not demonstrate a hydrophobic effect. In an ionic reaction of this type polar solvent effects could be very important. Thus the availability of another test—the salting out and salting in contrast—was critical in establishing that indeed hydrophobic packing in the transition state is involved.

We could not use guanidinium ion, since the benzoin condensation rate depends on pH and at high Gn concentrations the pH was affected. However, lithium salts with various anions had no such pH effect, and could be used to test for a hydrophobic effect on the rate. They were calibrated by examining their effect on the water solubility of benzaldehyde, a solute with a hydrophobic tail and a hydrophilic head group that we selected as an analog of the species in the reaction transition state (one such species is in fact benzaldehyde, the other its cyanohydrin anion). In water at 20 °C benzaldehyde was soluble to 60 mM, but adding 5 M LiCl decreased this solubility to 27 mM while in 5 M LiClO$_4$ it increased to 100 mM. This is the type of solubility effect that contrasts salting out and salting in behavior.

resembles
transition
state

benzoin

(4)

We then examined the effects of these salts (all at 5.0 M) on the pseudo-second-order (in benzaldehyde) rate constants for benzoin condensation with a standard concentration of KCN and standard conditions. With LiCl the rate increased by 3.8 fold, but with LiClO$_4$ the rate decreased by 3.4 fold (that is, the rate constant was divided by this factor). Relative to the rate in water alone, the observed second order rate constant with various salts was as follows: LiCl, 380%; LiBr, 135%; LiClO$_4$, 29%; LiI, 20%; KCl, 285%; CsCl, 124%; CsI, 185%. With lithium cation the rate decreases as the anion gets larger, as expected from previous studies of the salting in phenomenon. In the chloride series the rate also decreases as the cation gets larger, from Li to K to Cs. However, the CsI result is anomalous, and not yet understood.

An additional piece of evidence for the hydrophobic packing shown in equation 4 is the effect of added cyclodextrins on this reaction rate (*19*). The cavity of γ-cyclodextrin (cyclooctaamylose) is large enough to bind two benzene rings, but β-cyclodextrin (cycloheptaamylose) cannot bind both at once. As expected from this, γ-cyclodextrin is a catalyst of the reaction, with a 76% rate increase at 10 mM γ-CD. By contrast, β-CD is an inhibitor, 10 mM of it dropping the rate to 70% of that in water alone.

As further evidence, we made an enzyme mimic (*20*) by attaching a thiazolium salt to γ-cyclodextrin. The thiazolium ring can substitute for cyanide as a catalyst of the benzoin condensation. We found that this artificial enzyme was the best known catalyst for this reaction, as expected if the two benzene rings are both hydrophobically packed into the cyclodextrin cavity in the transition state.

We also examined (*19*) salt effects on the rate of the cyanide-catalyzed benzoin condensation in the two water-like solvents formamide and ethylene glycol, and in DMSO. In these solvents both LiCl and LiClO$_4$ slowed the rates, and by similar amounts. The rate in ethylene glycol with 1.0 M salt went to 76% with LiCl, and 62% with LiClO$_4$. In formamide with 2.0 M salt the rate went to 57% with LiCl, and 25% with LiClO$_4$. In DMSO with 1.0 M salts, the rate went to 30% with LiCl, and 31%

with LiClO$_4$. Again we do not see a salting in or salting out contrast in such solvents, as we do in water. The salt effects themselves probably reflect polar influences on the reaction rate, perhaps by coordination of Li$^+$ to anions.

This case makes it clear that a fast rate in water, or a slowing of a reaction by an antihydrophobic agent, cannot alone be used to diagnose the presence of hydrophobic acceleration in a reaction. Most reactions are subject to substantial polar effects—the Diels-Alder reaction being an exception—so in all cases the critical test is the demonstration of dichotomous rate effects from the inclusion of prohydrophobic and antihydrophobic additives. The test is easy, and it gives important and unique information about the shape of transition states in aqueous solution—at least about their exposed apolar surface area.

Quantitative Effects of Antihydrophobic Agents

A rate increase with a prohydrophobic agent, and a rate decrease with an antihydrophobic one, tell us that hydrophobic packing is involved in the transition state. Literally, they say that the transition state has less hydrophobic surface exposed to water solvent than do the reactants. However, it seemed to us that this could be made quantitative. That is, the magnitude of the rate effects should correlate with the amount of hydrophobic surface that becomes hidden in the transition state. We have made a start (21) on establishing this as a quantitative tool.

Consider first the effect of urea and of guanidinium chloride on solubility. We examined p-t-butylbenzyl alcohol (1) and also p-methylbenzyl alcohol (2). In both cases the addition of urea or of GnCl increased their water solubility, but now we want to focus on the extent of the increase. With 8 M urea the solubility of 1 increased by a factor of 3.3, and that of 2 by a factor of 2.5. Thus the effect on the solubility of 2 was 76% that on 1. Similarly, 6 M GnCl increased the solubility of 1 by a factor of 3.9, and of 2 by a factor of 2.8. The effect on 2 was 72% of the effect on 1, a similar ratio to that with urea. Since the hydrophobic surface of 1 is larger than that of 2, the difference is not surprising. More interesting, the exact quantitative relationship is essentially that which one would predict from the relative sizes of the two hydrocarbon segments of 1 and 2.

We performed (21) a standard calculation of the van der Waals surface areas of the hydrocarbon sections of 1 and 2.. That of toluene is 71% that of t-butylbenzene! This ratio is very much like the 76% and 72% numbers obtained above. It seems that the magnitude of the antihydrophobic effect on a solubility directly reflects the amount of hydrophobic surface that must be exposed to water.

1 R = t-Butyl   β-Cyclodextrin

2 R = Me

In this study we examined the effect of urea and of GnCl on the binding of a t-butylphenyl group into the cavity of β-cyclodextrin. With 8 M urea the binding constant decreased by a factor of 2.4 (compare the factor of 3.3 effect on the solubility of a substrate **1** carrying a t-butylphenyl hydrophobic group). With 6 M GnCl the binding of a t-butylphenyl group into β-cyclodextrin decreased by a factor of 4.3 (compare this with the factor of 3.9 for the increased solubility of **1** in water with the same additive). The similarity of the magnitudes of the effects on solubilities and on binding into a cyclodextrin cavity are reasonable if two things are true. First of all, removal of a t-butylphenyl group from contact with water can be accomplished more or less equally either by taking it entirely out into a separate phase or by binding it into a cyclodextrin cavity microphase. Secondly, the urea and GnCl affect the t-butylphenyl groups, but not the cyclodextrin cavity.

This last point adds to our evidence that the mechanism of salting in by agents such as urea and GnCl is by direct interaction with the hydrocarbon, not by a generalized change in the properties of the water. It seems reasonable that these agents could not easily bridge between the water solvent and the interior of a cavity, for geometric reasons. If instead these agents acted to modify the properties of water one might have expected that they would have decreased the hydrophobic energy of both the t-butylphenyl group and of the cyclodextrin cavity. If they had, the effects on binding constants should have been larger than those on solubilities.

We also examined the effects of urea and of GnCl on the binding of some dimeric substrates to a cyclodextrin dimer, in which two hydrophobic groups are simultaneously bound to the two cyclodextrin cavities. We saw that much larger effects were seen in these cases. If there were simply a doubling of the amount of hydrophobic surface that was buried when compound **3** binds to cyclodextrin dimer **4**, for instance, the free energy perturbation by urea or GnCl should have simply doubled. This would make the factor by which the binding constant changes be the square of that for the effect on single binding of **1** into a cyclodextrin cavity. Something close to this was seen when ethanol was used to modify binding constants, but a larger effect was seen with urea and with GnCl.

3

4

For instance, with 10% ethanol the binding constant of **1** into cyclodextrin decreased by a factor of 0.37, while that of **3** into **4**—which is 10^4 higher in pure water—decreased by a factor of 0.12 (the square of 0.37 is 0.14). Similarly with 20% ethanol the binding constant of **1** into cyclodextrin decreased by a factor of 0.17, while that of **3** into **4** decreased by a factor of 0.04 (the square of 0.17 is 0.03). However, with 8 M urea a 0.42 decrease in the **1**:cyclodextrin binding constant became a 0.05 factor with **3**:**4** (expected 0.18) while with 6 M GnCl a 0.23 factor for **1**:cyclodextrin became a 0.012 factor for **3**:**4** (expected 0.05). The extra decrease for the dimeric binding in these cases may reflect effects on the linking groups, which are present in the dimers but not in the monomeric systems.

It is too early to tell how well the expected quantitative relationships will be upheld. However, it is certainly clear that the effect of antihydrophobic agents increases as the buried hydrophobic area increases.

Other Studies on Cyclodextrin Dimers. The 10^4 higher binding constant of **3** into **4** than of **1** into cyclodextrin in pure water is one example of many we have reported (*22-25*) reflecting the chelate effect for such hydrophobic complexing. The free energy of binding for a dimeric case can be even more than twice that of the monomeric analog, since part of the chelate effect is the fact that translational entropy does not double in the dimeric case. Loss of translational entropy thus diminishes the monomeric binding constant but does not doubly diminish the dimeric one. Interestingly, we have recently done calorimetric studies on such binding (*25*), and find that this predicted entropy advantage for dimeric binding is not reflected in the experimental results.

In common with previous work, we saw that simple binding of monomeric substrate **5** into β-cyclodextrin was driven by enthalpy (ΔH° = -5.21 kcal/mole) and not by entropy ($T\Delta S^\circ$ = 1.06 kcal/mole), even though it is often found that hydrophobic binding is entropy driven. This is one of many examples that make it clear that the thermodynamic criterion for hydrophobic binding is not a reliable one. In the dimeric analogs **6-9**—with binding of **6** into **7**, **8** or **9**—the ΔH°'s went to -16, -14, and -15 kcal/mole respectively. However, they were compensated to some extent by unfavorable entropy changes, with $T\Delta S^\circ$'s of -6, -5, and -6 kcal/mole, respectively. The net result in these three cases was that dimeric binding was much stronger than was monomeric binding, but not by as much as double the monomeric free energy.

In reactions in solution there is often entropy-enthalpy compensation, leading to surprising thermodynamic results. As discussed at the beginning of this chapter, solvent molecules can be bound or released in order to lower the total free energy, but binding a solvent molecule can lower the enthalpy at an entropy cost. Thus the finding that the chelate effect in this system does not follow simple ideas about entropy advantages is not completely surprising.

Dimeric binding of substrates can also lead to strong catalyses in appropriate cases. For example, we have reported (*24*) the very fast hydrolysis of substrate **10** when it is bound to catalyst **11** so as to place the ester group directly above the catalytic metal ion. Such systems hold great promise for the future.

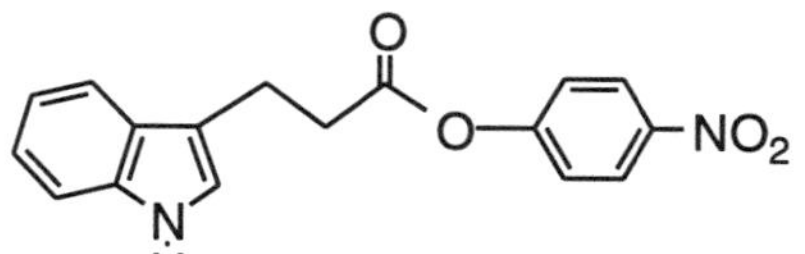

10

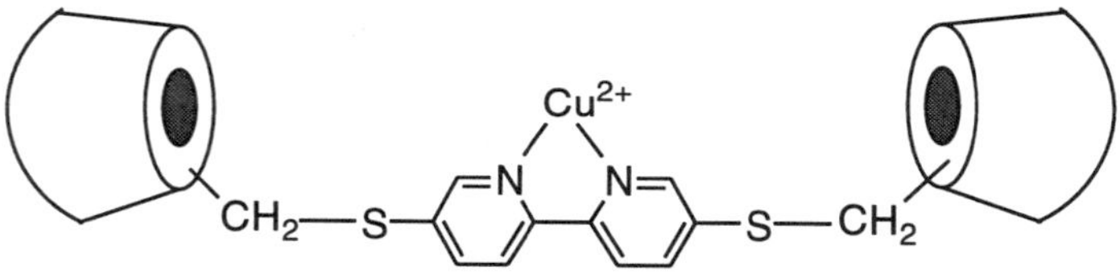

11

Hydrophobic Effects on Nucleic Acid Structures

The double helix is usually described in terms of its base-pairing interactions, by which the genetic code operates. The hydrogen bonds in such a structure may determine the specificity of binding, but much of the binding energy comes from hydrophobic stacking of the purines and pyrimidines in the core of the helix. As expected from this, antihydrophobic agents can disrupt the double helix just as they can denature proteins. The importance of this hydrophobic stacking interaction became clear to us in some studies we did (*26-29*) on an analog of normal DNA.

For other reasons we had prepared (*26*) (Figure 1) the isomer of deoxyadenosine and of thymidine in which the secondary hydroxyl group was on carbon 2', not the normal 3'. We constructed some 16-mers with these nucleosides to see how their properties compare with those of normal DNA, and saw striking differences (*27*). Mixed AT sequences showed no tendency at all to form a double helix with their complementary strands under conditions in which the analogous normal 3',5"-linked A and T sequences were strongly bound to each other. However, a 16-mer containing only the abnormal A was strongly bound to a 16-mer of abnormal T! Such a contrast in properties between mixed sequences and homopolymers is unprecedented for normal nucleosides.

Molecular models suggested what the difference was. In normal DNA—with its 3',5" phosphate links—the heterocyclic base groups in the core of the double helix can stack well, with little hydrophobic surface exposed to solvent. By contrast, in a computer model of a double helix constructed of our isomeric nucleosides—with their

2',5" phosphate links—the geometry is such that the bases do not stack so completely. There is exposed hydrophobic surface, and this diminishes the major stabilizing force for the normal double helix. Thus mixed sequences of our isomeric DNA do not form a double helix.

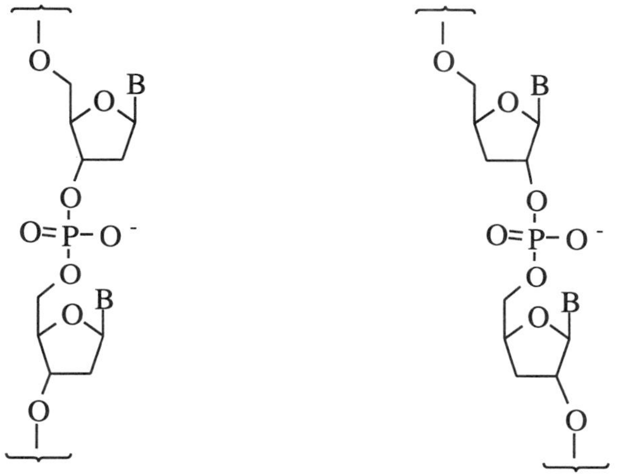

Normal DNA
linked to position 3

Our DNA isomer
linked to position 2

Figure 1. Normal DNA and its isomer.

What then can be the explanation of the finding that homopolymers—strands containing only A or only T—do indeed associate? We suggested (28) that they could be forming a triple helix, and physical studies (29) have now confirmed this. As the computer model shows, the extra strand in the triple helix helps to cover more of the hydrophobic surface in our system. We found that the base association is pyrimidine-purine-pyrimidine, as in triple helices of normal DNA with both Watson-Crick and Hogsteen base pairing interactions. The homopolymers can set up such a structure by using two pyrimidine strands and one purine strand, but mixed sequences cannot.

One might ask: why does Nature use DNA with 3',5" links, when one could imagine that 3-deoxynucleosides could also have been available in which 2',5" links are present? The answer seems to be that hydrophobic binding is a principal force in holding the double helix together, and only with the 3',5" links is this binding strong enough to permit the operation of a genetic system.

Conclusions

Water is an exciting solvent for organic chemistry, not just for biochemistry. The hydrophobic effect in water can greatly influence binding constants, rate constants, and selectivities. Its role can be diagnosed by the use of prohydrophobic and antihydrophobic additives. Quantitative studies suggest that the magnitudes of their effects may even be used to furnish a detailed picture of transition states for reactions in water.

Acknowledgments

Experimental and intellectual contributions by my cited coworkers, and support of this work by the National Institutes of Health and the Office of Naval Research, are gratefully acknowledged.

Literature Cited

1. Ben-Naim, A. *Hydrophobic Interactions*; Plenum Press: New York, NY, 1980.
2. Tanford, C. *The Hydrophobic Effect*; 2nd ed..; Wiley: New York, NY, 1980.
3. Leffler, J. E.; Grunwald, E. *Rates and Equilibria of Organic Reactions*, John Wiley & Sons: New York, NY, 1963.
4. McDevit, W. F.; Long, F. A. *J. Am. Chem. Soc.* **1952,** *74*, 1773-1777.
5. Dack, M. R. *Chem. Soc. Rev.* **1975,** *4* , 211-229.
6. Wetlaufer, D. B.; Malik, S. K.; Stoller, S.; Coffin, R. L. *J. Am. Chem. Soc.* **1964,** *86*, 508.
7. Rizzo, C. *J. Org. Chem.* **1992,** *57*, 6382-6384.
8. Breslow, R.; Guo, T. *Proc. Natl. Acad. Sci. USA* **1990,** *87*, 167-169.
9. Siskova, M.; Hejtamankova, J.; Bartovska, L. *Coll Czech. Chem. Commun.* **1985,** *50*, 1629-1635.
10. Breslow, R.; Guo, T. *J. Am. Chem. Soc.* **1988,** *110*, 5613-5617.
11. Rideout, D.; Breslow, R. *J. Am. Chem. Soc.* **1980,** *102*, 7816-7817.
12. Breslow, R.; Maitra, U.; Rideout, D. *Tetrahedron Lett.* **1983,** *24*, 1901-1904.
13. Breslow, R.; Rizzo, C. J. *J. Am. Chem. Soc.* **1991,***113*, 4340-4341.
14. Blake, J. F.; Jorgensen, W. L. *J. Am. Chem. Soc.* **1991,***113*, 7430-7432.
15. Grieco, P.; Garner, P.; He, Z. Tetrahedron Lett. **1983,** 24, 1897-1900, and later work.
16. Breslow, R.; Maitra, U. *Tetrahedron Lett.* **1984,** *25*, 1239-1240.
17. Schneider, H.-J.; Sangwan, N. K. *J. Chem. Soc. Chem. Commun.* **1986,** 1787.
18. Abraham, M. H. *J. Am. Chem. Soc.* **1982,** *104*, 2085.
19. Kool, E. T.; Breslow, R. *J. Am. Chem. Soc.* **1988,** *110*, 1596-1597.
20. Breslow, R.; Kool, E. *Tetrahedron Lett.* **1988,** *29*, 1635-1638.
21. Breslow, R.; Halfon, S. *Proc. Natl. Acad. Sci. USA* **1992,** *89*, 6916-6918.
22. Breslow, R.; Greenspoon, N.; Guo, T.; Zarzycki, R. *J. Am. Chem. Soc.* **1989,** *111*, 8296-8297.
23. Breslow, R.; Chung, S. *J. Am. Chem. Soc.* **1990,** *112*, 9659-9660 .
24. Breslow, R.; Zhang, B. *J. Am. Chem. Soc.* **1992,** *114*, 5882-5883.
25. Breslow, R.; Zhang, B. *J. Am. Chem. Soc.* **1993,** *115*, 9353-9354.
26. Rizzo, C. J.; Dougherty, J. P.; Breslow, R. *Tetrahedron Lett.* **1992,***33*, 4129-4132.
27. Dougherty, J. P.; Rizzo, C. J. ; Breslow, R. *J. Am. Chem. Soc.* **1992,** *114*, 6254-6255.
28. Breslow, R.; Xu, R. *Proc. Natl. Acad. Sci. USA* **1993,***90*, 1201-1207.
29. Jin, R.; Chapman, Jr., W. H.; Srinivasan, A. R.; Olson, W. K.; Breslow, R.; Breslauer, K. J. *Proc. Natl. Acad. Sci. USA* **1993,** *90*, 10568-10572.

RECEIVED May 4, 1994

Chapter 21

Enforced Hydrophobic Interactions and Hydrogen Bonding in the Acceleration of Diels—Alder Reactions in Water

Wilfried Blokzijl[1] and Jan B. F. N. Engberts

Department of Organic and Molecular Inorganic Chemistry, University of Groningen, Nijenborgh 4, 9747 AG Groningen, The Netherlands

Following pioneering work of Breslow, Grieco and others, we find that intermolecular Diels- Alder (DA) reactions of cyclopentadiene with alkyl vinyl ketones and 5-substituted-1,4 naphthoquinones as well as intramolecular DA reactions of N-furfuryl-N-alkylacrylamides are greatly accelerated in water as compared with traditional organic solvents. Iso-baric activation parameters in combination with thermodynamic quantities for transfer of reactants and activated complex from alcohols and alcohol-water mixtures to water indicate, that the decrease of the hydrophobic surface area of the reactants during the activation process ("enforced hydrophobic interactions") is an important factor determining the rate enhancement in water. A second factor involves hydrogen-bond stabilization of the polarized activated complex. Aggregation or stacking of the reactants do not play a role at the concentrations used for the kinetic measurements. In alcohol-water mixtures the rate accelerations are confined to the highly water-rich solvent composition range and result from favorable changes in both $\Delta^{\#}H^{\ominus}$ and $\Delta^{\#}S^{\ominus}$.

For a large variety of Diels-Alder (DA) reactions, both rate constants and stereospecificities are little or only moderately sensitive to changes in the nature of the reaction medium (1). However, in 1980 Breslow et al.(2) showed that, quite surprisingly at that time, DA reactions are dramatically accelerated in aqueous solution. Ever since, similar curious rate effects have been found for many other organic reactions including Claisen rearrangements (3), benzoin condensations (4) and aldol reactions of silyl enol ethers (5). A common aspect of these reactions is that the reactants are relatively apolar species. Particularly Grieco et al.(6) have shown that the remarkable rate enhancements in water are not just curiosities but that

[1]Current address: Unilever Research Laboratories, P.O. Box 114, 3130 AC Vlaardingen, The Netherlands

water can be a highly useful solvent in synthetic organic chemistry. Recently, the first extensive review of organic reactions in aqueous media has appeared (7). Several explanations have been presented to account for the high rate constants of DA processes in water. These include micellar-like catalysis (8), effects due to the high cohesive density of water (9), internal pressure (10), solvophobicity (11) and enhanced hydrogen bonding to the carbonyl function in the activated complex (12). Breslow (13) proposed that "hydrophobic packing" of diene and dienophile is the most likely explanation, largely based upon kinetic studies of DA reactions in water in the presence of "salting-in" and "salting-out" materials. Apart from aqueous solvent effects, DA reactions can be accelerated by Lewis-acid catalysis and catalytic antibodies (14) and by increased external pressure. Finally, we note that DA reactions are often speeded up in "organized aqueous media", such as microemulsions (15), micellar systems (16) and clay emulsions (17). These rate effects have been attributed to cage effects, pre-orientation of the reactants, microviscosity effects and local concentration effects. Interestingly, DA processes can also be quite fast in heterogeneous aqueous media (7).

In an endeavor to obtain a proper understanding of the reasons for the rate acceleration of DA reactions in water, we have performed a detailed kinetic study of the intermolecular DA reactions of cyclopentadiene (1) with alkyl vinyl ketones (2a,b) and 5-substituted-1,4-naphthoquinones (3a-c) in water, in alcohols and alcohol-water mixtures and in a series of organic solvents (18,19) (Scheme 1). Furthermore we have examined solvent effects on the intramolecular DA reaction of N-furfuryl-N-alkylacrylamides (18,20). Effects due to the hydrogen-bond acceptor capability of the dienophile have been probed via a comparison of the rates of the DA reaction of 1 with methyl vinyl ketone (2a) and methyl vinyl sulfone (4) (21). It will be shown that the rate acceleration of the DA reactions in water is governed by the decrease of the hydrophobic surface area of the reactants during the activation process ("enforced hydrophobic interactions"). A second factor involves hydrogen-bonding stabilization of the polarizable activated complex.

Kinetic Solvent Effects on Intermolecular DA Reactions in Water

The rate constants for the intermolecular DA reactions studied (Scheme 1) are spectacularly enhanced in water. For example, the second-order rate constant for the reaction of 1 with 3a is increased by a factor of 7600 going from n-hexane to water. For a series of 17 organic solvents, the Gibbs energy of activation for the reaction of 1 with 3c shows a trend with the solvent polarity as expressed in the Dimroth-Reichardt $E_T(30)$ value (Figure 1), but the reaction in water is much faster than expected on the basis of this trend. A similar result is obtained if $\Delta^{\#}G^{\circ}$ for the same reaction is plotted against the acceptor numbers of the solvents following a procedure advanced by Desimoni et al.(23). The following discussion will be focused on a quantitative analysis of rate constants and isobaric activation parameters of the DA reactions in water, in some monohydric alcohols and in binary mixtures of water with these alcohols. Relevant data are presented in Table I. It is clear that the substantial rate accelerations going from the alcohols to water are reflected in a more favorable magnitude of both $\Delta^{\#}H^{\circ}$ and $-T\Delta^{\#}S^{\circ}$. Second-order rate constants for the DA reaction of 1 with 3a in EtOH-H$_2$O, n-PrOH-H$_2$O, and t-BuOH-H$_2$O (Figure 2)

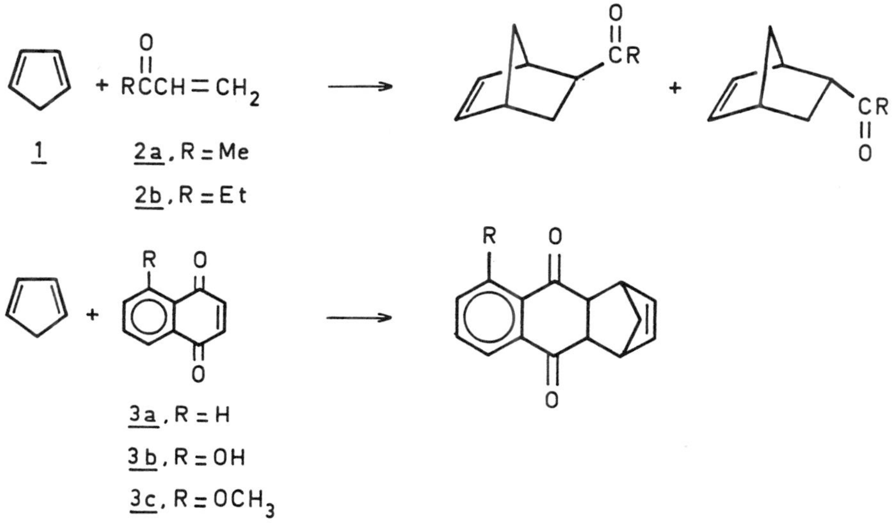

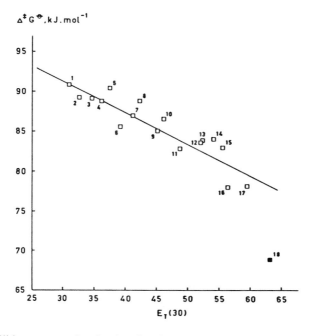

Scheme 1

Figure 1. Gibbs energy of activation for the Diels-Alder reaction of **1** with **3c** in n-hexane (1), tetrachloromethane (2), benzene (3), 1,4-dioxane (4), THF (5), chloroform (6), dichloromethane (7), acetone (8), DMSO (9), acetonitrile (10), 2-propanol (11), ethanol (12), N-methylacetamide (13), N-methylformamide (14), methanol (15), glycol (16), trifluoroethanol (17) and water (18) as a function of $E_T(30)$ at 25°C.

Table I. Second-order Rate Constants and Isobaric Activation Parameters for the Diels-Alder Reactions of Cyclopentadiene (1) with Dienophiles 2a, 2b, and 3a-c in Aqueous Solutions and Organic Solvents, at 25°C

D^a	Solvent	k_2 $dm^3 \, mol^{-1} \, s^{-1}$	$\Delta^{\#}G^{\ominus}$ $kJ \, mol^{-1}$	$\Delta^{\#}H^{\ominus}$ $kJ \, mol^{-1}$	$-T\Delta^{\#}S^{\ominus}$ $kJ \, mol^{-1}$
2a	Methanol	0.883×10^{-3}	90.46		
	Ethanol	0.868×10^{-3}	90.50		
	1-Propanol	0.912×10^{-3}	90.37	45.1(0.7)	45.3(0.7)
	Water	51.9×10^{-3}	80.35	39.4(0.7)	40.9(0.7)
	Water+SDS^b	45.6×10^{-3}	80.68		
	Water+SDS^c	40.5×10^{-3}	80.97		
	Water+CTAB^b	29.5×10^{-3}	81.75		
	Water+CTAB^c	20.8×10^{-3}	82.63		
2b	1-Propanol	0.762×10^{-3}	90.82	45.8(0.5)	45.0(0.6)
	Water	48.9×10^{-3}	80.50	41.5(0.6)	39.0(0.6)
3a	1-Propanol	17.6×10^{-3}	83.05	42.9(0.6)	40.1(0.6)
	Water	4.28	69.42	36.6(0.4)	32.8(0.5)
3b	1-Propanol	32.2×10^{-3}	81.54	44.2(0.5)	37.3(0.6)
	Water	4.33	69.39	44.2(0.8)	25.2(0.8)
3c	1-Propanol	14.9×10^{-3}	83.44	43.3(1.0)	40.2(1.0)
	Water	5.26	68.91	40.5(0.7)	28.4(0.7)

^{a} Dienophile. ^{b} 50 mmol in 1 kg of water. ^{c} 100 mmol in 1 kg of water.

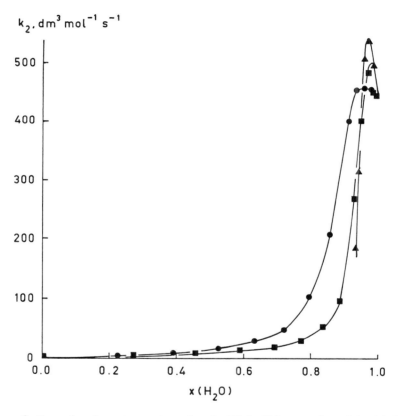

Figure 2. Second-order rate constants for the Diels-Alder reaction of **1** with **3a** in mixed aqueous solutions as a function of the mole fraction of water at 25°C; ●, ethanol; ■, 1-propanol; ▲, 2-methyl-2-propanol.

show that the large rate enhancements going to water are confined to the water-rich solvent composition range. This type of behavior is general for all DA reactions examined in this study. We note that for the reaction of **1** with **3a-c** even a slight increase of k_2 is found upon the addition of small amounts of n-PrOH and t-BuOH to water. This remarkable effect was not observed for the DA reactions of **1** with **2a-b**. The solvent has also a marked effect on the endo/exo product ratios for the reactions of **1** with **2a** and **2b** (Figure 3). Again the greatest solvent-dependence is found in the water-rich region. In mixed aqueous solvent systems with high mole fractions of water ($X(H_2O) = 0.95-1.00$), the effect of the cosolvent on the second-order rate constant for the DA reactions is modest (data are given in ref.(22)). Since the cosolvent is not aggregated in this dilute region, we conclude that the sum of the Gibbs energies for the pairwise interactions of the cosolvent with diene and dienophile almost equals the pairwise interaction of the cosolvent with the activated complex. Interestingly, $\Delta^{\#}H^{\circ}$ and $\Delta^{\#}S^{\circ}$ undergo large changes upon addition of e.g. small amounts of 1-PrOH (Figure 4). However, these changes are largely compensating, leading to modest changes in $\Delta^{\#}G^{\circ}$. For the reaction of **1** with **3c**, the rate-accelerating effect in water relative to pure 1-PrOH is almost entirely entropic in origin, whereas the rate accelerating effect at $X(H_2O) = 0.9$ (i.e. at 10 mol % of 1-PrOH) is overwhelmingly caused by a favorable enthalpy change ($\delta\Delta^{\#}H^{\circ} = 22$ kJ.mol^{-1} vs. $-\delta T\Delta^{\#}S^{\circ} = 10$ kJ.mol^{-1}).

Evidence against Homotactic or Heterotactic Association of Diene and Dienophile in Water

A possible explanation for the rate enhancement of the DA reactions in water could involve homo- or heterotactic association of diene and dienophile. These effects would lead to enhanced local concentrations of both reaction partners leading to acceleration of the bimolecular reaction. A similar effect operates on intermolecular DA reactions in aqueous solutions in the presence of cyclodextrins (13). Strong experimental evidence, however, argues against the operation of this entropy effect in the kinetic study of the DA reactions shown in Scheme 1:

(a) Vapor pressure measurements of **1** above its aqueous solution (25°C) in the concentration range 0-0.06 mol.kg^{-1} and, in the same concentration range, above a solution of **1** in EtOH-H$_2$O ($X(H_2O) = 0.90$) show (Figure 5) that the vapor pressure is a linear function of the molality of **1** (m_{CPD}) until at least $m_{CPD} = 0.03$ mol.kg^{-1}. Small deviations from linearity only occur near the solubility limit of **1**. We contend that **1** does not aggregate significantly in water at the concentrations used in our kinetic measurements ($2.10^{-3} - 2.10^{-2}$ M). In the EtOH-H$_2$O mixture, thermodynamically ideal behavior extends to even higher concentrations. Similar experiments with aqueous solutions of **2a-b** also indicated that these dienophiles do not aggregate in aqueous solution;

(b) In our kinetic measurements, the naphthoquinones **3a-c** were used in concentrations of ca. 5.10^{-5} M. Aggregation or stacking of **3a-c** at these low concentrations is very unlikely;

(c) Second-order rate constants for our DA reactions were determined using different excess concentrations of diene and dienophile (18,22). The rate constants were

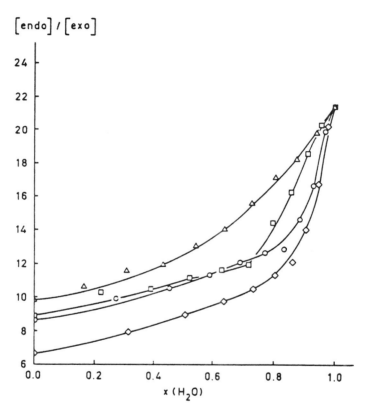

Figure 3. Endo/exo product ratio for the Diels-Alder reaction of **1** with **2a** in mixtures of water and methanol, ▲; ethanol, □; 1-propanol, ○; and 2-methyl-2-propanol, ◇ as a function of the mole fraction of water, at 25°C.

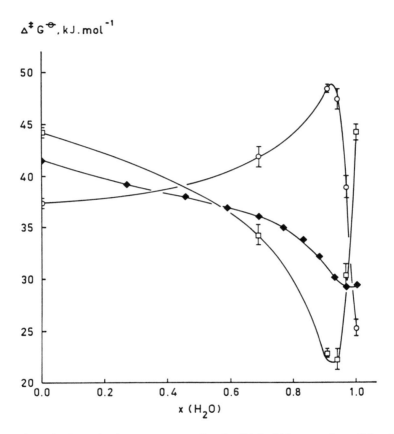

Figure 4. Isobaric activation parameters for the Diels-Alder reaction of **1** with **2a** as a function of the mole fraction of water in mixtures of water and 1-propanol; $\Delta^{\#}G^{\ominus}$ (◆), $\Delta^{\#}H^{\ominus}$ (□) and $-T\Delta^{\#}S^{\ominus}$ (○) at 25°C. The values of $\Delta^{\#}G^{\ominus}$ have been displaced downwards by 40 kJ mol^{-1} for clarity.

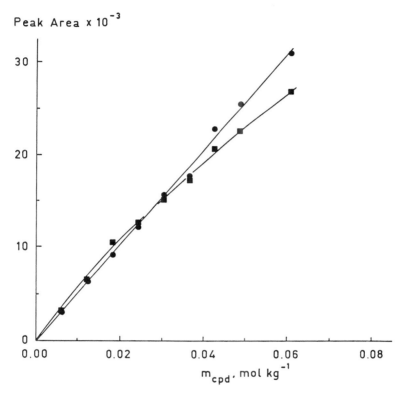

Figure 5. Plots of the peak area of cyclopentadiene, obtained after injection of a standard volume of vapor, withdrawn from the vapor above an aqueous solution of cyclopentadiene as a function of the molality of cyclopentadiene, at 25°C; solution of cyclopentadiene in pure water, ■ and solution of cyclopentadiene in an aqueous solution, containing 10 % (w/w) of ethanol, ●.

reproducible within small error limits which indicates that the reactants do not form aggregates under these conditions;

(d) The intramolecular DA reactions are also greatly accelerated in water. This observation, in combination with the fact that the conformation of the substrate in water is not significantly different from that in the organic solvents (*24*) is not easily reconcilable with reactant aggregation being the reason for the large reaction rates of the intermolecular DA reactions in water.

Pseudothermodynamic Analysis of Medium Effects on DA Reactions in Water and Alcohol-Water Mixtures. Enforced Hydrophobic Interactions and Hydrogen Bonding Effects

An adequate analysis of kinetic medium effects benefits greatly from an approach which allows a separation of the effect of the solvent on the initial state (IS; reactants) and the activated complex (AC). Therefore, standard Gibbs energies of transfer were measured for the IS, AC, and also the reaction product of the DA reactions of **1** with **2a** and **2b**. The transfer parameters for **1**, **2a-b**, and the respective reaction products were obtained from vapor pressure measurements (*18, 22*), those for the AC were calculated from the Gibbs energies of activation in combination with the Gibbs energies of transfer for the reactants. The data for the reaction of **1** with **2a** over the whole mole fraction range in 1-PrOH-H$_2$O are shown graphically in Figure 6. These results are very rewarding and the following conclusions can be drawn. First of all, the IS (**1** + **2a**) is greatly destabilized in the water-rich region and the effect is dominated by the hydrophobic nature of **1**. We note that the behavior of the IS and the reaction product is rather similar. By contrast, the standard chemical potential of the AC is much less dependent on X(H$_2$O) and can adapt itself remarkably well to the solvation changes over the whole mole fraction range ("chameleon behavior"). The data in Figure 6 provide strong evidence that the rate acceleration in water relative to the rate in 1-PrOH is mainly caused by destabilization of the IS. Since no IS aggregation is involved, we propose that, going from 1-PrOH to water, the stabilization of the transition state relative to the reaction partners is, at least in part, caused by the reduction of the hydrophobic surface area of the reaction partners during the activation process. Since this effect is dictated by the activation process of the DA reaction, the effect may be called "enforced hydrophobic interaction". This effect should be clearly distinguished from pairwise hydrophobic packing of diene and dienophile which may well involve a complex with a geometry quite different from that of the AC for the DA reaction. For a discussion of the factors contributing to the gain in Gibbs energy resulting from hydrophobic interaction, the reader is referred to a recent review (*25*). The favorable contraction of the hydrophobic volume of the reaction partners during activation is also expressed in the pronounced preference for the endo isomer if the mole fraction of water is increased in the alcohol-water mixtures (vide supra; Figure 3). The observed small increase of k$_2$ upon addition of small (1-2 mole %) quantities of 1-PrOH or t-BuOH to water (Figure 2) indicates that these relatively hydrophobic alcohols exert a competing favorable effect on the enforced hydrophobic interaction. Indeed, Ben-Naim (*26*) has shown that pairwise hydrophobic interactions are

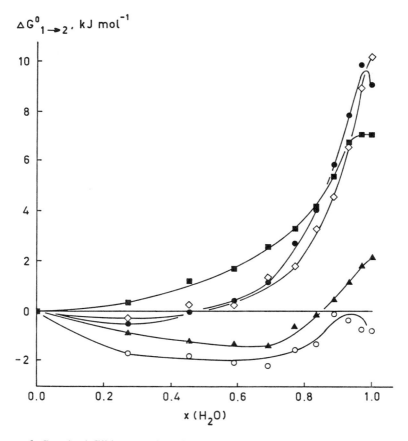

Figure 6. Standard Gibbs energies of transfer for reactants, activated complex and products of the Diels-Alder reactions of **1** with **2a** from 1-propanol to mixtures of 1-propanol and water as a function of the mole fraction of water at 25°C; **2a**, ▲ ; **1**, ■; initial state **(1+2a)**, ●; activated complex, ○; product, ◇.

augmented in the presence of small concentrations of an organic cosolvent. Returning now to the data points for the AC in Figure 6, we contend that the remarkable insensitivity of the standard potential calls for a second effect operating on the kinetic medium effect. We suggest that water induces a more polar activated complex through increased hydrogen-bonding to the polarized carbonyl moiety. This effect will induce an AC that can be optimally accommodated in the aqueous medium. Recent Monte Carlo simulations of the changes in Gibbs energy of solvation for the reaction of **1** with **2a** in n-propane, methanol and water yielded strong evidence for the strengthening of the two hydrogen-bonds to the carbonyl moiety during the activation process (*12, 27*). The effect on the AC is probably not of primary importance for the overall kinetic medium effect since the largest rate enhancement is found in water and not in the strongly hydrogen-bond donating solvent $(CF_3)_2CHOH$.

Substituent Effects

Substituent effects on the DA reaction of **1** with 5-substituted naphthoquinones (**3a-c**; **3d**, R = Me; **3e**, R = Ac; **3f**, R = NO_2) will be reported in a separate paper. We only note that no linear Hammett-type relation is obtained in water, most likely because of steric interaction between the 5-substituent and the carbonyl group (*21*). Nevertheless, substituent effects clearly decrease in the series n-hexane > MeCN > EtOH > CF_3CH_2OH > -$(CF_3)_2CHOH$ > H_2O, consistent with the idea that enforced hydrophobic interactions and hydrogen-bond stabilization of the AC both contribute to the rate acceleration in aqueous media.

Comparison of Methyl Vinyl Ketone and Methyl Vinyl Sulfone as the Dienophile

The proposed explanation for the high rates of DA reactions in aqueous solutions implies that the kinetic solvent effect will respond to changes in overall hydrophobicity of the IS and hydrogen-bond capability of the AC. This is borne out in practice by a comparison of kinetic solvent effects on the reaction of **1** with methyl vinyl keton (**2a**) and methyl vinyl sulfone (**4**) (*21*). Reaction rates for the reaction of **1** with **4** were determined in MeCN, EtOH, 1-PrOH, CF_3CH_2OH, $(CF_3)_2CHOH$ and water. It was found that the rate enhancement going from MeCN to water amounts to 71x for **4** as compared with 250x for **2a**. Combination of Gibbs energies of activation with Gibbs energies of transfer of **4** from CH_3CN to water led to the somewhat unexpected result that **4** is less hydrophobic than **2a**. The presence of two oxygen hydrogen-bond acceptor sites in the sulfone compared to one in the ketone may account for this result. It is obvious that the rate acceleration for a particular DA process in aqueous solution depends on a quite specific way on the nature of the diene and dienophile as far as hydrophobicity and hydrogen-bonding capability is concerned.

Intramolecular DA Reactions in Water

The intramolecular version of the DA reaction is over twenty years younger than its bimolecular counterpart. The first example was reported by Alder et al.(*28*). The reaction is of great utility in natural product synthesis (*29*). Because of its intramolecular nature, "entropic assistance" often leads to Gibbs energies of activation which are 22-30 kJ.mol^{-1} lower than those for analogous bimolecular DA reactions.

We have found that the intramolecular DA (IMDA) reaction of N-furfuryl-N-methylmaleamic acid (**5a**; Scheme 2) is also greatly accelerated in water (*18,20*).

Some further examples are listed in Table II (*20,22*). The rate enhancements in water are comparable to those for intermolecular DA processes in aqueous solution. ^{1}H-NMR studies (*20*) indicate that the high rate constants for IMDA reactions cannot be attributed to enhanced concentrations of the s-cis conformation in water. Apparently, the high rate of IMDA reactions in water can be rationalized by similar considerations as outlined above for the intermolecular DA reactions. A detailed account of IMDA reactions in water will be given elsewhere (*20*).

Conclusions

It appears that several previous explanations for the large rate enhancements of DA reactions in aqueous solution are ambiguous or even fallacious. For example, the effect cannot be rationalized in terms of the internal pressure since the internal pressure of water is small instead of large (*1,30*). Furthermore, micellar effects are also not operative since micelles retard instead of accelerate DA reactions as compared with the reaction in water (Table I). The present results indicate that two factors dominate the rate acceleration in water: (1) enforced hydrophobic interactions during the activation process and (2) hydrogen-bond stabilization of the polarizable activated complex. A quantitative assessment of the relative contributions of both

5a, R = methyl
5b, R = ethyl
5c, R = n-propyl
5d, R = n-octyl

Scheme 2

Table II. Rate Constants and Gibbs Energies of Activation for the IMDA
Reactions of 5a-d in a Series of Organic Solvents and in Water at 25°C

Compound	Solvent	Rate Constant s^{-1}	$\Delta^{\#}G^{\ominus}$ kJ mol^{-1}
5a	Ethanol	3.09×10^{-4}	93.1
5a	1-Propanol	2.90×10^{-4}	93.2
5a	Hexane	1.63×10^{-4}	94.6
5a	1,4-Dioxane	4.93×10^{-5}	97.6
5a	Dichloromethane	2.36×10^{-5}	99.4
5a	Acetonitrile	9.27×10^{-5}	96.0
5a	Trifluoroethanol	2.82×10^{-3}	87.6
5a	Water	2.50×10^{-2}	82.0
5b	Ethanol	4.10×10^{-4}	92.4
5b	1-Propanol	3.65×10^{-4}	92.6
5b	Water	2.94×10^{-2}	81.8
5c	Ethanol	4.77×10^{-4}	92.0
5c	1-Propanol	4.37×10^{-4}	92.2
5c	Water	1.80×10^{-2}	83.0
5d	Ethanol	4.85×10^{-4}	91.9
5d	1-Propanol	4.36×10^{-4}	92.2
5d	Water	1.40×10^{-2}	83.6

effects is difficult and will depend, *inter alia*, on the reference solvent employed for characterizing the hydrophobicity of diene and dienophile. Our experimental results are fully consistent with the elegant computer simulations performed by Jorgensen et al.(*12,27*).

Literature Cited

(1)   Reichardt, C. Solvents and Solvent Effects in Organic Chemistry; VCH: Cambridge, UK; 1990.
(2)   Rideout, D.C.; Breslow, R. *J.Am.Chem.Soc.* **1980**, *102*, 7816.
(3)   Grieco, P.A.; Brandes, E.B.; McCann, S.; Clark, J.D. *J.Org.Chem.* **1989**, *54*, 5849.
(4)   Kool, E.T.; Breslow, R. *J.Am.Chem.Soc.* **1988**, *110*, 1596.
(5)   Lubineau, A.; Meyer, E. *Tetrahedron* **1988**, *44*, 6065.
(6)   Grieco, P.A. *Aldrichim.Acta* **1991**, *24*, 59.
(7)   Li, C-J. *Chem.Rev.* **1993**, *93*, 2023.
(8)   Grieco, P.A.; Yoshida, K.; Garner, P. *J.Org.Chem.*, **1983**, *48*, 3137.

(9)   Grieco, P.A.; Larsen, S.D.; Fobare, W.F. *Tetrahedron Lett.*
        1986, *27*, 1975.
(10)   Lubineau, A.; Queneau, Y. *J.Org.Chem.* **1987**, *52*, 1001.
(11)   Sangwan, N.K.; Schneider, H-J. *J.Chem.Soc. Perkin Trans.2* **1989**, 1223.
(12)   Blake, J.F.; Jorgensen, W.L. *J.Am.Chem.Soc.* **1991**, *113*,
        7430.
(13)   Breslow, R. *Acc.Chem.Res.* **1991**, *24*, 159.
(14)   Hilvert. D.; Hill, K.W.; Nared, K.D.; Auditor, M.T.M.
        J.Am.Chem.Soc. **1989**, *111*, 9262.
(15)   Ramamurthy, V. *Tetrahedron* **1986**, *42*, 5753.
(16)   Sakellarou-Fargues, R.; Maurette, M-T.; Oliveros, E.; Riviere, M.;
        Lattes, A. *J.Photochem.* **1982**, *18*, 101.
(17)   Lazlo, P.; Lucchetti, J. *Tetrahedron Lett.* **1984**, *25*, 2147.
(18)   Blokzijl, W.; Blandamer, M.J.; Engberts, J.B.F.N. *J.Am.
        Chem.Soc.* **1991**, *113*, 4241.
(19)   Blokzijl, W.; Engberts, J.B.F.N. *J.Am.Chem.Soc.* **1992**, *114*, 5440.
(20)   Blokzijl, W.; Engberts, J.B.F.N. To be published.
(21)   Otto, S.; Blokzijl, W.; Engberts, J.B.F.N. To be published.
(22)   Blokzijl, W. *Ph.D.Thesis*, University of Groningen, 1991.
(23)   Desimoni, G.; Faita G.; Righetti, P.P.; Toma, L.
        Tetrahedron **1990**, *46*, 7951.
(24)   Jung, M.E.; Gervay, J. *J.Am.Chem.Soc.* **1991**, *113*, 224.
(25)   Blokzijl, W.; Engberts, J.B.F.N. *Angew.Chem., Int.Ed.*
        Engl. **1993**, *32*, 1545.
(26)   Ben-Naim, A. *Hydrophobic Interactions*, Plenum: New York,
        N.Y., 1980.
(27)   Jorgensen, W.L.; Blake, J.F.; Lim, D.; Severance, D.J.
        J.Chem.Soc., Far.Trans. **1994**, in press.
(28)   Alder, K.; Schumacher, M. *Fortschr.Chem.Org.Naturstoffe*
        1953, *10*, 66.
(29)   For a recent review, see: Roush, W.R. *Adv.Cycloaddit.*
        1990, *2*, 91.
(30)   Dack, M.J.R. *Chem.Soc.Rev.* **1975**, *4*, 211.

RECEIVED May 16, 1994

Chapter 22

Influence of Induced Water Dipoles on Computed Properties of Liquid Water and on Hydration and Association of Nonpolar Solutes

Daniel Van Belle, Martine Prévost, Guy Lippens, and Shoshana J. Wodak

Unité de Conformation des Macromolécules Biologiques, Université Libre de Bruxelles, CP 160/16, P2 avenue P. Héger, B–1050 Bruxelles, Belgium

The extended Lagrangian method is used to compute the induced polarization of water in molecular dynamics simulations.

In a first part, this method is applied in simulations of pure water, using the polarizable PSPC model. Thermodynamic, structural and dynamical properties are computed and compared to those obtained using the mean–field SPC. Simulations with the extended Lagrangian method are shown to reproduce results obtained with the more classical self–consistent iterative/predictive procedure. In particular, they confirm that the explicit representation of the induced polarization considerably improves the transport properties of the SPC model. They demonstrate in addition, that the major impact of the many–body effect is on the water molecule orientational times, whereas the static properties and the self–diffusion coefficient are much less affected.

In a second part, PSPC implemented with the extended Langragian method is used to investigate the effect of adding explicit polarization terms to the water potential on the structural and dynamic properties of aqueous methane solutions and on methane–methane association in water. It is shown that electronic polarization of the water molecules has a subtle though significant influence on the structure and dynamics of water molecules surrounding the hydrophobic solute. But its most remarkable influence, by far, is on the methane–methane potential of mean force, where it appears to abolish the much questioned solvent separated minimum obtained in many previous studies with nonpolarizable water models. Polarizable water models hence seem to yield an improved physical picture of pure water and should be an interesting tool for investigating the processes of hydrophobic association.

Following the pioneering work of Rahman and Stillinger (*1*), many water models have been developed for use in simulation studies (*2–12*). With only a few exceptions (*9–12*) these models neglect many body effects underlying induced polarization. They assume pair–wise additivity of the potential, which decreases substantially the time needed to compute the energies and forces in the system. A pair–wise treatment may be considered as an acceptable approximation as long as one deals with pure water. Indeed, in this case, the effect of polarization can be readily taken into account in an average way by enhancing the permanent dipole moment of water. While this has led, in general, to a satisfactory description of the structural properties of water, difficulties were encountered in reproducing concomitantly, both the thermodynamic and dynamic behavior (*2, 13*). Furthermore, in mixtures where the isotropy of the system is broken, the 'average' view of the water molecule is highly questionable. More accurate descriptions seem indeed to be required to evaluate the effect of hydrophobic or ionic solutes on the surrounding solvent (*9, 14*), or to adequately represent electrostatic interactions in highly heterogeneous systems such as proteins (*15, 16*).

Explicit treatment of electronic polarization effects (*5, 9, 10*) constitutes therefore a welcome refinement of the water–water interaction potential. Different methods (*5, 10*) have been proposed to reduce the computer time required for calculating the induced dipoles. The procedure derived by Sprik and Klein (*5*), specifically designed for use in molecular dynamics simulations, completely abandons the iterative schemes or matrix inversion required to solve the self–consistent equations. Instead, it follows an approach related to the constant pressure and constant temperature method that uses the extended Lagrangian formalism (*17*). The induced dipole on each molecule is treated as an additional degree of freedom, and the Lagrangian of the system is 'extended' with a fictitious kinetic term associated with this extra variable. The equations of motion of this variable, derived by the standard procedure of the Lagrangian theory (*18*), are integrated by standard numerical procedures used in molecular dynamics programs.

In this study, the extended Lagrangian method of Sprik & Klein (*5*) is applied to simulations of pure liquid water and of methane–water solutions, using PSPC (*10*), a polarizable variant of the classical three–center water model SPC (*2*). The present approach differs from the original procedure by keeping the induced dipoles centered on the water oxygens, rather than distributing them on additional sites, which results in an appreciable gain in computer time. From the generated trajectories structural and dynamic properties of liquid bulk water and of water around the hydrophobic solute are computed, and compared to those obtained from simulations of the identical systems carried out using the mean-field SPC water model. Lastly, we also compare the potentials of mean force between two methane molecules dissolved in water, calculated using the extended–Lagrangian PSPC and the classical SPC respectively.

These comparative analyses allow us to investigate the influence of induced polarization of water on its bulk properties, as well as on hydrophobic hydration and hydrophobic association.

Methods

Computation of induced polarization. Assuming a scalar polarizability and linear polarization, the induced dipole moment of the kth water molecule is given by:

$$\mathbf{p}_k = \alpha_k \, \mathbf{E}_k \qquad (1)$$

where $\mathbf{p}_k$ is the induced dipole moment, α_k is the polarizability and $\mathbf{E}_k$ is the total electric field measured at the water oxygen.

The induced polarization is computed using the approach described in detail in ref. 5, 16. Only its main features are therefore outlined below.

The dipole moment $\mathbf{p}$ is treated as an additional (internal) degree of freedom of the water molecule. Using the extended Lagrangian formalism (17), the system is represented in a new phase space defined by the cartesian coordinates of each water molecule, its induced dipole and the conjugated momenta of both. For each extra degree of freedom k, a potential energy term is added to the system (19). This term contains two contributions, the energy of the induced dipole given an external field, and the energy for creating the induced dipole. The extra degrees of freedom also have a kinetic term, $K_p = \dfrac{1}{2} \sum_k m_p \, \dot{\mathbf{p}}_k^2$, associated with them, where $\dot{\mathbf{p}}_k$ is the time–derivative of the induced dipole and m_p is an 'inertial factor' associated with the extra dynamical variables whose dimensions are those of mass.charge^{-2}.

With the standard procedures of Lagrangian mechanics the following equation of motion is then derived:

$$m_p \, \ddot{\mathbf{p}}_k = \mathbf{E}_k - \frac{\mathbf{p}_k}{\alpha_k} \qquad (2)$$

where $\ddot{\mathbf{p}}_k$ is the second time derivative of the induced dipole. Integration of equation 2 which is readily performed by standard numerical integration algorithms such that of Verlet (20), yields the value of $\mathbf{p}_k$ which is then used to compute the electrostatic terms and the forces that depend on the induced dipoles. The dipole inertial factor determines the time–scale of the response of the dipoles to fluctuations of the electric field produced by the motion of the particles. A fixed value of 0.5 gr/eu^2 was used for this factor throughout the simulations, following a calibration against simulations of the same PSPC system, but where the full iterative procedure is used to evaluate the many–body effects (22).

This procedure for computing the many–body polarization effects is extremely efficient, with an increase of no more than a factor of 2 in computer time, essentially due to the overhead associated with computing energies and forces between the dipoles.

Water models, and the methane–water potential. Two water models are used here. The classical SPC model (2) and the polarizable PSPC model derived from it by Ahlström et al. (10). Both are members of the three–center models class, with a single Lennard–Jones interaction positioned at the oxygen and three charges positioned at the nuclear coordinates. The parameters for both models are summarized in Table 1 of ref. 22.

The methane is taken as non polarizable, and interacts with the water molecules through a single Lennard–Jones potential (23), with the following parameters, obtained by the Lorentz–Berthelot combination rules: $\sigma_{MW} = 3.448\,\text{Å}$, $\epsilon_{MW} = -0.214$ kcal/mol for the classical SPC system and $\sigma_{MW} = 3.497\,\text{Å}$, $\epsilon_{MW} = -0.195$ kcal/mol for the polarizable PSPC system.

Molecular dynamics calculations. The integration algorithm used for computing positions and dipoles is the 'velocity version' of the Verlet algorithm (RATTLE) (*21*). The molecular dynamics simulations are performed in the isovolumic–isothermal N,V,T ensemble, sampled by the extended Lagrangian Nosé method (*17*), with routines implemented in the BRUGEL package (*24*).

The pure liquid simulations are carried out on 216 water molecules at a density of 1 g/cm^3, confined to a cubic cell with periodic boundaries. The simulations of methane in water consider 341 water molecules at a density of 1 g/cm^3, plus two methane atoms, held fixed at their contact distance (4 Å) (*25*). In both systems, geometrical constraints are applied to the bond distances and angles of the water molecules following the method of Ryckaert (*26*), and an integration time–step $\Delta t = 2. \cdot 10^{-15}$ s is used. The systems are equilibrated during 20 ps and simulations are carried out for 50 ps.

A site–site spherical cut–off of 8.5 Å is applied to all the terms in the potential. The water–water electrostatic potential term is multiplied over the entire distance range (between 0 and the cutoff distance) by a termination function $S(r)$ (*27, 28*):

$$S(r) = 1 - 2\frac{r}{r_c} + (\frac{r}{r_c})^2 \qquad (3)$$

where r is the site–site distance ($r < r_c$) and r_c is the cut–off distance.

This termination scheme has been extensively tested on SPC liquid water, against the Ewald summation, leading to quite satisfactory results on thermodynamic, structural and transport properties (*28*). Results on the latter 2 properties are illustrated in Figures 1 and 2.

Methane–Methane potential of mean force (pmf) calculations. The free energy profile is calculated as a function of the intermolecular distance between the methane molecules using the free energy difference perturbation technique which leads to (*29, 30*):

$$\Delta A\, (r\text{–}{>}r + \Delta r) = -k_BT \ln <e^{-\Delta U/k_BT}>_r \qquad (4)$$

where the angle brackets with the subscript r represent an ensemble average, computed with the intermolecular distance between the two methanes held fixed at a distance r, and ΔU is the potential energy difference obtained by changing the intermolecular distance from r to r + Δr.

To evaluate the potential of mean force between two methanes dissolved in water a total trajectory of 670 ps has been computed. This trajectory consists of 17 individual molecular dynamics runs, each performed at a given methane–methane separation distance r, with r varying from 8.0 to 3.75 Å, in steps of 0.250 Å. Each run consists of 10 ps equilibration followed by 20 ps to calculate the averages. In order to improve convergence, in the distance interval of 7.0–5.25 Å the trajectories used for averaging were extended to 40 ps. At the methane–methane distance of 4 Å, an additional 60 ps molecular dynamics simulations have been performed for the purpose of computing the average properties.

The uncertainties associated with the calculated free energy differences were evaluated using the procedure described in ref. *25*.

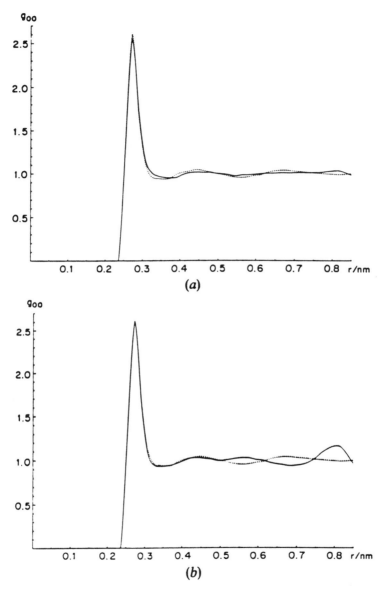

Figure 1. Oxygen–oxygen Pair Correlation Functions in Simulations of Pure SPC Water Using Different Termination Functions.
The figure illustrates results obtained in molecular dynamics simulations [27], using the SPC model either with Ewald's summation (dashed curve) or with 2 different terminations functions: (a) the one in equation 3, and (b) $S(r) = 1-(r/r_c)^2 + (r/r_c)^4$ from ref. *40*.
(Reproduced with permission from ref. *28*. Copyright 1990 Taylor & Francis Ltd.)

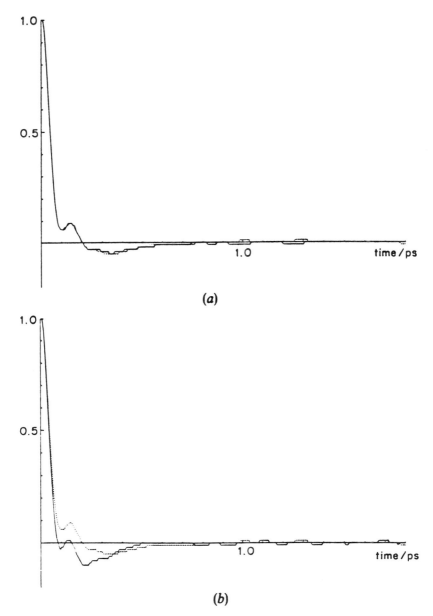

(a)

(b)

Figure 2. Normalized Center of Mass Velocity Autocorrelation Function for Simulations with Pure SPC Water Using Different Termination Functions. The figure illustrates results obtained in molecular dynamics simulations [27], using the SPC model either with Ewald's summation (dashed curve) or with 2 different terminations functions: (a) the one in equation 3, and (b) $S(r) = 1-(r/r_c)^2 + (r/r_c)^4$ from ref. *40*.
(Reproduced with permission from ref. *28*. Copyright 1990 Taylor & Francis Ltd.)

A comparison between the mean–field SPC and polarizable PSPC water models: simulation studies of pure liquid water (*22, 31*)

Structural and thermodynamic properties. Values of thermodynamic and average properties for SPC and PSPC models calculated from our simulations are listed in Table I. Overall, the properties derived from our simulations show a good agreement with the results previously described on the same PSPC model (*10*), although a slight lowering (0.1 Debye) of the magnitude of the average dipole moment is observed in our simulation. This could be due to the use of the termination function which is applied to all the terms involving interactions between and with permanent partial charges. It has been shown that this termination scheme behaves as if the partial charges are replaced by a distance dependent but smaller effective charge (*28*), with a net result of reducing the electrostatic energy and weakening the forces and electric fields in the system. The weakened fields, will in turn reduce the response of the polarizable particles, thereby leading to smaller induced dipoles.

Table I. Thermodynamic and Average Properties of the Simulated PSPC and SPC Systems

	SPC	*PSPC**
Temperature (K)	300	300
Polarization energy (kcal/mol)	0.88[(d)]	–2.91 ± 0.05
Vaporization enthalpy (kcal/mol)[(a)]	9.10 ± 0.05	9.10 ± 0.05[(b)]·
Induced dipole moment (Debye)[(c)]	0.42	1.0 ± 0.01
Total dipole (Debye)	2.27	2.8 ± 0.01
Angle between induced and permanent dipoles	–	22° ± 0.3

SOURCE: adapted from ref. *22*
*values obtained by Van Belle *et al.* (*22*)
(a) experimental value is 9.91 kcal/mol
(b) with $3/2k_BT$, the internal energy of the isolated molecule subtracted for comparison
(c) the induced dipole moment is calculated as the difference between the actual dipole moment and the gas–phase value (1.85 Debye).
(d) point dipole creation energy only.

In agreement with Ahlström *et al.* (*10*), we find that the induced dipoles are nearly parallel to the permanent dipoles. The average angle between the two vectors is 22°, and the component of the induced dipoles along the permanent ones is equal to 0.91 Debye.

The radial distribution functions g_{OO} for SPC and PSPC calculated from our simulations are very similar (Figure 3), with the positions of the first peaks at about the same distance (2.75 Å). Unlike Ahlström *et al.* (*10*), we observe a lowering of the first maximum in PSPC relative to SPC. We also observe a slight lowering of the second maximum of g_{OH} in PSPC, although in our case, the first maxi-

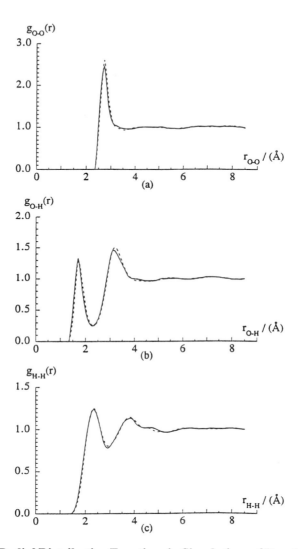

Figure 3. Radial Distribution Functions in Simulations of Pure Water, with and without Induced Polarization.
Solid curve: results obtained using PSPC polarizable model and the extended Lagrangian method. Dashed curve: non–polarizable SPC model. (a) oxygen–oxygen radial distribution function, (b) oxygen–hydrogen intermolecular radial distribution function, (c) hydrogen–hydrogen intermolecular radial distribution function.
(Reproduced with permission from ref. 22. Copyright 1992 Taylor & Francis Ltd.)

mum has exactly the same height. The g_{HH} functions are almost identical up to 4 Å where a slight maximum appears at 4.75 Å in the PSPC model (22).

Dynamical properties. Table II quotes the diffusion coefficients D, derived using the Einstein relation, and the experimental value measured at 300 K. We see that the diffusion coefficients computed for PSPC are about half the value obtained for mean field SPC which brings it in excellent agreement with the experimental value. This agreement is in fact slightly better than the value previously obtained with PSPC under different simulation conditions (10). We thus confirm that explicitly including induced polarization effects significantly slows down particle motion, as seen from the mean square displacement time correlation function (Figure 4).

Table II. Transport Properties of the Pure Water Liquid

	SPC	PSPC[(a)]	Exp.
D (10^{-9} m^2/s)	4.6 ± 0.2	2.4 ± 0.2	2.3[(b)]
D_α (10^{-9} m^2/s)	3.8 ± 0.4	1.9 ± 0.4	–
D_β (10^{-9} m^2/s)	4.7 ± 0.4	2.8 ± 0.4	–
D_γ (10^{-9} m^2/s)	4.2 ± 0.4	2.2 ± 0.3	–
$\tau_\alpha^{(1)}$ (ps)	2.4 ± 0.1	4.8 ± 0.3	$3.8 - 5.0$[(c)]
$\tau_\alpha^{(2)}$ (ps)	1.2 ± 0.1	2.1 ± 0.2	–
$\tau_\beta^{(1)}$ (ps)	1.6 ± 0.1	2.3 ± 0.2	–
$\tau_\beta^{(2)}$ (ps)	0.9 ± 0.1	1.7 ± 0.2	2.0[(d)]
$\tau_\gamma^{(1)}$ (ps)	2.4 ± 0.1	3.2 ± 0.3	–
$\tau_\gamma^{(2)}$ (ps)	1.3 ± 0.1	2.3 ± 0.2	–

SOURCE: adapted from ref. 22
Indices α, β, γ are defined in Figure 4 of ref. 22, indices (1) and (2) refer to the first and second order Legendre polynomials respectively.
(a) values obtained by Van Belle et al. (22)
(b) experimental value published by Krynicki et al. (32)
(c) experimental value from Eisenberg et al. (33)
(d) experimental value from Halle et al. (34)

Analysis of the time autocorrelation functions of individual components of the center of mass velocity along the α,β,γ directions of the molecular reference frame (22), shows that in both models, the water molecules move slowest in the in-plane α direction and faster along the out of plane β direction (Table II), where the correlation functions are more dominated by the lower frequency.

To probe yet another aspect of the dynamical properties, the reorientation correlation functions of the first and second order Legendre polynomials are also calculated:

$$< P_1(\mathbf{e}^x(t) . \mathbf{e}^x(0)) > = < (\mathbf{e}^x(t) . \mathbf{e}^x(0)) > \qquad (5)$$

$$< P_2(\mathbf{e}^x(t) \cdot \mathbf{e}^x(0)) > = \frac{1}{2} < 3(\mathbf{e}^x(t) \cdot \mathbf{e}^x(0))^2 - 1 > \qquad (6)$$

where $\mathbf{e}^x$ are unit vectors defined in the above mentioned molecular reference frame.

The corresponding decay times (labeled $t_x^{(1,2)}$), given in Table II, are quantities which can be compared with measurements of dielectric relaxation (*29*) and relaxation of the Zeeman magnetization (*35*). They were calculated by a least square fit, assuming an exponential decay for the reorientation correlation functions defined by equations 5 and 6. The short times (up to 0.5ps), corresponding to the librational motion of the water molecule in its neighbors cage, were not included in the fit.

Comparing the results for the SPC and PSPC models, the change in these properties is striking: the PSPC model reorients much more slowly than the mean–field SPC equivalent, the agreement with the experimental values has thus been encouragingly improved.

More recently, the T_1 relaxation time of the proton Zeeman magnetization has been computed from molecular dynamics simulations using both models (*31*). The values obtained for $1/T_1$ were $1/9.4$ s^{-1} and $1/5.4$ s^{-1}, for the mean–field and polarizable PSPC models respectively. Although the PSPC value is closer to the experimental measure of $1/3.1$ s^{-1} than that obtained with SPC, it still exceeds it by 75%. The decomposition of $1/T_1$ into intra– and inter– molecular contributions gives an indication to where the problem lies. With PSPC, the inter–molecular contribution is 11.69×10^{-2} s^{-1}, in good agreement with experimental measures of the corresponding quantity by Goldammer and Zeidler (*36*) (11×10^{-2} s^{-1}). This is due, at least in part, to its well behaved diffusion coefficient. On the other hand, the intra–molecular contribution for PSPC (6.73×10^{-2} s^{-1}) is much smaller than the experimental measure (*36*) (17×10^{-2} s^{-1}). Clearly, the latter contribution is determined by the $1/r^6$ dependence of the intra–molecular H–H distance. Based on recent experimental measures of the water molecule bond distance and bond angle (*37*), this distance should be 1.51 Å, while that in both SPC and PSPC models is larger (1.63 Å), enough to account for a drop of 60% in the intramolecular contribution to the relaxation rate. It is therefore concluded that geometrical rather than dynamic factors are responsible for the observed discrepancies. A subsequent parametrization of a new water model should therefore not only focus on the charge distributions but also on the experimentally determined geometrical parameters of the water molecule in the liquid state.

Molecular dynamics study of methane hydration and methane association in a polarizable water phase (*25*).

Structural and static properties. Methane–water radial distribution functions (rdf) computed from simulations with SPC and the polarizable PSPC models are shown in Figure 5.

The methane–water functions display appreciable structure with two maxima at 3.7 and 6.5 Å approximately. The maxima of the first and second peaks occur at slightly larger distances for the polarizable water model, consistent with

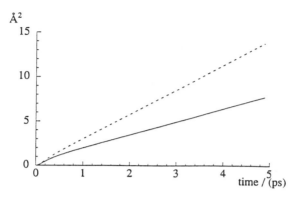

Figure 4. Mean Square Displacement Time Correlation Function.
Solid curve: PSPC. Dashed curve: SPC.
(Reproduced with permission from ref. *22*. Copyright 1992 Taylor & Francis Ltd.)

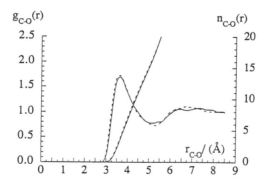

Figure 5. Methane–oxygen Pair Correlation Function and Corresponding Coordination Numbers, Computed with and without Induced Polarization of Water.
Plain curve: polarizable PSPC system computed using the extended Lagrangian procedure. Dashed curve: mean–field SPC system.
The computed uncertainty amounts to ± 0.05 at the first maximum; the rdf's are computed by averaging the results from the two methanes molecules.
(Reproduced with permission from ref. *25*. Copyright 1993 American Chemical Society).

a larger methane–water Lennard–Jones σ_{MW} parameter (0.05 Å) associated with this model. The heights of the first peaks are almost identical for the SPC and PSPC water models but some differences appear at larger distances. In SPC the first minimum is somewhat deeper and the second maximum is somewhat higher and better defined. In addition, the broader second peak in PSPC is 'split' into two peaks in SPC. All this indicates that SPC displays more structure than its polarized equivalent.

Integration of the rdf's for both models, yields 3 neighbors at the first maximum and 11 neighbors to the point where the radial distribution function crosses unity. Integration up to the minimum yields 16 neighbors at 5.1 Å and 18 neighbors at 5.3 Å for the polarizable and permanent systems respectively. The differences at longer distances are probably not significant due to the fact that the minimum in the rdf of the polarizable system is quite flat (Figure 5). Interestingly the same rdf's computed for simulations of one methane in a box of 215 water molecules (*25*), show a higher first peak for PSPC than SPC, yielding 4.8 neighbors (at 3.7 Å) and 3.5 neighbors (at 3.6Å) for the two models respectively.

The solvent–solvent rdf's averaged over all the molecules of the system (not shown) (*25*) are identical to those computed from simulations of pure water (*10*, *22*), indicating that, on the whole, the solvent structure is not perturbed by the presence of the hydrophobic solutes considered in this study.

In addition to structural properties, we also analyze the magnitude of the induced water dipoles around the apolar solute. The results show that water molecules in contact with the hydrophobic species are less polarized than their counterparts in the bulk. The average drop in polarization magnitude is of 10% at 3.0 Å, 6% at 3.5 Å and 4% at 4 Å, resulting in a total average dipole moment of 2.6 D, 2.7 D, and 2.75 D, at distances of 3 Å, 3.5 Å and 4.0 Å respectively. This is well understood since the presence of the apolar solute tends to lower the local density of polarizing dipoles. This effect is however limited to the neighborhood of the methane and is completely damped at distances of about 5 Å after the first peak of the solute–solvent pair correlation function. A much more significant drop in the magnitude of induced water dipoles of 25% has been computed for water molecules near a hydrophobic wall (*14*), suggesting that the influence of the hydrophobic solute on the water dipole moment probably depends on the size of the solute.

The average angle between the induced and the permanent dipole components also shows a change due to the presence of the apolar solute. The induced dipole is less parallel to the permanent dipole at 3 Å from the solute center (24° ± 0.3) than at 5 Å (21° ± 0.3).

One can conclude from these results, that the electrostatic interactions between water molecules in contact with the methanes are weakened in the polarizable model, while they would obviously remain unchanged in SPC.

An analysis of the water–water rdf's for molecules in the first solvation shell of the methane molecules (see legend to Table III for detail) confirms this conclusion (*22*). It shows indeed that the height of the first peak drops further, relative to that in the bulk, for PSPC than for SPC, as would be expected from a weakening of the electrostatic interactions (*27*, *28*) between PSPC water molecules close to the hydrophobic solute.

Table III. Self–diffusion Coefficient and Reorientational Times of Water in the First Solvation Shell of Methanes

	SPC methane	SPC*	PSPC methane	PSPC*
D $(10^{-9}$ m^2/s)	2.9 ± 0.4	4.6 ± 0.2	2.0 ± 0.4	2.4 ± 0.2
$\tau_\alpha^{(1)}$ (ps)	5.6 ± 0.8	2.4 ± 0.1	6.4 ± 0.9	4.8 ± 0.3
$\tau_\beta^{(1)}$ (ps)	3.0 ± 0.8	1.6 ± 0.1	2.9 ± 0.8	2.3 ± 0.2
$\tau_\gamma^{(1)}$ (ps)	4.2 ± 0.6	2.4 ± 0.1	3.1 ± 0.5	3.2 ± 0.2

SOURCE: adapted from ref. 25
A water molecule is considered as belonging to the first solvation shell if it is within a radius r from the center of the methane, and has not left this perimeter during the simulation for a continuous period longer than 10% of a maximum correlation time s. In all cases r is fixed to 4 Å and s to 4 ps.
* denotes pure liquid values. Indices α, β, γ correspond to the molecular axes as defined in ref. 22. Index (1) refers to the first order Legendre polynomial. The computed uncertainties are the standard errors calculated by breaking up the total trajectory into blocks of 10 ps (25).

Dynamic properties – Diffusion and reorientational times. The effect of the hydrophobic solute being largest on the water molecules which are in its immediate vicinity, the analysis of the dynamic properties focused on the first solvation shell (defined as in legend of Table III). The translational diffusion coefficient D, and the orientational correlation times $\tau_x^{(1)}$ in the molecular frame α, β, γ, calculated as described above, are listed in Table III, where the pure liquid water values are provided for comparison (22, 28).

We see that overall, the presence of methane tends to slow down the dynamical properties, in agreement with the well known 'freezing effect' of the hydrophobic solutes on the surrounding water molecules (38). This effect is somewhat more pronounced in the SPC system than with PSPC, possibly due to the smaller induced dipoles of the PSPC water molecules in contact with the methane, resulting in weaker mutual interactions and forces. This will in turn increase the diffusion coefficient and accelerate the reorientation of the molecules, counterbalancing somewhat the 'freezing effect'.

The dynamical properties calculated considering all the water molecules in the system were identical to those obtained previously for pure water (22, 28). However, since the effects observed here are sometimes of comparable magnitude to the computed uncertainties (Table III), much longer molecular dynamics simulations would be needed to confirm these findings.

Potential of mean force for the methane–methane interaction in water. The pmf's computed as a function of the inter–methane distance for the SPC and PSPC systems respectively, are shown in Figure 6, where the zero energy level is taken at the minimum of each curve.

We see that for both water models, the minima occur at 4 Å, corresponding to the methane–methane contact distance. This near perfect coincidence is the consequence of the methane–methane energy parameters being the same in both

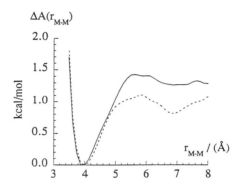

Figure 6. Potential of Mean Force of Two Methanes on Water as a Function of the Intermolecular Distance.
Plain curve: polarizable PSPC system. Dashed curve: mean–field SPC system.
The statistical uncertainty for each point relative to the adjacent points ranges from 0.01 to 0.1 kcal/mol, depending on the free energy difference itself. The total uncertainty on both ends of the curve amounts to 0.35 kcal/mol (*22*)·
(Reproduced with permission from ref. *25*. Copyright 1993 American Chemical Society).

systems. At larger intermolecular distance however, the curves display quite a different shape. The height of the energy barrier for separating the methanes is 1.5 kcal/mol in the polarizable PSPC system but only 1.0 kcal/mol in the permanent SPC one. This result is significant, given that the statistical uncertainty for each pair of adjacent points along the curve in the interval of 4–6Å does not exceed 0.01 kcal/mol. We see furthermore that the permanent model displays a well–defined minimum at 6.9 Å corresponding to the solvent–separated configuration. This distance corresponds exactly to a linear alignment between the centers of the two methanes with that of an intervening water molecule. The energy barrier to squeeze this water molecule out is 0.35 kcal/mol. In the polarizable system, this second minimum is very ill defined, the height of the barrier being reduced to 0.15 kcal/mol only. The fact that the water molecule situated between the two methanes has a smaller induced dipole can qualitatively explain this difference.

The results we obtain with the permanent water model are in very good agreement with those of Jorgensen et al. (23), although the position of the minimum corresponding to the solvent–separated configuration is shifted to larger distances in our case, which is probably due to differences in the water models used.

The raised energy barrier for separating the two methanes in the polarizable system could be explained by the traditional view of the hydrophobic effect, by considering the differences in hydration properties between the aggregated and non–aggregated species. Indeed, the number of water neighbors of the isolated methane was shown to be larger in the polarizable model (4.8) than in the permanent one (3.5) (25), while the number of neighbors of the dimer in contact is equal in both models (3.0). We can thus deduce that the polarizable system orders less water molecules than the permanent model, when the two methanes are in contact than when they are apart.

We thus see that the polarizable model lowers the probability for the complex to dissociate and raises the energy level of the solvent–separated configuration, while in the same time decreasing the energy barrier of squeezing out the water molecule from the solvent–separated configuration, a net effect that should favor aggregation relative to the permanent model.

Conclusions

The extended Lagrangian method adapted from Sprik and Klein (5) is shown to be very effective in computing induced polarization effects in pure liquid water. Applied to the PSPC model it yields structural, thermodynamics and transport properties in good agreement with those computed previously by iterative/predictive algorithms, at only a factor of 2 in computational overhead, compared to the classical SPC. The extended Lagrangian computations furthermore yield translational diffusion coefficients for pure water that are in better agreement with the experimental values than those reported previously.

Using the extended Lagrangian method, the influence of including electronic polarization of water on the hydration and association properties of methane solutions, was also analyzed. Water polarization is shown to have perceptible effects on methane hydration and a marked effect on methane association.

Polarizable water molecules close to the hydrophobic solute experience a loss in their dipole moment, which results in weaker electrostatic interactions with neighboring molecules and hence to less structure than in the bulk. These effects are however small, and decrease with increasing distance to the solute. It is not surprising therefore that both water models yield very similar hydration free energies for methane (unpublished results), and it is encouraging that the computed values are in good agreement with the experimental measure (*39*).

In accord with previous findings we see that the major influence of the nonpolar solute on the water phase is to slow down the motion of the waters surrounding it; the effect to which the important unfavorable entropic contributions to hydrophobic solvation have been attributed (*38*). Due to weakened electrostatic interactions, polarizable water surrounding the solute, being able to re-orient faster than non polarizable water, will have their motion less affected.

The most striking effect of including water polarization, is on the computed methane–methane potential of mean force. The energy barrier to separate the two methanes is higher, while that for squeezing out the water from the solvent-separated configuration is lower, nearly abolishing the minimum corresponding to the solvent–separated configuration. It is possible therefore that including electronic polarization in water simulations yields an improved physical description of the system, in agreement with the proposal of Berne and Wallqvist (*39*), and should be instrumental in future studies of hydrophobic aggregation in water. It must be noted however, that methane polarization has not been taken into account in the present study. Its contribution to the methane–methane potential of mean force in our system still needs to be investigated.

Finally, it should be mentioned that the extended Lagrangian method could be generalized, without major difficulties, to tensorial polarizabilities and implemented for more complex heterogeneous systems such as solvated biological molecules.

Acknowledgments

Daniel Van Belle acknowledges support from the European Communities BRIDGE Action Programme (BIOT–CT91–0270), Martine Prévost is research associate with the National Fund for Scientific Research (Belgium), and Guy Lippens is senior research assistant with the National Fund for Scientific Research (Belgium). This text presents research results of the Belgian programme on Interuniversity Poles of Attraction initiated by the Belgian State, Prime Minister's Office, Science Policy Programming. The scientific responsibility is assumed by its authors.

Literature Cited

(1)   Rahman, A.; Stillinger, F. H. *J. Chem. Phys.* **1971**, *55*, 3336.
(2)   Berendsen, H. J. C.; Postma, J. P. M.; van Gunsteren, W. F.; Hermans, J. *Intermolecular Forces*; Pullmann, B.: Dordrecht, The Netherlands, 1981; 331.
(3)   Rullmann, J. A. C.; van Duijnen, P. Th. *Mol. Phys.* **1988**, *63*, 451.
(4)   Jorgensen, W. L.; Chandrasekhar, J.; Madura, J. D.; Impey R. W.; Klein, M. L. *J. Chem. Phys.* **1983**, *79*, 926.
(5)   Sprik, M.; Klein, M.L. *J. Chem. Phys.* **1988**, *89*, 7556.
(6)   Matsuoka, O.; Clementi, E.; Yoshimine, M. *J. Chem. Phys.* **1976**, *64*, 1351.

(7)   S. Kuwajima S.; Warshel, A. *J. Phys. Chem.* **1990**, *94*, 460.
(8)   Wallqvist, A.; Ahlström P.; Karlström, G. *J. Phys. Chem.* **1990**, *94*, 1649.
(9)   M. Sprik, M. *J. Phys. Chem.* **1991**, *95*, 2283.
(10) Ahlström, P.; Wallqvist, A.; Engström S.; Jönsson, B. *Molec. Phys* **1989**, *68*, 563.
(11) Barnes, P.; Finney, J. L.; Nicholas, J. D.; Quinn, J. E. *Nature* **1979**, *282*, 459.
(12) Wojcik M.; Clementi, E. *J. Chem. Phys.* **1986**, *84*, 5970.
(13) Impey, R.W.; Madden, A.; Mc Donald, I.R. *Molec. Phys*. **1982**, *46*, 513.
(14) Wallqvist, A. *Chem. Phys. Lett.* **1990**, *165*, 437.
(15) A. Warshel A.; Levitt M. *J. Mol. Biol.* **1976**, *103*, 227.
(16) Van Belle, D.; Couplet, I.; Prévost, M.; Wodak, S. J. *J. Molec. Biol.* **1987**, *198*, 721.
(17) Nosé, S. *Molec. Phys.* **1984**, *52*, 255.
(18) Goldstein, H. *Classical Mechanics*; Addison–Wesley, 1959.
(19) Böttcher, C. J. F. *Theory of Electric Polarisation*, 2nd edit.; Elsevier Publishing Company: Amsterdam, The Netherlands, 1973.
(20) Verlet, L. *Phys. Rev.* **1967**, *159*, 98.
(21) Swope, W.C.; Andersen, H.C.; Berens, H.; Wilson, K.R. *J. Chem. Phys.* **1982**, *76*, 637.
(22) Van Belle, D.; Foeyen, M.; Lippens, G.; Wodak, S.J. *Mol. Phys.* **1992**, *77*, 239.
(23) Jorgensen, W. L.; Buckner, J. K.; Boudon, S.; Tirado–Rives, J. *J. Chem. Phys.* **1988**, *89, 3742*.
(24) Delhaise, P.; Van Belle, D.; Bardiaux, M.; Alard, P.; Hamers, P.; Van Cutsem, E.; Wodak, S. *J. Molec. Graphics* **1985**, *3*, 116.
(25) Van Belle, D.; Wodak, S. J. *J. Am. Chem. Soc.* **1993**, *115*, 647.
(26) Ryckaert, J. P.; Cicotti, G.; Berendsen, H. J. C. *J. Comput. Phys.* **1977**, *23*, 327.
(27) Pettitt, B. M.; Rossky, P. J. *J. Chem. Phys.* **1982**, *77*, 1451.
(28) Prévost, M.; Van Belle, D.; Lippens, G.; Wodak, S. J. *Molec. Phys.* **1990**, *71*, 587.
(29) Mc Quarrie, D. *Statistical Mechanics*; Harper and Row Publishers: New York, USA, 1976.
(30) Tobias D. J.; Brooks, C. L. III. *Chem. Phys. Lett.* **1987**, *142*, 472.
(31) Lippens, G.; Van Belle, D.; Wodak S. J.; Jeener, J. *Mol. Phys.* **1993**, *80*, 1469.
(32) Krynicki, K.; Green, C. D.; Sawyer, D.W. *Discuss. Faraday Soc.* **1978**, *66*, 199.
(33) Eisenberg, D.; Kauzman, W. *The Structure and Properties of Water*; Oxford University Press: Oxford, UK, 1969; 207.
(34) Halle, B.; Wennerström, H. *J. Chem. Phys.* **1982**, *75*, 1928.
(35) Ribeiro, A. A.; King, R.; Restico, C.; Jardetzky, O. *J. Am. Chem. Soc.* **1980**, *102*, 4040.
(36) E. Goldammer, E.; Zeidler, M.D. *Ber. Bunseng. Phys. Chem.* **1969**, *73*, 4.
(37) Thiessen, W. E.; Narten, A. M. *J. Chem. Phys.* **1982**, *77*, 2656.
(38) Frank, H. S.; Evans, W. J. *J. Chem. Phys.* **1945**, *13*, 507.
(39) Berne B.J.; Wallqvist, A. *Chemica Scripta* **1989**, *29A*, 85.
(40) Brooks C.L. III; Pettitt, B.M.; Karplus M. *J. Chem. Phys.* **1985**, *83*, 5897.

RECEIVED April 25, 1994

Chapter 23

Hydrophobic Interactions from Surface Areas, Curvature, and Molecular Dynamics

Use of the Kirkwood Superposition Approximation To Assemble Solvent Distribution Functions from Fragments

Robert B. Hermann[1]

Lilly Research Laboratories, Eli Lilly and Company, Indianapolis, IN 46285

A previously developed methodology for the calculation of hydrophobic interactions based on resolving solution-free energies into solvent cavity potentials and solute-solvent interaction energies is investigated further. The aromatic systems benzene and toluene now are well fitted along with the aliphatic compounds. The inclusion of the local mean curvature of the accessible surface is examined. The solvent effect on cyclohexane dimerization is found to be unfavorable, in accord with previous estimates. The calculations rely on a molecular dynamics determined solute-solvent interaction energy. However, it is possible to build the solute-solvent distribution function for a molecule or dimer from molecular fragment distribution functions via the Kirkwood superposition approximation. Distribution functions built from such transferable solvent distribution functions, in the case of methane dimer and ethane, give encouraging results for the solute-solvent interaction energy.

Approximate intermolecular potentials and semi-empirical methods are widely used today in the calculation of molecular and bulk thermodynamic properties. Solvent effects models including hydrophobic interaction models are useful in conjunction with such methods for the semi-empirical calculation of free energies and dissociation constants in solution *(1-11)*. In this regard, the development of a working model for the calculation of hydrophobic interactions is continued here.

In a previous paper *(4)*, a method was developed which allows one to calculate the hydrophobic interactions between small hydrocarbon molecules. In that method, two important quantities are solvent cavity potentials and solute-solvent interaction energies. Solute-solvent interaction energies are found from molecular dynamics simulations, and cavity potentials are found by calibrating a parameterized version of scaled particle theory with experimental hydrocarbon solvation energies.

[1]Current address: Department of Chemistry, Indiana University/Purdue University at Indianapolis, 402 North Blackford Street, LD3326, Indianapolis, IN 46202–3274

0097–6156/94/0568–0335$08.72/0

Here this methodology is applied to several other interesting cases. In addition, the use of local curvature *(12)* is investigated rather than just the use of inherent curvature of spherical cavities as was done before. Some current methods for continuum solvation effects treat different regions of the accessible surfaces *(13)* with different parameters. Local curvature considerations are important for treating portions of accessible surfaces *(12)*, e.g. in the vicinity of hydrophobic groups, since these have their own associated curvature.

In addition to the development of a means to obtain cavity energies, a simpler method currently under development to obtain interaction energies within the scope of continuum methods is outlined; namely, an attempt is made to construct a solvent distribution function for a larger solute from the distribution functions of smaller molecules, or fragments. While the solvent distribution function of the fragments may be obtained from molecular dynamics calculations, the solvent distribution function associated with the larger system is obtained by putting these fragment distribution functions together from a set of rules.

Theory

The solubility of hydrocarbons in water has been treated by breaking down the solvation-free energies into two main contributions--the cavity potential and solute-solvent interaction energy *(14-20)*. The following equation for dilute solutions relates the solvation potential to Henry's law constant:

$$kTlnK_H = \mu^* + kT \ln N_0kT/V \tag{1}$$

where K_H is Henry's law constant, k is Boltzman's constant, T is the temperature, N_0 is Avogadro's number and V is the molar volume of the solvent.

The method as developed in a previous paper *(4)* and outlined here is based on the idea that the solvation potential μ^* can be expressed as

$$\mu^* = \mu_c + e_i \tag{2}$$

where e_i is defined as the following average:

$$e_i = < \sum_{j=1}^{N_w} U_{oj} > \tag{3}$$

where N_w is the number of water molecules and the solute subscript is 0. μ_c, the cavity potential, is then defined by the difference

$$\mu_c = \mu^* - e_i \tag{4}$$

By this definition, μ_c contains all entropic contributions to μ^* and e_i contains none. The quantity e_i may be found then by averaging molecular dynamics runs using a hydrocarbon whose coordinates are fixed and TIP3 water *(21)*. Fixed conformations of hydrocarbons are used and each must be calculated individually. The energy was calculated out to 15 Å so that a correction

$$E_{corr} = 4\pi\rho_1 \int_{R_c} U_{ow}r^2dr \tag{5}$$

where $R_c = 15$ Å must be applied. The results of the molecular dynamics calculations are shown in Table I. The column labeled e_i is the sum of the dynamics result plus E_{corr}.

The calculation of the cavity potential is done by first measuring the accessible surface area *(13)* of the molecule or system. The solute-solvent thermal radii *(22)* are used which simplifies the parameter choice problem somewhat. After carrying out such calculations for 13 hydrocarbons, the resulting calculated cavity potentials were fitted to the following equation *(4)* suggested by scaled particle theory *(17, 18, 23, 24)*:

$$\mu_c = aA(1 + b/r + c/r^2) \tag{6}$$

where r is found from the accessible surface area $A = 4\pi r^2$, and a is approximated by the surface tension of water *(25)*, 103.5 cal Å^{-2}, b is 3.09 Å and c is 3.25 Å^2. The parameters a, b and c are, in general, temperature dependent; however, they have been found by calibrating against solubility data at a particular temperature *(4)*. To calculate hydrophobic interactions between pairs of hydrocarbon molecules, the molecular dynamics calculations are performed on the (associated) pair to get e_i and the cavity potential μ_c for the pair is found from the area of the pair and Equation 6. The curvature is taken into account through the term b/r. Table II lists the areas and Table III gives the resulting cavity potentials. Figure 1 shows the cavity potentials of the monomers plotted against the accessible surface areas. The cavity potential as given by calibrating Equation 6 is dependent of the nature of the molecules occupying the cavity. The above calibration should represent cavity potentials for aliphatic systems generally.

Combined Treatment of Aliphatic and Aromatic Systems

It would be desirable to treat aliphatic and aromatic systems the same way, i.e. with the same set of cavity parameters. Benzene using OPLS parameters *(26)* gives an interaction energy of -14.132 kcal mol^{-1}. This value, together with the accessible surface area of 230.5 Å^2 and an experimental solvation energy of -883 cal mol^{-1}*(27)* produces a cavity potential from Equation 6 much too high to fit on the cavity potential vs. area curve of Figure 1.

Presumably because of its larger quadrupole moment, benzene interacts more strongly with water than an aliphatic compound of similar molecular weight. Because of this extra electrostatic interaction, the cavity formed by aromatic systems is perturbed relative to the cavity formed by aliphatic molecules. The resulting cavity energy would be significantly higher in energy per unit area from the aliphatic cavities. Therefore, the parameters for Equation 6, defining the cavity potential for a solute, would not apply to aromatic systems.

A solution to this problem applied here was to carry out the molecular dynamics simulations on benzene using aliphatic charges rather than aromatic charges but otherwise using aromatic OPLS parameters. This insures that the cavity surface has a similar structure to that formed by the aliphatic molecules, so that the cavity energy per unit area is comparable. When evaluating the interaction energy e_i afterward by averaging over all the configurations, the benzene molecule is then given the aromatic charges.

The energy is found from

$$e_i = <\sum_{j=1}^{N_w} U_{0j}>_{al} \tag{7}$$

Table I. Molecular Dynamics Interaction Energies and Related Data

System	$<E>(MD)$[a] $(kcal\ mol^{-1})$	e_i $(kcal\ mol^{-1})$	Run Length (ps)	Waters
Methane	-3.172	-3.216	180	926
Ethane	-5.186	-5.264	180	960
Propane	-7.086	-7.198	300	1025
trans n-Butane	-8.670	-8.815	210	1074
gauche n-Butane	-8.702	-8.847	180	1136
Isobutane	-8.654	-8.799	180	1111
trans n-pentane	-10.630	-10.808	180	1118
gauche n-pentane[b]	-10.242	-10.420	180	1167
Neopentane	-9.873	-10.052	180	1171
Cyclopentane	-9.786	-9.954	180	1100
Dimethylbutane	-11.578	-11.790	180	1223
Cyclohexane	-11.120	-11.321	180	1165
Dimethylpentane	-12.914	-13.159	180	1247
Cycloheptane	-12.618	-12.852	180	1207
Isooctane	-14.151	-14.430	300	1282
Benzene[c]	-11.664	-11.841	300	1045
Toluene[c]	-13.325	-13.536	270	1137
CH_4 - CH_4 0.0 Å	-7.070	-7.159	300	963
CH_4 - CH_4 1.54 Å	-6.380	-6.469	240	1014
CH_4 - CH_4 3.45 Å	-5.766	-5.855	210	1082
CH_4 - CH_4 4.0 Å[d]	-5.850	-5.939	248	1082, 335
CH_4 - CH_4 5.0 Å	-5.866	-5.955	240	1136
CH_4 - CH_4 6.0 Å	-6.081	-6.170	300	1150
CH_4 - CH_4 7.16 Å	-6.383	-6.472	270	1160
CH_4 - CH_4 8.0 Å	-6.363	-6.452	270	1239
CH_4 - C_2H_6	-7.800	-7.922	210	1096
C_2H_6 - C_2H_6	-9.340	-9.496	180	1096
i-C_4H_{10} - i-C_4H_{10}	-14.989	-15.279	180	1294
n-C_5H_{12} - n-C_5H_{12}	-17.757	-18.114	180	1330
C_6H_{12} - C_6H_{12}[e]	-18.900	-19.303	300	1358
C_6H_{12} - C_6H_{12}[f]	-19.959	-20.361	270	1376
C_6H_6 - C_6H_6	-20.911	-21.265	270	1356
$(C_2H_6)_3$	-13.270	-13.504	300	1220
$(C_2H_6)_4$	-15.864	-16.176	330	1321

[a]Aliphatic monomers from Ref. 4.
[b]Conformation with one gauche interaction.
[c]Interaction energy calculated from Equation 7.
[d]48 ps with 1082 waters and 200 ps with 335 waters.
[e]Parallel or stacked configuration. See Figure 2b.
[f]Perpendicular or "T" configuration. See Figure 2c.

Table II. Geometric Features of the Cavity Surfaces

System[a]	Accessible Surface Area[b] ($\AA^2$)	Average Mean Curvature[c,e] $\overline{K_m}$ ($\AA^{-1}$)	Average Squared Mean Curvature[d,e] $\overline{K_m^2}$ ($\AA^{-2}$)
Methane	135.5	0.312	0.099
Ethane	173.2	0.278	0.081
Propane	204.4	0.259	0.071
trans n-Butane	235.3	0.244	0.064
gauche n-Butane	231.1	0.243	0.064
Isobutane	229.6	0.243	0.063
trans n-Pentane	266.1	0.232	0.060
gauche n-Pentane[f]	262.0	0.230	0.059
Neopentane	250.7	0.232	0.059
Cyclopentane	236.8	0.240	0.063
Dimethylbutane	273.2	0.223	0.055
Cyclohexane	258.4	0.230	0.057
Dimethylpentane	306.2	0.215	0.053
Cycloheptane	278.5	0.221	0.053
Isooctane	321.5	0.208	0.050
Benzene	230.5	0.243	0.064
Toluene	259.4	0.232	0.060
CH_4 - CH_4 0.0 $\AA$	148.1	0.295	0.088
CH_4 - CH_4 1.54 $\AA$	167.1	0.281	0.081
CH_4 - CH_4 3.45 $\AA$	207.5	0.259	0.072
CH_4 - CH_4 4.0 $\AA$	222.5	0.253	0.071
CH_4 - CH_4 5.0 $\AA$	244.0	0.245	0.071
CH_4 - CH_4 6.0 $\AA$	270.9	0.239	0.076
CH_4 - CH_4 7.16 $\AA$	271.7	0.285	0.088
CH_4 - CH_4 8.0 $\AA$	271.0	0.304	0.096
CH_4 - C_2H_6	244.1	0.242	0.066
C_2H_6 - C_2H_6	266.8	0.232	0.061
i-C_4H_{10} - i-C_4H_{10}	358.5	0.199	0.050
n-C_5H_{12} - n-C_5H_{12}	393.4	0.191	0.045
C_6H_{12} - C_6H_{12} [g]	407.0	0.186	0.047
C_6H_{12} - C_6H_{12} [h]	430.7	0.185	0.045
C_6H_6 - C_6H_6	379.3	0.198	0.052
$(C_2H_6)_3$	357.2	0.205	0.055
$(C_2H_6)_4$	383.7	0.194	0.048

[a]See text for dimer configurations.

[b]Calculated using MOLAREA, Quantum Chemistry Program Exchange 225, Indiana University, Bloomington, IN.

[c]Mean curvatures at a point were measured by the method of Nicholls et al.; Ref. 12. A 2.8 radius probe was used.

[d]The mean curvature at a point was squared to get the squared mean curvature at that point.

[e]Averages are for the entire surface.

[f]Only one gauche interaction.

[g]Parallel or stacked configuration. See Figure 2b.

[h]Perpendicular or "T" configuration. See Figure 2c.

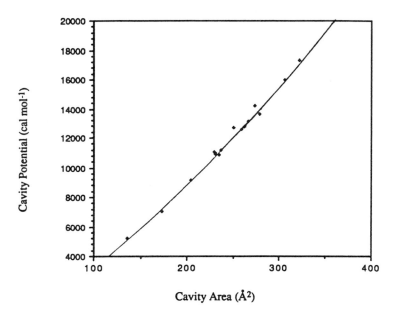

Figure 1. Cavity potentials from equation 4 for aliphatic hydrocarbons, toluene and benzene.

where the subscript al indicates that the configurational average is over the aliphatically charged cavity and not the aromatic charged cavity. U_{oj} is the complete aromatic molecule-water molecule interaction potential.

Thus the difference between the aromatic and aliphatic charge-solvent interaction is treated as a first order perturbation of a zeroth order dynamics distribution function. With the e_i calculated in this way, benzene and toluene are fitted reasonably well, as can be seen from Figure 1. For toluene, OPLS aromatic parameters are used for the phenyl ring and OPLS aliphatic parameters *(4)* are used for the methyl group. The charges on the methyl group carbon and the phenyl group carbon to which it is joined are then adjusted to give a dipole moment of 0.37 Debye *(28)* with the positive end of the dipole in the direction of the methyl group.

Inclusion of Local Curvature

In the calculation of μ_c in Equation 6, the curvature of the cavity is taken into account implicitly since it is found from the accessible surface area of the molecule as the inverse of the radius of the sphere which has the same area as the accessible surface. This corresponds to the second term in Equation 6 involving the parameter b. The assumption is made that the energy of the cavity formed by the molecule is the same as that for a spherical cavity of equal surface area *(4, 20, 29)*. This equivalent sphere method of representing the cavity is necessary if the liquid state theory used to calculate the cavity energy is applicable only to spherical cavities. In the equivalent sphere approximation, it is assumed that when the sphere is deformed to the accessible surface of the molecule, regions that were deformed such that curvature increased were somewhat compensated for by corresponding regions where curvature is thereby necessarily decreased or negative; since the cavity potential depends on curvature, such free energy changes tended to cancel. The good correlation with the data indicates that this assumption is reasonable at least for the accessible surfaces of small hydrocarbons. This equivalent sphere approximation was used previously in connection with Barker Henderson perturbation theory based on a Percus-Yevick zeroth order hard sphere representation and produced results in qualitative agreement with this paper *(20, 29)*.

Nicholls et al. *(12)*, in their treatment of hydrophobic interactions through interfacial tension, have demonstrated the effect of local curvature on the interfacial free energy. They have used the accessible surface area and interfacial tension together with a modification due to local curvature.

In this paper, the detailed curvature of the accessible surface cavity is similarly considered. The two descriptions of curvature of a surface are the mean curvature and the Gaussian, or total curvature *(30)*. It is this mean curvature that is of interest here. The mean curvature K_m at a point P on the smooth cavity surface is given by

$$K_m = (1/r_1 + 1/r_2)/2 \tag{8}$$

where r_1 and r_2 are the principal radii at a point P on the surface. For a sphere of radius r, this reduces to $K_m = 1/r$.

The average value of the mean curvature is the value of the mean curvature at each point on the surface, averaged over the entire surface. This will be designated $\overline{K_m}$. Another quantity which will be of interest is the average value of the square of the mean curvature $\overline{K_m}^2$. The curvature at each point on the surface is squared to get the mean curvature squared at that point. This is then averaged over the surface.

The mean curvature is measured at a point on the surface in the manner of Nicholls et al. *(12)*. They calculate the local curvature at a point on the accessible surface of a molecule by determining the accessible surface of a (spherical) water molecule at that point. For a planar accessible surface, one half of the water molecule is accessible but varies more or less for curved surfaces. Calculations of this accessibility determines the approximate curvature at that point on the surface. Table II gives the necessary geometrical parameters to be used in subsequent calculations on molecules and complexes. Five thousand points per spherical surface were used in the calculation in this paper.

Two alternatives to Equation 6 are now considered, as possible methods for the treatment of local curvature. These possibilities are given by Equations 9 and 10. In both cases in the curvature term, $1/r$ is replaced by $\overline{K_m}$, the average mean curvature.

$$\mu_c = aA(1 + b'\overline{K_m} + c'/r^2) \tag{9}$$

$$\mu_c = aA(1 + b''\overline{K_m} + c''\overline{K_m^2}) \tag{10}$$

In Equation 10, the last term involves the average of the square of the mean curvature instead of $1/r^2$. The results for the inclusion of local curvature, based on Equations 9 and 10 is given in Table III along with the results of Equation 6. The agreement with experimental solubilities is only slightly better than the equivalent sphere method of Equation 6. The aromatic systems, given the special treatment mentioned above, can in all cases also be accommodated with good agreement.

The parameter a is the surface tension of water and is the same for Equations 6, 9 and 10. The parameters b' and c' for Equation 9 were found to be 2.77 and 2.39, respectively while b" and c" for Equation 10 were found to be 3.09 and 3.25, respectively. The parameters for Equation 10 are the same as for Equation 6.

$\overline{K_m^2}$ and the Dissociation of Dimers

In some cases, e.g. the dissociation of dimers, the solute system can consist of more than one molecule, thereby requiring two or more distinct cavities, i.e. two or more separate surfaces. In order to treat such composite solutes involving multiple surfaces, Equation 10, rather than Equation 6 or 9 is necessary. This is easily demonstrated for a collection of spherical cavities, as follows:

First of all, for a cavity having a single spherical surface, the contribution to the potential due to the last term of Equation 10 is

$$Aac''\overline{K_m^2} = 4\pi ac''. \tag{11}$$

This is independent of the cavity radius and is related to the work of introducing a single point particle into the solvent. In scaled particle theory, the cavity potential is derived from the process of introducing a point particle into the solvent, and then allowing it to grow up to the size of the final desired (spherical) particle. Introducing a point particle increases the chemical potential, and such a process should be associated with each distinct spherical surface. Equations 6 and 9, as in scaled particle theory, can account for the introduction of only one point particle.

The important difference is that in addition to being able to describe a single surface, Equation 10, unlike Equations 6 or 9 can also describe several spherical surfaces. If the total area of the system is $A = 4\pi r^2$ and the system is made up of n molecules, then

$$A = 4\pi(r_1^2 + r_2^2 + \cdots r_n^2) \tag{12}$$

where the molecules 1, 2, $\cdots$ n have the spherical curvatures

$$K_m = 1/r_1, 1/r_2, \cdots 1/r_n. \tag{13}$$

Table III. Solvation Potentials and Cavity Potentials[a]

System[b]	μ_C[c]	$\mu^*(exp)$[d]	Equation 6 μ_C	Equation 6 μ^*[e]	Equation 9 μ_C	Equation 9 μ^*[e]	Equation 10 μ_C	Equation 10 μ^*[e]
Methane	5190	1974	5053	1837	5029	1813	5042	1826
Ethane	7069	1805	7231	1968	7237	1973	7258	1994
Propane	9143	1946	9171	1973	9098	1900	9150	1953
trans n-Butane[f]	10968	2051	11189	2373	11032	2216	11116	2300
gauche n-Butane[f]	10691	2051	10912	2064	10934	2087	10932	2084
Isobutane	11029	2230	10811	2013	10902	2103	10852	2053
trans n-Pentane[f]	13117	2346	13278	2469	12939	2131	13141	2332
gauche n-Pentane[f]	12840	2346	12997	2576	12932	2511	13003	2583
Neopentane	12693	2641	12221	2170	12347	2296	12249	2198
Cyclopentane	11169	1216	11288	1334	11319	1366	11322	1368
Dimethylbutane	14245	2456	13767	1978	13909	2119	13801	2012
Cyclohexane	12554	1234	12748	1427	12829	1509	12723	1402
Dimethylpentane	16008	2849	16083	2923	15923	2764	16100	2941
Cycloheptane	13655	803	14129	1276	14267	1415	14122	1270
Isooctane	17354	2924	17177	2747	17233	2803	17268	2839
Benzene	10957	-883	10869	-971	10924	-916	10896	-944
Toluene	12771	-765	12812	-724	12702	-834	12801	-729
CH_4 - CH_4 0.0 Å			5760	-1398	5897	-1262	5747	-1412
CH_4 - CH_4 1.54 Å			6868	399	6937	469	6834	365
CH_4 - CH_4 3.45 Å			9372	3516	9205	3350	9313	3458
CH_4 - CH_4 4.0 Å			10344	4405	9974	4035	10282	4343
CH_4 - CH_4 5.0 Å			11769	5814	11252	5298	12008	6053
CH_4 - CH_4 6.0 Å			13604	7433	12588	6418	14271	8101
CH_4 - CH_4 7.16 Å			13663	7191	9057	2586	11455	4983
CH_4 - CH_4 8.0 Å			13612	7160	7518	1066	10399	3947
CH_4 - C_2H_6			11779	3857	11418	3496	11796	3874
C_2H_6 - C_2H_6			13323	3827	12984	3488	13339	3842
i-C_4H_{10} - i-C_4H_{10}			19865	4585	19747	4468	20277	4998
n-C_5H_{12} - n-C_5H_{12}			22457	4343	22321	4207	22718	4604
C_6H_{12} - C_6H_{12}[g]			23482	4179	23530	4227	24290	4987
C_6H_{12} - C_6H_{12}[h]			25279	4918	24903	4542	25732	5371
C_6H_6 - C_6H_6			21406	141	20857	-408	21918	653
$(C_2H_6)_3$			19774	6270	19095	5591	20142	6638
$(C_2H_6)_4$			21730	5554	21445	5269	22106	5930

[a]In cal mol^{-1}.

[b]For dimer configuration, see text.

[c]Except for n-butane and n-pentane, this is found from $\mu_C = \mu^* - e_i$, where μ^* is the experimental solvation energy (column 2) and e_i is from Table I.

[d]Aliphatic entries previously listed in Ref. 4; Benzene and Toluene values from Ref.27.

[e]Calculated using μ_C from the preceding column and e_i from Table I.

[f]For treatment when more than one conformation is present, see Ref. 4.

[g]Parallel or stacked configuration. See Figure 2b.

[h]Perpendicular or "T" configuration. See Figure 2c.

The average squared mean curvature consists of these n terms, each weighted with their corresponding areas and divided by the total area A:

$$\overline{K_m^2} = \{4\pi r_1^2(1/r_1)^2 + 4\pi r_2^2(1/r_2)^2 + \cdots + 4\pi r_n^2(1/r_n)^2 \}A^{-1} \qquad (14)$$

so that

$$Aac''\overline{K_m^2} = 4\pi ac''n \qquad (15)$$

It can be seen that the quantity $4\pi ac''$ enters in once for each distinct spherical surface in the system. In this manner, Equation 10 can represent n spherical surfaces, or cavities. The expectation is that Equation 10 will represent any number of distinct solute surfaces. This is critical, for example, in following dimer dissociation such as methane-methane (see below). In order to get the correct description of a dimer in the limit of large separation, Equation 10 is necessary.

Applications to Selected Molecular Pairs and Aggregates

The numerical results for the molecular systems below are shown in Tables III and IV. The choice of dimer configuration shown in Figure 2 is somewhat arbitrary and usually depends in part on maximum contact. The remaining dimer configurations were defined in ref. *(4)*. Once the general dimer configuration was chosen, the energy was minimized using MM2 or MMP2 *(31)* on the isolated system, with the constraint that the chosen relative orientation be maintained. The solvation energies calculated for these dimers are for these fixed configurations and are given in Table III. The binding energies in Table IV are found from these fixed configuration solvation energies and the solvation energies of the monomers. Comparing the solvation energies of these selected dimers, while incomplete in the sense that all possible configurations were not considered, gives some insight as to whether the solvation effect augments or hinders association in solution.

Cyclohexane Dimer. Two configurations were treated, as shown in Figures 2b and 2c. The relative positions were minimized using MM2, giving a 4.55 Å between ring centers for parallel and 4.65 Å between centers for the perpendicular configuration. The parallel configuration of the isolated dimer had a binding energy of -3.29 kcal mol^{-1}, while the perpendicular configuration has a binding energy of -1.93 kcal mol^{-1}. The calculated solvation free energies associated with each indicate less association in water than in the gas phase. The values for both the equivalent sphere method and the explicit local curvature methods are qualitatively similar.

The MM2 energy and solvation energy together were then minimized, varying only the intermolecular distance (between ring centers). For the parallel configuration, a new minimum was found at 4.45 Å, favoring a greater association by 190 cal mol^{-1} over the results of Table IV. In the case of the perpendicular configuration, a similar minimization placed the molecules .2 Å closer and decreased the energy by 172 cal mol^{-1}. Subtracting this from the results in Table VI, solvent effects still favor dissociation of the dimer.

Based on the results for the one parallel conformation only, the equivalent sphere method predicts the dimer is favored 7 to 1 in the gas phase over solution, i.e. the solvent effect favors dissociation. Second virial coefficient data *(32)* and osmotic coefficients calculated for cyclohexane has suggested a dimer concentration three times greater in the gas phase than in solution *(33)*. There is a reduction in both cavity area and cavity energy upon association. Usual methods for calculations of changes in hydrophobic interactions would give a reduction in area and would, therefore, incorrectly predict a greater association in solution.

Table IV. Calculated Hydrophobic Interactions[a]

Pair		(Equation 6) Binding potential[b]		(Equation 9) Binding potential[b]		(Equation 10) Binding potential[b]		Reduction in area
CH_4 - CH_4	0.0 Å	-5072	(-5347)	-4888	(-5211)	-5064	(-5360)	123.0
	1.54	-3274	(-3549)	-3157	(-3479)	-3288	(-3583)	103.9
	3.45	-158	(-432)	-276	(-598)	-195	(-490)	63.5
	4.0	731	(456)	409	(86)	690	(394)	48.5
	5.0	2140	(1866)	1672	(1349)	2400	(2105)	27.0
	6.0	3760	(3485)	2793	(2470)	4448	(4152)	.1
	7.16	3517	(3243)	-1040	(-1363)	1330	(1034)	-.7
	8.0	3486	(3212)	-2560	(-2883)	285	(-1)	0.0
CH_4 - C_2H_6		52	(77)	-289	(-283)	53	(95)	64.6
C_2H_6 - C_2H_6		-109	(217)	-458	(-122)	-147	(232)	79.6
i-C_4H_{10} - i-C_4H_{10}		559	(125)	262	(8)	891	(538)	100.0
n-C_5H_{12} - n-C_5H_{12}		-664	(-348)	-306	(-484)	-226	(-87)	136.1
C_6H_{12} - C_6H_{12}[c]		1324	(1711)	1209	(1759)	2183	(2519)	110.0
C_6H_{12} - C_6H_{12}[d]		2064	(2451)	1525	(2075)	2567	(2904)	86.3
C_6H_6 - C_6H_6		2081	(1909)	1425	(1360)	2541	(2420)	81.7
C_2H_6 - $(C_2H_6)_3$		-2683		-2295		-2703		146.7

[a]Binding potentials in cal mol^{-1}. Areas in $Å^2$.

[b]Binding potential is dimer solvation potential minus the solvation potential of two monomers from Table III. The first binding potential is based on the monomer calculated energies while the numbers in parentheses are based on the experimental monomer energies. Solute-solute interaction not included.

[c]From Figure 2b.

[d]From Figure 2c.

Benzene Dimer. Table IV shows the results when benzene and the dimer are treated as discussed above. For both monomer and dimer, the molecular dynamics was carried out on the aliphatic charge model as outlined above and the interaction energy evaluated afterward with aromatic charges as shown in Equation 7.

The dimer configuration that was suggested by Jorgensen and Severance *(26)* was used here and shown in Figure 2d. The distance between ring centers is 4.99 Å measured as indicated above. As Table IV shows, the solvation energy favors dissociation. Again, this is qualitatively in accord with second virial coefficient data *(32)* and estimates of the osmotic coefficient *(33, 34, 35)*. Based on this single configuration only, the solvent effect in all three cases is somewhat overestimated.

The MMP2 energy and solvation energy were minimized by varying the intermolecular distance along a line through the center of one ring and perpendicular to the plane of the other ring. The distance was thereby reduced by .2 Å and the energy by 95 cal mol^{-1}. In spite of the small change, the overall effect of solvation is found to favor dissociation.

Ethane Dimer, Isobutane Dimer, n-Pentane Dimer and Methane-Ethane. Calculations for the equivalent sphere method were reported previously *(4)*, but are included for comparison. The dimer configuration chosen *(4)* was one of the more

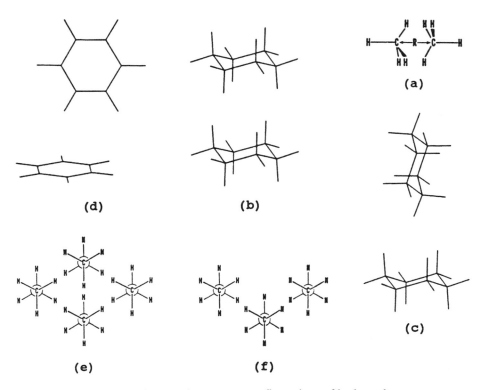

Figure 2. Dimer and aggregate configurations of hydrocarbons.

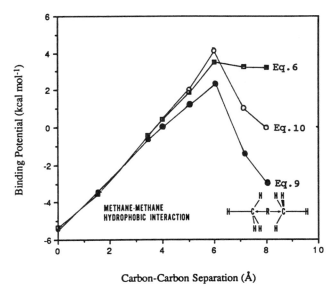

Figure 3. Methane-methane hydrophobic interaction according to Equations 6, 9, and 10.

closely packed possibilities to demonstrate an upper limit to the solvation effect. A slightly lessened contribution of the solvent effect toward dissociation was found upon the inclusion of local curvature as in Equation 9, but not in Equation 10. In all cases, a reduction in accessible surface area was observed upon association.

Ethane Aggregation. To demonstrate that under certain conditions, the solvation effect can favor strong association, the interaction of an ethane molecule with an aggregate of three strategically placed ethanes was considered. Table IV gives the relevant binding energies. The configurations are shown in Figures 2e and 2f. It can be seen, in accord with previous results *(29)*, that solution favors association by 2.3-2.7 kcal mol^{-1}. In this case, like the others above, association results in a reduction in area.

Methane-Methane. A range of intermolecular distances is considered for the methane-methane interaction. A plot of the association energy, based on the three methods for the calculation of the cavity energy, is shown in Figure 3. The advantage of the third term in Equation 10 involving the average squared mean curvature can now be easily seen. The solvent effect correctly goes to zero for large separations, as can be seen from the curve representing Equation 10. Equation 6, on the other hand, implies too low a curvature resulting in a high cavity energy upon separation. The last term in Equation 9 corresponds to only one point particle and gives too low an energy upon greater separation.

Beginning with the MM2 vacuum minimized distance of 3.45 Å between the methane molecules, a new minimum at 3.35 Å was obtained by minimizing again but including the solvent effect. This resulted in a small increase in dimer stability of 55 cal mol^{-1} over the results of Table IV. These results are for a closest approach interaction as shown in Figure 2a and neglect other methane-methane orientations at contact.

Modeling the Interaction Energy e_i.

A further simplification would be desirable so that a molecular dynamics calculation does not have to be done on each system for which e_i is needed. It would be useful to model the interaction so that with a set of rules and parameters, e_i for a molecule could be found, to sufficient accuracy. In order to obtain e_i, the solvent distribution function must be known so that e_i may be calculated from the integral over the intermolecular potential weighted with the solvent distribution function *(36)*, $\rho^{(2)}$:

$$e_i = \int \sum_{k=1}^{m} U_k \rho^{(2)} d\tau \qquad (16)$$

where U_k is the interatomic potential between the kth solute atom and a water molecule, m is the total number of solute atoms and $\rho^{(2)} = d_s g^{(2)}$ is the water distribution about the solute atom, where $g^{(2)}$ is the solute-solvent correlation function, d_s is the solvent density and $d\tau$ indicates integration over the relevant coordinates.

The method under investigation is based on the idea of obtaining the solvent distribution function for a molecular "fragment", e.g. a methane molecule, and then just as these fragments may be combined to build a large molecule, the distribution functions associated with the fragments are combined to produce the large molecule distribution function. The discussion below is in terms of the correlation function $g^{(n)}$ rather than the distribution function $\rho^{(n)}$.

Methane Dimer-Water Correlation Functions. It is first instructive to examine the solute-solvent correlation functions of some systems for which there are molecular dynamics results. In Figures 4 and 5, the distribution of water oxygens about the methane dimer (inter-carbon distance = 7.16 Å) as determined by a molecular dynamics calculation is plotted. These figures are different views of a cylindrical plot in which the methane-methane axis is taken as the cylinder axis x. The water molecule oxygen positions obtained from 270 ps dynamics run sampling 13,500 configurations using Amber 3.0A *(37)* were allocated to $2\pi r \times .25 \times .25$ Å^3 bins over three dimensional space. Adding the waters around the cylinder axis in this manner gives a representation of the correlation function in two dimensional space. The .25 Å grid spacing in the x and r directions can be seen from the figures. The intensity of the correlation function is then plotted in the z (upward) direction.

Referring to Figure 4, the position of the methanes is easily seen to be where the correlation function is zero. The methanes are oriented as shown in Figure 2a and a slight egg-shaped asymmetry may be noticed due to the axial hydrogens of the methanes pointing away from the center.

The first peak in the correlation function is clearly indicated as a wall of oxygen atom density around the methanes. Since methane is not exactly cylindrically symmetrical, the first peak may be slightly low due to being slightly smoothed out around the three non-axial hydrogens of each methane. A slight trough lies right behind the wall. After a very slight peak after this, the density beyond is fairly uniform.

Examination of Figures 4 and 5 indicate interesting secondary features. First, a build-up of water between the methanes can be seen. The presence of a water molecule between two hydrophobic solutes has been pointed out previously *(38, 39, 40, 41, 42)*. Second, a slight deepening of the trough of the correlation function behind the peak, can be seen equidistant from the methanes. This deepening can be seen in the center of Figure 5. Finally, there are two slight dips in the wall on either side of the peak, approximately at the position where the trough would intersect the crest. These features may be interpreted as an interference pattern between two methane centered methane-water correlation functions if a separate methane water correlation function is assumed to be present around each methane. However, the two assumed functions multiply rather than add. Figure 6 shows a simple methane-water correlation function, from a 510 ps molecular dynamics run on methane.

Figures 7 to 9 show results of other dynamics runs. In general, there is an increase of density of water between the methanes where possible. In Figure 8 at 3.45 Å, a slight increase in density can still be seen. Figure 9 shows the molecules coalesced to the unnatural distance of 1.54 Å.

Kirkwood Superposition Approximation. In order to produce similar results from a synthetic approach, it is necessary to make use of the Kirkwood superposition approximation *(43, 36)*. The superposition approximation gives the three body correlation function $g^{(3)}(123)$, in this case two methanes, designated 1 and 2, and a water molecule, designated 3, as a product of three two body correlation functions:

$$g^{(3)}(123) = g^{(2)}(12)g^{(2)}(13)g^{(2)}(23) \tag{17}$$

If the methanes are held fixed, then the superposition approximation may be written in unsymmetrical form *(43, 36)*:

$$g^{[3]}(123) = g^{(2)}(13)g^{(2)}(23) \tag{18}$$

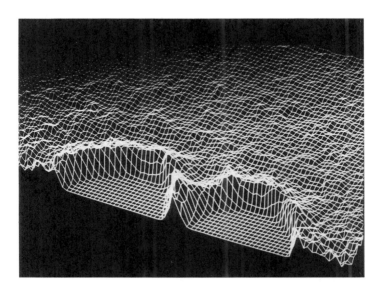

Figure 4. Methane-methane-water correlation function for carbon-carbon distance of 7.16 Å from a 270 ps molecular dynamics run.

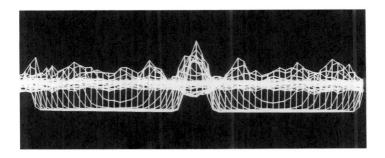

Figure 5. Front view of Figure 4. Carbon-carbon axis horizontal.

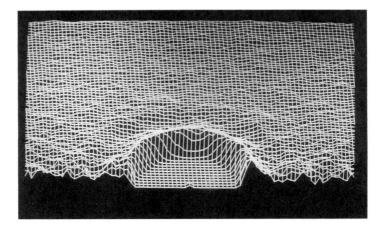

Figure 6. Methane-water correlation function from a 510 ps molecular dynamics run. Blip in front center indicates position of carbon atom.

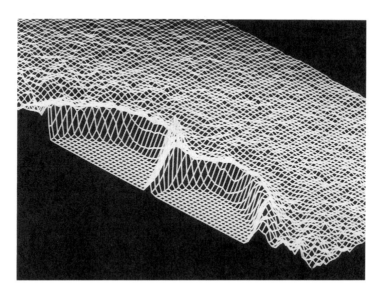

Figure 7. Methane-methane-water correlation function for carbon-carbon distance of 6 Å from a 300 ps molecular dynamics run.

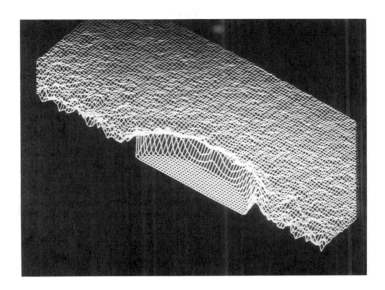

Figure 8. Methane-methane-water correlation function for carbon-carbon distance of 3.45 Å from a 210 ps molecular dynamics run.

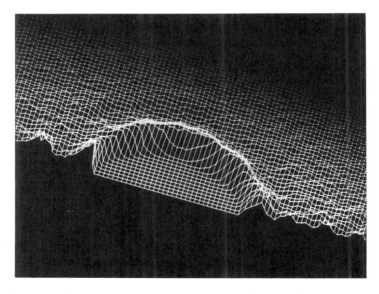

Figure 9. Methane-methane-water correlation function for carbon-carbon distance of 1.54 Å from a 240 ps molecular dynamics run.

Figure 6 is regarded as $g^{(2)}(13)$ and the symmetry operation $x \rightarrow -x$ is performed on this to get $g^{(2)}(23)$.

If these correlation functions, namely $g^{(2)}(13)$ and $g^{(2)}(23)$, are translated relative to each other so that the carbons are 7.16 Å apart, and then the product $g^{(2)}(13)$ $g^{(2)}(23)$ is formed, the result is shown in Figure 10. The product $g^{(2)}(13)$ $g^{(2)}(23)$ reproduces the true function (Figure 4) rather well, except for possibly an extra large build-up between the methanes. Of course, sampling on or very near the x axis is much less adequate than at larger r values.

The same procedure of obtaining a correlation function from the product of two methane water correlation functions was applied to the other methane-methane separations. The results are shown in Figures 11, 12, and 13. In all cases, the superposition of the peaks of the correlation functions lead to some build-up between the methanes, even in the 1.54 Å and 3.45 Å cases.

In Figure 13 for the 1.54 Å case, the crest-trough cancellation of the product $g^{(2)}(13)g^{(2)}(23)$ in the region where the water molecule is in line with and outside of the two methanes is apparent. This is intuitively incorrect; it is unlikely that the presence of one methane can change the water distribution on the far side of the other methane. This failure does not occur at the large methane-methane separations because there the crest only overlaps the other $g^{(2)}$ where it is equal to 1. At the smaller R values, it becomes an important difficulty. The failure of the superposition approximation for overlapping cavities has been discussed by Pratt and Chandler (44).

In building the function $g^{[3]}(123)$ from the superposition approximation, the shape of the function between the methanes is qualitatively correct, although in some cases too large. In the outer regions, it can lead to unphysical crest-trough canceling. Because of these considerations, the treatment is modified in the following way. In the region between the methanes, the superposition approximation is applied. On the other side of the methanes, the correlation function pertaining solely to that methane is retained:

(Domain 1)
$$g^{[3]}(123) = g^{(2)}(13) \cdot g^{(2)}(23) \quad \text{for} \quad x_1 < x < x_2 \tag{19}$$

(Domain 2)
$$g^{[3]}(123) = g^{(2)}(13) \quad \text{for} \quad x < x_1 \tag{20}$$

(Domain 3)
$$g^{[3]}(123) = g^{(2)}(23) \quad \text{for} \quad x > x_2 \tag{21}$$

Two cases were considered: I) x_1 and x_2 are given by the methane carbon positions, and II) they are each .75 Å inside the C-C distance. It was found that placing the domain boundary directly at the carbon atoms as in method I still gives too much of the field to the superposition approximation. Therefore, in method II, the boundary is moved so that it is about .75 Å (one half a van der Waals radius) inside the space between the two fragments. The result for the 1.54 Å case is shown in Figure 14.

In addition to a reasonable appearance of the correlation function, it is necessary that reasonable interaction energies may be calculated. In Table V, the energies for the several approximations are compared to the dynamics results. Method II gives good results for all practical distances. Even in the case of 1.54 Å, which is important for building a larger molecule from fragments, the agreement is satisfactory. At the unimportant distance of 0 Å, there is an ambiguity in how a correlation function can be constructed, due to the lack of symmetry.

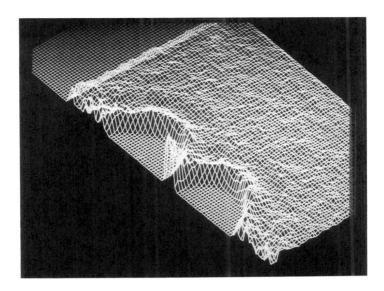

Figure 10. Methane-methane-water correlation function for carbon-carbon distance of 7.16 Å built from the superposition approximation.

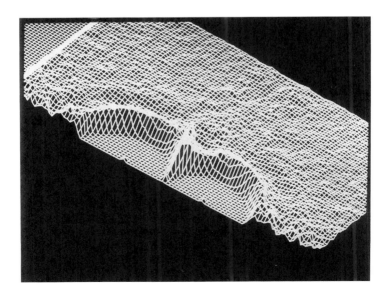

Figure 11. Methane-methane-water correlation function for carbon-carbon distance of 6 Å from the superposition approximation.

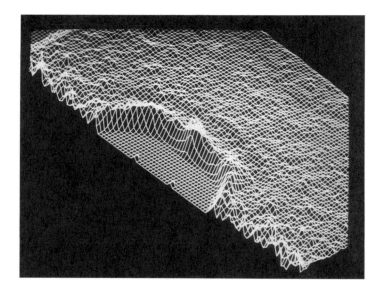

Figure 12. Methane-methane-water correlation function for carbon-carbon distance of 3.45 Å from the superposition approximation.

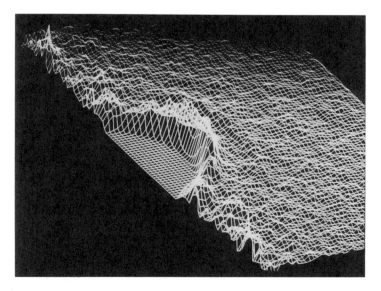

Figure 13. Methane-methane-water correlation function for carbon-carbon distance of 1.54 Å from the superposition approximation.

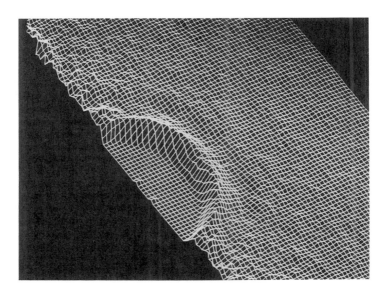

Figure 14. Methane-methane-water correlation function for carbon-carbon distance of 1.54 Å using Method II from Equations 19, 20 and 21.

The energies in Table V were calculated from the tabular form of $g^{[3]}$ as shown in the figures. The energy for the two methanes at infinite separation calculated in this manner agrees well with the direct dynamic results in Table I.

Table V. Methane Dimer-Water Interaction Energies Using Fragment Constructed Distribution Functions[a]

Carbon-Carbon Interatomic Distance	Results from MD	Superposition Approximation	Method I	Method II
0.0 Å	-7.070	-8.70	-5.61	-5.61
1.54	-6.380	-7.03	-6.73	-6.19
3.45	-5.766	-5.61	-5.79	-5.90
5.0	-5.866	-5.80	-5.73	-5.72
6.0	-6.081	-6.15	-6.11	-6.10
7.16	-6.383	-6.70	-6.61	-6.59
8.0	-6.355	-6.38	-6.35	-6.34
∞	-6.344	-6.37	-6.37	-6.37

[a]Distances in Angstroms. Energies in kcal mol^{-1}.

Application to Ethane. To form the function for ethane, two methane correlation functions are brought together from the other direction than in the cases above, to a distance of 1.54 Å. The result of the superposition approximation produced crest trough cancellations in the wrong regions. The energy was -4.87 kcal mol^{-1} in fair agreement with the result in Table I (-5.19 kcal mol^{-1}). Method II produced a correlation function in good agreement with the authentic function, similar to Figure 14. The energy for Method II was -4.81 kcal mol^{-1}.

Discussion

The methodology as developed so far has several positive attributes. It is in agreement with current calculations that show that the solvent effect favors methane association (40, 41, 42). It is also in agreement with current estimates that benzene and cyclohexane dimerize to a greater extent in the gas phase than in solution (33, 35). The treatment is not confined to aliphatic systems. Aromatic hydrocarbons may be treated along with aliphatic hydrocarbons with the same cavity potential parameters.

Molecules studied here produce a decrease in surface area and usually a decrease in cavity potential on dimerization. Only in the case of Equation 10 for the high energy contact dimer of cyclohexane and the benzene dimer is the cavity potential slightly higher than the monomer pair. Thus, the free energy of association is not directly related to reduction in area or cavity potential. While the direct surface area methods will always predict a solvent effect favoring dimerization, the method of this paper can predict dimer dissociation. However, it is possible by burying enough surface area as shown in Figures 2e and 2f to get significant binding.

It has been demonstrated that the equivalent sphere method of determining the cavity curvature gives results similar to the use of the explicit treatment of average mean curvature using a 2.8 Å radius sphere as a probe (12) in determining the curvature. For the simple hydrocarbon series solubilities, the inclusion of local curvature results in only a small improvement over the equivalent sphere method.

As a molecular complex separates into monomers, the free energy of the system goes to the correct separation limit only if Equation 10 is used. In the case of dimers, it may be seen that the hydrophobic interaction results for Equation 9 are mostly lower, and the Equation 10 results mostly higher than the results for the equivalent sphere method.

The average squared mean curvature is much more dependent on the detailed shape of a surface than is the average mean curvature. In the method above of treating the curvature of cavity surfaces in which measurements of $\overline{K_m^2}$ of the accessible surface are made with a 2.8 Å spherical probe, it should be noted that the calculated value of $\overline{K_m^2}$ is sensitive to the size of the probe.

Compared to potential of mean force calculations *(40, 41, 42)*, the barrier for pulling apart two molecules from a dimer configuration to separate monomers is too high. A second minimum is present in some *(40, 41, 42)* but not all *(45)* PMF calculations. One possible reason for the higher barrier is that the accessible surface is too convoluted in the barrier region and does not represent very well the first solvation shell.

In an effort to calculate e_i generally, it appears that it may be possible to build distribution functions for larger systems to a good approximation from transferable solvent distribution functions of fragments, or small molecules. A method is developed here in which novel features such as build-up and interference effects appear automatically between fragments and in the vicinity of their union. Table V shows that the molecular dynamics results for e_i are reasonably well reproduced by Equations 19-21.

Such functions built in this manner could be incorporated into continuum solvation models. The representation of the correlation functions by analytical functions would be a simplification of the treatment.

Acknowledgments

The author wishes to acknowledge discussions with Dr. James Metz and Dr. Bo Saxberg of Lilly Research Laboratories. The author also wishes to thank Professor W. Jorgensen for the use of the OPLS aliphatic parameters.

Literature Cited

1. Cramer, C. J.; Truhlar, D. G. *J. Am. Chem. Soc.* **1991**, *113*, p. 8305.
2. Cramer, C. J.; Truhlar, D. G. *Science* **1992**, *256*, p. 213.
3. Hermann, R. B. *J. Phys. Chem.* **1972**, *76*, p. 2754.
4. Hermann, R. B. *J. Comp. Chem.* **1993**, *14*, p. 741.
5. Honig, B.; Sharp, K.; Yang, A. *J. Phys. Chem.* **1993**, *97*, p. 1101.
6. Ooi, T.; Oobatake, M.; Némethy, G.; Scheraga, H. A. *Proc. Natl. Acad. Sci. U.S.A.* **1987**, *84*, p. 3086.
7. Still, W. C.; Tempczyk, A.; Hawley, R. C.; Hendrickson, T. *J. Am. Chem. Soc.* **1990**, *112*, p. 6127.
8. Kang, Y. K.; Gibson, K. D.; Némethy, G.; Scheraga, H. A. *J. Phys. Chem.* **1988**, *92*, p. 4739.
9. von Freyberg, B; Braun, W. *J. Comp. Chem.* **1993**, *14*, p. 510.
10. Warshel A. *J. Phys. Chem.* **1979**, *83*, p. 1640.
11. Wesson, L.; Eisenberg, D. *Protein Science* **1992**, *1*, p. 227.
12. Nicholls, A.; Sharp, K. A.; Honig, B. *Proteins: Struct. Funct. Genet.* **1991**, *11* p. 281.
13. Lee, B.; Richards, F. M. *J. Mol. Biol.* **1971**, *55*, p. 379.
14. Eley, D. D. *Trans. Faraday Soc.* **1939**, *35*, p. 1281.
15. Ulig, H. H. *J. Phys. Chem.* **1937**, *41*, p. 1215.
16. Choi, D. S.; Jhon, M. S.; Eyring, H. *J. Chem. Phys.* **1970**, *53*, p. 2608.

17. Pierotti, R. A. *J. Phys. Chem.* **1963**, *67*, p. 1840.
18. Pierotti, R. A. *J. Phys. Chem.* **1965**, *69*, p. 281.
19. Sinanoglu, O. In *Molecular Associations in Biology*; Pullman, B., Ed.; Academic Press: New York, N.Y., 1968; p. 427.
20. Hermann, R. B. *J. Phys. Chem.* **1975**, *79*, p. 163. Erratum, *J. Phys. Chem.* **1975**, *79*, p. 3080.
21. Jorgensen, W. L.; Chandrasekhar, J.; Madura, J. D.; Impey, R. W.; Klein, M. L. *J. Chem. Phys.* **1983**, *79*, p. 926.
22. Potsma, J. P. M.; Berendsen, J. C.; Haak, J. R. *Faraday Symp. Chem. Soc.* **1982**, *17*, p. 55.
23. Reiss, H.; Frisch, H. L.; Lebowitz, J. L. *J. Chem. Phys.* **1959**, *31*, p. 369.
24. Reiss, H. *Adv. Chem. Phys.* **1965**, *9*, p. 1.
25. *CRC Handbook of Chemistry and Physics*; Lide, D. R., Ed.; Chemical Rubber Publishing Co.: Cleveland, OH, 1991.
26. Jorgensen, W. L.; Severance, D.L. *J. Am. Chem. Soc.* **1990**, *112*, p. 4768.
27. Ben Naim, A.; Marcus, Y. *J. Chem. Phys.* **1984**, *81*, p. 2016.
28. *The Chemists Companion: A Handbook of Practical Data, Techniques and References*; Gordon, A. J.; Ford, R. A., Eds.; John Wiley & Sons: New York, N.Y., 1972.
29. Hermann, R. B. In *Seventh Jerusalem Symposium on Molecular and Quantum Pharmacology*; Bergman, E; Pullman, E., Eds.; D. Reidel: Dordrecht, Holland, 1974; p. 441.
30. DoCarmo, M. P. *Differential Geometry of Curves and Surfaces*; Prentice Hall: Englewood Cliffs, N.J., 1976.
31. Burkert, U.; Allinger, N. L. *Molecular Mechanics*: ACS Monograph 177; American Chemical Society, Washington, D.C., 1982.
32. Dymond, J. H.; Smith, E. B. *Virial Coefficients of Gases: A Critical Compilation*; Oxford University Press: New York, N.Y., 1980; 2nd Ed.
33. Wood, R. H.; Thompson, P. T. *Proc. Natl. Acad. Sci. U.S.A.* **1990**, *87*, p. 946.
34. Tucker, E. E.; Lane, E. H.; Christian, S. D. *J. Solution Chem.* **1981**, *10*, p. 1.
35. Rossky, P. J.; Friedman, H. L. *J. Phys. Chem.* **1980**, *84*, p. 587.
36. Hill, T. *Statistical Mechanics*; McGraw-Hill: New York, N.Y., 1956.
37. Weiner, S. J.; Kolmann, P. A.; Nguyen, D. T.; Case, D.A. *J. Comp. Chem.* **1986**, *7*, p. 230.
38. Geiger, A.; Rahman, A.; Stillinger, F. H. *J. Chem. Phys.* **1978**, *70*, p. 263.
39. Pangali, C.; Rao, M.; Berne, B. J. *J. Chem. Phys.* **1979**, *71*, p. 2982.
40. Jorgensen, W. L.; Buckner, J. K.; Boudon, S.; Tirado-Rives, J. *J. Chem. Phys.* **1988**, *89*, p. 3742.
41. Ravishanker, G; Mezei, M.; Beveridge, D. L. *Faraday Symp. Chem. Soc.* **1982**, *17*, p. 79.
42. Smith, D. E.; Haymet, A. D. J. *J. Chem. Phys.* **1993**, *98*, p. 6445.
43. Kirkwood, J. G. *J. Chem. Phys.* **1935**, *3*, p. 300.
44. Pratt, L. R.; Chandler, D. *J. Chem. Phys.* **1977**, *67*, p. 3683.
45. van Belle, D.; Wodak, S. J. *J. Am. Chem. Soc.* **1993**, *115*, p. 647.

RECEIVED April 25, 1994

BIOPOLYMERS AND INTERFACES

Chapter 24

Treatment of Hydration in Conformational Energy Calculations on Polypeptides and Proteins

Harold A. Scheraga

Baker Laboratory of Chemistry, Cornell University,
Ithaca, NY 14853–1301

Calculations of the conformations of proteins in aqueous solution require a treatment of the influence of water on the dissolved solute molecule. Because present-day computers cannot accommodate a complete search of conformational space if the surrounding water molecules are treated explicitly, resort is had to solvent-shell or solvent-exposed surface-area models, with parameters for the component amino acids obtained primarily from experimental data on free energies of hydration but also from Monte Carlo or molecular dynamics simulations of solutions of small-molecule solutes. These models are presented here, and some of the thermodynamic parameters, viz. those for hydrophobic interactions, are rationalized in terms of theoretical treatments that yield results that have been checked by experiment.

In order to understand the molecular mechanisms of chemical processes in aqueous solution (*1*), it is necessary to have potential functions for the solute molecule and for its interactions with the surrounding water molecules (*2,3*). Water has a considerable influence on the structure and reactivity of solute molecules. For example, in many protein structures, positively-charged arginine side chains are close to each other, and the strong electrostatic repulsion that would ensue in vacuum is modified by bridging water molecules, as demonstrated by semi-empirical quantum mechanical calculations on a model of two guanidinium ions surrounded by several water molecules (*4*).

For an understanding of the role of hydration, one approach uses some type of quantum mechanical calculation, treating the solvent as a polarizable continuum (*5*), to obtain free energies of hydration. A recent example is the analysis of heterocyclic tautomerization in aqueous solution (*6*). Alternatively,

0097–6156/94/0568–0360$08.00/0

use is made of molecular mechanics, with the solvent molecules treated explicitly (*7-10*), or implicitly by either a solvent-shell model (*11*) or by a solvent-exposed surface-area model (*12,13*), especially for large-molecule solutes of arbitrary shape. Integral equation methods have also been applied to treat solvation; for example, use has been made of XRISM (extended reference interaction-site model) methods to calculate pair correlation functions of protein atoms with surrounding water molecules, and thereby extract solvation free energies for each residue (*14,15*). In addition, for treating electrostatic interactions of such large molecules in solution, the solvent is considered to be a continuum dielectric, and the non-linear Poisson-Boltzmann equation is solved numerically (*16,17*).

For the particular application to aqueous solutions of macromolecules such as proteins, the quantum mechanical approach is not feasible, and molecular mechanics with an explicit treatment of the water molecules, as indicated schematically in Figure 1, is possible only if limited changes in the conformation of the solute are of interest. If, however, large changes in

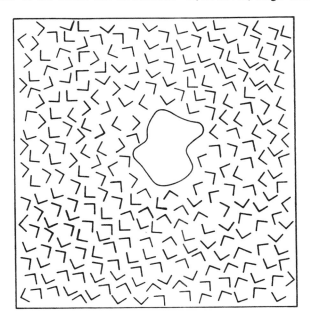

Figure 1. Schematic representation of a macromolecule solute in a box of water molecules.

protein conformation are allowed as, for example, in studying protein folding in aqueous solution, then the model of Figure 1 cannot be treated by present-day computers, and resort must be had to the solvent-shell model or the solvent-exposed surface-area model. For either approach, it is necessary to have solvation free energies for the components of the amino acids from which the protein is constituted. While such thermodynamic quantities can be obtained for small-molecule solutes by quantum mechanical calculations (*6,18,19*),we have used experimental free energies of hydration in accord with a similar use of experimental data to parameterize our potential function (*20*) for amino acid and polypeptide solutes. Therefore, in this paper, we will first describe the solvent-shell and solvent-exposed surface-area models, and then their parameterization. Finally, some applications to aqueous systems will be discussed.

Solvent-Shell Model

Our earliest use of the solvent-shell model (*21*) was made in 1967, and has been implemented more recently with improvements in the geometry of overlapping spheres (*22-27*) and with more recent experimental data on free energies of hydration (*11,28-32*).

Figure 2 illustrates the free energy penalty involved due to the removal of water when the van der Waals sphere of a solute group overlaps the

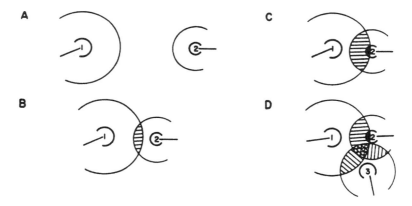

Figure 2. A schematic drawing illustrating various stages of the approach of two solute groups 1 and 2. A, No overlaps. B, Overlap of the hydration spheres of groups 1 and 2. C, Overlap of the van der Waals sphere of group 2 with the hydration sphere of group 1. D, Overlap of three hydration spheres of solute groups 1, 2 and 3. (Reproduced with permission from reference 33. Copyright 1981 New York Academy of Sciences.)

hydration shell of another solute group. In Figure 2B, the free energy of hydration of groups 1 and 2 is unaltered; in Figure 2C, however, while the hydration volume of group 2 is unaltered, that of group 1 is reduced by the densely shaded volume of overlap. Figure 2D illustrates a multiple-site interaction (*33*).

This model has been used to compute the contribution of hydration in molecular mechanics calculations on polypeptides and proteins in water. As an example, we cite the role of water in the helix-coil transition of polyamino acids (*34*), such as poly-L-valine, where the helix content *increases* as the temperature is raised because of the well-known behavior of the temperature dependence of hydrophobic interactions (in this case, between the nonpolar side chains of the helix). As another example, the experimental change in energy to convert the triple-helical collagen-like poly(Gly-Pro-Pro) to the single-chain statistical-coil form is 1.95 kcal/mol per Gly-Pro-Pro tripeptide unit (*35*); a theoretical value, making use of the shell model to treat hydration, is 2.4 kcal/mol (*36*), in fair agreement with the experimental value. If hydration is not included, the theoretical value is much higher, viz., 5.0 kcal/mol (*36*).

As a third example, Han and Kang (*37*) have recently used the hydration shell model to compute the ratio of cis-to-trans isomers of N-acetyl-N'-methylamides of Pro-X dipeptides, where X is a series of amino acids. These authors report good agreement with experimental data for this ratio and also for the theoretical propensity of the Pro-X dipeptides to adopt β-bend conformations.

Solvent-Exposed Surface-Area Model

Alternatively, solvent-exposed surface area is used as a basis to assess the hydration contribution, using surface (instead of volume) free-energy density parameters (see Figure 3). Various models are used to compute the solvent-

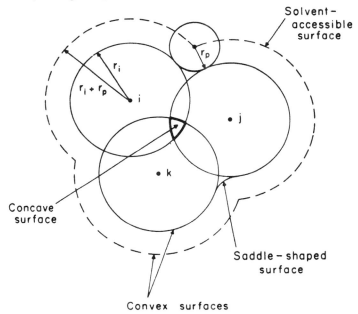

Figure 3. The solvent-accessible surface (shown as a dashed line) for atoms i, j, and k with a probe radius of r_p. (Reproduced with permission from reference 13. Copyright 1992 John Wiley.)

exposed surface area (13,38) in order to compute the free energy of hydration (39-41) of a solute molecule. Vila et al (42) have evaluated a number of models and several sets of hydration parameters. From a computational point of view, the model of Perrot et al (13) is much faster than that of Connolly (38), and Williams et al (43) have used it to show that computed structures of bovine pancreatic trypsin inhibitor, with hydration included with a solvent-exposed surface-area model, show better agreement with the X-ray structure than do those in which hydration is not taken into account. The increased speed of the algorithm of Perrot et al (13) makes it a practical one for inclusion (together with a molecular mechanics algorithm) in a procedure to minimize the sum of the energy and hydration free energy of a macromolecule. However, such minimization algorithms, which involve the computation of first derivatives, sometimes encounter discontinuities in the first derivatives of the accessible surface area and, hence, in the hydration free energy. Wawak et al (44) have recently shown that there are only two situations in which such discontinuities arise, and work is currently in progress to surmount these problems. Work is also in progress (32) to improve the hydration parameters for polypeptides by making use of a recent compilation of free energies of hydration of small molecules (31).

By calculating the free energy of solvation for the residues of melittin with both the accessible surface-area approach and the XRISM method, Kitao et al showed that the two methods agree qualitatively, but that the XRISM approach consistently overestimates the solvation free energy (15).

Rationalization of Some Hydration Parameters

While our approach is based on the use of experimental free energies of hydration to parameterize either a solvent-shell or solvent-exposed surface-area model, it is of interest to conclude this article with a theoretical rationalization (and corresponding experimental verification) of some of these parameters, viz., as an example, those involving the hydrophobic interaction between nonpolar groups in water.

Our earlier theoretical treatment of hydrophobic interactions (45,46) was based on a statistical mechanical theory of the thermodynamic properties of liquid water (47) and of aqueous solutions of hydrocarbon (48). The free energy of formation of a pair of nonpolar groups making maximum contact in water was expressed as

$$\Delta G^\circ = a + bT + cT^2 \qquad (1)$$

with a corresponding expression for the volume change, ΔV°. Near room temperature, $\Delta V^\circ > 0$, $\Delta G^\circ < 0$, $\Delta S^\circ > 0$ and $\Delta H^\circ > 0$. The unfavorable enthalpy of formation is more than counterbalanced by the positive entropy, resulting in a favorable free energy arising from changes in the surrounding water structure as the two nonpolar groups come into contact.

As an illustration of the effect of temperature, the thermodynamic parameters are given as a function of temperature in Table I for the leucine-isoleucine hydrophobic interaction. These parameters were obtained from equation 1, with a = 7290, b = 47.8 and c = 0.0660 for $\Delta G°$ in cal/mol.

Table I. Thermodynamic Parameters for the Formation of a Leucine-Isoleucine Hydrophobic Interaction of Maximum Bond Strength (45)

t °C	$\Delta G°$ kcal/mol	$\Delta H°$ kcal/mol	$\Delta S°$ e.u.
0	-0.8	2.3	11.6
10	-0.9	2.0	10.4
20	-1.0	1.6	9.2
25	-1.0	1.4	8.5
30	-1.1	1.2	7.8
40	-1.2	0.8	6.5
50	-1.2	0.4	5.1
60	-1.3	-0.1	3.6
70	-1.3	-0.6	2.1

Source: Reprinted from reference 45. Copyright 1962 American Chemical Society.

Table II illustrates the range of the theoretical thermodynamic parameters at 25°C for several pairs of interacting nonpolar side chains of amino acids in maximum contact.

Table II. Theoretical Thermodynamic Parameters for Hydrophobic Interaction at 25°C (45)

Side Chains	$\Delta G°$ kcal/mol	$\Delta H°$ kcal/mol	$\Delta S°$ e.u.
Alanine-alanine	-0.3	0.4	2.1
Isoleucine-isoleucine	-1.5	1.8	11.1
Phenylalanine-leucine	-0.4	0.9	4.7
Phenylalanine-phenylalanine	-1.4	0.8	7.5

Source: Reprinted from reference 45. Copyright 1962 American Chemical Society.

A variety of experiments have been carried out to verify these thermodynamic parameters. One of these involves the dimerization of carboxylic acids in aqueous solution. From the increase in the observed dimerization constant with increasing chain length, it has been suggested (49) that, in aqueous solution, the dimers are side-by-side rather than cyclic as observed in the gas phase. Thus, dimerization involves single hydrogen bonds between the carboxyl groups, and hydrophobic interactions between the nonpolar portions. The experimental and theoretical values of $\Delta G°$ for the hydrophobic interaction are compared in Table III.

Table III. Free Energy of Hydrophobic Interaction in the Dimerization
of Carboxylic Acids at 25°C

	$\Delta G°$ (kcal./mole)				
	Experimental[a]			Calculated[b]	
Side-Chain	(1)	(2)	(3)	(4)	(5)
CH_3^-	-0.79	-0.95	-0.80	-0.70	-0.70
$CH_3CH_2^-$	-1.03	-1.06	-1.09	-0.90	-1.00
$CH_3CH_2CH_2^-$	-1.31	-1.47	-1.41	-1.15	-1.35
$C_6H_5CH_2^-$	-	-	-1.57	-1.45	-1.63

[a]These three columns correspond to three different sets of experimental data
(see Reference 49).
[b]These two columns correspond to two slight variations in the theory
(see Reference 49).

Similar binary complexes were examined by fluorescence quenching of
phenols with carboxylates, from which the thermodynamic parameters for the
hydrophobic interaction were extracted (50). These are shown in Table IV
together with the theoretical values.

As a final example, we cite some experimental data for the volume
decrease when nonpolar groups are added to water. The data in Table V
were obtained from experiments on a homologous series (51) in which the
data for the first member of the series were subtracted in order to obtain the
contribution of the nonpolar group (52). The agreement between
experimental and theoretical data is fairly good, especially at the higher
temperatures. Of course, the volume change accompanying the hydrophobic
interaction will have the opposite, i.e. positive, sign. Such increases in volume
are observed in association reactions (53) where hydrophobic interactions may
be involved.

From the examples cited above, it appears that the theory of reference
45 provides a good rationalization of the origin of the thermodynamic
parameters for hydrophobic interactions between nonpolar groups in water.
The theory of reference 45 was based on a model in which partial clathrate-
like cages of water molecules surround the nonpolar group. This model has
subsequently been validated by numerous Monte Carlo simulations of aqueous
solutions of hydrocarbons (54,55); these include, e.g., simulations of methane
in water, in which $C\cdots O$ pair correlation functions (compared to $O\cdots O$ pair
correlation functions in pure water) indicate the presence of clathrate-like
structures.

Table IV. Thermodynamic Parameters for Pairwise Hydrophobic Interaction between the Nonpolar Side Chains of Sodium Carboxylates and Phenolic Compounds at 25° (50)

System	Experimental data			Theoretical data		
	$\Delta G°$	$\Delta H°$	$\Delta S°$	$\Delta G°$	$\Delta H°$	$\Delta S°$
	cal/mol	cal/mol	eu	cal/mol	cal/mol	eu
Phenol-acetate	-374[a]	100 ± 450[a]	2.1 ± 1.8[a]	-349	230	1.9
-propionate	-451	180 ± 330	2.6 ± 1.4	-426	700	3.8
-isobutyrate	-414	860 ± 280	4.8 ± 1.3	-465	940	4.7
-butyrate	-649	940 ± 240	5.8 ± 1.1	-735	670	4.7
Xylenol-acetate	-479	470 ± 380	3.6 ± 1.7	-472	550	3.4
-propionate	-552	880 ± 390	5.2 ± 1.7	-580	750	4.5
-isobutyrate	-693	980 ± 590	6.1 ± 2.3	-687	1050	5.8
-butyrate	-761	350 ± 400	4.2 ± 1.6	-786	780	5.3

[a]Probable error. The probable error in $\Delta G°$ is ± 100 cal/mol.
Source: Reprinted from reference 50. Copyright 1968 American Chemical Society.

**Table V. Volume Decrease (in c.c./mole) Accompanying the Addition of
Nonpolar Groups to Water (51)**

| | $-\Delta V$ | | | | | | | |
| | 0°C | | 20°C | | 40°C | | 50°C | |
Group	Exp.	Theor.	Exp.	Theor.	Exp.	Theor.	Exp.	Theor.
CH_3	1.1	1.9	1.2	1.5	1.2	1.3	1.3	1.1
C_2H_5	1.9	3.7	2.1	3.1	2.2	2.5	2.3	2.3
C_3H_7	2.9	5.6	3.1	4.6	3.3	3.8	3.5	3.4

Some Applications of the Foregoing Theory of Hydrophobic Interactions

Two applications of the foregoing theory (45) are presented here. The first
one treats the entropy changes accompanying association reactions of proteins
(56). If hydration is neglected, and the entropy change, ΔS_{assoc}, is attributed
only to loss of translational and rotational freedom, then $\Delta S_{assoc} = -122$ e.u.
for a molecule the size of insulin. However, by taking hydration (and internal
degees of freedom of the insulin dimer) into account, the computed value of
ΔS_{assoc} is $-10(\pm 8)$ e.u., in agreement with the experimental value (56).

As a second application, nucleation (or chain-folding-initiation) sites
were computed for initiating protein folding (57). The model was based on
hydrophobic interactions, to remove the nonpolar groups from contact with
water. The free energy change was calculated for the conversion of extended
chain segments into hairpin-like structures with hydrophobic interactions
between the side chains. By considering all segments along the whole chain,
the most likely chain-folding-initiation sites were computed for several
proteins. The computed sites for bovine pancreatic ribonuclease A (57) were
subsequently verified by experiment and also by an alternative treatment
based on triangular contact maps (58).

Concluding Remarks

While the data in Tables III-V provide a rationalization of the early theoretical treatment of the hydrophobic interaction, and the subsequent application of simulation methods to aqueous systems, reliance is being placed, at
least in the near future, on two alternative approaches to obtain the thermodynamic parameters for the hydration of both polar and nonpolar groups.
These are the use of experimental data, on the one hand, and Monte Carlo
or molecular dynamics simulations on the other. These will improve the data
for use in either a solvent-shell or a solvent-exposed surface-area model. As
potential functions improve, so will the simulation results. Finally, if

computer power increases by several orders of magnitude, it may then become possible to search the conformational space of a macromolecule solute by using explicit water molecules, as in Figure 1, instead of relying on these volume or surface models.

Literature Cited

1. Dzingeleski, G. D.; Wolfenden, R. *Biochemistry* **1993**, *32*, 9143.
2. Scheraga, H. A. *Accts. Chem. Res.* **1979**, *12*, 7.
3. Némethy, G.; Peer, W. J.; Scheraga, H. A. *Ann. Rev. Biophys. Bioeng.* **1981**, *10*, 459.
4. Magalhaes, A.; Maigret, B.; Hoflack, J.; Gomes, J. N. F.; Scheraga, H. A. *J. Protein Chem.* **1994**, in press.
5. Grant, J. A.; Williams, R. L.; Scheraga, H. A. *Biopolymers* *30*, **1990**, 929.
6. Cramer, C. J.; Truhlar, D. G. *J. Am. Chem. Soc.* **1993**, *115*, 8810.
7. Rapaport, D. C.; Scheraga, H. A. *J. Phys. Chem.* **1982,** *86*, 873.
8. Jorgensen, W. L.; Ravimohan, C. *J. Chem. Phys.* **1985**, *83*, 3050.
9. Kollman, P. A.; Merz, K. M., Jr. *Acct. Chem. Res.* **1990**, *23*, 246.
10. Straatsma, T. P.; McCammon, J. A. *J. Chem. Phys.* 1991, *95*, 1175.
11. Kang, Y. K. Gibson; K. D.; Némethy; G.; Scheraga, H. A. *J. Phys. Chem.* **1988**, *92*, 4739.
12. Perrot, G.; Maigret, B. *J. Mol. Graph.* **1990**, *8*, 141.
13. G. Perrot, G.; Cheng, B.; Gibson, K. D.; Vila, J.; Palmer, K. A.; Nayeem, A.; Maigret, B.; Scheraga, H. A. *J. Comput. Chem.*, **1992**, *13*, 1.
14. Lee, P. H.; Maggiora, G. M. *J. Phys. Chem.* **1993**, *97*, 10175.
15. Kitao, A.; Hirata, F.; Gō, N. *J. Phys. Chem.* **1993**, *97*, 10231.
16. Jayaram, B.; Sharp, K. A.; Honig, B. *Biopolymers* **1989**, *28*, 975.
17. Vorobjev, Y. N.; Grant, J. A.; Scheraga, H. A. *J. Am. Chem. Soc.* **1992**, *114*, 3189.
18. Kweon, G. Y.; Scheraga, H. A.; Jhon, M. S. *J. Phys. Chem.* **1991**, *95*, 8964.
19. Chipot, C.; Angyan, J. G.; Maigret, B.; Scheraga, H. A. *J. Phys. Chem.* **1993**, *97*, 9797.
20. G. Némethy, G.; Gibson, K. D.; Palmer, K. A.; Yoon, C. N.; Paterlini, G.; Zagari, A., Rumsey, S.; Scheraga, H. A. *J. Phys. Chem.* **1992**, *96*, 6472.
21. Gibson, K. D.; Scheraga, H. A. *Proc. Natl. Acad. Sci., U.S.A.* **1967**, *58*, 420.
22. Kang, Y. K.; Némethy, G.; Scheraga, H. A. *J. Phys. Chem.* **1987**, *91*, 4105.
23. Kang, Y. K.; Némethy, G.; Scheraga, H. A. *J. Phys. Chem.* **1988**, *92*, 1382.
24. Gibson, K. D,; Scheraga, H. A. *J. Phys. Chem.* **1987**, *91*, 4121.
25. Gibson, K. D,; Scheraga, H. A. *J. Phys. Chem.* **1987**, *91*, 6326.

26. Gibson, K. D.; Scheraga, H. A. *Mol. Phys.* **1987**, *62*, 1247.
27. Gibson, K. D.; Scheraga, H. A. *Mol. Phys.* **1988**, *64*, 641.
28. Kang, Y. K.; Némethy, G.; Scheraga, H. A. *J. Phys. Chem.* **1987**, *91*, 4109.
29. Kang, Y. K.; Némethy, G.; Scheraga, H. A. *J. Phys. Chem.* **1987**, *91*, 4118.
30. Kang, Y. K.; Némethy, G.; Scheraga, H. A. *J. Phys. Chem.* **1987**, *91*, 6568.
31. Abraham, M. H.; Whiting, G. S.; Fuchs, R.; Chambers, E. J. *J. Chem. Soc., Perkins Trans. II* **1990**, 291.
32. Augspurger, J. D.; Scheraga, H. A., work in progress.
33. Paterson, Y.; Némethy, G.; Scheraga, H. A. *Annals N. Y. Acad. Sci.* **1981**, *367*, 132.
34. Gō, M.; Scheraga, H. A. *Biopolymers* **1984**, *23*, 1961.
35. Shaw, B. R.; Schurr, J. M. *Biopolymers* **1975**, *14*, 1951.
36. Némethy, G.; Scheraga, H. A. *Biopolymers* **1989**, *28*, 1573.
37. Han, S. J.; Kang, Y. K. *Int. J. Peptide Protein Res.* **1993**, *42*, 518.
38. Connolly, M. L. *J. Appl. Cryst.* **1985**, *18*, 499.
39. Eisenberg, D.; McLachlan, A. D. *Nature* **1986**, *319*, 199.
40. Ooi, T.; Oobatake, M.; Némethy, G.; Scheraga, H. A. *Proc. Natl. Acad. Sci., U.S.A.* **1987**, *84*, 3086.
41. Ooi, T.; Oobatake, M.; Némethy, G.; Scheraga, H. A. *Proc. Natl. Acad. Sci., U.S.A.* **1987**, *84*, 6015.
42. Vila, J.; Williams, R. L.; Vasquez, M.; Scheraga, H. A. *Proteins: Structure, Function, and Genetics* **1991**, *10*, 199.
43. Williams, R. L.; Vila, J.; Perrot, G.; Scheraga, H. A. *Proteins: Structure, Function, and Genetics* **1992**, *14*, 110.
44. Wawak, R. J.; Gibson, K. D.; Scheraga, H. A. *J. Math. Chem.* **1994**, in press.
45. Némethy, G.; Scheraga, H. A. *J. Phys. Chem.* **1962**, *66*, 1773.
46. Némethy, G.; Scheraga, H. A. *J. Phys. Chem.* **1963**, *67*, 2888.
47. Némethy, G.; Scheraga, H. A. *J. Chem. Phys.* **1962**, *36*, 3382.
48. Némethy, G.; Scheraga, H. A. *J. Chem. Phys.* **1962**, *36*, 3401.
49. Schrier, E. E.; Pottle, M.; Scheraga, H. A. *J. Am. Chem. Soc.* **1964**, *86*, 3444.
50. Kunimitsu, D. K.; Woody, A. Y.; Stimson, E. R.; Scheraga, H. H. *J. Phys. Chem.* **1968**, *72*, 856.
51. Friedman, M. E.; Scheraga, H. A. *J. Phys. Chem.* **1965**, *69*, 3795.
52. Scheraga, H. A. *Ann. N. Y. Acad. Sci.* **1965**, *125*, 253.
53. Lauffer, M. A. *The Molecular Basis of Neoplasia*; University of Texas Press, Austin, Texas, **1962**, 180.
54. Swaminathan, S.; Beveridge, D. L. *J. Am. Chem. Soc.* **1977**, *99*, 8392.
55. Owicki, J. C.; Scheraga, H. A. *J. Am. Chem. Soc.* **1977**, *99*, 7413.
56. Steinberg, I. Z.; Scheraga, H. A. *J. Biol. Chem.* **1963**, *238*, 172.
57. Matheson, R. R., Jr.; Scheraga, H. A. *Macromolecules* **1978**, *11*, 819.
58. Némethy, G.; Scheraga, H.A. *Proc. Natl. Acad. Sci., U.S.* **1979**, *76*, 6050.

RECEIVED June 17, 1994

Chapter 25

Hydrophobic–Hydrophilic Forces in Protein Folding

A. Ben-Naim

Department of Physical Chemistry, Hebrew University of Jerusalem, Jerusalem 91904, Israel

The total force exerted on each atom, or group of atoms, of a protein can be split into two parts: A direct force originating from the protein itself and an indirect part, induced by the solvent.

We focus on the solvent induced part of the force exerted on a typical hydrophobic (say methyl) and hydrophilic (say hydroxyl) groups. It is argued that the solvent induced force between hydrophilic groups are stronger than the corresponding forces between hydrophobic groups. Some recent simulations of the indirect force confirm this conclusion. The implication of these findings to biochemical processes is discussed.

1. Introduction

The problem of "how" and "why" protein folds into an almost unique three dimensional structure has become a central problem in theoretical biochemistry.[1] Most attempted answers to this question focus on the characterization of the structural intermediates in the protein folding pathway.[2,3] A complete knowledge of all structural intermediates will give us an answer to the question of "how" protein folds, leaving the "why" part of the question unanswered.

Clearly, the "why" question is a more fundamental one in the problem of protein folding. This question can be further subdivided into two questions: One, why protein folds along a particular pathway? Second, why the protein eventually attains a specific three dimensional structure?

To answer the first question we need to know all the forces that operate on a protein at any given configuration. These forces will determine the trajectory of motion of the protein within its configurational space.[4] Clearly, because of the incessant thermal fluctuations, a protein cannot follow a precise trajectory. Instead, we can think of a corridor of trajectories (the width of which depending on the temperature) along which the protein folds. Knowledge of the precise forces at each moment will also tell us "how" protein is most likely to fold, i.e., what are all the intermediates in the folding pathway. We shall henceforth refer to these forces as the *dynamic force*.

0097–6156/94/0568–0371$08.00/0

To answer the second question we need not know the dynamic force at each stage of the folding pathway, nor do we need to know the structural intermediates. All we need is to examine the difference in the Gibbs energy of the initial (unfolded) and the final (folded) states. We shall henceforth refer to this change in Gibbs energy as the *thermodynamic force*. In a recent review article, entitled "Dominant forces in protein folding"[5] only thermodynamic forces are discussed, almost nothing is said on the dynamic forces.

In this paper we shall address ourselves mainly to the examination of *dynamical forces*, and see if at the present stage of our knowledge we can answer the question of which forces are the dominant ones. Once we know the dynamical forces we could, in principle, tell "how" the protein will proceed from one intermediate structure to another, leading to the ultimate 3-D structure. The latter might not coincide with the most stable state, i.e., the state of the lowest Gibbs energy.

In the next section, we present a simple illustration of the difference between the dynamic and the thermodynamic force. We then focus on the former, which consists of two parts, direct forces originating from the protein itself, and indirect, or solvent-induced forces. In section 3 we present an exact argument showing why forces between hydrophilic HΦI groups are expected to be much stronger than those between hydrophobic HΦO groups. We then show that the same argument, though not exactly the same, can be applied to real aqueous solutions.

2. Thermodynamic versus dynamic force

There seems to be some confusion regarding the concepts of thermodynamic and dynamic forces. The difference between the two can be demonstrated by a very simple example, figure 1.

Consider first two particles in vacuum constrained to move along the line connecting their centers. Figure 1-a shows the intermolecular potential energy as a function of the distance R between the two particles.

For any initial distance, say R_1 or R_2, the system will tend to reach the equilibrium distance R_e. We say that there is a *thermodynamic* force driving the system from any initial state (R) to the final state R_e whenever

$$U(R_e) - U(R) < 0 \qquad (2.1)$$

In this particular example the actual or the *dynamic* force is attractive at $R = R_1$, but repulsive at $R = R_2$. However, since in both cases the dynamic force *drives* the system towards R_e, we can loosely say that both the thermodynamic and the dynamic force *drives* the system *towards* the same goal, R_e. In this sense, and in this particular example, we may interchange the two concepts of forces.

Such an interchange is not valid in general. We demonstrate this point in figure 1-b. Suppose that the same two particles, constrained to move along the same line, are immersed in a solvent. If we fix the temperature T and the pressure P of the solvent, then the criterion for a thermodynamic force (2.1) is replaced by

$$G(R_e) - G(R) < 0 \qquad (2.2)$$

where $G(R)$ is the Gibbs energy of the system having the two particles at a fixed distance R.

Indeed, for any initial distance R, the *thermodynamic* force, drives the system into the state of lowest Gibbs energy. However, in contrast to the previous example, the actual or the dynamic force at each initial distance *does not* drive the system towards R_e. Thus at R_1, the dynamic force is attractive, driving the system towards

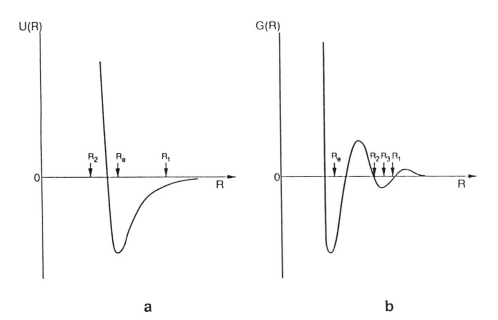

a   b

Figure 1
    The Gibbs energy function $G(R)$ for two simple solutes as a function of R. a)
In vacuum, where $G(R) = U(R)$, and b) In a solvent. R_e is the location of the lowest
minimum of $G(R)$. The average force between the two solutes is attractive at $R = R_1$,
repulsive at $R = R_2$ and zero at $R = R_3$ or at $R = R_e$.

R_e, but at R_2 the dynamic force is repulsive. This means that if we release the two particles held initially at a fixed distance R_2, the dynamic force will *on the average* drive the two particles *away* from each other, i.e., in opposite direction to the thermodynamic force. Similarly, starting from an initial distance at R_3, the dynamic force is zero.

Thus, knowledge of the thermodynamic force does not tell us anything on the dynamical force and vice versa.

There is another important difference between the two cases in figure 1. In 1-a, on releasing the two particles at $R = R_1$, the force is attractive and the particles will move towards each other. On the other hand, releasing the two particles at $R = R_1$ in a solvent, figure 1-b, the *average* force is attractive, hence it is only on the average that the two particles will be moving towards each other. In the latter case, the solvent-induced force (see section 3) is a result of the fluctuating forces exerted by many solvent molecules in the surroundings of the two particles at $R = R_1$. Therefore the instantaneous motion at $R = R_1$ could be in any direction. Likewise, at $R = R_2$, the instantaneous force could be either attractive or repulsive. It is these fluctuating forces that eventually will drive the system from $R = R_2$ to $R = R_e$ in spite of the fact that the *average* force is repulsive at $R = R_2$. Since the fluctuating forces depend on temperature, it is in principle possible that they will not be strong enough to move the system across the barrier from $R = R_2$ to $R = R_e$. In such a case, the system will be trapped in some metastable state, say at $R = R_3$.

The distinction between the thermodynamic force and the dynamic force can be easily extended to more complex systems such as proteins. Let $\mathbf{R}^M = \mathbf{R}_1 \cdots \mathbf{R}_M$ be a specific configuration of a protein (where $\mathbf{R}_i$ is the locational vector of the i-th nucleus). We say that there exists a thermodynamic force leading from $\mathbf{R}^M$ to $\mathbf{R}'^M$ whenever

$$G(\mathbf{R}'^M) - G(\mathbf{R}^M) < 0 \qquad (2.3)$$

For an isothermic - isobaric system, the condition (2.3) is the generalization of (2.2). The actual average motion of the protein, released from an initial configuration $\mathbf{R}^M$ is determined by the gradient of the function $G(\mathbf{R}^M)$. Normally, we are interested in the actual force operating on a specific nucleus, say the $i = 1$ nucleus, at the initial configuration $\mathbf{R}^M$. This force is obtained from the gradient of $G(\mathbf{R}^M)$ with respect to $\mathbf{R}_1$. Thus we write

$$\mathbf{F}_1 = - \nabla_1 G(\mathbf{R}^M) \qquad (2.4)$$

where $\nabla_1 = (\dfrac{\partial}{\partial x_1}, \dfrac{\partial}{\partial y_1}, \dfrac{\partial}{\partial z_1})$ is the gradient operator with respect to $\mathbf{R}_1$ (normally we also require that this will be consistent with the conservation of all bond lengths and angles of the protein).

As in the example depicted in figure 1-b, starting from an initial configuration $\mathbf{R}^M$, we might have a thermodynamic force leading to some equilibrium configuration $\mathbf{R}_e^M$. However, neither the instantaneous (fluctuating) force nor the average force at $\mathbf{R}^M$ will necessarily be in the direction towards $\mathbf{R}_e^M$. Whether the system will eventually reach $\mathbf{R}_e^M$ or not depends on whether there are high barriers that cannot be overcome by the fluctuating forces.

3. Direct versus indirect forces

Consider again a macroscopic system consisting of water molecules at a given T and P, and one protein molecule at some fixed configuration $\mathbf{R}^M = \mathbf{R}_1 \cdots \mathbf{R}_M$, where $\mathbf{R}_i$ is the locational vector of the i-th nucleus. We are interested in the force operating on a selected nucleus, say $i = 1$, given the configuration $\mathbf{R}^M$, and averaged over all possible configurations of the solvent molecules. This force is obtained from the gradient of the function $G(R^M)$ with respect to R_1.[4,6]

Using classical statistical mechanics we can always write $G(\mathbf{R}^M)$ as[4]

$$G(\mathbf{R}^M) = U(\mathbf{R}^M) + \delta G(\mathbf{R}^M) \qquad (3.1)$$

where $U(\mathbf{R}^M)$ is the *direct* potential energy originating from the protein itself and $\delta G(\mathbf{R}^M)$ includes all solvent effects. The function $U(\mathbf{R}^M)$, although a very complicated function, is the relatively simpler of the two functions on the right hand side of (3.1). It includes all interactions between bonded and non-bonded nuclei that constitute the protein.

The force on the nucleus 1 is obtained from (3.1) as

$$\mathbf{F}_1 = -\nabla_1 U(\mathbf{R}^M) - \nabla_1 \delta G(\mathbf{R}^M) = \mathbf{F}_1^D + \mathbf{F}_1^{SI} \qquad (3.2)$$

where $\mathbf{F}_1^D$ is the direct force on 1 due to all other nuclei at $\mathbf{R}_2 \cdots \mathbf{R}_M$. The second term is the indirect, or the solvent-induced (SI) force, originating from the presence of the solvent. The components of $\mathbf{F}_1^D$ are fairly well understood; these forces are between bonded nuclei, van der Waals, electrostatic and hydrogen bonds between non-bonded nuclei, etc. On the other hand, the solvent induced force depends on the interactions between the protein and the solvent as well as between solvent molecules. This is an extremely complex function to analyze analytically. Undoubtedly, this is the reason why the components contributing to $\mathbf{F}_1^{SI}$ were never studied systematically.

We first note that assuming classical statistical mechanics we may write the identity[4]

$$\delta G(\mathbf{R}^M) = \Delta G^*(\mathbf{R}^M) - \Delta G^*(\infty) \qquad (3.3)$$

where $\Delta G^*(\mathbf{R}^M)$ is the solvation Gibbs energy of the protein at a fixed configuration $\mathbf{R}^M$, and $\Delta G^*(\infty)$ is the solvation Gibbs energy of all nuclei at infinite separation from each other. Since at the latter configuration there are no net forces between the nuclei, we have

$$\mathbf{F}_1^{SI} = -\nabla_1 \delta G(\mathbf{R}^M) = -\nabla_1 \Delta G^*(\mathbf{R}^M) \qquad (3.4)$$

From (3.4) it follows that a *solvent-induced force is exerted on 1, if and only if, a small change in the location of 1 produces a change in the solvation Gibbs energy of the system.*

It is intuitively clear that the changes in the solvation Gibbs energy referred to above must be "near" R_1. In order to make this statement more precise, we use a statistical mechanical expression which is very convenient for analyzing the nature of this solvent induced force, namely,[4]

$$\mathbf{F}_1^{SI} = \int -\nabla_1 U(\mathbf{R}_1, \mathbf{X}_W) \rho(\mathbf{X}_W | \mathbf{R}^M) d\mathbf{X}_W \qquad (3.5)$$

Here, $-\nabla_1 U(\mathbf{R}_1, \mathbf{X}_W)$ is the *direct* force exerted by a water molecule at $\mathbf{X}_W$ on the nucleus at $\mathbf{R}_1$, and $\rho(\mathbf{X}_W | \mathbf{R}^M)$ is the conditional density of water molecules at $\mathbf{X}_W$, given a protein at a fixed configuration $\mathbf{R}^M$. The integration in (3.5) extends over all possible configurations of a water molecule. For latter use we shall refer to $-\nabla_1 U(\mathbf{R}_1, \mathbf{X}_W)$ as the direct force exerted by W on 1, and to $\rho(\mathbf{X}_W | \mathbf{R}^M)$ as the probability density of finding W at $\mathbf{X}_W$.

From the structure of the integrand in (3.5), it is clear that a direct force $-\nabla_1 U(\mathbf{R}_1, \mathbf{X}_W)$ will be non zero only if $\mathbf{X}_W$ is not too far from $\mathbf{R}_1$. Let us denote by $D(\mathbf{R}_1)$ the region from which a water molecule can exert a significant force on 1, we rewrite (3.5) as

$$\mathbf{F}_1^{SI} = \int_{D(\mathbf{R}_1)} -\nabla_1 U(\mathbf{R}_1, \mathbf{X}_W) \rho(\mathbf{X}_W | \mathbf{R}^M) d\mathbf{X}_W \qquad (3.6)$$

within the region $D(\mathbf{R}_1)$ the local solvent density is affected only by those nuclei of the protein that are not too far from $\mathbf{R}_1$. We denote by $\mathbf{R}^n$ those nuclei that, at a specific configuration $\mathbf{R}^M$, are close enough to the region $D(\mathbf{R}_1)$ so that they can affect the local density of water molecules at $\mathbf{X}_W$. Thus, instead of (3.6) we can write

$$\mathbf{F}_1^{SI} = \int_{D(\mathbf{R}_1)} -\nabla_1 U(\mathbf{R}_1, \mathbf{X}_W) \rho(\mathbf{X}_W | \mathbf{R}^n) d\mathbf{X}_W \qquad (3.7)$$

Note that $\mathbf{R}^n$ must include $\mathbf{R}_1$, but otherwise the other nuclei in $\mathbf{R}^n$ need not be close to $\mathbf{R}_1$ along the sequence of the amino acids.

4. Classification of the solvent induced forces

From the analysis made in the previous section it follows that a substantial force $\mathbf{F}_1^{SI}$ will be realized whenever both factors in the integral (3.7) are large. This requires that $\mathbf{X}_W$ be in the region $D(\mathbf{R}_1)$ and that there is a subset of nuclei at $\mathbf{R}^n$ that produces a large local density of water molecules in $D(\mathbf{R}_1)$. Clearly, the magnitude of the force $\mathbf{F}_1^{SI}$ depends on the *type* of the nucleus at $\mathbf{R}_1$ as well as on the *number* and the types of nuclei at $\mathbf{R}^n$. It is impossible at present to analyze all possible cases of nuclei at $\mathbf{R}^n$. Instead, we explore only some representative examples.

Instead of having *any* nucleus at $\mathbf{R}_1$, we examine two cases; a hydrophobic (HΦO) and a hydrophilic (HΦI) group. As a typical HΦO group we choose a methyl group (viewed as a single entity), and as a typical HΦI group we choose a hydroxyl or a carbonyl group (again viewed as a single entity). We could also have added a charged group, but as we shall soon see, even with this limited number of representative groups; HΦO and HΦI we have a multitude of possible forces to be studied. A partial list is the following:

A HΦO group at $R_1 \cdots$ and one HΦO group at R^{n-1}

$\cdots$ and one HΦI group at R^{n-1}

$\cdots$ and two HΦO groups at R^{n-1}

$\cdots$ and two HΦI groups at R^{n-1}

$\cdots$ one HΦO and one HΦI group at R^{n-1}

$\cdots$ three HΦO groups at $R^{n-1} \cdots$

A HΦI group at $R_1 \cdots$ and one HΦO group at R^{n-1}

$\cdots$ and one HΦI group at R^{n-1}

$\cdots$ and two HΦO groups at R^{n-1}

$\cdots$ and two HΦI groups at R^{n-1}

$\cdots$ one HΦO and one HΦI group at R^{n-1}

$\cdots$ three HΦO groups at $R^{n-1} \cdots$

By $\mathbf{R}^{n-1} = \mathbf{R}_2, \mathbf{R}_3 \cdots \mathbf{R}_n)$ we mean $\mathbf{R}^n$ excluding $\mathbf{R}_1$.

In the next section we explore some of the cases listed above. We shall first present an exact analysis for a simple solvent i.e., water at very low density. In section 6 we make a plausible analysis for the case of water at liquid densities.

5. Nearly ideal-gas water as a solvent

We consider here a very simple case where we have besides group 1 at R_1 only one group at $\mathbf{R}_2$. The solvent density is very low $\rho_W \to 0$ so that to first order in the solvent density, the solvent induced force is

$$\mathbf{F}_1^{SI} = -(\rho_W/8\pi^2) \int \nabla_1 U(\mathbf{R}_1, \mathbf{X}_W) \exp[-\beta U(\mathbf{X}_W|\mathbf{R}_1, \mathbf{R}_2)] d\mathbf{X}_W \quad (5.1)$$

where $\beta = (k_B T)^{-1}$, k_B being the Boltzmann constant and T the absolute temperature. $-\nabla_1 U(\mathbf{R}_1, \mathbf{X}_W)$ as before, is the force exerted on 1 at $\mathbf{R}_1$ by a water molecule at $\mathbf{X}_W$. The conditional probability of finding a water molecule in an element of configurational volume $d\mathbf{X}_W$ is now

$$Pr(\mathbf{X}_W) d\mathbf{X}_W = \frac{\rho_W d\mathbf{X}_W}{8\pi^2} \exp[-\beta U(\mathbf{X}_W|\mathbf{R}_1, \mathbf{R}_2)] \quad (5.2)$$

where $U(\mathbf{X}_W|\mathbf{R}_1, \mathbf{R}_2)$ is the direct interaction energy between W at $\mathbf{X}_W$ and the two groups at $\mathbf{R}_1$ and $\mathbf{R}_2$.

In this particular example the solvent induced force in (5.1) may be calculated exactly for any given pair of pair potentials $U(\mathbf{R}_1, \mathbf{X}_W)$ and $U(\mathbf{R}_2, \mathbf{X}_W)$. We shall not need to make such a calculation here. Instead, we shall examine only the change of the force caused by replacing a HΦO group by a HΦI group.

Also for the present qualitative discussion, we use a simple form of a solute-solvent pair potential,[4], i.e.,

$$U(\mathbf{R}_i, \mathbf{X}_W) = U^{LJ}(\mathbf{R}_i, \mathbf{X}_W) + U^{HB}(\mathbf{R}_i, \mathbf{X}_W) \quad (5.3)$$

where U^{LJ} is a Lennard-Jones type of interaction and U^{HB} is the hydrogen-bonding (HB) part of the potential. The latter is non zero only when the group at R_i can form a HB with a water molecule. We shall not need any details on the form of these two parts, we note only that at room temperature $\beta U^{LJ} \approx -1$ at the minimum of the potential function, whereas $\beta U^{HB} \approx -10$, when the distance and orientations are such that a HB can be formed between the pair at $\mathbf{R}_i$ and $\mathbf{X}_W$.

We now compare three different cases:

a) A HΦO group at $\mathbf{R}_1$ and one HΦO group at $\mathbf{R}_2$. In this case, only the LJ part of the potential is operative and we have

$$\mathbf{F}_1^{SI} = -(\rho_W/8\pi^2) \int \nabla_1 U^{LJ}(1, W) \exp[-\beta U^{LJ}(1, W) - \beta U^{LJ}(2, W)] d\mathbf{X}_W \quad (5.4)$$

where we used a shorthand notation for the configuration of each pair of particles. We shall refer to this force as the "normal" force. It is "normal" in the sense that only βU^{LJ} appears everywhere - no contributions due to HB'ing. Thus we write

$$F_1^{SI}(H\Phi O - H\Phi O) = F_1^{SI} \text{ (normal)} \quad (5.5)$$

b) A HΦO group at $\mathbf{R}_1$ and one HΦI group at $\mathbf{R}_2$. In this case the direct force is the same as in (5.4), but the local density due to the group at $\mathbf{R}_2$ changes by a factor of $\exp[-\beta U^{HB}(2, W)] \approx \exp[10] = 2.2 \times 10^4$. Thus although the region from which a HB may be formed is quite small ($\Delta \mathbf{X}_W/8\pi^2$ is of the order of 5×10^{-3} cm^3/mol)[4] there will be a net increase in the local density of water molecules at $\mathbf{X}_W$ due to the HB'ing capability of the HΦI group.

We may roughly write for this case

$$F_1^{SI}(H\Phi O - H\Phi I) = -(\rho_W/8\pi^2) \int \nabla_1 U^{LJ}(1, W) EXP[-\beta U^{LJ}(1, W)$$

$$\beta U^{LJ}(2, W) - \beta U^{HB}(2, W)] d\mathbf{X}_W \quad (5.6)$$

Since the additional factor in the integrand of (5.6) is greater than one, we can write

$$|F_1^{SI}(H\Phi O - H\Phi I)| > |F_1^{SI}(normal)| \tag{5.7}$$

c) A $H\Phi I$ group at $\mathbf{R}_1$ and a $H\Phi I$ group at $\mathbf{R}_2$. In this case there are two changes in the integrands of (5.4). First, the forces exerted by W on 1 have now two components.

$$\nabla_1 U(1, W) = \nabla_1 [U^{LJ}(1, W) + U^{HB}(1, W)] \tag{5.8}$$

second, there are configurations from which a water molecule can form simultaneously two HB's with 1 and 2. Clearly, the configurational space from which this can be done is quite small (this was estimated to be of the order of $\Delta X_W'/8\pi^2 \approx 2.8 \times 10^4 \text{cm}^3/\text{mol})^4$ but now the exponential term can get as large as $\exp[20] \approx 4.8 \times 10^8$, due to the formation of two HB's. Therefore the net solvent-induced force on 1 is expected to be much larger than in the previous case. We thus write

$$|F_1^{SI}(H\Phi I - H\Phi I)| >> |F_1^{SI}(H\Phi O - H\Phi I)| > |F_1^{SI}(normal)| \tag{5.9}$$

To summarize, when we replace one $H\Phi O$ group at R_2 by a $H\Phi I$ group, we increase the local density of water molecules by a factor of about $\exp[10]$, at some configurations X_W. The direct force exerted by one W at X_W does not change. When we replace the two groups by $H\Phi I$ groups, both the direct force $\nabla_1 U(1, W)$, and the local density of water molecules will change, giving rise to large solvent induced forces.

6. Solvent induced force in real liquid water

Considering again the three cases listed above in real liquid water, say, at room temperature and at 1 Atm., we find that the changes from one case to another are less dramatic than the previous case, but the order of the forces, as presented in (5.9) is maintained.

The general expression for the force (3.6) is the same as before, with a reinterpretation of the local density as

$$\rho(X_W|\mathbf{R}_1, \mathbf{R}_2) = \rho_W\, g(X_W|\mathbf{R}_1, \mathbf{R}_2)/8\pi^2 \tag{6.1}$$

(In (5.1) we have used the low density limit of $g(X_W|\mathbf{R}_1, \mathbf{R}_2)$.)

The number density ρ_W in (6.1) is the density of liquid water at room temperature $\rho_W = 5.5 \times 10^{-2}$ mol/cm^3, and $g(X_W|\mathbf{R}_1, \mathbf{R}_2)$ is the pair correlation function between a water molecule and the pair-of-groups at $\mathbf{R}_1, \mathbf{R}_2$ viewed as a single entity (note that $g(X_W|\mathbf{R}_1, \mathbf{R}_2)$ is different from the triplet correlation function $g(X_W, \mathbf{R}_1, \mathbf{R}_2)$).[4,7]

The direct force $\nabla_1 U(\mathbf{R}_1, X_W)$ is the same as before. For the $H\Phi O - H\Phi O$ case, $g(X_W|\mathbf{R}_1, \mathbf{R}_2)$ is of the order of 2-3. This may be referred to as the "normal" case for the liquid density. For the $H\Phi O - H\Phi I$ case, we have estimated[4,7] that $g(X_W|\mathbf{R}_1, \mathbf{R}_2)$ is of the order of 110. This is considerably small than in the gaseous phase, but still considerably larger than the "normal" case. Finally, for the $H\Phi I - H\Phi I$ case, we have estimated[4,7] $g(X_W|\mathbf{R}_1, \mathbf{R}_2)$ to be of the order of 2.6×10^3 (for those configurations from which W can form simultaneously two HB's with 1 and 2).

We therefore conclude, based on theoretical arguments,[4,7,8] that although the changes are more modest than in the gaseous phase discussed in the previous section, the order of the forces, as presented in (5.9), is preserved.

This conclusion was recently confirmed by molecular dynamics simulations. The average forces between two water molecules were calculated and found to be stronger than the corresponding force between two nonpolar solutes.[9]

7. Conclusion

In assessing the relative contributions of the various effects involved in the stabilization of proteins, one need to have first a complete inventory of all possible effects and, second, an idea of the order of magnitude of each of these contributions.

Clearly, if we are going to analyze all possible amino acid residues, we shall have to deal with a huge number of effects, on which we have at present neither experimental nor theoretical information. Therefore we limit ourselves to only two representative groups that occur in proteins, the HΦO group (represented by, say, methyl groups) and the HΦI group (represented by hydroxyl or carbonyl groups). Even with this reduced classification, one must distinguish between different specific effects such as solvation, pairwise correlations, triplet correlations, etc.

The current prevailing opinion is that the "HΦO effect" is the most important one in protein stabilization.[5] This is based on Kauzmann's original idea,[10] that the free energy of transferring of a HΦO group into the interior of the protein can be estimated from the reaction

$$CH_4 \text{ (water} \rightarrow \text{organic liquid)} \tag{7.1}$$

which is quite large. On the other hand, the free energy of the "HΦI effect" has been considered in the context of the following "reaction"

$$C=O \cdots W + NH \cdots W \rightarrow C=O \cdots HN + W \cdots W \tag{7.2}$$

The dots "$\cdots$" indicated hydrogen bonding (HB). The latter equation suggests that the number of HB's broken on the lhs of (7.2) is the same as the number of HB's formed on the rhs (7.2). If we assume that the HB strength between the various groups are nearly the same, one arrives at the conclusion that HB's do not contribute significantly to the protein folding process.

Therefore, the current view tends to dismiss the contribution of the HB'ing, while at the same time attributing a greater importance to the HΦO effect. A recent examination of the various HΦO and HΦI effects shows that Kauzmann's model of transfer of free energy (7.1) is not adequate for the protein folding, and that the presentation of the HΦI effect in the sense of (7.2) leads to erroneous conclusions.[11,12]

In the above discussion we have used the terms HΦO and HΦI "effects", with reference to equations (7.1) and (7.2). Actually there are more than one HΦO and HΦI effects. In order to make more meaningful statements, we must specify more precisely which effects are we comparing. If we do a one-to-one comparison of the various effects, we find the following:

1) HΦO versus HΦI loss of solvation

This involves the complete loss of the solvation of, say, methyl or a hydroxyl group when it is transferred into the interior of the protein. For a HΦO group, we find that the loss of the (conditional) solvation Gibbs energy is negative and on the order of -0.5 kcal/mol.[2] For the loss of the solvation of a hydroxyl group the corresponding value is about +6.7 kcal/mol. Thus, for this particular effect, the HΦI effect is almost an order of magnitude larger than the corresponding HΦO effect. Even when taking into account the formation of internal HB's within the interior of the protein (as most HΦI groups actually do), one still finds that the HΦI effect due to the loss of the solvation is far larger than the corresponding HΦO effect.

2) Pairwise HΦO versus HΦI interactions

The solvent induced interaction between two HΦI groups at some specific configurations was recently estimated to be on the order of -3 kcal/mol.[4,11] The corresponding value for HΦO groups such as methyl or ethyl groups is on the order of -0.3 to -0.6 kcal/mol.

3) Triplet and higher order solvent induced interactions

Although very little is known on these interactions, a recent calculation indicated that triplet and quadruplet solvent induced interactions between HΦI groups are quite large[7] - probably much larger than the corresponding effects for triplet and quadruplet HΦO groups.

4) The solvent induced forces between HΦI groups are larger than the corresponding forces between HΦO groups - as we have discussed in sections 5 and 6, and confirmed by simulations.[9]

By comparing one-by-one effects, we see that any of the HΦI effects listed above is much stronger than the corresponding HΦO effects. This still leaves open the question of the overall or the accumulative effect of all groups in a specific protein. For instance, the interaction free energy between a pair of HΦI groups could be 5-6 times larger than for a pair of HΦO groups. If the protein has, say, ten times more pairs of interacting HΦO groups than HΦI groups, of course the overall effect of the HΦO groups will be larger. However, given that the HΦO side chain consistency of proteins are about 30-50% of the total side chains, and remembering that each backbone residue has two HΦI groups (the C=O and NH of the amide group), it is unlikely that the overall combined effects of all HΦO groups will produce a dominant driving force for the folding process. This still leaves the possibility that in some specific, but probably rare proteins, the combined HΦO effect might be dominant in the protein folding process. To the best of my knowledge this has never been demonstrated conclusively for any single protein.

We therefore conclude that various HΦO effects and their importance to protein folding were overestimated. We believe that the various HΦI effects play a more important role in the driving forces for protein folding as well as for other biochemical processes.

Acknowledgment
This work was partially supported by the Basic Research Foundation administered by the Israel Academy of Sciences and Humanities.

References

1)   Creighton, T.E., J. Phys. Chem. **89**, 2452 (1985).
2)   Creighton, T.E., *Proteins, Structure and Molecular Principles*, Freeman, New York (1984).
3)   Jaenicke, R., Prog. Biophys. Molec. Biol. **49**, 117 (1987).
4)   Ben-Naim, A., *Statistical Thermodynamics for Chemists and Biochemists*, Plenum Press, N.Y. (1992).
5)   Dill, K.A., Biochemistry **29**, 7133 (1990).
6)   Hill, T.L., *Statistical Mechanics*, McGraw-Hill, New York (1956).
7)   Ben-Naim, A., J. Chem. Phys. **93**, 8196 (1990).
8)   Ben-Naim, A., J. Phys. Chem. **94**, 6893 (1990).
9)   Durell, S.R., Brooks, B.R. and Ben-Naim, A., J. Phys. Chem. (in press, 1994).
10)   Kauzmann, W., Adv. Protein Chem. **14**, 1 (1959).
11)   Ben-Naim, A., Biopolymers **29**, 567 (1990).
12)   Ben-Naim, A., J. Phys. Chem. **95**, 1437 (1991).

RECEIVED April 25, 1994

Chapter 26

Molecular Dynamics Simulations of DNA and Protein–DNA Complexes Including Solvent

Recent Progress

D. L. Beveridge, K. J. McConnell, R. Nirmala, M. A. Young, S. Vijayakumar, and G. Ravishanker

Department of Chemistry and Program in Molecular Biophysics, Wesleyan University, Middletown, CT 06459

The results of three new molecular dynamics (MD) trajectories for the d(CGCGAATTCGCG) double helix in water, including one of 1 nanosecond (ns) duration, and an MD study of the λ repressor-operator complex are described. The DNA simulations form the basis for a detailed analysis of the progress of the trajectory over time and the dynamics of axis bending. The results indicate that the ns dynamical trajectory progresses through a series of three substates of B form DNA, with lifetimes of the order of hundreds of picoseconds (ps). Evidence for an incipient dynamical structure is presented. To validate the simulation, the calculated axis bending is compared with that observed in corresponding crystal structure data. The results indicate that, for this system at least, significant new dynamical behavior is introduced in the ns regime, that previously reported calculations at the ps level did not reveal. Simulation of the protein-DNA complex and the independent components thereof have been carried out for 100-320 ps to date. Some preliminary results from the protein-DNA complex and indications of the extent of structural reorganization on complex formation are presented.

Molecular Dynamics (MD) computer simulations have been used in a number of recent studies (*1*) to investigate the nature of DNA fine structure, axis bending and helical flexibility implicated in important protein-DNA interactions (*2*). We report herein progress in the development of an accurate theoretical model of the structure and motions of the DNA double helix in water in the unbound state, and in a complex with a regulatory protein. The systems under consideration are the oligonucleotide duplex of sequence d(CGCGAATTCGCG), which contains the target sequence for the restriction enzyme *Eco RI* endonuclease, and the λ repressor-operator protein-DNA complex. The results are being compared in detail with experimental data from X-ray crystallography to validate the theoretical model.

Previous studies from this laboratory have addressed various aspects of water structure around DNA (3), in particular the spine of hydration in the minor groove (4) and comparison with crystallographic ordered water sites (5). These studies suggest considerable localization of ordered water in the minor groove and a more disordered water in the major groove, and that both effects could contribute to the overall energetics of stabilization. It must be noted that the thermodynamics for the 'spine of hydration' is not unequivocally established but there is increasing evidence suggesting a stabilizing role (5). Here, we consider the influence of solvation on the behavior of the structural dynamics of DNA and of protein-DNA complexes, including waters explicitly.

MD on the d(CGCGAATTCGCG) Duplex

MD on DNA including solvent is computationally quite intensive, and trajectories published to date consider time scales of only a few hundred ps (6,7). In order to examine the sensitivity of results to trajectory length and choice of starting structure, we have recently performed a series of three simulations on the d(CGCGAATTCGCG) duplex, two for 500 ps and one extended to 1000 ps, or 1 ns (8). These simulations are based on the GROMOS86 force field (9), which has been setup for the simulation of biomolecules in aqueous environment (10), and the SPC model of water (11). The GROMOS force field has been extensively employed for simulation of proteins and DNA in this laboratory (6,12,13) and others (14-17). The SPC model of water is a non-polarizable effective pair potential for liquid water and is similar to other widely used models such as SPC/E (18), F3C (19), TIPS (20,21) and its variants, in terms of geometrical and nonbonded properties (22). All of the above models are rigid with the exception of F3C, which is flexible. The water models vary in their ability to reproduce experimentally observed properties of water. The radial distribution function agrees well up to the first solvation shell for all of these models and up to the third (~ 7 Å) for SPC and F3C (22). Calculated diffusion constant (Å^2/ps) for the SPC model (0.36) is comparable to that of TIP3P (0.40) but slightly higher than the experimental value (0.23) at 300 K (22). The dipole moment for the SPC model is 2.27 D, compared to 1.85 for the isolated molecule and the density at 300 K, and at 1 atmosphere pressure turns out to be 0.97 g/cm^3 compared to an experimental value of 0.995 at 305 K (18). Overall, it is apparent that the SPC model of water and the GROMOS force field provide a reasonable model for simulation of proteins, DNA and protein-DNA complexes.

Using GROMOS unmodified, irreversible base pair opening events were observed. Such behavior was subsequently noted in MD studies of DNA based on other force fields as well (23,24). Base pair opening is observed experimentally in DNA via NMR spectroscopy (25). However, the experiments are carried out on the ms time scale and the results indicate the phenomena to be infrequent. This led us to suspect that the opening events seen in our previous MD are deficiencies in the potentials and not accurate theoretical descriptions of the phenomena. A subsequent simulation employed a harmonic restraint function with GROMOS86 to assure that Watson-Crick base pairing was maintained intact. The results were found to be consistent with available experimental data within reasonable limits of expectation, both in comparisons with crystal structure data (6), and with 2D-NOESY build up curves obtained from NMR spectroscopy (26,27).

Subsequent studies of hydrogen bond interaction energies in Watson-Crick base pairs revealed that GROMOS underestimates the magnitude of these interactions. A hydrogen bond potential in the form proposed by Tung et al. (28) was then added to our implementation of the GROMOS force field. The resulting base pair interaction energies compared closely with corresponding experimental data and corresponding *ab initio* quantum mechanical calculations(Gould, I.; Kollman, P., submitted). The simulations described herein are based on the GROMOS force field with this modification. Unlike the hydrogen bonds, no additional potential has been applied for stacking interactions. The current studies also utilize a longer range switching function, which feathers the truncation of potentials over the length scale from 7.5 to 11.5 Å, which eliminates the tendency of charged groups to cluster at the cutoff limit when potentials are truncated too abruptly, a behavior recently noted by Auffinger et al. (29). Further concerns about the accuracy of force field in ionic interactions and the convergence in the dynamical behavior of mobile counterions have led us to follow Tidor et al. (30) to treat the effect of counterions implicitly in this set of studies, using reduced charge of -0.24 eu on the phosphate groups, consistent with Manning's theory of counterion condensation (31).

Solvation is carried out by placing the solute in a box of a given shape and filling with water to obtain an overall density of 1 gm/cc. The shape of the box is chosen, that requires the least number of waters for solvation of a given thickness. The box is then divided into uniform grids and a water molecule is placed in each grid that has no solute atoms. The initial placement is relaxed by a Monte Carlo (MC) simulation employing the Metropolis algorithm, wherein the solute is held rigid and waters are allowed to equilibrate around the solute. The MC simulations are carried out until the total and solute-solvent interaction energies converge. The simulations on the dodecamer duplex DNA are carried out using free MD, surrounded by ~3500 water molecules in a hexagonal prism elementary cell of constant volume, treated under periodic boundary conditions to model dilute aqueous solution. Velocities are rescaled when necessary (owing to conformational transitions) to produce an average kinetic energy corresponding to 300 K.

The three new simulations on the d(CGCGAATTCGCG) duplex were performed under protocols identical except for starting structure and length. The starting configurations were a) the canonical B80 (32) fiber form of DNA (500 ps trajectory), b) the Drew-Dickerson (33) crystal form (500 ps trajectory) and c) the protein bound form observed in the complex of d(CGCGAATTCGCG) with the restriction enzyme *Eco RI* endonuclease (34) (1 ns trajectory). The canonical form is a regular Watson-Crick double helix with a straight axis. The crystal form shows axis bending at or near the interfaces of the CG and AT tracts, and the *Eco RI* form shows, in addition to deformations at nearly the same positions as in the uncomplexed form, an extreme kink in the middle region of the structure at the A6 and T7 steps. The simulations were performed using the program WESDYN (35) and analyzed by means of various utilities available in MD Toolchest (36). All three simulations were found to converge to essentially similar MD behavior, indicating the results, at least for this system, are not sensitive to the choice of starting structure. The extreme kink in the *Eco RI* dodecamer was found to relax within the first 20 ps of MD, indicating the protein-bound form to be a strained conformation rather than a metastable intermediate. This results support those of a current study by Rosenberg and coworkers (*37*) based on the AMBER force field.

The MD based on the *Eco RI* form was extended to 1 ns, and forms the basis for this more detailed analysis. The stability and convergence behavior of the simulation was monitored by one-dimensional (1D) and two-dimensional (2D) root mean square deviation (RMS) maps. The latter is especially informative, since the extent of similarity among all structures in the trajectory is depicted graphically (Figure 1). Essentially, the 2-D RMS provides a measure of the deviation of every structure in the trajectory with respect to all others. Thus, a square matrix of RMS values are generated wherein the upper and bottom halves are symmetrical and, therefore, only one half of the matrix is presented. The diagonal elements represent deviation of a structure with respect to itself and consequently have an RMS value of zero. The off-diagonal elements provide a measure of the similarity of any two structures from different time points in the trajectory and their gray scaling reflects the extent to which they are similar. The results indicate that after an initial equilibration period, the MD structure resides for ~300 ps in a form ~4.5Å RMS from canonical B form DNA. The dynamical structure then makes a distinct transition to a new form, still in the B family but somewhat more distant (~7.5Å RMS) from the canonical form, where it remains for 180 ps. A rapid (~1.5 ps) reversible base pair opening event occurs in this structure at the T7 step, concomitant with displacements in helicoidal roll and twist. Then the dynamical structure transits to a third form, where it resides at the termination of the run. The third form, as evidenced by a cross peak in the 2-D RMS map, bears a strong resemblance to the first, indicating that the MD results appear to describe an incipient dynamical equilibrium among putative dynamical substates of the B family. The lifetime for a given structural form of DNA suggests that a minimal trajectory length of at least 200-300 ps would be necessary in order to fully relax the DNA in solution, given our methods and protocol.

Validation of the MD results was pursued by a comparison of calculated and observed helix bending characteristics. To analyze axis bending in a given structure, the magnitude β and angular direction of bending α are computed from deviations in the helicoidal parameters roll and tilt. The values of β and α for a given step can be projected onto a polar plot or "bending dial", seen in perspective at the bottom of Figure 2a. The detailed analysis of multiple sequences can be carried out by superimposing results from individual structures on a single bending dial for each base pair step in a DNA sequence. The use of bending dials to analyze DNA crystal structures and MD simulations has been described previously by Young et al. (*38*).

The bending dials for 20 reports of d(CGCGAATTCGCG) duplex structures in the Nucleic Acids Data Bank (*39*) are shown in comparison with the MD results in Figure 2b. The results show graphically that bending towards the minor groove occurs at the G2-C3 step, followed in the succeeding step C3-G4 by a bend towards the major groove. The origin of the bending in the helicoidal parameters is inter-base pair roll. A similar effect is seen in the other flanking sequence; these are the upper and lower roll points identified by Dickerson and coworkers (*40*). The axis deformations from the 1 ns MD trajectory are shown in Figure 2c, and indicate that

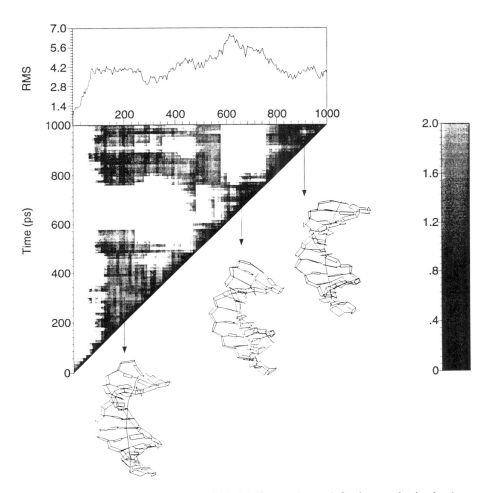

Figure 1. 1D RMS map (top) and 2D RMS map (center) for base pairs in the 1 ns MD simulation of the d(CGCGAATTCGCG) duplex. Shaded areas in the 2D RMS map indicate regions of RMS deviation < 2 Å. The diagonal elements represent the RMS deviation of a given structure with respect to itself (at each time point) and the off-diagonal elements represent similarity of structures from different time points in the trajectory, the gray scale reflecting the extent of the RMS deviation between them. Average MD structures for each of three putative substates are shown at bottom.

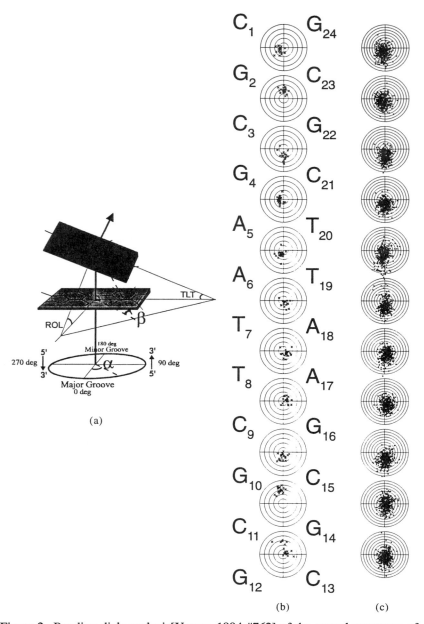

Figure 2. Bending dials analysis[Young, 1994 #762] of the crystal structures of the d(CGCGAATTCGCG) duplex resident in the NDB[Berman, 1992 #745], compared with corresponding results from a 1 ns MD simulation[Ravishanker, 1994 #809]. a) Definition of bending dial (seen in perspective view at bottom) in terms of base pair roll and twist; b) Bending dials analysis for 20 crystal structures, and c) Bending dials analysis for the MD structures. Each ring represents a 5° increment in axis deflection. To make trends in the analysis more discernible, crystal structure dials extend to 25°, while MD dials extend to 45°.

the natural roll points are found at G2-C3 steps and the C9-G10 steps. The deformation in each case is a bend towards the major groove. The adjoining step does not distinctly show the corresponding roll into the minor groove as observed crystallographically, a discrepancy between the calculated and observed results. The effect of crystal packing, which can be significant in DNA oligonucleotides (*41-43*), is not considered in the calculations. Thus roll bends corresponding to experiment are found in the MD at the agreement level indicated, and not unreasonable considering the calculated results are for an aqueous solution model. The interesting question of the role of crystal packing effects on axis bending will the subject of a forthcoming paper in which we describe an MD performed on the crystallographic unit cell. Analysis of the water structure in this system will also be the subject of a subsequent paper.

We find that MD models of DNA are increasing in their accuracy and utility in investigating the fundamental nature of DNA structure and motions. Detailed analysis of the progress of the simulations reveals that the trajectories progress through a series of substates with lifetimes of several hundred ps, and provides evidence for the presence of an incipient dynamic equilibrium among substates on the ns time scale. The stability of the MD calculations over such a long duration suggest our model of DNA is reasonable and is encouraging to note the feasibility of studying dynamical processes occurring in the ns regime. It is likely that further detail will emerge as MD simulations are extended to successive decades of time and new motional regimes are encountered.

MD Simulation Studies of the λ Repressor-Operator Complex

The preceding series of MD simulations on DNA as well as several previously published MD simulations on proteins (*12,14*) demonstrate that MD models based on the GROMOS force field are reasonably accurate when applied to these systems independently. A logical next step is to apply these models for studying the intricate interactions involved in the formation of a protein-DNA complex. Protein-DNA complexes present an additional level of complexity to MD simulation methods, since the intermolecular interactions are likely to be governed by a highly subtle balance of forces. We seek to evaluate the accuracy with which calculations on interacting macromolecules can be carried out, using the λ repressor-operator complex as a prototype case.

A number of characteristic structural elements have now been recognized in proteins interacting with DNA, including the helix-turn-helix (HTH) motif, zinc fingers and leucine zippers (*44*). The *modus operandi* of binding motifs involving recognition and regulation of genetic biochemical processes have been found to vary from one complex to another (*2*). Further, regions of protein structure that are contiguous to the above motifs, some dynamically labile, have been found to contribute significantly to the binding. There is no clear indication of the structural changes, particularly in the DNA, that occur on complex formation. Collective efforts in the fields of molecular biology, biochemistry and molecular biophysics are now directed towards determining the relationship between the structural details of protein-DNA complexes and the origin of the remarkable specificity of their interactions.

Understanding the specificity of a regulatory protein to it's various cognate operator sequences involves interpretive knowledge of the relative contributions of all significant factors to the free energy of binding relative to each other and to random sequence DNA. Genetic and chemical protection experiments at the molecular level and crystallography at or near the atomic level provide a knowledge of the direct interactions involved, but not the energetics (beyond the idea that if a structure occurs, it must have a favorable free energy). Direct measurement of free energy via equilibrium binding studies provides information on the energetics, but unequivocally linking the energetics to structure is difficult, especially when subtle non-local changes are involved. Theoretical methods can, in principle, provide the link between structure and energetics. In addition, the dynamical aspects of protein-DNA complexes, not accessible to experiment, can be explored via MD studies, forming a basis for interpretation of experimental results. However, MD simulations are still limited by assumptions inherent in the underlying molecular force field, the sensitivity of results to simulation protocols and truncation of potentials, and the limited time frame that can be reasonably simulated on systems of this size. Detailed study of a prototype system is necessary to determine the capabilities and limitations of MD applied to protein-DNA interactions.

We describe herein separate MD studies of the repressor protein from the bacteriophage λ, its cognate duplex DNA OL1, of sequence d(TATCACCGCCAGTGGTA), and the repressor-operator complex. This system is of historic interest in the elucidation of the 'genetic switch' from lysogenic to lytic phases of the phage (45) and has been studied extensively from diverse points of view. The availability of crystal structures of the complex (46,47) and that of the unbound protein (48) makes this system ideally suited for inquiring into the overall accuracy and utility of MD simulation applied to a regulatory protein-DNA complex. All three simulations were carried out with water included explicitly. For the simulation of the free DNA and the complex an hexagonal prism cell was employed and the simulation of the protein employed a simple cubic cell. Simulation of the free DNA, free protein and the complex required 4411, 6576 and 7133 waters, respectively, which provided at least 9.0 Å solvation in all systems. The effect of counterions was introduced via the use of a reduced charge of -0.24 on phosphates fully exposed to solvent. Sequestered phosphates are set to -1.0, and partially exposed phosphates have their charges scaled proportional to their solvent accessibility. All other protocols were similar to that of the preceding three simulations on the dodecamer duplex DNA. The results are analyzed with respect to the intrinsic stability of the model system, the dynamical structure of the protein-DNA interface, adaptive changes in structure on complex formation, and solvent effects.

The initial structure for the protein, DNA, and the complex are taken from crystal structure data (46,47). We consider our MD results from this study thus far to be quite preliminary, especially in the light of the results presented in the previous section of this article. A combined 2-D RMS plot comparing the calculated dynamical structure of the uncomplexed protein and that of the protein in the

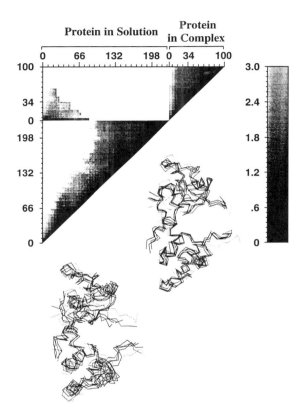

Figure 3. 2D RMS map comparing structures from the MD trajectory of the free and complexed protein, in solution. Both axes represent structures from the MD trajectory at 1.0 ps intervals and averaged over 5.0 ps blocks. Simulation of the free protein in solution was carried out for 230 ps and that of the complex for 100 ps. Cross peaks indicate similarity between two structures either from a given trajectory or between the two trajectories and the shading reflects the extent of similarity (RMS deviation over backbone atoms). For each simulation, the starting structure and four snapshots from the MD trajectory, at equally spaced intervals, are overlaid to indicate the dynamical flexibility. The gray scaling varies from light to dark as the simulation progresses.

complex is shown in Figure 3. An α carbon overlay of five snapshots from each trajectory indicate the dynamic range covered by the MD. Although the overall structure of the proteins in the two simulations are similar, the absence of RMS cross peaks within 3.0 Å indicate the structures are diverging, as the simulation progresses. The MD structures of the uncomplexed protein also indicate that the C-terminal helices, located at the protein dimer interface in the complex, are substantially destabilized. This result is consistent with the experimental observation that the free repressor does not form dimers in solution (48). Despite the unwinding of the C-terminal helices, the calculated % helicity for the uncomplexed protein is in good agreement with experimental measurements (49). Also, both the N-terminal arms of the dimer show significant changes with respect to the crystal conformation and appear to flop back on the protein. In the simulation of the complex, the dimer interface is more stable than in the uncomplexed protein, and the structural rearrangement of the arms remain localized within the major groove area.

A comparison of structures from the trajectory of the uncomplexed DNA and that of the DNA in complex with the protein is presented in Figure 4. The simulation on the uncomplexed DNA was carried out to a trajectory length of 320 ps and the simulation of the protein-DNA complex to about 100 ps at the time of this writing. The isolated DNA remains within the B family of structures, although there appears to be some slight fraying at the ends. In general, a slight expansion of the major groove and a compression of the minor groove at all base steps is evident, except for a three base pair region in the middle segment of the DNA. The overall dynamics of the complex is shown in Figure 5. The calculations indicate the complex to be stable up to this point. The dynamical structure features a compression of the major groove, beginning at both ends of the DNA, and an expansion in the base steps between the two DNA binding pockets. In the minor groove, there is an expansion in the protein binding regions and at the ends. We now plan to extend each of these simulations as far as is necessary to characterize the dynamical stability, and then analyze the dynamical structures and the molecular motions in detail.

Summary and Conclusion

We have reviewed herein MD simulations on DNA oligonucleotides and a protein-DNA complex recently performed in this laboratory. MD simulation is a potentially powerful approach to the study of problems in this area, since full details of the molecular motions are obtained at a level inaccessible to any other method, theoretical or experimental. However, due to the empirical nature of the underlying molecular force field, the methods require extensive independent validations before the results obtained can be considered reliable and accurate. The results to date are encouraging but suggest a significantly new dynamical behavior in the ns regime. This may be a consequence of the particular flexibility of the DNA double helix, but more likely due to relaxation processes occurring over a longer time scale. Further investigations are clearly necessary to delineate the capabilities and limitations of MD methodology applied to this class of problems.

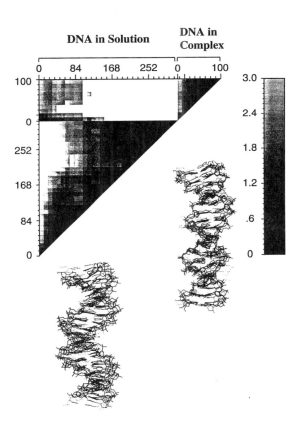

Figure 4. 2D RMS map comparing structures from the MD trajectory of the free and complexed DNA, in solution. Simulation of the uncomplexed DNA was carried out for 320 ps. See legend of Figure 3 for additional details.

Figure 5. Overlay of the stick figure of the DNA and α-carbon drawing of the protein from the simulation of the complex. The crystal structure and four snapshots from the MD trajectory, taken at equally spaced intervals, are shown. The gray scaling varies from light to dark as the simulation progresses.

Acknowledgments

This project is supported by NIH Grants # GM 37909 from the National Institutes of General Medical Sciences, and RR 07885 from the Division of Research Resources. K. J. McConnell is supported by an NIH Molecular Biophysics Training Grant # GM 08271. Cray C90 and YMP computer facilities for this project were made available to us by the Pittsburgh Supercomputer Center and the NCI Frederick Cancer Research and Development Center of NIH.

Literature Cited

(1)   Beveridge, D. L.; Swaminathan, S.; Ravishanker, G.; Withka, J. M.; Srinivasan, J.; Prevost, C.; Louise-May, S.; LAngley, D. R.; DiCapua, F. M.; Bolton, P. H. In *Water and Biological Molecules*; E. Westhof, Ed.; The Macmillan Press, Ltd: London, 1993; pp 165.

(2)   Steitz, T. A. *Quart. Rev. Biophys.* **1990**, *23*, 205.

(3)   Westhof, E.; Beveridge, D. L. In *Water Science Reviews*; F. Franks, Ed.; Cambridge University Press: Cambridge, 1990, Vol. 5; pp 24.

(4)   Subramanian, P. S.; Ravishanker, G.; Beveridge, D. L. *Proc. Natl. Acad. Sci. (USA)* **1988**, *85*, 1836.

(5)   Subramanian, P. S.; Beveridge, D. L. *J. Biomol. Struct. & Dyn.* **1989**, *6*, 1093.

(6)   Swaminathan, S.; Ravishanker, G.; Beveridge, D. L. *J. Am. Chem. Soc.* **1991**, *113*, 5027.

(7)   Pohorille, A.; Ross, W. S.; Tinoco Jr., I. *Int. J. Supercomp. App.* **1990**, *4*, 81.

(8)   Ravishanker, G.; Nirmala, R.; McConnell, K. J.; Young, M. A.; Beveridge, D. L. **1994**, *manuscript in preparation.*

(9)   van Gunsteren, W. F.; Berendsen, H. J. C. GROMOS, University of Groningen: Groningen, The Netherlands, 1986.

(10) van Gunsteren, W. F.; Berendsen, H. J. C. *Angew. Chem. Int. Ed. Engl.* **1990**, *29*, 992.

(11) Berendsen, H. J. C.; Postma, J. P. M.; van Gunsteren, W. F.; Hermans, J. In *Intermolecular Forces*; B. Pullman, Ed.; D. Reidel: Dordrecht, 1981; pp 331.

(12) Harte Jr., W. E.; Swaminathan, S.; Beveridge, D. L. *Proteins: Structure, Function and Energetics* **1992**, *13*, 175.

(13) Vijayakumar, S.; Visveshwara, S.; Ravishanker, G.; Beveridge, D. L. *Biophysical J.* **1993**, *65*, 2304.

(14) van Gunsteren, W. F. *Curr. Opin. Struct. Biol.* **1993**, *3*, 277.

(15) Hermans, J.; Berendsen, H. J. C.; van Gunsteren, W. F.; Postma, J. P. M. *Biopolymers* **1984**, *23*, 1513.

(16) Berendsen, H. J. C.; van Gunsteren, W. F.; Swinderman, H. R. J.; Geurtsen, R. G. *Ann. N. Y. Acad. Sci.* **1986**, *482*, 269.

(17) Brunne, R. M.; van Gunsteren, W. F.; Bruschweiler, R.; Ernst, R. R. *J. Am. Chem. Soc.* **1993**, *115*, 4764.

(18) Berendsen, H. J. C.; Grigera, J. R.; Straatsma, T. P. *J. Phys. Chem.* **1987**, *91*, 6269.

(19) Levitt, M.; Sharon, R. *Proc. Natl. Acad. Sci. (USA)* **1988**, *85*, 7557.

(20) Jorgensen, W. L. *J. Am. Chem. Soc.* **1981**, *103*, 335.

(21) Jorgensen, W. L.; Chandrasekar, J.; Madura, J.; Impey, R.; Klein, M. L. *J. Chem. Phys.* **1983**, *79*, 926.

(22) Daggett, V.; Levitt, M. *Ann. Rev. Biophys. Biomol. Struct.* **1993**, *22*, 353.

(23) Miaskiewicz, K.; Osman, R.; Weinstein, H. *J. Am. Chem. Soc.* **1992**, *115*, 1526.

(24) Hirshberg, M.; Sharon, R.; Sussman, J. L. *J. Biomol. Struct. Dyn.* **1988**, *5*, 965.

(25) Van de Ven, J. M.; Hilbers, C. W. *Eur. J. Biochem.* **1988**, *178*, 1.

(26) Withka, J. M.; Swaminathan, S.; Beveridge, D. L.; Bolton, P. H. *Science* **1991**, *225*, 597.

(27) Withka, J. M.; Swaminathan, S.; Beveridge, D. L.; Bolton, P. H. *J. Am. Chem. Soc.* **1991**, *113*, 5041.

(28) Tung, C. S.; Harvey, S. C.; McCammon, A. J. *Biopolymers* **1984**, *23*, 2173.

(29) Auffinger, P.; Nirmala, R.; Ravishanker, G.; Beveridge, D. L. **1994**, *manuscript in preparation.*

(30) Tidor, B.; Irikura, K. K.; Brooks, B. R.; Karplus, M. *J. Biomol. Struct. Dyn.* **1983**, *1*, 231.

(31) Manning, G. S. *Quart. Rev. Biophys.* **1978**, *11*, 179.

(32) Arnott, S.; Chandrasekeharan, R.; Birdsall, D. L.; Leslie, A. G. W.; Ratliffe, R. L. *Nature* **1980**, *283*, 743.

(33) Drew, H. R.; Wing, R. M.; Takano, T.; Broka, C.; Tanaka, S.; Itakura, K.; Dickerson, R. E. *Proc. Natl. Acad. Sci. (USA)* **1981**, *78*, 2179.

(34) McClarin, J. A.; Frederick, C. A.; Wang, B.-C.; Greene, P.; Boyer, H. W.; Grabel, J.; Rosenberg, J. M. *Science* **1986**, *234*, 1526.

(35) Ravishanker, G. WESDYN 2.0, Wesleyan University: Middletown, CT, 1994.

(36) Ravishanker, G.; Beveridge, D. L. MD TOOLCHEST, Wesleyan University: Middletown, CT, 1994.

(37) Kumar, S.; Kollman, P. A.; Rosenberg, J. M. *J. Biomol. Struct. Dyn.* **1994**, *in press*.

(38) Young, M. A.; R., N.; Srinivasan, J.; McConnell, K. J.; Ravishanker, G.; Beveridge, D. L. In *Structural Biology: State of the Art: Proceedings of the Eighth Conversation*; E. R.H. Sarma and M. H. Sarma, Ed.; Adenine Press: Albany, N.Y., 1994; Vol. 2; pp 197.

(39) Berman, H. M.; Olson, W. K.; Beveridge, D. L.; Westbrook, J.; Gelbin, A.; Demeny, T.; Hsieh, S.-H.; Srinivasan, A. R.; Schneider, B. *Biophys J.* **1992**, *63*, 751.

(40) Fratini, A. V.; Kopka, M. L.; Drew, H. R.; Dickerson, R. E. *J. Biol. Chem.* **1982**, *257*, 14686.

(41) Dickerson, R. E.; Goodsell, D. S.; Kopka, M. L.; Pjura, P. E. *J. Biomol. Struct. Dyn.* **1987**, *5*, 557.

(42) Shakked, Z. *Curr. Opin. Struct. Biol.* **1991**, *1*, 446.

(43) Srinivasan, A. R. *Biophys. Chem.* **1992**, *43*, 279.

(44) Branden, C.; Tooze, J. *Introduction to Protein Structure*; Garland Publishing, Inc.: New York and London, 1991.

(45) Ptashne, M. *A Genetic Switch*; Cell Press/Blackwell Scientific Publications: Cambridge, MA., 1986, pp 192.

(46) Jordan, S. R.; Pabo, C. O. *Science* **1988**, *242*, 893.

(47) Clarke, N. D.; Beamer, L. J.; Goldberg, H. R.; Berkower, C.; Pabo, C. O. *Science* **1991**, *254*, 267.

(48) Pabo, C. O.; Lewis, M. *Nature* **1982**, *298*, 443.

(49) Thomas, G. J., Jr.; Prescott, B.; Benevides, J. M. *Biochemistry* **1986**, *25*, 6768.

RECEIVED April 25, 1994

Chapter 27

Interaction of a Model Peptide with a Water–Bilayer System

A. Pohorille and M. A. Wilson

Department of Pharmaceutical Chemistry, University of California, San Francisco, CA 94143

We discuss a molecular dynamics study of the alanine dipeptide at the interface between water and a glycerol-1-monooleate (GMO) bilayer. The dipeptide is interfacially active and incorporates into the bilayer without disrupting its structure. The interfacial region that is readily penetrated by the dipeptide spans the entire head group portion of the bilayer. The polar groups of the alanine dipeptide mostly remain well solvated by water and the oxygen atoms of GMO, and conformations of the dipeptide are characterized by (ϕ, ψ) angles typical of α-helix and β-sheet structures. When the molecule is deeper in the bilayer, the C_{7eq} state also becomes stable. The barrier to the isomerization reaction at the interface is lower than in bulk phases. After 7 ns of trajectories, the system is not fully equilibrated, due to slow collective motions involving GMO head groups. These result in decreased mobility and lower rates of isomerization of the dipeptide at the interface.

Peptides with affinity for membrane surfaces exhibit a broad range of cellular activities. They are hormones, antibiotics, toxins, and fusogenic and signal peptides (*1-3*). Their biological functions are determined by their secondary structure at the membrane surface. This secondary structure is often different at the membrane–water surface than in aqueous solution. Recent studies of the conformational changes induced by the membrane surface in synthetic, model molecules revealed two remarkable features of peptide–membrane interactions. First, short peptides, which are disordered in water, can acquire secondary structure at the water–membrane interface. Second, polarities rather than the specific identities of residues determine the secondary structure of the peptide at membrane surfaces. Peptides with alternating polar/nonpolar residues form β-sheets (*4,5*), while α-helices have polar and nonpolar amino acids arranged so that they match the period of this structure (*5-7*). This indicates that the amphiphilic nature of

0097–6156/94/0568–0395$08.00/0

the water–membrane interface, where polar and nonpolar phases exist in direct proximity, affects the conformational preferences of peptides.

Since experimental methods for studying membrane–bound peptides suffer from several difficulties, such as limited residence times of these peptides at the interface, poor water solubilities of some amino acid sequences, and inability to obtain crystal structures in most cases of interest, molecular–level computer simulations can potentially play a significant role in determining how peptides interact with water–membrane interfaces and how these interactions depend on amino acid sequence. Despite the obvious biological importance of these problems, no such simulations have been reported so far, probably due to difficulties in proper treatment of highly anisotropic interfacial environments. However, recent progress in our understanding of the structure and properties of aqueous interfaces (8), and water–membrane systems in particular (9-14) render realistic simulations of small peptides at membrane surfaces possible.

In this paper, we present results of molecular dynamics simulations of the alanine dipeptide, which is probably the simplest model peptide, at the interface between water and a lipid membrane formed by glycerol-1-monooleate (GMO). This work builds on our previous studies of the water–GMO membrane system (14) and the alanine dipeptide at the water–hexane interface (15). While this peptide is obviously too short to form a secondary structure, it has been considered a good model for examining conformational preferences of the protein backbone. Several computational studies have show that these preferences depend on the environment. The conformations of the alanine dipeptide corresponding to the local energy minima in the gas phase (16,17) and in hexane (15) are stabilized by intramolecular hydrogen bonds. In water, the hydrogen bonds are destabilized by hydration effects and the most stable conformations are characterized by ϕ and ψ angles typical of α-helices and β-sheets (17-19). At the water–hexane interface both types of conformations are populated (15).

Based on our results for the alanine dipeptide in the water–hexane system, we expect that this molecule exhibits interfacial activity at the water–membrane interface. One of our objectives is to determine how this complex interface influences the conformational preferences of the dipeptide backbone. Further, we want to obtain a microscopic description of the environment surrounding the dipeptide and its orientational preferences at the interface. Finally we would like to gain information about the relative time scales of processes affecting the peptide, such as diffusion and conformational transition of the solute, and relaxation of the solvent. These issues are addressed in the Results and Discussion section which follows the description of the system and the methods used.

Method

The system under study consisted of two alanine dipeptide molecules, 72 GMO molecules, and 2274 water molecules in the simulation box of 36.94 Å × 36.94 Å × 150 Å. The chemical structures of the dipeptide and the GMO molecule with the numbering system used in the text are shown in Figure 1a. The head group of GMO contains atoms 18–27; the rest of the molecule constitutes the hydrocarbon tail (atoms 1–17). The GMO molecules were arranged as a planar bilayer perpendicular to the z-direction of the simulation box. The xy-dimensions of the system were set to reproduce the experimental surface density of molecules in the bilayer (*20*). The membrane was located between two water lamellae, each containing 1137 water molecules, as is schematically represented in Figure 1b.

The arrangement of the system was the same as that used in our molecular dynamics simulations of the water–GMO bilayer (*14*). In those calculations, it was found that the width of the membrane, measured as the distance between peaks in the density of the head–group atoms on both sides of the bilayer, is 30 Å. This density decreased to 10% of its maximum value at ±19.8 Å from the middle of the bilayer. The width of each water lamella was 26 Å, which correspond to approximately 8 layers of water. On the bilayer side, the water density decreased to 90% and 50% of the value in bulk liquid at ±19.0 and ±16.5 Å, respectively.

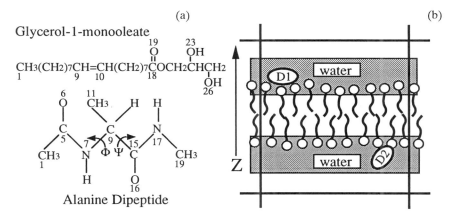

Figure 1. (a) Chemical structures of GMO and the alanine dipeptide. The dipeptide angles ϕ and ψ are marked; (b) Schematic of the water–bilayer system. The alanine dipeptide molecules are shown in the interfacial region.

These results guided our choice for the initial locations of the two alanine dipeptide molecules in the system. They were placed in aqueous solution close to the interface with the bilayer, such that their α carbon atoms (C(9)) were at $z = \pm 24$ Å from the middle of the bilayer. The locations of the dipeptide molecules in the system is schematically depicted in Figure 1b. Three initial configurations were constructed by placing the alanine dipeptides in configurations of the water–bilayer system drawn from the equilibrium distribution obtained in our previous work (*14*) and removing any water molecules which overlapped the solutes. The x, y-coordinates of C(9) and the molecular orientation of the peptide were chosen at random. The initial conformations of the alanine dipeptide corresponded to the energy minima found for this molecule at the water–hexane interface (*15*)— α_R (-60,-40), β (-60,150), C_5 (-160,150) and C_{7eq} (-60,60), where the ϕ and ψ angles for each conformation are given in parentheses. Other conformations which are also stable at the water–hexane interface, but are separated from this set of conformations by fairly high barriers, such as α_L (60,60) and C_{7ax} (60,-60), were not considered in this work.

Potential Energy Functions. The potential energy functions for GMO and water molecules were the same as those used for the water–bilayer system (*14*). Water molecules were described by the TIP4P model (*21*). In the GMO molecule, all CH_n groups were represented as united atoms, whereby the hydrogen atoms were not explicitly considered and the carbon atom was assigned the mass of the whole group. All bond lengths and bond angles were kept rigid. In this approximation, the potential energy was expressed as a sum of torsional potentials of GMO molecules, van der Waals energies between atoms (or groups) and Coulomb energies between partial charges on water molecules and GMO head groups. The details of the potential energy functions, the atomic partial charges and the van der Waals and torsional parameters were given previously (*14*).

The alanine dipeptide was represented by a flexible model with full atomic detail. The potential energy function contained contributions from bond stretching and bending of planar angles in addition to torsional, van der Waals and Coulomb terms. This potential has been described in our study of the dipeptide at the water–hexane interface (*15*). The relative free energies of stable conformations of the alanine dipeptide in the gas phase and in water, calculated in that work, are in close agreement with the free energies recently obtained by Tobias and Brooks (*17*). The dipeptide–water and dipeptide–GMO potential energy functions were evaluated from the Lorentz–Bertholet combination rules (*22*).

Molecular Dynamics. Three molecular dynamics trajectories, each 1 ns long, were generated, starting from the configurations described above. Assuming that interactions between the solute molecules across the bilayer can be neglected, six independent trajectories for the alanine dipeptide were obtained.

The equations of motion were integrated using the Verlet algorithm with a time step of 2.5 fs. The temperature of the system was 300 K. Minimum image periodic boundary conditions were applied in all three dimensions. The bond lengths and planar angles of the GMO and water molecules were kept fixed by using the SHAKE algorithm (*23*). No constraints were applied to the degrees of freedom of the dipeptide. The intermolecular interactions were truncated smoothly between distances of 7.5 and 8.0 Å, as measured between centers of charge–neutral groups (*14*).

Free Energy Calculations. The stable conformations of the alanine dipeptide in the ϕ, ψ range studied here, α_R, β and C_{7eq} are characterized by approximately the same values of ϕ. Thus, the free energy profiles for transitions between these conformations can be described as a function of only ψ. These profiles, $A(\psi)$, were obtained from the probability distributions, $P(\psi)$, of finding the molecule in a conformation with by angle ψ:

$$A(\psi) = -kT \ln P(\psi) \tag{1}$$

where k is the Boltzmann constant and T is temperature.

As described in the next section, a sufficient number of transitions was observed in some cases to permit a direct calculation of $P(\psi)$ from the molecular dynamics trajectory (after equilibration). In other instances, additional molecular dynamics calculations were needed in which ψ was restricted to different, but overlapping ranges. Four such additional trajectories, 0.35 ns long, were obtained in each instance. In these trajectories the angle ϕ remained unconstrained. To ensure uniform sampling of each region and, therefore, improve statistical precision of the resulting $P(\psi)$ an additional external potential $U_{ext}(\psi)$ was sometimes added. Then, $A(\psi)$ in a given range of ψ is obtained as:

$$A(\psi) = -kTlnP(\psi) - U_{ext}(\psi) \qquad (2)$$

and the full free energy profile is constructed by requiring that $A(\psi)$ be a continuous function of ψ (*24*).

Results and Discussion

Interfacial Activity of the Alanine Dipeptide. The time evolution of the positions of the six alanine dipeptide molecules along the direction perpendicular to the interface is shown in Figure 2a. These positions are defined as the absolute values of the z-coordinate of atom C(9) measured from the mid–plane of the bilayer. The dipeptide molecules move from water to the interface in the first 0.3–0.5 ns of the molecular dynamics trajectories. Each then explores a wide interfacial region occupied by GMO head groups and water molecules, which extends approximately from $|z|=10$ Å to $|z|=21$ Å. There are, however, differences in the degree to which the molecules penetrate the head group portion of the bilayer. In particular, molecule 6 appears to be located, on average, deeper in the bilayer than other molecules. As will be shown further in the paper, this molecule behaves somewhat differently than molecules 1-5 and, therefore, will be discussed separately. A question may be raised whether molecule 6 has reached its equilibrium position or, alternatively, will continue to move deeper into the bilayer. This issue was resolved by performing an additional 1.3 ns of molecular dynamics trajectory for the system containing this molecule. No deeper penetration of the solute into the membrane was observed during this trajectory.

In Figure 2b we present the free energy profile of the alanine dipeptide in the interfacial region estimated from the probability distribution along z averaged for molecules 1-5 and, separately, from the distribution for molecule 6. We note three important features of these results: First, the alanine dipeptide is interfacially active. This conclusion is further supported by a series of short molecular dynamics trajectories in which the dipeptide was forced into the hydrocarbon core of the bilayer. When the external force was removed and some equilibration was performed, the dipeptide returned to the interface in less than 0.05 ns. A similar interfacial activity is exhibited by the dipeptide at the water–hexane interface (*15*). It was found that this molecule is stabilized at the interface, relative to aqueous and hexane solutions, by 5.5 and 10.0 kcal/mol, respectively. Thus, it appears that, in general, the alanine dipeptide binds to interfaces between water and nonpolar phases.

Second, the interfacial region that is readily accessible to the dipeptide is remarkably broad. The averaged free energy for molecules 1-5 is almost constant in the range $13 < |z| < 20$ Å. The results for molecule 6 show that penetration into the bilayer by another 2 Å is also probable. This is different from the situation at the water–hexane interface, where the free energy along z exhibits a much sharper minimum near the plane where the densities of water and hexane decrease to the half of their bulk values. The observed width of the free energy profile at the water–membrane interface does not appear to be unique to the alanine dipeptide. A very similar profile was found for the anesthetically active molecule, 1,1,2-trifluoroethane (*25*). These results suggest that small, neutral solutes active at interfaces between water and nonpolar media experience considerably greater freedom of movement at the water–membrane interface than at the water–oil interface.

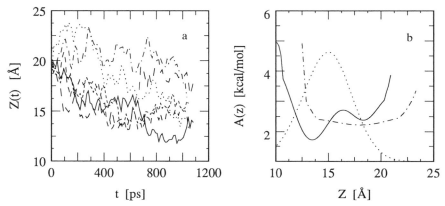

Figure 2. (a) Trajectories of the dipeptide molecules perpendicular to the membrane surface. Run 6 is the solid line; (b) Free energy profiles of the alanine dipeptide as a function of z, for runs 1–5 (dash–dotted line) and run 6 (solid line). The density profile of the GMO head groups (dotted lines) is shown as a reference. The center of the bilayer is located at $z = 0$.

Third, we note that differences between the free energy profiles along z for molecules 1–5 and molecule 6, which did not disappear during the course of the simulations, probably indicate the existence of slow relaxation modes in the solvent which were not fully equilibrated. Other consequences of this effect and structural differences in the environment around molecules 1–5 and molecule 6 will be discussed farther in the paper.

Conformations of the Alanine Dipeptide at the Membrane–water Interface. The trajectories of the six alanine dipeptide molecules yield somewhat different results regarding conformational equilibria and transitions. Molecules 1-3 remained in their initial (after equilibration) α_R or β conformations during the entire trajectory. In contrast, molecules 4–6 underwent isomerization. For molecules 5 and 6, several transition between different conformations were observed, so that it was possible to estimate the free energy profile for the isomerization reaction from the probability distribution of finding different values of ψ during the trajectory. For molecules 3 and 4 this profile was obtained from a series of calculations in which ψ was constrained in different ranges, as described in the method section. The results are summarized in Table I and the four free energy profiles are shown in Figure 3.

The observed differences between the trajectories indicate that the system does not fully equilibrate in the course of the simulations. Nevertheless, several conclusions can be drawn from the results. The conformations α_R and β, which are stable in aqueous solution, are also stable at the water–membrane interface. The free energy difference between α_R and β does not change significantly upon transferring the dipeptide from water to the interfacial region. At the interface, this difference ranges from 0.0 to 1.0 kcal/mol for molecules 3–5, compared to 0.4 kcal/mol in the aqueous solution (*15*). The position of the free energy barrier separating the two stable conformations is also similar to that for the dipeptide in water. However, the height of the barrier at the interface is 1.6 kcal/mol (measured from the α_R state), only about half of that in aqueous solution. A statistically significant difference in the relative stabilities of the α_R and β conformations between molecule 3 and molecules 4 and 5 is a consequence of slowly relaxing modes in the system. This point is further underscored by the fact that molecule 6 exhibits a different free energy profile. This profile resembles that obtained previously for the alanine dipeptide at the water–hexane interface (*15*). In addition to the α_R and β conformations, C_{7eq} is also stable. This corresponds to a free energy minimum in hexane but not in water. The barriers are again markedly lower than those in bulk aqueous and hexane phases.

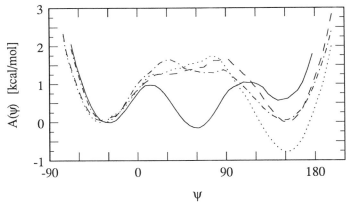

Figure 3. Free energy profiles for $\alpha_R \rightleftharpoons \beta$ isomerization at the membrane interface for molecules 3 (dashed line), 4 (dotted line), 5 (dash–dotted line) and 6 (solid line).

The observed differences in the calculated free energy profiles raise an important question whether an internal solute degree of freedom (angle ψ) is a proper choice for the reaction coordinate. In general, such a choice is satisfactory when other degrees of freedom of the system, in particular those associated with the solvent, equilibrate rapidly, compared to the relaxation time of the reaction. Then, the reaction can be considered as proceeding along the reaction coordinate in the potential of mean force exerted by other degrees of freedom. However, if some of these degrees of freedom relax slowly the rate of the reaction depends on their specific values. This is apparently the case here and, therefore, the calculated free energy profiles depend not only on ψ but also on some "hidden variable(s)" which describe the slow relaxation of the system and can facilitate or, alternatively, impede the isomerization reaction. Examples of such structural "variables" would be water molecules protruding unusually deep into the bilayer to hydrate the solute, or GMO molecules tightly bound to the dipeptide. Such a phenomenon was observed by Benjamin (*26*) in his study of the transfer of an ion from 1,2-dichloroethane to water. He found that the reaction proceeds readily only on fairly rare occasions when a chain of water molecules (a "water finger") develops between the aqueous phase and the ion. Despite extensive analysis, we were unable to identify any such specific interactions in the present case. This points to the participation of collective modes in the slow relaxation of the system. This conclusions is reinforced by the analysis of the solvation of the dipeptide, the structure of the membrane around this molecule, and dynamic properties of the system, presented below.

Table I. Summary of the Results for Conformational Equilibria of the Alanine Dipeptides at the Water–Bilayer Interface

	molecules:					
	1	2	3	4	5	6
conformations[a]	α_R	β	β	α_R,β	α_R,β	C_{7eq},α_R,β
$\Delta A(\alpha_R - \beta)$ [b]	d	d	-0.8	0	0	0.5 (-0.1)[e]
$\Delta A^{\ddagger}$ [c]	d	d	1.7	1.6	1.6	1.2

[a] conformations observed in the trajectories with no constraints on ψ; [b] the free energy difference between α_R and β conformations, in kcal/mol.; [c] activation free energy for the $\alpha_R \rightleftharpoons \beta$ isomerization reaction; [d] not calculated; [e] the free energy difference between α_R and C_{7eq} in parenthesis.

Hydration of Alanine Dipeptide at the Interface.   Earlier studies of the alanine dipeptide in aqueous solution (*17-19,27*) stressed that the α_R and β conformations are strongly stabilized by water relative to the conformations containing intramolecular hydrogen bonds. Since the same conformations are also stable at the water–bilayer interface, it raises a question to what extent the alanine dipeptide remains solvated at the surface of the membrane. To answer this question we compared the calculated radial distribution functions between oxygen atoms of water and oxygen atoms, methyl groups and nitrogen atoms of the dipeptide in aqueous solution and at the water–bilayer interface. The distribution functions for the water oxygen atoms and the oxygen atoms of the dipeptide are shown in Figure 4a. From the average of runs 1–5, it is clear that the peptide becomes partially dehydrated upon transfer from bulk water to the interface. Integration of the oxygen–oxygen radial distribution functions to the first minimum yields hydration numbers of 2.2 and 1.6 for the alanine dipeptide in aqueous solution and at the water–bilayer interface, respectively. A similar pattern of dehydration is observed for the methyl groups (Figure 4b) and nitrogen atoms (not shown).

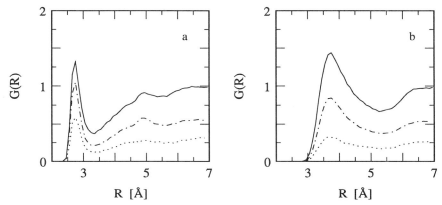

Figure 4. (a) Radial distribution function between the water oxygen and the alanine dipeptide oxygen atoms in bulk water (solid line), averaged over molecules 1-5 (dash–dotted line), and for molecule 6 (dotted line); (b) the same as in (a), but for the water oxygen atoms with the dipeptide methyl carbon atoms.

At the interface the alanine dipeptide is surrounded not only by oxygen atoms from water but also from the GMO head groups. Integration of the radial distribution function between oxygen atoms in the dipeptide and the head groups yields a solvation number of 0.4. This yields a total solvation number for the oxygen atoms in the dipeptide of 2.0, a value close to that in aqueous solution. Thus, the loss of hydration around the dipeptide at the interface is almost fully offset by its interactions with the oxygen atoms in GMO head groups. While some differences were observed in the degree to which the five molecules were hydrated, their *total* solvation numbers were almost identical, irrespective of the dipeptide conformation.

The radial distribution functions for molecule 6 differ markedly from those for molecules 1-5. As can be seen from Figure 4, molecule 6 undergoes extensive dehydration. The hydration and total solvation numbers for oxygen atoms of this molecule are 0.9 and 1.3, respectively.

These results provide a link between solvation of the alanine dipeptide at the interface and its conformational preferences. Molecules 1–5 remain in locally polar environments, and their conformational preferences for the α_R and β structures are similar to those of the dipeptide in water. In contrast, molecule 6 looses about 40% of its solvation. This degree of desolvation is similar to that at the water–hexane interface. As a consequence, the conformational preferences of the dipeptide in these two cases are similar. In particular, a considerable population of the C_{7eq} conformation is present in the absence of strong solvation. Desolvation of the polar oxygen atoms of molecule 6 destabilizes this molecule at the interface, However, this is accompanied by a compensating gain in stability resulting from the removal of methyl groups from water.

Arrangement of the Alanine Dipeptide and the Neighboring GMO Molecules at the Interface. In this section we address two questions: (1) what is the orientation of the alanine dipeptide at the interface, and (2) how are the nearby head groups arranged to accommodate the solute? The orientational preferences of the dipeptide can be described by the probability distribution of finding different values of the angles between the normal to the interface and three vectors – the vector from C(1) to C(19), pointing along the backbone of the molecule, and the carbonyl bond vectors C(5)–O(6) and C(15)–O(16), pointing approximately perpendicular to the backbone.

The orientational distribution of the angle between C(1)–C(19) and the normal, shown in Figure 5a, is fairly broad. Even broader distributions are obtained for the other angles, indicating that the dipeptide exhibits considerable orientational freedom at the interface. Orientations in which the dipeptide lies parallel to the interface are preferred, because perpendicular arrangements would require dehydration of one of the oxygen atoms, destablizing these arrangements. Similar preferences were also observed at the water–hexane interface (*15*).

Similar information about the arrangement of the alanine dipeptide at the water–bilayer interface is conveyed by the density profiles of oxygen atoms, nitrogen atoms and methyl groups of this molecule in the direction perpendicular to the interface, shown in Figure 5b. As we can see, they do not differ dramatically from one another. One feature worth noting is that the hydrophobic methyl groups penetrate somewhat deeper into the nonpolar core of the bilayer than do the polar oxygen and nitrogen atoms.

The fact that near the interface the dipeptide does not lie with its hydrophilic groups pointing toward water and its hydrophobic groups pointing toward the membrane can be understood by noting that this molecule is not amphiphilic. Although the dipeptide contains both polar and nonpolar groups, they cannot be segregated into different environments due to constraints imposed by the chemical structure of the molecule. Instead, a compromise is reached, whereby the nonpolar groups are partially removed from water while the polar groups retain most of their solvation shell. This can be accomplished in a relatively broad range of orientations. In contrast, it has been shown both experimentally (*28*) and computationally (*29,30*) that typical amphiphilic molecules, such as phenol and *p-n*-pentylphenol, become quite rigid at the water liquid–vapor interface, so that simultaneously the hydrocarbon part is excluded from the water and the polar alcohol group remains hydrated.

When the dipeptide moves into the interfacial region, the GMO molecules must undergo some structural changes to accommodate it. Extensive analysis of the structure of the GMO molecules near the dipeptide reveals that these changes are very subtle. Most of the structural properties of the bilayer, including those of the hydrocarbon core, are not affected by the presence of the dipeptide. It therefore appears that the dipeptide is able to slip into the head group region without causing any major structural perturbations to the bilayer. The transfer of the dipeptide into the interfacial region becomes possible through a series of small adjustments of the orientations and conformations of both the solute and the neighboring GMO molecules. It is possible that this is a generally preferred way in which the bilayer interface interacts with fairly small, neutral species, as large structural perturbations would tend to destablize the bilayer. However, this conclusion does not apply to ions, which induce large changes in the structure of the membrane and the interfacial water (*25*).

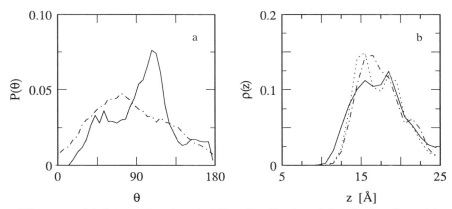

Figure 5. (a) Orientational probability distribution of the alanine dipeptide vectors C(1)–C(19) (solid line), and C=O (dash–dotted line). with respect to a vector normal to the bilayer surface pointing from the bilayer into the water phase; (b) Density profiles for the dipeptide methyl groups (solid line), oxygen atoms (dash-dotted line), and nitrogen atoms (dotted line) at the membrane interface. The center of the bilayer is located at $z = 0$, and the water lies to the right ($z > 16$).

Dynamics of the Alanine Dipeptide at the Interface.   The different results on the conformational equilibria of the alanine dipeptide obtained from different MD trajectories indicate that on nanosecond time scales the system is not in equilibrium with respect to slow solvent relaxation. On the other hand, the broad distributions of orientations and positions of the dipeptide along the z-direction of the bilayer suggest that some molecular motions are fairly well equilibrated on the same time scale. The existence of different types of motions in membranes which relax on different time scales has recently been discussed in the context of computer simulations of a dipalmitoylphosphatidylcholine bilayer (*13*). In that study, it was suggested that reorientational dynamics of carbon atoms is fast, while lateral diffusion in the head group region is slow.

Based on our results the following picture may be proposed: The alanine dipeptide is in a "cage" of several neighboring GMO molecules. It possesses considerable orientational and positional freedom as long as its movement requires only reorientation or small structural adjustment of the neighboring GMO molecules. Any large motions which require reorganization of the local environment would be impeded by the GMO "cage". The head group region should be especially rigid while the environment deeper in the bilayer might be more flexible. The dipeptide molecules which penetrate deeper into the bilayer would therefore exhibit more conformational freedom. Note, that the transition between α_R and β conformations is associated with a fairly substantial molecular rearrangement in which ψ changes by about 180°.

One measure of fast dynamics in the system is the average displacement correlation function of the dipeptide. This function is linear on the time scale of 1-5 ps, allowing us to calculate the coefficient of self–diffusion from its slope. In the direction parallel to the bilayer, this coefficient equals 0.05×10^{-5} cm^{2}sec^{-1}. By comparison, the self–diffusion coefficients of the dipeptide in water, hexane and at the water–hexane interface are 0.085, 0.34, and 0.16×10^{-5} cm^{2}sec^{-1}, respectively. Thus, self–diffusion of the alanine dipeptide in all those environments is faster than that at the water–bilayer interface. It appears that solute diffusion at the membrane surface over longer time scales is even slower. This issue is currently being investigated and the results will be published separately.

If the idea that conformational transitions in the alanine dipeptide cause substantial disruptions in the "cage" of the nearby GMO molecules is correct, then a decrease in the rate of isomerization due to solvent effects should be observed. In the transition state theory approximation, every trajectory which reaches the transition state leads to the formation of product. However, when the solvent reorganization lags behind the evolution of the solute along the reaction coordinate, solvent molecules tend to pull the solute back to the reactant state. A convenient way to analyze this effect in isomerization reactions is by calculating the reactive flux correlation function (31). A conventional method for obtaining the reactive flux is to monitor a series of short trajectories initiated from the transition state. In the present case, this approach would require additional, extensive calculations. However, by taking advantage of fairly frequent conformational transitions in molecules 5 and 6, we have estimated the dynamic solvent effect from the existing trajectories. The trajectory was monitored for 2 ps after it crossed the transition state. After several recrossings, the system stabilized in either the reactant or the product well. The observation resumed upon the next crossing of the transition state, etc. The exact length of a single observation had no effect on the results.

A total of 167 transition events between the α_R and β (or C$_{7eq}$) conformations were observed, which approximately matches the predictions of transition state theory (32). Of these, 76 started in the α_R state and 91 started in the β or C$_{7eq}$ state. During each event, the system crossed the transition state 9 times on average. Only 29 events were reactive, i.e. led to the isomerization, yielding a transmission coefficient of 0.17. This value is markedly lower than that obtained in our study of the isomerization of 1,2-dichloroethane at the water–hexane interface (15). Thus, as expected, the solvent exerts dynamical effects on the alanine dipeptide and lowers the isomerization rate.

Conclusions

The alanine dipeptide exhibits a free energy minimum at the water–membrane interface, even though it does not have well–defined hydrophobic and hydrophilic parts and, therefore, should not be considered as amphiphilic. Such interfacial activity appears to be a general feature of molecules which possess sufficient hydrophilicity and lipophilicity. The interface "as seen" by the dipeptide extends over a wide region in which the molecule can move without free energy cost. In contrast, the interfacial region in the water–hexane system is substantially narrower. This is probably due to the fact that in the water–membrane system the environment changes from polar to nonpolar progressively in the head group region while the water–hexane interface is quite sharp.

Over most of the interfacial region the polar groups of the dipeptide remain well solvated by water molecules and the oxygen atoms of the GMO head groups. This probably determines that the α_R and β conformations, which are stable in aqueous solution, are also preferred at the water–membrane interface. Only when the dipeptide penetrates into the bilayer side of the head group region does partial desolvation occur and the C_{7eq} conformation, stabilized by an intramolecular hydrogen bond, becomes significantly populated. A distinguishing effect of the interface is that the free energy barriers to isomerization are considerably reduced compared to those in bulk water and hexane phases.

Accurate determination of conformational equilibria and barriers from molecular dynamics simulations appears to be quite difficult due to slowly relaxing motions in the system which impede equilibration. Their presence was supported by calculations of the dipeptide mobility and dynamic solvent effects on the isomerization reaction. The exact nature of these motions is not well understood but most likely they are associated with collective reorganization of membrane head groups. A better understanding of these slow relaxation phenomena is needed before proceeding to large–scale molecular simulations of longer peptides which exhibit secondary structure at the water–membrane interface.

Acknowledgments. This work was supported by NASA–UCSF Consortium Agreement No. NCC 2–772 and NASA–UCSF Joint Research Interchange No. NCA 2–792. Computer facilities were provided by the National Aerodynamics Simulator (NAS) and by the National Cancer Institute.

Literature Cited

1 Kaiser, E. T.; Kezdy, F. J. *Ann. Rev. of Biophys. Biophys. Chem.* **1987**, *16*, 561–582.

2. Sargent, D. F.; Schwyzer, R. *Proc. Natl. Acad. Sci. USA* **1986**, *83*, 5774–5778.

3. White, J. M. *Annu. Rev. Physiol.* **1990**, *52*, 675–697.

4. Ono, S.; Lee, S.; Mihara, H.; Aoyagi, H.; Kato, T.; Yamasaki, N. *Biochim. Biophys. Acta* **1990**, *1022*, 237–244.

5. DeGrado, W. F.; Lear, J. D. *J. Am. Chem. Soc.* **1985**, *107*, 7684–7689.

6. Parente, R. A.; Nadasdi, L; Subbarao, N. K.; Szoka, F. C. *Biochemistry* **1990**, *29*, 8713–8719.

7. Kato, T.; Lee, S.; Ono, S.; Agawa, Y; Aoyagi, H.; Ohno, M; Nishino, N. *Biochim. Biophys. Acta* **1991**, *1063*, 191–196

8. Pohorille, A.; Wilson, M. A. *J. Mol. Struct. (Theochem)* **1993**, *284*, 271–298.

9. Raghavan, K.; Reddy, M. R.; Berkowitz, M. L. *Langmuir* **1992**, *8*, 233–240.

10. Damodaran, K. V.; Merz, K. M.; Garber, B. P. *Biochemistry* **1992**, *31*, 7656–7664.

11. Heller, H.; Schaefer, M.; Schulten, K. *J. Phys. Chem.* **1993**, *97*, 8343–8360.

12. Bassolino-Klimas, D.; Alper, H. E.; Stouch, T. R. *Biochemistry* **1993**, *32*, 12624–12637.

13. Venable, R. M.; Zhang, Y; Hardy, B. J.; Pastor, R. W. *Science* **1993**, *262*, 223–226.

14. Wilson, M. A.; Pohorille, A. *J. Am. Chem. Soc.* **1994**, *116*, 1490–1501.

15. Pohorille, A; Wilson, M. A. In *Reaction Dynamics in Clusters and Condensed Phases*; Jortner, J.; Levine, R. D. and Pullman, B., Eds.; The Jerusalem Symposia on Quantum Chemistry and Biochemistry; Kluwer, Dordrecht, 1993, Vol. 26; pp 207–226.

16. Czerminski, R.; Elber, R. *J. Chem. Phys.* **1990**, *92*, 5580–5601.

17. Tobias, D. J.; Brooks, C. L. *J. Phys. Chem.* **1992**, *96*, 3864–3870.

18. Ravishanker, G.; Mezei, M.; Beveridge, D. L. *J. Comput. Chem.* **1986**, *7*, 345–348.

19. Anderson, A.; Hermans, J. *Proteins* **1988**, *3*, 262–273.

20. White, H. S. *Biophys. J.* **1978**, *23*, 337–347.

21. Jorgensen, W. L.; Chandrasekhar, J.; Madura, J. D.; Impey, R. W.; Klein, M. L. *J. Chem. Phys.* **1983**, *79*, 926–935.

22. Jorgensen, W. L.; Madura, J. D.; Swenson, C. J., *J. Am. Chem. Soc.* **1984**, *106*, 6638–6646.

23. Ciccotti, G.; Ryckaert, J. P. *Comput. Phys. Rep.* **1986**, *4*, 345–392.

24. Chandler, D. *Introduction to Modern Statistical Mechanics*; Oxford: New York, NY, 1987; pp 168–175.

25. Pohorille, A.; Wilson, M. A. *Origins of Life and Evolution of the Biosphere* **1994**, in press.

26. Benjamin, I. *Science* **1993**, *261*, 1558–1560.

27. Pettitt, B. M.; Karplus, M. *J. Phys. Chem.* **1988**, *92*, 3994–3997.

28. Hicks, J. M.; Kemnitz, K.; Eisenthal, K. B.; Heinz, T. F. *J. Phys. Chem.* **1986**, *90*, 560–564.

29. Pohorille, A.; Benjamin, I. *J. Chem. Phys.* **1992**, *94*, 5599–5605.

30. Pohorille, A.; Benjamin, I. *J. Phys. Chem.* **1993**, *97*, 2664–2670.

31. Chandler, D. *J. Chem. Phys.* **1978**, *68*, 2959–2970.

32. Pechukas, P. In *Dynamics of Molecular Collisions; Part B*, W. H. Miller, Ed.; Plenum: New York, NY, 1976.

RECEIVED April 25, 1994

Chapter 28

A Molecular Model for an Electron-Transfer Reaction at the Water–1,2-Dichloroethane Interface

Ilan Benjamin

Department of Chemistry, University of California, Santa Cruz, CA 95064

The structure, energetics and dynamics involved in a model electron transfer reaction at the interface between water and 1,2-dichloroethane are investigated by molecular dynamics computer simulation. The molecular model provides information about the spatial localization of the redox pair and its orientation at the interface, which is generally missing from other treatments of the problem. Although each center is mainly solvated by one of the liquids, due to surface roughness, significant contributions from both liquids to the electrostatic potential fluctuations at each site are observed. The solvent free energy curves for the reaction are computed by an umbrella sampling procedure and are shown to be in very good agreement with the linear response approximation. Solvent relaxation following photochemically induced "vertical" electron transfer is shown to contain components from both liquids on different time scales.

Electron transfer (ET) reactions at the interface between two immiscible electrolyte solutions (ITIES) are of considerable interest in many diverse areas such as electrochemistry, hydrometallurgy and biophysics. Despite the importance of these reactions, our understanding of the basic molecular factors which control the thermodynamics and dynamics of ET at liquid/liquid interfaces is limited compared to the enormous progress that has been made in the study of ET reactions in bulk solution, at metal and semiconductor interfaces and in biomolecules. This situation is a result of the lack of experimental information about the structure and dynamics of the liquid-liquid interface on the relevant distance and time scales. Indeed, most of the experimental data about this system have been obtained using classical electrochemical techniques (1-5) which only indirectly probe processes at the interface, and which depend on relatively crude models for the extraction of the relevant kinetic information.

0097–6156/94/0568–0409$08.00/0

This situation is beginning to change with the development and application of new experimental techniques which have the potential of directly probing the interfacial region. Information about the roughness of the oil-water interface has been obtained by specular reflection of neutrons (6) and measurements of reorientation dynamics using fluorescence anisotropy decay have provided insight into the molecular nature of the roughness of water/alkane interfaces (7). The ability of optical second harmonic generation (SHG) to selectively study the orientation and adsorption of molecules at the water/CCl$_4$, water/decane (8) and water/1,2-dichloroethane (DCE) (9) interfaces has been demonstrated. Corn and coworkers have also used optical SHG to probe photoinduced electron transfer at the water/DCE interface (10, 11). Nanometer-sized tip electrodes, which have been used to carry out electrochemical measurements at the solution-polymer interface (12) may have the potential to probe ET at ITIES.

There have been very few theoretical treatments of electron transfer reactions at the liquid-liquid interface, and most of these were limited to a continuum model description of the liquids. Kharkats and Vol'kov (13) calculated the reorganization free energy for electron transfer at a continuum electrostatic model of a liquid-liquid interface. Kuznetsov and Kharkats (14) also used the continuum electrostatic model to calculate the reorganization free energy and work terms for different orientations of the redox couple at the interface. Marcus has presented an electrostatic model for the reorganization free energy at the liquid-liquid interface (15) and he has obtained an expression for the rate constant in the non-adiabatic limit for a reaction between a donor in one phase and an acceptor in another phase (16, 17).

The above treatments did not consider several important issues, some of which are impossible to address using a continuum solvent model: (i) A key issue is the applicability of the linear response approximation, which is at the heart of the reorganization free energy calculations; (ii) A knowledge of the interface structure is a fundamental requirement for any model of ET at the interface, because this structure determines the spatial location and distribution of the reactants; (iii) The dynamical response of the solvents at the interface may be of special importance for recent photochemically induced electron transfer experiments at the liquid-liquid interface (10). Several theoretical and experimental studies have indicated that a continuum model and the linear response approximation may not be appropriate for a description of the solvent dynamical response in bulk solution (18-22).

We have previously used a molecular model for an ET reaction at the interface between a model non-polar liquid and dipolar liquid and have considered the applicability of the linear response for the calculations of the solvent free energy (23) and the solvent dynamical response (Benjamin, I. *Chem. Phys.*, in press.) In this paper, we use a recently developed realistic model for the interface between water and 1,2-dichloroethane (DCE) (24) to examine the issues listed above for an electron transfer at this (experimentally popular) interface. In addition to the examination of the solvent free energy and the solvent dynamical response, we discuss the spatial localization of the reactants at the interface, a property for which experimental information is beginning to emerge (9).

The Neat Interface

The system includes 108 DCE molecules and 343 water molecules in an elongated parallelepiped of cross section 21.7Å × 21.7Å parallel to the *XY* plane. The water potential energy surface is described using a flexible SPC model (*25*) including the intramolecular potential of Kuchitsu and Morino (*26*). The DCE intermolecular potential is modeled using four Lennard-Jones centers with point charges and parameters which are selected to give reasonable structural and thermodynamical properties of this liquid. The DCE model is also fully flexible, with harmonic bond stretch, angle bend and 3-term harmonic series for the Cl-C-C-Cl torsional mode. The interaction between the water and the DCE is calculated using the combination rule for mixtures (*27*). More details about the potential parameters and the preparation and equilibration of the system can be found elsewhere (*24*).

The integration of the equations of motion is done using the velocity version of the Verlet leap-frog algorithm with a time step of 0.5 fs. Periodic boundary conditions in the *X* and *Y* directions produce, on average, a flat liquid-liquid interface perpendicular to the long axis of the box (*Z* direction). The properties of the neat interface have been described in detail (*24*). An important result of that study was the observation that the interface between the water and DCE is molecularly sharp, but it is quite rough due to constantly moving "fingers" of water and DCE. However, the long time average of the density of each liquid as a function of location along the interface normal is relatively smooth (especially for water), as can be seen from Figure 1.

The Redox Couple at the Interface

Interfacial electron transfer may occur at different molecular geometries relative to the interface. The electron may hop between two centers which are located on one side of the interface, as in the case of electron transfer in monolayers adsorbed on water surfaces (*28*). In many cases of experimental and technological interest, the two reactants are in different media. For example, the electron donor (D) may be in the organic phase and the electron acceptor (A) in the aqueous phase. The products of the charge separation reaction $DA \rightarrow D^+A^-$, (which may be initiated photochemically), can then be removed from the interface by an applied electric field or by having significantly different adsorption free energy at the interface.

A similar situation is considered in this paper and is shown schematically in Figure 1. We model each redox site as a united atom with a point charge (which may be 1, −1 or zero). The combination rules are used to determine the interaction between each united atom and the two liquids. This interaction consists of a Lennard-Jones term, which is independent of the electronic state of the solute, and an electrostatic term. The Lennard-Jones parameters ε_s and σ_s of the two centers are identical and selected to be 0.1 kcal/mol and 5Å - typical of a relatively large organic group.

When the redox pair is in the uncharged state, the two redox centers are quite hydrophobic and tend to be expelled from the water phase into the DCE phase. However, each member of charged pair is hydrophilic, and tends to migrate into the bulk water phase. Since we are interested in the interfacial ET, we restrict the center of mass of the pair to be inside a slab of fixed width centered at the interface ("interface zone"). In addition, since the electronic coupling between the two electronic states is a rapidly varying function of the distance between them, we also fix the interatomic distance between the redox pair to be 6Å using a SHAKE algorithm (29). The redox pair is free to move and rotate inside the "interface zone". Figure 2 displays the probability distribution for the position of the pair's center of mass. The tendency of the D^+A^- pair to migrate into the water phase and the DA pair toward the organic phase is clearly evident. Figure 2 also presents the probability distribution of the location of the fictional state $D^{+\frac{1}{2}}A^{-\frac{1}{2}}$, which approximately corresponds to the charge distribution of the transition state for the ET reaction. The nearly flat line reflects a balance between the hydrophobicity of the large Lennard-Jones center and the (mild) hydrophilicity due to the partial charges. Control of the adsorption/desorption tendency of the solute (or the free energy for adsorption) by means of changing charges and sizes has been recently explored experimentally at the water surface (30).

Of particular importance is the orientation of the redox couple relative to the interface. If one assumes that the two liquids are separated by a sharp boundary, (as most continuum models do), then the two reactants can approach each other via a limited range of solid angles (15) (for the case in which the two reactants are each restricted to being in one of the two immiscible liquids). In the other extreme, if the interface is regarded as a narrow homogeneous mixture of the two liquids, then every angle of approach is possible. It is difficult (if not impossible) to decide between these two possibilities on the basis of direct experimental data, but resolution of this issue is of fundamental importance to the theoretical framework necessary for extracting rate information. Our molecular model, in general, is more in line with the picture of a flat, sharp interface. This is shown in Figure 3, which gives the orientational probability distribution for the redox couple relative to the interface normal. When two spheres of diameter σ are a distance R apart across the interface ($R \geq \sigma$), the maximum angle between the vector connecting the two spheres and the interface normal would be $\theta_{max} = \cos^{-1}(\sigma/R)$, which in our case is $\theta_{max} = \cos^{-1}(5/6) \approx 34°$. Although a significant portion of the population is indeed below this angle (especially for the uncharged state), the larger angles of approach indicate that the reactants are able to approach each other over a wider angle due to the interface roughness. There is a significant barrier to complete rotation of the redox couple, as is evident in the small probability of angles above 90°. Also note that the observation of orientations larger than the maximum angle consistent with a sharp boundary is especially notable for the D^+A^- case. This suggests that surface roughness is more pronounced in the presence of ions.

The modification in liquid structure due to the presence of the ions is a well known problem with continuum models in bulk liquid, and attempts have been made to account for this short-range effect by introducing a non-local dielectric response. We conclude this section by providing one example of the type of problem one can encounter with the use of a (local) dielectric response. Consider the electrostatic potential at the location of an ion (charge Q, diameter σ) immersed in liquid A (dielectric constant ε_A) due to the polarization induced by the field of this ion in liquid B (dielectric constant ε_B), which forms a sharp boundary with liquid A. It is given by the "image" term:

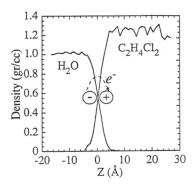

Figure 1. Density profiles of water and 1,2-dichloroethane at the interface at 300K.

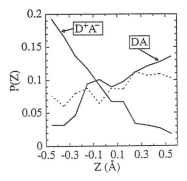

Figure 2. Probability distribution of the position of the center of mass of the redox pair in states with different charge distributions. The dotted line represents the "transition state" with the $D^{\frac{1}{2}}A^{-\frac{1}{2}}$ charge distribution.

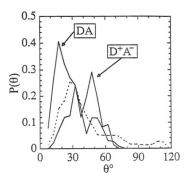

Figure 3. Probability distribution of the angle θ between the redox pair's "bond" and the normal to the interface for the charged and neutral pairs and for the $D^{\frac{1}{2}}A^{-\frac{1}{2}}$ pair (dotted line).

$$U = \frac{Q(\varepsilon_A - \varepsilon_B)}{2\varepsilon_A(\varepsilon_A + \varepsilon_B)d}, \tag{1}$$

where d is the distance from the center of the ion to the interface (31). For the geometries that are relevant to ET reactions where the electronic coupling is an exponential function of the distance between the ions, we may assume that the ion touches the interface, and thus $d = \frac{1}{2}\sigma$. The potential that is used in calculating the Born free energy also includes the potential due to the polarization produced by the ion in liquid A (which is $2Q/(\varepsilon_A\sigma)$ and will be referred to below as the "self term"), as well as the potential due to the other ion. We note that for the ion in the water phase (next to DCE), the image term is a factor of about 2.6 smaller than the "self term" (using $\varepsilon \approx 80$ for water and $\varepsilon \approx 10$ for DCE, which are the values obtained from the molecular model described previously).

The electrostatic potentials (due to the solvents only) at the location of each of the two ions determined from the molecular dynamics simulation are given in Table I. They are calculated by dividing the ensemble average of the electrostatic potential energy by the charge of the ion (which could be ± 1 and $\pm \frac{1}{2}$), or by using a test charge which is not included in the Hamiltonian in the case of the neutral pair.

Table I. Solvent Contributions to the Electric Potential at the Redox Sites

Site	DCE potential (Volts)	H_2O potential (Volts)
D	+0.11	−0.08
A	+0.08	−0.23
$D^{1/2}$	−0.64	−0.69
$A^{-1/2}$	+0.39	+1.38
D^+	−1.14	−1.46
A^-	+0.40	+3.69

The electrostatic potential produced by the water at the location of the anion A^- is a factor of 9 larger than the potential produced by the DCE. One obvious reason for this deviation from the 2.6 factor given by the electrostatic model is the strong, short-range hydrogen bonding interactions between the water and the ion. (Selecting a larger radius for the ion in the water phase would make only a small difference in these numbers). When the charge on the ion is reduced by a factor of 2, the water contribution to the electrostatic potential goes down by almost a factor of 3, whereas the DCE contribution is unchanged. The continuum model predicts both contributions to scale linearly with the charge. Thus, because of error cancellation, the total (water + DCE) electrostatic potential at the location of the anion approximately scales linearly with Q.

A different picture emerges for the cation (which is located in the DCE phase). The electrostatic potential at the location of the cation D^+ produced by the DCE is almost identical to the potential produced by the water. This discrepancy with the continuum model can again be traced to the short-range specific interactions of the water. When the water is specifically oriented to solvate the anion, with the hydrogen pointing toward it, the oxygen may be able to partially stabilize the nearby cation. In addition, due to surface roughness, water molecules may be able to approach the cation and favorably interact with it. Further support for this is provided by the fact that the DCE potential at the location of the cation (−1.14V) is much smaller than the potential when the cation is in the bulk of DCE (−2.78V -- which is obtained from an independent simulation of the cation in bulk DCE).

A detailed examination of the scaling of solvent electrostatic potentials and free energies with solute charge for a single ion in bulk polar solvents has been presented by Fonseca, Ladanyi and Hynes using the RHNC integral equation approach (32). These authors have shown that the deviation from linear behavior can be understood in terms of cooperative contributions of the solvent dipole and quadrupole.

We finally note that the difference between the electrostatic potential at the two sites, (which gives the total potential energy of the two ions), scales linearly with the charge on the redox pair even better than when the potential on one site is considered (6.69V for D^+A^- compared with 3.1V for the $D^{\frac{1}{2}}A^{-\frac{1}{2}}$, a factor of 2.15). We will return to this point below.

Solvent Free Energy for the Electron Transfer Reaction

In this section, we present the calculations of the diabatic free energies for the model electron transfer at the liquid-liquid interface, and we compare the results with a dielectric continuum model described by Marcus (15). The actual calculation of the rate constant for a given reaction depends on a knowledge of the electronic coupling between the initial and final electronic states. This rate has been estimated by Marcus for the one reaction for which experimental data are available (3).

In a discussion of the solvent free energy which governs the electron transfer kinetics, it is useful to define a solvent coordinate by

$$\Delta E(\mathbf{r}) = V_{D^+A^-} - V_{DA}, \tag{2}$$

where $V_{D^+A^-}$ is the potential energy of the system when the redox couple is charged, and V_{DA} is the potential energy of the system for the same solvent configuration $(\mathbf{r})$ when the redox couple is uncharged (22, 23, 33-37). The equilibrium average of ΔE is large and negative when the system is in the D^+A^- state, and near zero when it is in the DA state. The fluctuations in ΔE determine the probability for the two potential energies of the two electronic states to intersect and thus give rise to an electron transfer (Condon approximation). The fluctuations in ΔE when the system is in one of the two states are quantified by the corresponding free energies:

$$G_{D^+A^-}(x) = -\beta^{-1}\ln\left\{<\delta(\Delta E(\mathbf{r}) - x)>_{D^+A^-}\right\}, \tag{3}$$

$$G_{DA}(x) = -\beta^{-1} \ln \left\{ <\delta(\Delta E(\mathbf{r}) - x)>_{DA} \right\}, \tag{4}$$

where $\beta = (kT)^{-1}$ and $<>_{D^+A^-}$, $<...>_{DA}$ are the ensemble averages in the states D^+A^- and DA, respectively. The intersection between $G_{D^+A^-}$ and G_{DA} determines the activation free energy for the ET reaction. This intersection usually occurs at values of the reaction coordinate x which correspond to a very rare equilibrium fluctuation in ΔE, and thus they cannot be calculated directly from equations 3 and 4. To obtain values of $G_{D^+A^-}$ and G_{DA} for large deviations from equilibrium, one can selectively increase the sampling frequency in a given narrow interval of x by adding a biasing potential (which is later removed) to the Hamiltonian. Accurate values of the free energy in several overlapping "windows" can then be matched to produce the full free energy curve (23, 35, 37).

The free energy curves for the D^+A^- and DA states when the reaction free energy is zero are shown in Figure 4. They are calculated as explained before, using 10 overlapping "windows" of 50 ps each. Each curve can be fitted to a parabola with excellent accuracy, even for free energies as high as 100 kcal/mol. As assumed in Marcus' theory, the curvature of the two paraboli are nearly equal. For our model, the reorganization free energy, (defined as the free energy difference between the minimum of one electronic state and the free energy of the non-equilibrium polarization one gets after a "vertical" transition from the minimum of the other state), is $\lambda = 80 \pm 2$ kcal/mol. It is of interest to compare this with the prediction of the continuum electrostatic model. The reorganization free energy for an electron transfer across the interface between two dielectric media is given by Marcus (15). The expression for λ in the case of two ions of equal diameter σ separated by a distance R such that the center of mass is at the interface and the inter-ionic vector forms an angle θ with respect to the interface normal, is given by:

$$\frac{\lambda}{332}(\text{kcal/mol}) = \frac{1}{\sigma}\left[2 - \frac{1}{\varepsilon_A} - \frac{1}{\varepsilon_B} \right] \tag{5}$$

$$+ \frac{1}{R}\left[\frac{1}{2\cos\theta} \frac{\varepsilon_B - \varepsilon_A}{\varepsilon_B + \varepsilon_A}\left(\frac{1}{\varepsilon_A} - \frac{1}{\varepsilon_B} \right) + \frac{2}{\varepsilon_B + \varepsilon_A} - 1 \right],$$

where ε_A and ε_B are the static dielectric constants of the two liquids, and the infinite frequency dielectric constant is set to one (consistent with our electronically non-polarizable liquid models). Substituting the parameters relevant to our system (given above), and taking θ to be 45° (which is approximately the average orientation taken from Figure 3), gives $\lambda = 74$ kcal/mol.

The reasonable agreement with the λ calculated by the molecular dynamics and the nearly identical parabolic representation of the free energy curves are quite surprising given that the detailed calculation of the electric potential discussed above and the results regarding the structure of the interface are at odds with a simple continuum model. One possible explanation may be related to the fact, mentioned above, that although the electric potentials near the individual ions do not scale with the ion's charge according to the linear relation expected from the continuum model, the total electric potential for the whole system does. However, more work is necessary to understand why the continuum model and the linear response approximation work so well in this system. A calculation of the reorganization free energy as a function of ion size and location relative to the interface would allow for a more stringent test of the continuum model.

Solvent Dynamics

In this section we briefly consider the solvent dynamical response to a photoinduced electron transfer reaction at the water/DCE interface. Recently, this reaction has been probed at the water/DCE interface using optical second harmonic generation (*10*). Although the time resolution of that work was not sufficient to follow the solvent response, the topic is important enough to warrant theoretical examination at this point. The importance of the solvent dynamical response has been demonstrated for electron transfer reactions in bulk liquids (*22, 38-40*) and many of the issues discussed in these investigations may be relevant to interfacial ET.

Recently we have demonstrated that the rate of solvent relaxation following the creation of a large dipole across a model interface between non-polar and dipolar liquids is slower than the relaxation in the bulk (Benjamin, I. *Chem. Phys.*, in press). It was found that this is a direct consequence of the structural constraint imposed by the interface. Below we briefly consider charge separation (DA $\rightarrow$ D$^+$A$^-$) and charge recombination (D$^+$A$^-$ $\rightarrow$ DA) processes at the water/DCE interface.

To study the charge recombination process, the system which was equilibrated in the D$^+$A$^-$ state is transferred at time $t = 0$ to the DA electronic state. The solvent must adjust to the new charge distribution, and the system relaxes to a new equilibrium appropriate to the DA state. The charge separation process is similarly investigated by considering a sudden change from the DA to the D$^+$A$^-$ state. For both reactions, the solvent coordinate is monitored as a function of time. Each reaction is repeated 100 times using independent initial conditions (which were generated from the equilibrium trajectories discussed in the previous section).

The rate of the solvent relaxation is most easily examined using the following non-equilibrium correlation function:

$$C(t) = \frac{\bar{x}(t) - \bar{x}(\infty)}{\bar{x}(0) - \bar{x}(\infty)}, \tag{6}$$

which may be directly related to experimental data such as the time-dependent Stokes shift (*18*). In equation 6, $\bar{x}(t)$ is the average over the ensemble of independent trajectories of the solvent coordinate defined in equation 2, and $\bar{x}(\infty)$ is the equilibrium value of the solvent coordinate in the final state. Since the solvent coordinate contains contributions from both liquids, it is of interest to examine the decay of the individual components of x, even though these two components cannot be separately observed experimentally.

Figure 5 provides the total correlation functions for the charge recombination ($C_r(t)$) and charge separation ($C_s(t)$), as well as the correlation of the two individual components for the two liquids. The total correlation functions exhibit a non-

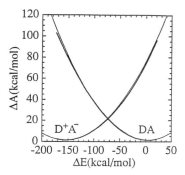

Figure 4. Solvent reorganization free energy curves for the charged solutes (D^+A^-) and the neutral solutes (DA) at the water/1,2-dichloroethane interface. Thick line - free energy calculations using umbrella sampling; thin line - the best fit to a parabola.

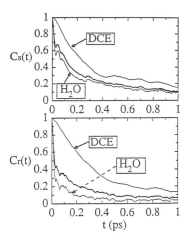

Figure 5. Non-equilibrium solvent relaxation correlation function. The top panel corresponds to the charge-separation reaction, and the bottom panel corresponds to the charge-recombination reaction. The lines labeled DCE and H_2O refer to the relaxation of the individual liquids, and the unlabeled thick line is the total relaxation.

exponential decay which includes a very rapid component (complete in 50 fs), which is responsible for about 60% of the total decay in the charge recombination reaction, but accounts for only 30% of the total decay in the charge separation reaction. The rest of the decay is non-exponential and about a factor of 2 faster than in the charge recombination process. An examination of the separate contributions of the two liquids shows that the initial fast decay is due to the water, as the DCE component does not include a fast initial decay and in fact is quite similar for the two reactions. The time scale over which the initial water component decays suggests that water librational motion is the main molecular mode responsible for the early relaxation. The tracking of the full response and the water response seen in Fig. 5 even at longer times is due to the fact that most of the contribution to the electrostatic potential at the location of the ions comes from the water (see Table I). It is reasonable to expect that in systems where the charge is buried inside a large hydrophobic group (which is more likely to be shielded from water at the interface), the response would track that of the DCE solvent. Thus monitoring the solvent response to a photochemical ET at a liquid-liquid interface would make it possible to determine the location of the redox pair and the extent to which each liquid participates in the solvation. A possible explanation for the slower response of the water to charge separation than to charge recombination at the interface (which is observed to a lesser degree in bulk water), is that at the interface, more structural rearrangement is required to solvate the newly created large dipole because the interfacial water is oriented at the interface (*24*). Additional "maneuvering" is necessary to satisfy the constraints imposed by the DCE and the D^+A^- pairs. Similar results have been obtained and analyzed in great detail for another liquid-liquid interface model (Benjamin, I. *Chem. Phys.*, in press). Further particulars about the structural changes accompanying the non-equilibrium charge transfer, as well as a full comparison with the calculation in bulk water, will be reported elsewhere.

Summary and Conclusions

A molecular model for an electron transfer reaction at the water/1,2-dichloroethane interface is developed and used to examine some basic issues about liquid/liquid electron transfer, some of which have not been addressed in existing treatments.

It has been established that the interfacial region is narrow even at the molecular scale. However, changes in interfacial liquid structure which are necessary in order to accommodate and solvate an ion, as well as surface roughness (induced by thermal fluctuations), cast doubt on any attempt to describe the interface as a mathematically sharp line separating two distinct dielectric media. Thus, electric potentials induced by the liquids at the interface do not agree with predictions of continuum models. This may limit the applicability of these models for calculating concentration profiles and work terms that are necessary for evaluating the observed rate constant for interfacial ET.

Nevertheless, and quite surprisingly, the solvent reorganization free energy for the redox pair at the interface is in close agreement with the prediction of the continuum model. The full free energy curves, which have been calculated by umbrella sampling, can be fitted by almost perfect and identical curvature paraboli, which is in agreement with the linear response approximation.

The solvent dynamical response relevant to fast photoinduced ET is non-exponential and contains a significant inertial component due to water libration modes and slower components from both water and dichloroethane intermolecular modes. The slower solvent response to charge separation than to charge recombination observed in the molecular dynamics calculations may be possible to observe experimentally.

Acknowledgments

This work was supported by the National Science Foundation (CHE-9221580).

Literature Cited

(1)   Girault, H. H. J.; Schiffrin, D. J. In *Electroanalytical Chemistry*; A. J. Bard, Editor.; Dekker: New York, 1989; p 1.

(2)   *The Interface Structure and Electrochemical Processes at the Boundary Between Two Immiscible Liquids*; V. E. Kazarinov, Editor.; Springer: Berlin, 1987.

(3)   Geblewicz, G.; Schiffrin, D. J. "Electron transfer between immiscible solutions. The hexacyanoferrate-lutetium biphthalocyanine system," *J. Electroanal. Chem.* **1988**, *244*, 27.

(4)   Maeda, K.; Kihara, S.; Suzuki, M.; Matsui, M. "Voltammetric interpretation of ion transfer coupled with electron transfer at a liquid liquid interface," *J. Electroanal. Chem.* **1991**, *303*, 171.

(5)   Cheng, Y. F.; Schiffrin, D. J. "Electron transfer between bis(pyridine)meso-tetraphenylporphyrinato iron(II) and ruthenium(III) and the hexacyanoferrate couple at the 1,2-dichloroethane water interface," *J. Electroanal. Chem.* **1991**, *314*, 153.

(6)   Lee, L. T.; Langevin, D.; Farnoux, B. "Neutron reflectivity of an oil-water interface," *Phys. Rev. Lett.* **1991**, *67*, 2678.

(7)   Wirth, M. J.; Burbage, J. D. "Reorientation of acridine orange at liquid alkane water interfaces," *J. Phys. Chem.* **1992**, *96*, 9022.

(8)   Grubb, S. G.; Kim, M. W.; , Th. Raising; Shen, Y. R. "Orientation of molecular monolayers at the liquid-liquid interface as studied by optical second harmonic generation," *Langmuir* **1988**, *4*, 452.

(9)   Higgins, D. A.; Corn, R. M. "2nd harmonic generation studies of adsorption at a liquid-liquid electrochemical interface ," *J. Phys. Chem.* **1993**, *97*, 489.

(10)   Kott, K. L.; Higgins, D. A.; McMahon, R. J.; Corn, R. M. "Observation of photoinduced electron transfer at a liquid-liquid interface by optical second harmonic generation," *J. Am. Chem. Soc.* **1993**, *115*, 5342.

(11)   Brown, A. R.; Yellowlees, L. J.; Girault, H. H. "Photoinduced electron-transfer across the interface between 2 immiscible electrolyte solutions.," *J. Chem. Soc. Faraday Trans.* **1993**, *89*, 207.

(12)   Mirkin, M. V.; Fan, F-R. F.; Bard, A. J. "Direct electrochemical measurments inside a 2000-angstrom thick polymer film by scanning electrochemical microscopy," *Science* **1992**, *257*, 364.

(13)   Kharkats, Yu. I.; Vol'kov, A. G. "Interfacial catalysis: Multielectron reactions at the liquid-liquid interface," *J. Electroanal. Chem.* **1985**, *184* , 435.

(14)   Kuznetsov, A. M.; Kharkats, Y. I. In *The Interface Structure and*

Electrochemical Processes at the Boundary Between Two Immiscible Liquids; V. E. Kazarinov, Editor.; Springer: Berlin, 1987; p 11.

(15)   Marcus, R. A. "Reorganization free energy for electron transfers at liquid-liquid and dielectric semiconductor-liquid interfaces," *J. Phys. Chem.* **1990**, *94*, 1050.

(16)   Marcus, R. A. "Theory of electron-transfer rates across liquid-liquid interfaces," *J. Phys. Chem.* **1990**, *94*, 4152.

(17)   Marcus, R. A. "Theory of electron-transfer rates across liquid-liquid interfaces. 2. Relationships and Application," *J. Phys. Chem.* **1991**, *95*, 2010.

(18)   Simon, J. D. "Time-resolved studies of solvation in polar media," *Acc. Chem. Res.* **1988**, *21*, 128.

(19)   Maroncelli, M.; Fleming, G. R. "Computer simulation of the dynamics of aqueous solvation," *J. Chem. Phys.* **1988**, *89*, 5044.

(20)   Bagchi, B. "Dynamics of solvation and charge transfer reactions in dipolar liquids," *Annu. Rev. Phys. Chem.* **1989**, *40*, 115.

(21)   Fonseca, T.; Ladanyi, B. M. "Breakdown of linear response for solvation dynamics in methanol," *J. Phys. Chem.* **1991**, *95*, 2116.

(22)   Carter, E. A.; Hynes, J. T. "Solvation dynamics for an ion pair in a polar solvent: Time dependent fluorescence and photochemical charge transfer," *J. Chem. Phys.* **1991**, *94*, 5961.

(23)   Benjamin, I. "Molecular dynamics study of the free energy functions for electron transfer reactions at the liquid-liquid Interface," *J. Phys. Chem.* **1991**, *95*, 6675.

(24)   Benjamin, I. "Theoretical study of the water/1,2-dichloroethane interface: Structure, dynamics and conformational equilibria at the liquid-liquid interface," *J. Chem. Phys.* **1992**, *97*, 1432.

(25)   Berendsen, H. J. C.; Postma, J. P. M.; Gunsteren, W. F. van; Hermans, J. In *Intermolecular Forces*; B. Pullman, Editor.; D. Reidel: Dordrecht, 1981; p 331.

(26)   Kuchitsu, K.; Morino, Y. "Estimation of anharmonic potential constants. II. Bent XY_2 molecules," *Bull. Chem. Soc. Jpn.* **1965**, *38*, 814.

(27)   Hansen, J. -P.; McDonald, I. R. *Theory of Simple Liquids;* Academic Press: London, 2nd edition, 1986. p 179.

(28)   Charych, D. H.; Majda, M. "Electrochemical investigations of the lateral diffusion and electron hopping in langmuir monolayers at the air-water interface.," *Thin Solid Films* **1992** , *210*, 348.

(29)   Ciccotti, G.; Ryckaert, J. P. "Molecular dynamics simulation of rigid molecules," *Computer Physics Reports* **1986**, *4*, 345.

(30)   Castro, A.; Bhattacharyya, K.; Eisenthal, K. B. "Energetics of adsorption of neutral and charged molecules at the air water interface by 2nd second harmonic generation - Hydrophobic and solvation effects," *J. Chem. Phys.* **1991**, *95*, 1310.

(31)   Landau, L. D.; Lifshitz, E. M. *Electrodynamics of Continuous Media;* Pergamon: Oxford , 1980. p 40.

(32)   Fonseca, T.; Ladanyi, B. M.; Hynes, J. T. "Solvation free energies and solvent force constants," *J. Phys. Chem.* **1992**, *96*, 4085.

(33)   Warshel, A. "Dynamics of reactions in polar solvents. Semiclassical trajectory studies of electron-transfer and proton-transfer reactions," *J. Phys. Chem.* **1982**, *86*, 2218.

(34)   Hwang, J. K.; Warshel, A. "Microscopic examination of free-energy relationships for electron transfer in polar solvents," *J. Am. Chem. Soc.* **1987**, *109*, 715.

(35)   Kuharski, R. A.; Bader, J. S.; Chandler, D.; Sprik, M.; Klein, M. L.; Impey, R. W. "Molecular model for aqueous ferrous-ferric electron transfer," *J. Chem. Phys.* **1988**, *89*, 3248.

(36)   Carter, E. A.; Hynes, J. T. "Solute-dependent solvent force constants for ion pairs and neutral pairs in polar solvent," *J. Phys. Chem.* **1989**, *93*, 2184.

(37)   King, G.; Warshel, A. "Investigation of the free energy functions for electron transfer reactions," *J. Chem. Phys.* **1990**, *93*, 8682.

(38)   Maroncelli, M.; MacInnis, J.; Fleming, G. R. "Polar solvent dynamics and electron-transfer reactions," *Science* **1989**, *243*, 1674.

(39)   Barbara, P. F.; Jarzeba, W. "Ultrafast photochemical intramolecular charge and excited state solvation," *Adv. Photochem.* **1990**, *15*, 1.

(40)   Weaver, M. J. "Dynamical solvent effects on activated electron-transfer reactions: principles, pitfalls, and progress," *Chem. Rev.* **1992**, *92*, 463.

RECEIVED April 5, 1994

Author Index

Affiliation Index

Subject Index

Production: Meg Marshall
Indexing: Deborah H. Steiner
Acquisition: Rhonda Bitterli
Cover design: Alan Kahan

Printed and bound by Maple Press, York, PA